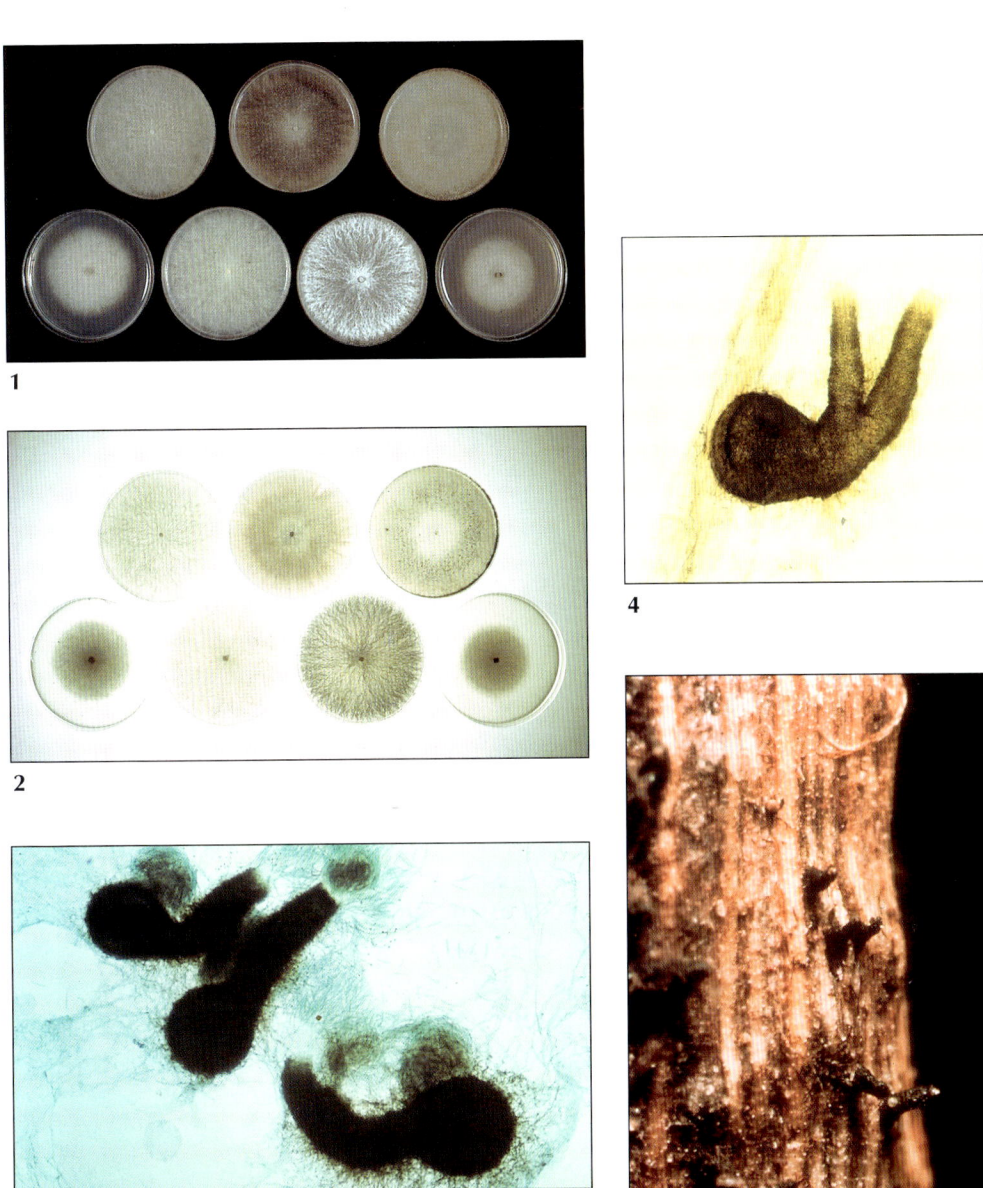

Plate 1. Take-all fungi and some related species. **1, 2.** Colony variants of fungi in the *Gaeumannomyces–Phialophora* complex on Lilly and Barnet/Weste and Thrower medium photographed against dark and light backgrounds. References to this semi-synthetic medium are given in Holden and Hornby (1981). The fungi are, from left to right, top row first: *G. graminis* var. *tritici* (the wheat take-all fungus), *G. graminis* var. *avenae* (the oat take-all fungus), *G. graminis* var. *tritici*, *P. graminicola*, *P. radicicola* sensu Cain, *Phialophora* sp. (lobed hypopodia), *P. graminicola* (see Section 5.3.1). (IACR-Rothamsted.) Aspects of the teleomorph of *G. graminis* var. *tritici* are shown in the next three photographs (for sizing cf. Figs 2.10 and 2.11). **3.** Perithecia and ascospores stained by cotton blue in lactophenol. (IACR-Rothamsted.) **4.** An atypical perithecium of isolate Og12 produced in a rotting test (Hornby and Holden, 1981). (IACR-Rothamsted.) **5.** Parts of perithecia, mostly necks, protruding from a wheat stem. (AgrEvo.)

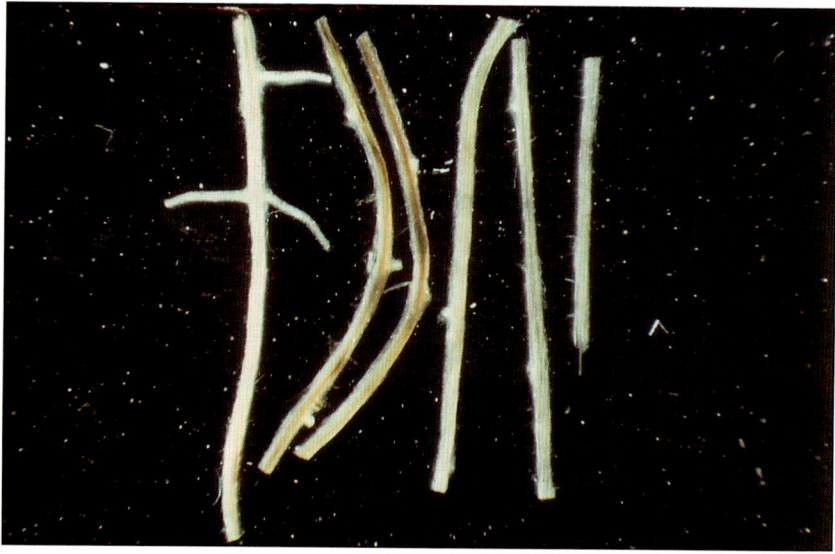

Plate 2. Wheat seedling diseases. **6.** A plant severely infected with *G. graminis* var. *tritici*, and showing symptoms of seedling blight, taken from an early-sown, second wheat crop after oilseed rape in late November. (IACR-Rothamsted.) **7.** A sequence of segments, approximately 2–3 cm long, cut from a detached lateral root, showing vascular discoloration caused by *G. graminis* var. *tritici*. The left-hand segment is the proximal segment. (IACR-Rothamsted.)

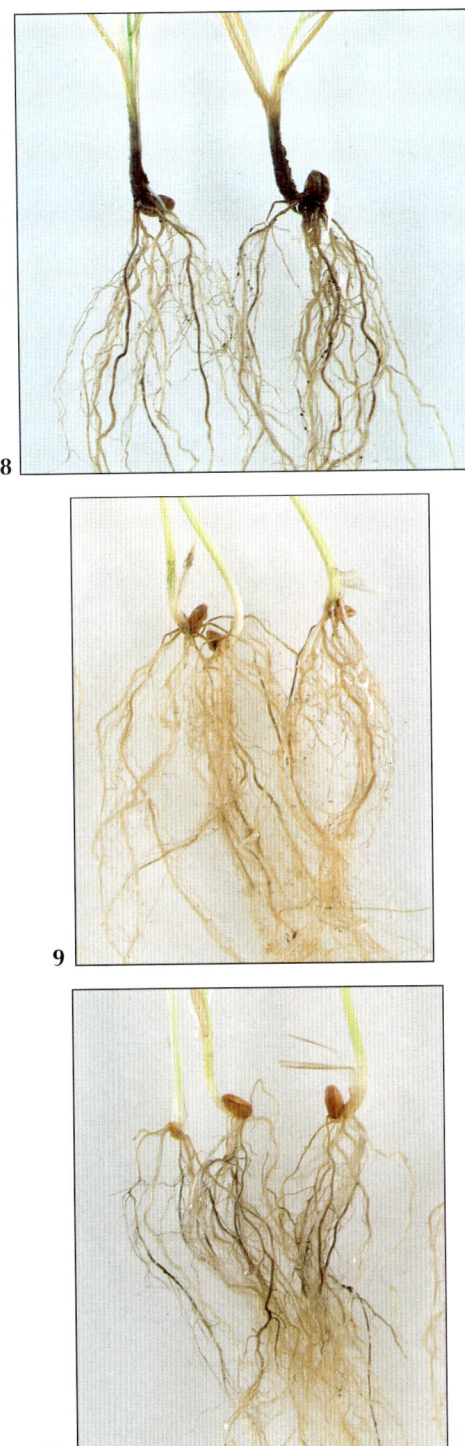

Plate 3. Superficial similarities in the discoloration of roots by some members of the *Gaeumannomyces–Phialophora* complex in pathogenicity tests (see Section 5.4.1); wheat seedlings displayed under water. **8.** *G. graminis* var. *tritici*. **9.** *G. graminis* var. *graminis*. **10.** *G. cylindrosporus* (*P. graminicola*). (IACR-Rothamsted.)

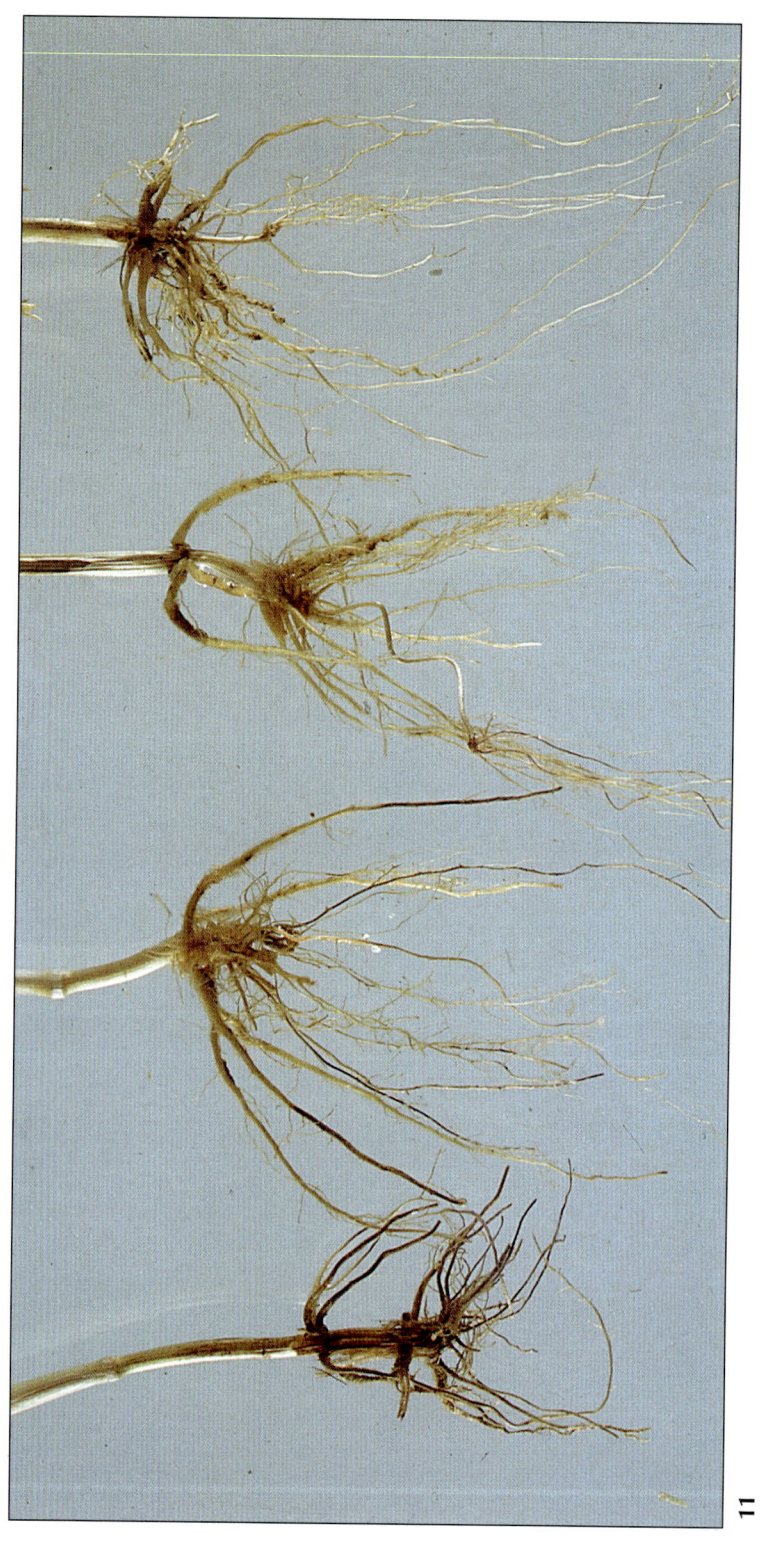

Plate 4. Take-all infection categories of wheat plants at about the time of anthesis. **11.** Root systems and shoot bases illustrating severe, moderate, slight and healthy categories used in assessments of take-all severity (see Sections 6.2 and 6.3.3). (IACR-Rothamsted.)

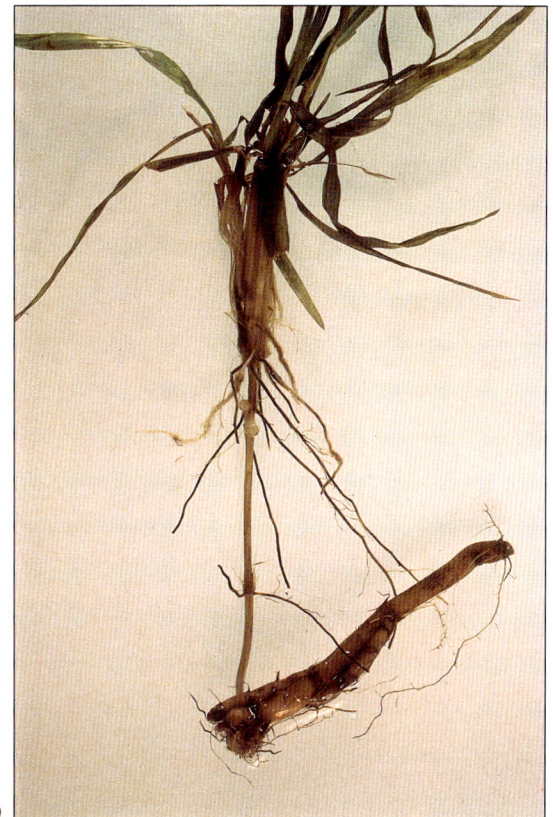

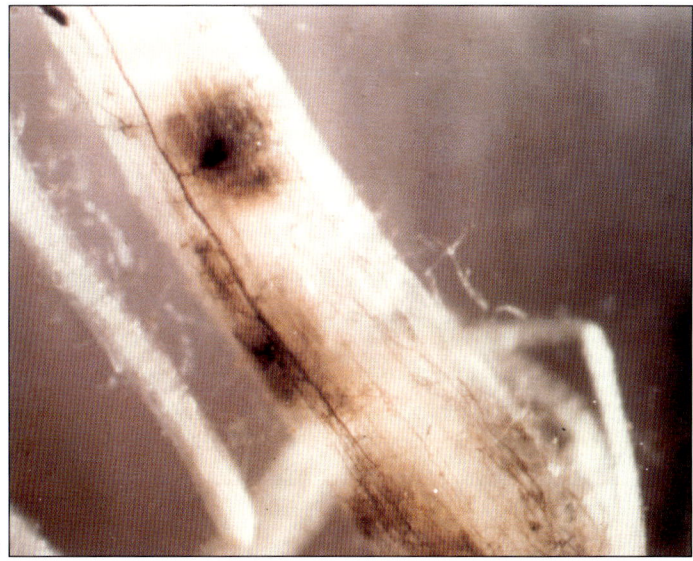

Plate 8. The take-all fungus on alternative hosts. **19.** Couchgrass infected with *G. graminis* var. *tritici* showing lesions on the rhizome and vascular discoloration of lateral roots; taken from a wheat field on 3 May 1995. (IACR-Rothamsted.) **20.** A root of oilseed rape (cv. Darmor; glucosinolate rating 5, where 9 = very low) infected with *G. graminis* var. *tritici* in a sand culture. Small lesions are associated with hyphal branches originating from a runner hypha on the root surface. Magnification 59X. (C.M.P. Davis.)

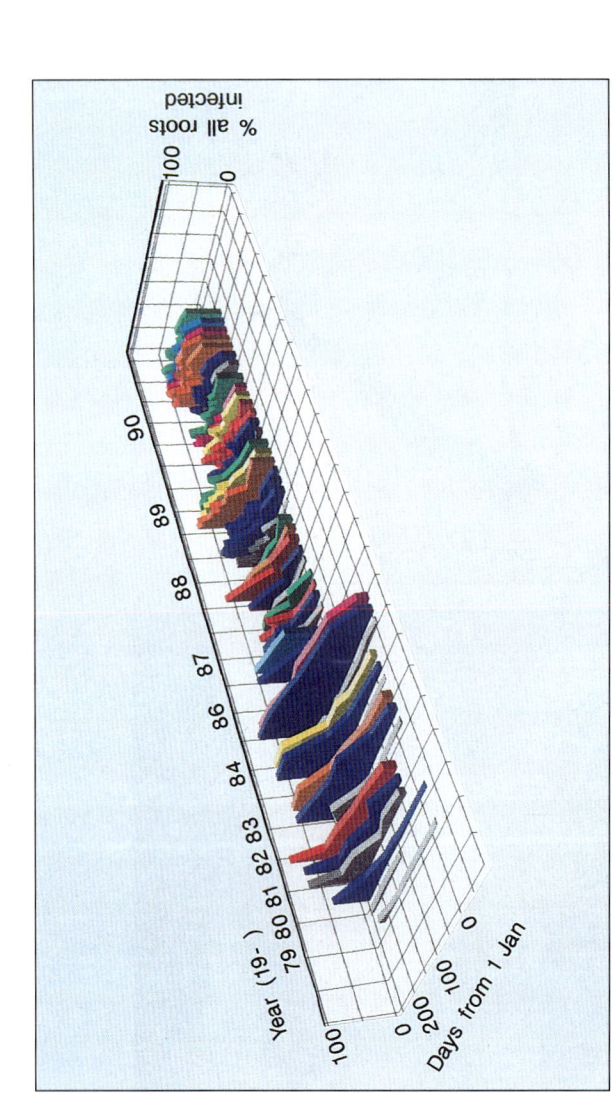

Plate 7. Epidemiology. **18.** Take-all in CS212, a phased sequence experiment on winter wheat at Rothamsted: 3-D representation of DPCs for disease measured as percentage of all roots infected from 1979 to 1990. Days = number of days from 1 January (the range shown is approximately a few weeks after emergence to a few days before harvest); sequences and sampling dates are explained in Fig. 2.14. TAD was assumed to be operative in fourth and subsequent wheat crops.

Colour	Grey	Blue	Brick red	Orange	Yellow	Magenta	Light blue	Green
Wheat crop	1st	2nd	3rd	4th	5th	6th	7th	9th, 10th, 11th, 12th and 13th in successive years

16

17

Plate 6. Take-all patches in wheat crops. **16.** Extensive patchiness in a third consecutive winter wheat crop at Woburn, Bedfordshire, 1984. (G.L. Bateman.) **17.** Wheat fields with take-all patches in Rio Grande do Sul, Brazil, 1981. Extensive patches show up as light areas in the field with contour banks. The isolated *Ilex paraguariensis* trees at the edge of this field were left to provide leaves for *maté* (a drink made from an infusion of leaves). (D. Hornby.)

Plate 5. Take-all patches in winter wheat crops. **12.** A very small patch at GS 75 in a crop at Rothamsted, 1992. A discoloured flag leaf in the foreground shows characteristic rolling. (IACR-Rothamsted.) **13.** A small patch soon after ear emergence, showing stunted plants. (AgrEvo.) **14.** A bare patch attributed to take-all in a severely affected crop showing extensive patchiness. (AgrEvo.) **15.** An extensive patch of early whiteheads at GS 71 in Stubbings field at Rothamsted in 1987. The patch coincides with a gully that was waterlogged during the winter, suggesting thas such large-scale disease patterns may be determined by site factors. By GS 83, whiteheads covered 80% of the field. (IACR-Rothamsted.)

TAKE-ALL DISEASE OF CEREALS
A Regional Perspective

TAKE-ALL DISEASE OF CEREALS
A Regional Perspective

By

D. HORNBY

Incorporating contributions from:

G.L. Bateman, R.J. Gutteridge, P. Lucas, A.E. Osbourn,
E. Ward and D.J. Yarham

CAB INTERNATIONAL

CAB INTERNATIONAL
Wallingford
Oxon OX10 8DE
UK

Tel: +44 (0)1491 832111
Fax: +44 (0)1491 833508
E-mail: cabi@cabi.org

CAB INTERNATIONAL
198 Madison Avenue
New York, NY 10016-4314
USA

Tel: +1 212 726 6490
Fax: +1 212 686 7993
E-mail: cabi-nao@cabi.org

© CAB INTERNATIONAL 1998. All rights reserved. No part of this publication may be reproduced in any form or by any means, electronically, mechanically, by photocopying, recording or otherwise, without the prior permission of the copyright owners.

A catalogue record for this book is available from the British Library, London, UK.

Library of Congress Cataloging-in-Publication Data
Hornby, D. (David)
    Take-all disease of cereals : a regional perspective / by D. Hornby; incorporating contributions by G. L. Bateman . . . [et al.].
      p. cm.
   ISBN 0–85199–124–6 (alk. paper)
   1. Take-all disease. I. Title.
  SB608.G6H65. 1998
  633.1′04943– –dc21   98–4514
                                                                                        CIP

ISBN 0 85199 124 6

Typeset in Photina by AMA Graphics Ltd, UK
Printed and bound in the UK at the University Press, Cambridge

Contents

Contributors	xi
Preface	xiii
Acknowledgements	xv
Abbreviations, Acronyms and Notes	xvii
Introduction	xix

1 Past and Present — 1
  1.1 Historical Background — 1
  1.2 Take-all in the UK Today — 18
      1.2.1 Importance — 18
      1.2.2 Control — 27
      1.2.3 Changes: evidence, causes and implications — 28
  1.3 Trends in Cereal Production and Take-all Research — 31
  1.4 Take-all Outside the UK — 34

2 Disease and Epidemiology — 47
  2.1 Preamble — 47
      2.1.1 Concerning the host — 47
      2.1.2 Concerning interactions — 48
  2.2 Disease Cycle — 49
  2.3 Disease–Environment Interactions — 56
      2.3.1 Introduction — 56
      2.3.2 Weather and seasonal effects (British Isles) — 58
      2.3.3 Temperature and water potential — 61
      2.3.4 Soil type: physicochemical aspects — 62
      2.3.5 Nitrate leaching — 63

2.4	Suppressive and Conducive Soils		67
	2.4.1 Recent research in the UK: take-all decline (TAD)		67
	2.4.2 Recent research in France: nitrogen form and bacteria		72
	2.4.3 Research in Australia: pathogen suppression during saprophytic growth		72
	2.4.4 Recent research in the USA		75
	2.4.5 Synthesis: mechanisms and regional differences		76
2.5	The Disease Progress Curve		76
	2.5.1 Description		77
	2.5.2 Interpretation		78
	2.5.3 Analysis		79
	2.5.4 Some practical problems		79
	2.5.5 Conclusions and future research		81
2.6	Modelling		82
	2.6.1 Qualitative models		83
	2.6.2 Mathematical models: temporal		84
	2.6.3 Selecting disease variables		87
	2.6.4 Mathematical models: spatial		88
	2.6.5 Empirical statistical models		90
	2.6.6 Host growth and yield		91
	2.6.7 Simulation of patch development: cellular automaton		92
	2.6.8 A study of the effects of crop succession and cultural practices: combining linear and non-linear models		96
2.7	Comparisons with Other Regions		100
3	**Take-all and Cereal Production Systems**		**103**
3.1	Factors that Affect Take-all		103
3.2	Host Resistance		103
	3.2.1 Exploiting different cereal species		105
	3.2.2 Cultivar resistance and possible mechanisms		109
	3.2.3 Novel breeding programmes		111
3.3	Rotations		112
	3.3.1 Changes in rotational practice		112
	3.3.2 Effects of soil type and climate on rotational options		118
	3.3.3 Exploiting TAD		119
	3.3.4 Effects of different break crops		121
3.4	Crop Nutrition		129
	3.4.1 Nitrogen		130
	3.4.2 Other major nutrients		134
	3.4.3 Minor nutrients		137
3.5	Husbandry		142
	3.5.1 Sowing date		142
	3.5.2 Method of cultivation		147
	3.5.3 Quality of cultivation		148
	3.5.4 Soil pH		151
	3.5.5 Herbicides		151

	3.6	Effects of Take-all on the Sensitivity of Crop Yield to Climate	153
	3.7	Comparisons with Cereal Production Systems in Other Regions	159

4 Strategies for Management — 165

4.1	Agronomic Practices		165
	4.1.1	Host nutrition	165
4.2	Chemical Control		166
	4.2.1	Fumigants	167
	4.2.2	Fungicides	167
	4.2.3	Herbicides and plant growth regulators	177
4.3	Biological Control		178
	4.3.1	Natural biological control	179
	4.3.2	Biological control by introduced organisms	182
	4.3.3	Bacteria as biological control agents	184
	4.3.4	Fungi	190
	4.3.5	Other organisms	194
	4.3.6	Prospects	195
4.4	Integrated Control		196
	4.4.1	Recent research in the UK and France	196
4.5	Decision Making and Forecasting		199
	4.5.1	Short-term: before sowing	199
	4.5.2	Short-term: during the crop	202
	4.5.3	Longer term	203
4.6	Recommendations for Take-all Control		204
4.7	Take-all Control in a Worldwide Context		206

5 The Pathogens and Related Fungi — 212

5.1	Introduction		212
5.2	The *Gaeumannomyces–Phialophora* Complex		213
5.3	Isolation and Culture Maintenance		223
	5.3.1	Isolation	223
	5.3.2	Maintenance	224
	5.3.3	Culture collections	225
5.4	Identification by Conventional Methods		226
	5.4.1	Pathogenicity tests, infection structures and perithecia	226
	5.4.2	Selective and other media	228
5.5	Identification by Molecular Methods		232
	5.5.1	DNA methods: background	232
	5.5.2	DNA methods: identification of *G. graminis*	234
	5.5.3	Serology	240
	5.5.4	Protein electrophoresis	240
	5.5.5	Prospects and implications	241
5.6	Population Studies		244

5.7	Genetics of *G. graminis*	245
	5.7.1 The need for mutants	245
	5.7.2 Development of techniques for mutagenesis	246
	5.7.3 Sexual crosses	246
	5.7.4 Molecular biology of *G. graminis*	247
5.8	Pathogenicity	248
	5.8.1 Host specificity	248
	5.8.2 Factors implicated in pathogenicity	249
	5.8.3 Specificity for oats: avenacin and avenacinase	250
5.9	Viruses of *G. graminis*	252
5.10	Regional Approaches	253

6 The Disease: Field Techniques — 254

6.1	Introduction	254
6.2	Assessing, Monitoring and Surveying Disease in Non-experimental Fields	254
6.3	Field Experiments	258
	6.3.1 Design	258
	6.3.2 Sampling	261
	6.3.3 Assessing disease	261
	6.3.4 Patterns of disease	266
	6.3.5 Inoculum	269
6.4	Artificial Inoculation	271
	6.4.1 Background	271
	6.4.2 Latest research in the UK	274
6.5	Disease–Yield Relationships	277
	6.5.1 Theoretical background	277
	6.5.2 Practical approaches to assessing losses in grain yield and quality	282
	6.5.3 Recent research in the UK	284
6.6	Comparisons of Regional Approaches to Field Experimentation	288

7 The Future — 291

7.1	The Uncertainties	291
	7.1.1 The cereal production system	291
	7.1.2 Climate change	292
	7.1.3 Unforeseen developments and applications	293
7.2	Resources	296
7.3	Research	296
	7.3.1 Establishing the importance of take-all	297
	7.3.2 Improving forecasting and risk assessment	299
	7.3.3 Understanding take-all biology	300
	7.3.4 Controlling take-all	301
7.4	Summary	306

References 309

Further Reading 342
    Pathogen and Related Fungi (includes virus infection) 342
        Soil and rhizosphere 343
        In vitro 344
        Pathogenicity 345
        Detection (includes identification and quantification) 346
    Disease (includes root infection) 347
        Occurrence and surveys 347
        Effect on host and yield 348
        Grass and turf diseases 349
    Epidemiology (includes environmental and soil factors) 351
        Alternative hosts 351
        Mathematics and statistics 351
        Fertilizers 351
        Crop sequence studies 352
        Other agronomic factors (e.g. tillage) 354
    Control (includes management) 356
        Biological: unclassified 357
        Biological: natural 359
        Biological: introduced organisms 359
        Fungicides 363
        Host resistance 364

Index 365

Contributors

Geoffrey L. Bateman
IACR-Rothamsted, Harpenden, Herts AL5 2JQ, UK.
Contributed to Chapters 1–7

Richard J. Gutteridge
IACR-Rothamsted, Harpenden, Herts AL5 2JQ, UK.
Contributed to Chapters 1–6

David Hornby
22 Chartwell Drive, Luton, Beds LU2 7JD, UK. (Formerly of: IACR-Rothamsted, Harpenden, Herts AL5 2JQ, UK.)
Contributed to Chapters 1–7

Philippe Lucas
Station de Pathologie Végétale, Centre de Recherches de Rennes, INRA, B.P. 29, F-35653 Le Rheu cedex, France.
Contributed to Chapters 1–4

Anne E. Osbourn
The Sainsbury Laboratory, John Innes Centre, Colney Lane, Norwich NR4 7UH, UK.
Contributed to Chapter 5

Elaine Ward
IACR-Rothamsted, Harpenden, Herts AL5 2JQ, UK.
Contributed to Chapter 5

David J. Yarham
Mill House, Croxton, Fulmodeston, Fakenham, Norfolk NR21 0NP, UK. (Formerly of: ADAS, Block C, Government Buildings, Brooklands Avenue, Cambridge CB2 2BL, UK.)
Contributed to Chapters 4 and 5

Preface

This book grew out of a Home-Grown Cereals Authority Review on take-all (Hornby and Bateman, 1991), the preparation of which had been an initiative of the IACR/ADAS/Universities Cereal Root Pathology Group. The group maintains about 12 members and was formed in 1988 to encourage cooperation, collaboration and integration among researchers receiving government funding for work on cereal root diseases. University members have represented Birmingham and Cambridge and in recent years the membership was widened to include representatives from The Sainsbury Laboratory at Norwich.

Many of the contributors have been long and actively associated with take-all in one way or another. Because of a threat to the quality and continuity of take-all research from an increasing shift to short-term contracts and projects and decreasing core funding, it seemed particularly important to tap this combined experience before it was lost to posterity. Besides, 'one of the depressing facts about science is the amount of worthy information that is collected, but then forgotten because it does not become integrated into general theory' (Berry, 1993). In the event, the decision proved to be providential, for the writing of this book took place against upheavals, unprecedented in decades of take-all research, caused by major restructuring in several of the organizations traditionally engaged in take-all research, swingeing cuts in funding and staff losses. Consequently, the book has not quite the breadth of coverage that was originally envisaged, but nonetheless the contributors believe that it serves as a useful record of the work done in northern Europe, particularly in the UK and in France, in the last quarter of a century and that it puts this into a worldwide perspective.

The first six of the seven chapters develop a modern concept of take-all by providing a general background and discussion of the problems that face farmers and researchers and the issues that arise from them. In some chapters or sections the local experience is discussed separately and in considerable detail, followed by comparisons with the approaches used elsewhere, usually in other continents where the conditions for cereal growing are quite different. Where this separation has been difficult to sustain without becoming stultified, topics have been discussed in a more general form, but still with emphasis on regional differences. In some subject areas, such as the development of molecular biological methods, the experiences and the approaches to research have been more international, and the discussion of these reflects this.

In the first chapter the history of take-all in the UK has been traced and set against the development of cereal production in this century in the belief that this provides clues to why the severity and impact of the disease have been perceived as erratic. Research aimed at finding means of controlling or managing take-all is described in the fourth chapter and recommendations for future research are made in the seventh chapter. Failure to solve the take-all problem with simplistic approaches has contributed to the image of the disease as intractable. Whether or not there is a case for continuing such simplistic approaches, it has to be conceded that in the past the search for them has shed new light on how take-all works.

Because the regional approach adopted in this book is based mostly on experiences in the British Isles and France, some topics considered of importance elsewhere are not dealt with in detail in the body of the text. To make up for this an appendix containing a supplementary bibliography with entries grouped by topic has been included. Although most bibliographic searching, including that for the text, had been completed by April 1997, the supplementary bibliography includes entries from subsequent searches up to October 1997.

David Hornby
Luton
30 October 1997

Acknowledgements

The contributors thank P.K. Leech for further analysis of data from the Survey of Fertiliser Practice, D.E. Mathre for information about take-all in Montana, R.W. Polley for information from winter wheat surveys in England and Wales and R. Webster for help with kriging. They are also indebted to the following, who read and commented on drafts of the chapters indicated: J.F. Antoniw (Chapter 2), G.V. Dyke (Chapter 2) and A.R. Werker (Chapters 2 and 6).

IACR-Rothamsted kindly made available office, computer and library facilities to D. Hornby during the writing of this book. O. Andrade provided him with computer facilities and much appreciated help with the index at INIA Carillanca, Temuco, Chile.

Hoechst Schering AgrEvo GmbH generously provided funds to cover the costs of the coloured plates and full-colour cover of the book.

Abbreviations, Acronyms and Notes

Abbreviations and Acronyms

ADAS	Agricultural Development and Advisory Service (UK)
AFRC	Agricultural and Food Research Council (UK), became the BBSRC in 1994
AUDPC	area under the DPC
BBSRC	Biotechnology and Biological Sciences Research Council (UK)
BCA	biological control agent
CAP	Common Agricultural Policy
cfu	colony-forming units
CSG	Chief Scientist's Group (UK)
CSL	Central Science Laboratory, MAFF (UK)
DM	dry matter
DPC	disease progress curve
EHF	Experimental Husbandry Farm
EU	European Union
Gga	*Gaeumannomyces graminis* var. *avenae*
Ggg	*Gaeumannomyces graminis* var. *graminis*
Ggt	*Gaeumannomyces graminis* var. *tritici*
GS	growth stage
IACR	Institute of Arable Crops Research (UK)
INRA	Institut National de la Recherche Agronomique (France)
MAFF	Ministry of Agriculture, Fisheries and Food (UK)
NIAB	National Institute of Agricultural Botany (UK)
Pg	*Phialophora graminicola*
PDA	potato-dextrose agar
RES	Rothamsted Experimental Station (UK)
SED	standard error of difference
TAD	take-all decline (explanation in Sections 2.4 and 4.3.1)
TAR	take-all rating (Dyke and Slope, 1978)
TI	take-all index

UK	United Kingdom, comprising Great Britain (England, Wales and Scotland) and (since 1922) Northern Ireland
WB	winter barley
WW	winter wheat

Notes

British Isles This refers to the geographical region occupied by the UK (see Abbreviations) and Ireland.

Cross-references These are given as numbers with or without parentheses, e.g. '(see 4.2.1)' refers to Chapter 4, Section 2, subsection 1.

Further information Much additional background information on take-all is available in Asher and Shipton (1981) and Kollmorgen (1985).

Growth stages Unless otherwise specified, the growth stage (GS) terminology for cereals used in this book is the decimal code of Zadoks *et al.* (1974). Occasional references are made to an earlier growth stage key proposed by Feekes (Large, 1954). A comparison of the Zadoks *et al.*, Feekes and Haun scales is made in Cook and Veseth (1991) and an expansion of the definitions of the Zadoks *et al.* decimal code with stylized drawings is available in Tottman *et al.* (1979).

Oxide terminology Regulations still require this old terminology for plant nutrients in fertilizers. K_2O and P_2O_5 are examples which appear in descriptions of work on K and P in 3.4.2.

Quintal As in Table 3.2, cereal crop yields may be given in quintals per hectare ($q\ ha^{-1}$) internationally (Kent and Evers, 1994):
   mass per unit area of $1\ q\ ha^{-1}$ is numerically equivalent to $kg\ 100\ m^{-2}$
   $1\ q = 100\ kg$
   $1\ ha = 100\ ares = 10{,}000\ m^2$
   $1\ t\ ha^{-1} = (1\ q\ ha^{-1}) \times 10$

Units of fertilizer Occasionally reference is made to older works in which the amount of a particular nutrient is given in units. A unit is 1.12 lb (i.e. 1% of 1 cwt) per acre, which is equal to $1.25\ kg\ ha^{-1}$.

Introduction

Over the years, the focus of research on take-all has swung from one research group to another in different parts of the world, particularly Australia, Europe and North America. Quite often these groups have differed in what they considered important, but this has not dampened a widespread tendency to generalize their results, leading to an acceptance that lessons learned in one area are likely to be applicable elsewhere. If this were true, one may be forgiven for wondering why, after so many years of research into take-all, we have failed to find much that remains universally applicable after critical examination. Fortunately, the message is striking home in some areas, such as biological control, where the idea of a 'customized' approach with a local basis has come to the fore (see 4.3.3).

Cosmopolitan generalization in the take-all literature has not helped in understanding the disease as it occurs in Britain and it is time to try another approach which recognizes and emphasizes the importance of regional differences. A section entitled 'A brief history of take-all research' in Stelljes and Hardin (1995) is a potted version of the experience of researchers in the Pacific Northwest of the USA with a narrow focus on biological control and scant reference to activities elsewhere. A European history of take-all would have quite a different emphasis.

Regional differences are not unique to take-all. Important work on the infection of wheat by *Fusarium* spp. has been carried out in the USA, but because of differences in cultivation and in populations of pathogens the results are not easily applied to European conditions (Colbach, 1995; Colbach *et al.*, 1996). Similarly, the fungus *Cochliobolus sativus* Ito and Kuribay is widespread and causes, or is implicated in, cereal infections variously known as common root rot, dryland root rot (Cook and Veseth, 1991), foot and root rot, foot rot and spot blotch. Whereas the disease it causes is important in warmer wheat-growing countries, it is relatively unimportant in Europe, especially where there is little moisture stress (Jones and Clifford, 1978), and it is infrequent on wheat, barley and rye in Britain (Gair *et al.*, 1983).

Unjustified generalization of regional findings and simplification of concepts in take-all research have created expectations of control and management that continue to be unfulfilled. The waste in time, effort and resources arising from the naïve and uncritical adoption of ideas from other regions might be largely avoided if there were an agreed global framework within which to report take-all results. This would have immediate application in areas such as: (i) reconciling data from the high input systems of Britain with data from the low input systems in places like Western Australia; (ii) understanding the discrepancies in reported behaviour of host, pathogen and disease; and (iii) assisting in unravelling complex natural phenomena such as take-all decline (TAD).

Regional research programmes with different interests and emphases, inadequate survey results and the plain absence of good data can all lead to presumptions about regional differences in take-all. Differences claimed on this basis may in time prove to be fallacious. Real differences are likely to exist because of differences in climate, husbandry (e.g. fertilizer practice, cultivars, rotations, fungicide usage), soil type, etc. One difference arises out of Britain's natural advantage for growing high-yielding wheat (Bingham et al., 1991). Because, it was claimed, the British farmer has to contend with a wider range of diseases, difficulties in achieving adequate grain quality for bread-making and greater land costs than his competitors, he needs to exploit the benefits of the climate to the full by farming intensively rather than extensively.

Regions may be large scale, e.g. wheat-growing regions on different continents, or smaller scale, e.g. contrasting wheat-growing regions in England (see Table 1.4). The universal set of attributes of take-all, comprised of attributes reported in the world literature (e.g. response to NaCl fertilizer, control by triadimenol fungicide, etc.) and as yet unrecognized attributes, should be exhibited by take-all everywhere if regional differences do not exist. The literature suggests this is not so (Fig. I.1) and this theme is taken up again in Section 4.7. Unfortunately many recognized attributes have not been assessed adequately in all regions and therefore it is not known if they are strongly regional. Examples are responses to phosphorus fertilizer (greater in regions where soils are deficient in phosphorus) and to fungal BCAs such as Ggg (greater in Australia?). There are, therefore, insufficient data to draw Fig. I.1 so that the relative areas have a significance other than illustrating an idea. However, should this approach prove useful, then a challenge to future researchers might be to catalogue the attributes in relation to regions so that such diagrams could come to illustrate regional differences and similarities more accurately. This would contribute a more rational basis for transferring, or not transferring, control strategies and devising and reviewing research proposals. Figure I.1 is in essence a Venn diagram, illustrating the relationships between sets. (Using set algebra, the total number of attributes in the three regions is expressed by $n(A \cup E \cup U)$ and the number of those common to all three regions by $n(A \cap E \cap U)$, where A = Australia, E = Europe and U = USA.)

In this book, much of the experience relating to take-all in northern Europe since the publication of Asher and Shipton's (1981) book, *Biology and Control of Take-all*, is gathered together. This experience has been gained against a background of growing interest in topics such as biotechnology, biological control, molecular biology and the increasing use of computers within the discipline. On a wider scale, a rising tide of technological innovation has been revolutionizing public attitudes to agriculture. Increased food productivity is seen to have 'costs' that affect food safety, animal welfare, environmental sustainability, security of employment and social justice. These concerns have now been given 'academic coherence' in the new field of agricultural bioethics (Mepham *et al.*, 1995). The impact of all these innovations is to some extent reflected in the way take-all research has developed globally in the last 15 years (Fig. I.2). Consequently it is

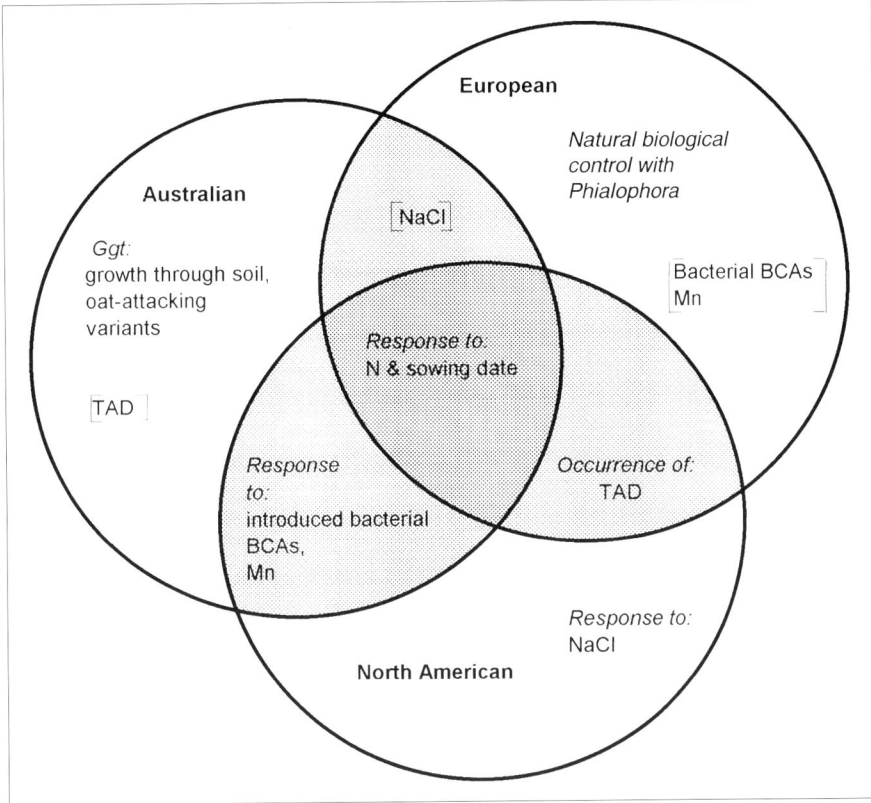

Fig. I.1. A way of considering regional differences in the disease caused by the wheat take-all fungus (*Gaeumannomyces graminis* var. *tritici*, Ggt).
The rectangle represents the universal set of attributes of take-all; within this, three subsets (circles) that characterize the disease in three regions (Australia, Europe and USA) are shown. The area outside any of these represents attributes that may occur only in other wheat-growing regions, which are omitted for clarity. The circles overlap so that each is divided into four areas: one with attributes in common with the other two regions (darkest shading); two, each with attributes in common with one other region (light shading); and one with attributes unique to the region (unshaded). A few selected attributes, which currently have strong regional associations and seem to indicate real regional differences, have been placed on the diagram. They are all discussed in the book and they are: the occurrence of the natural biological control phenomena, TAD and *Phialophora* (a delay in the onset of severe take-all attributed to the grassland *Phialophora, P. graminicola*); response to sowing date, or to N, Mn or NaCl fertilizers, or to introduced bacterial BCAs; and the pathogen able to grow through soil or infect oats. Some areas are distinguished by the absence of, or weak manifestation of, particular attributes (examples given within square brackets).

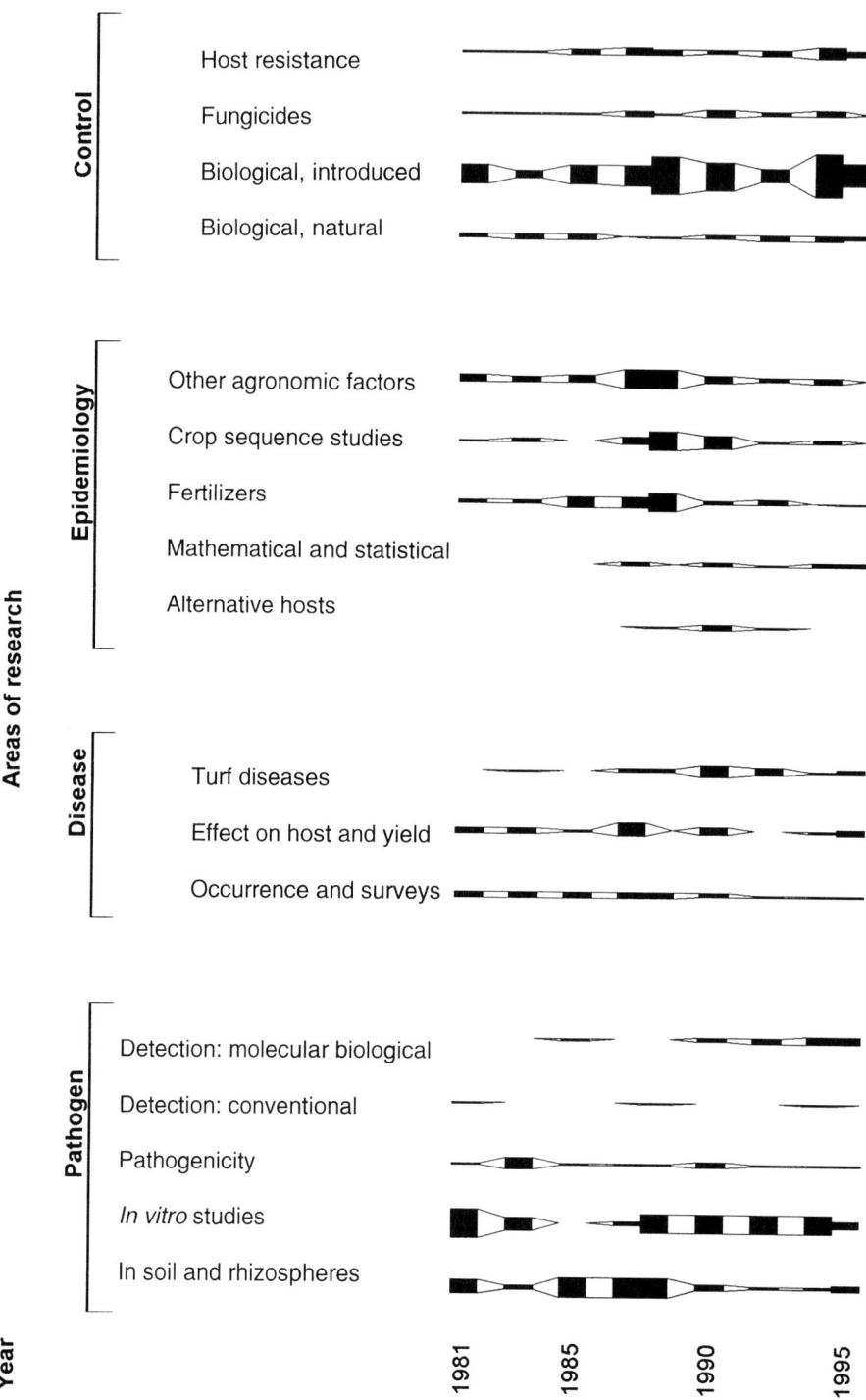

important to compare and contrast northern European experience with experiences reported from other wheat-growing regions. The intention is to provide a useful perspective for further research and development.

Fig. I.2 (opposite). Trends in world take-all research since 1980.
The widths of the kites indicate relative activity amongst topics and through years within topics, based on numbers of publications in selected years (source: bibliographic database on take-all, Rothamsted, February 1996). Black bands represent 9 years for which counts were obtained; in these years there was a total of 455 publications, 25 of which are not included in the figure because they were too general to categorize. White bands are interpolations for years without counts.
There is a certain arbitrariness in the groupings. For instance, 'Control' excludes control by fertilizers, rotation and other husbandry – these topics are included in 'Epidemiology'. Also, each paper was allotted to only one category, and where the contents were relevant to one or more categories the choice was made on what seemed to be the dominant topic.
Additional information about topics allotted to the subgroups:

Pathogen	In soil and rhizosphere	Inoculum, survival, growth through soil, rhizosphere studies, ecology.
	In vitro studies	Mycology, taxonomy, biochemistry, physiology, genetics, molecular genetics, viruses.
Disease	Effect on host and yield	Effect on yield, effect on host, host nutrient studies, infection process.
Epidemiology	Alternative hosts	Other cereals, grasses, weeds, volunteers (self-sown cereals).
	Mathematical and statistical	Field experiment design and analysis, modelling.
	Fertilizers	Fertilizers, minor elements, soil nutrients.
	Other agronomic factors	Cultivations, irrigation, sowing date, soil pH, herbicides.
Control	Biological, natural	TAD, *Phialophora*, natural fumigation.
	Biological, introduced	All work on introduced organisms, including mechanisms and strain improvement.
	Fungicides	Soil, seed and general.

For the major groupings given, the order of numbers of publications is control > pathogen studies > epidemiology > disease studies. The topic that has dominated take-all research has been biological control by introduced organisms, whereas establishing the importance of the disease has received least attention. The search for antagonists has continued to range widely, involving earthworms, a sterile fungus, Ggg and bacteria, including the popular fluorescent pseudomonads. In the last few years there have been rumours, news and some reports of chemical companies testing promising fungicide treatments for take-all. Little of this seems yet to have reached the literature and apparently, on the basis of publications, fungicide work still receives much less attention than biological control. However, some funding agencies cut their support for take-all research in the belief that the solution to the problems of take-all lay with these new treatments.

Past and Present

1.1 Historical Background

The name take-all was in use in Australia by 1870 for a devastating disease of cereals that had been recognized as early as 1852 in South Australia (Butler, 1961). For several decades there was a contentious debate concerning the cause of the disease and contemporary comment could be hard on those lacking 'scientific consideration' who presumed to set themselves up with an answer (Müller, 1873). Lawes and Gilbert (1870), ignorant of the cause of the disease, commented from Rothamsted on Australian experiences as follows:

> Take-all ... appears to flourish under as wide a range of circumstances as to soil, and a much wider as to climate .... it occurs on the more as well as the less fertile soils, and on newly broken-up as well as on exhausted land.

These observations are still true today, when the disease is regarded as the most damaging root disease of wheat worldwide (Heim *et al.*, 1986; Huber and McCay-Buis, 1993).

An early suggestion that a fungus 'Zenodochius Cerealium' or 'Xenodochus Cerealium' caused take-all was ridiculed (Müller, 1873), but from 1890 more and more reports attributed take-all to a fungus (Walker, 1981). However, the causal agent, an ascomycete fungus, was not fully proved until the first quarter of this century, when it was known under the misapplied name, *Ophiobolus graminis* (Sacc.) Sacc. About that time the first unequivocal reports of take-all were made in the UK (Table 1.1), although earlier, less certain reports do exist (Yarham, 1981). Until the name *Gaeumannomyces graminis* (Sacc.) Arx & Olivier var. *tritici* Walker, referred to throughout this book by the abbreviation Ggt, was published in 1972 (Walker, 1972) there had been no valid name for the wheat take-all fungus.

There is no comprehensive bibliography of take-all, but published items alone are likely to exceed 2000. A large majority of over 1000 references in Asher and Shipton's (1981) book concerned the disease or its causal agent and, at the end of 1995, the bibliographic database on take-all held at Rothamsted contained more than 700 references dated after 1980.) Also, many unpublished reports reside in

Table 1.1. A chronological impression of take-all in Britain, based on observations gleaned from published literature, disease intelligence reports and personal communications and relating mainly to winter wheat.

Harvest Year(s)	Notes	References
1912	First unequivocal records of take-all on wheat and oats in UK.	Massee (1912) Board of Agriculture and Fisheries (1915)
1924	First record in Hertfordshire, from the Plant Pathology Laboratory, Harpenden (MAFF).	Gregory (1951)
1933–1942	Period containing 2 years with severe, 5 with moderate and 3 with slight take-all.	Moore (1948)
1935	Disease more widespread than previously.	Garrett (1937)
1936	Take-all less prevalent than in 1935.	Samuel (1937)
1937	Most serious outbreaks to date had been on lighter chalk soils in Wiltshire, Hampshire, Cambridge, Norfolk and Yorkshire. This year more widespread than other years, except 1935.	Garrett (1937)
1938	Take-all scarcer than usual.	Buddin and Garrett (1941)
1940	Few cases.	Buddin and Garrett (1941)
1943–1948	Period contained 3 'take-all' years.	Moore (1948)
1946	Take-all occurred widely.	Moore (1948)
1948	**More widespread and destructive since first recognized 35 years ago.**	Moore (1948)
1955–1956	Take-all appeared to remain a problem in Wiltshire and Hampshire, but not in Cambridgeshire or Norfolk.	Smith (1960)
1975	Generally not a take-all year.	ADAS National Disease Intelligence Report – 28 July 1975. Cereal Disease Report No. 13
1976	Take-all obvious, level of infection generally mild.	Plant Disease Intelligence No. 14, 5/7/76
1977	Take-all levels higher than for many years. Take-all assessed for first time. On roots of > 50% of samples, most in the slight category; most severe in Wales.	Plant Disease Intelligence No. 16, 1/8/77 King (1977), also see Section 6.2.
1978	Season very favourable to build-up of Ggt, but disease not as severe as expected.	ADAS *Winter Wheat – Disease Intelligence Report and E. Region Trial Results – 1978/79*. PP 79/40, 31/12/79
1979	AUTUMN High levels of inoculum in the soil; breakdown inhibited by very dry conditions. WINTER Cold; inoculum breakdown inhibited. SPRING Wet; ideal for infection. SUMMER Mid-summer drought, stress. Mid-June patches. Mid-July, **worst epidemic for many years**.	Ditto

Year	Notes	Reference
1980	Take-all above average. AUTUMN Favourable for early sowing, subsequent mild, 'open' weather. SPRING Take-all patches occurred earlier (mid-April) than 1978 and 1979, crops growing away. Less serious losses than 1979.	ADAS Disease Report No. 20, 11/8/80 D.J. Yarham notes
1981	Take-all causing concern. Take-all mentioned in relation to premature ripening and whiteheads. More severe in winter barley than usual. Second wheats often worse than third wheats. AUTUMN Mild, 'open' weather; infections. WINTER Eyespot lesions. SPRING Wet March–May favoured build-up, waterlogging aggravated effects. Patches appearing in April, common late May. SUMMER Dry after ear emergence, moisture stress, premature death, extensive patches, serious yield losses.	ADAS Crop Pest and Disease Intelligence, Disease Report No.16, 13 July ADAS Crop Pest and Disease Intelligence, Disease Report No.19, 3 August Summary 1980/81 PP 82/21 May 82 D.J. Yarham notes Wheat Diseases Harvest Year 1981
1982	Epidemic much later than in 1981; losses 'will be' less severe. Occasional yellow patches on early-sown crops before Christmas. Dry weather in late spring checked, continued build-up in wet June. Not as severe as 1981. Premature ripening in many second and occasionally third wheats.	PP 82/32 ADAS
1983	Patches merged into general drought effects, indications that infections *less* severe than 1982. WINTER Mild, early, encouraged build-up. SPRING Build-up in early part, ameliorated by cool, wet period. SUMMER Severe take-all in several second and third wheat crops.	PP 83/45 PP 84/3, 18 Jan 1984
1984	AUTUMN Infection widespread and severe; common in early-sown crops. WINTER Disease decreased. SPRING Disease decreased in the early part, resurgence later. SUMMER Resurgence, patches in second and third wheats. *Not as serious as 1983.* Few reports of serious damage.	PP85/2

Table 1.1. *Continued*

Harvest Year(s)	Notes	References
1985	AUTUMN Well established in early-sown crops in mid-November. WINTER Levels high in some crops at end of winter. SPRING Disease declined to low. SUMMER Disease increased, patches in second and third wheats at the end of June. Wet weather reduced drought stress – plants withstood relatively high levels of attack without serious loss of yield. Conditions very conducive to infection, high proportion of crops affected by late July–early August.	ADAS Eastern Regional Disease Intelligence & Trials Summary 1984/85 Take-all – the 1985 epidemic PP85/69
1986	Generally less take-all than 1985 and 1984; few reports of severe infection even in second–fourth wheats.	PP 87/2
1987	Take-all a major problem in winter wheat. Very severe in second and third wheat crops, particularly on waterlogged clays and light soils that had puffy seedbeds. Aggravating eyespot and sharp eyespot. Second and third wheats suffering from take-all and sooty moulds. '... **a particularly bad year for take-all** ...'.	CI 87/25 17/7/87 CI 87/26 24/7/87 CI 87/27 31/7/87 D.J. Yarham: letter to Agricultural Advisory Officers
1988	AUTUMN Wet with high soil temperatures. Early infection slight. WINTER Mild and wet, favoured development of take-all. Patches beginning to show, generally associated with poor soil conditions and acid patches. Much disease in many crops; wet weather enabling plants to withstand this. Patches less obvious than in 1987, but often more disease as illustrated by the following data from Boxworth EHF:	CI 87/39 6/11/87 CI 88/6 26/2/88 CI 88/19 3/6/88 CI 88/25 14/7/88 ADAS Cropping Services Crop Intelligence Reports. Eastern Region CI 88/26 22/7/88

	% Plants (roots) infected in		
Date	1st WW after rape	2nd WW after rape	Continuous wheat
10/7/87	–	54(11)	50(12)
15/7/88	36(6)	80(34)	88(31)

Becoming more obvious (cool wet weather).

		D.J. Yarham letter to D. Hornby 10/10/89
1989	WINTER Late, severe attacks.	
	SPRING Further development favoured. Then became dry, generally favoured crop more than disease. Serious epidemics in certain areas only, e.g. peats, Norfolk Fens.	
	Severe take-all began to show in occasional crops, including some in Essex and Kent.	CI 89/6 10/2/89
	Disease showing in second and third wheats throughout the Eastern Region, aggravating moisture stress.	CI 89/21 29/5/89
	Infection seemingly favoured by a mild winter and a wet spring; then dry weather accelerated symptoms.	CI 89/23 9/6/89
	In many crops with leaf tipping.	CI 89/24 16/6/89
	Showing in many second and third wheats.	CI 89/27 7/7/89
	Poor grainfill, whiteheads common and causing concern.	CI 89/32 11/8/89
	Take-all showing as yellowing, particularly in Northamptonshire.	CI 8944 15/12/89
1990	Patches in some early wheat and barley crops in the Eastern Region.	CI907 23/2/90
	Mild winter favoured take-all in early-sown crops, the disease already severely restricting crop growth in affected areas.	CI9012 30/3/90
	Take-all aggravating drought stress in many second and third wheats.	CI9021 8/6/90
1991	Severe take-all in a crop in Norfolk.	CI9114 2/5/91
	Disease in a few second and third wheats, sometimes exacerbated by areas of low pH.	CI9115 10/5/91
	Drought stress most obvious in third wheats, crops beginning to show effects of take-all.	CI9118 31/5/91
	Take-all very common and contributing to whiteheads commonly associated with severe eyespot and sharp eyespot.	CI9126 19/7/91
	Severe take-all in many crops, particularly third wheats and crops on poor soils; whiteheads common; premature ripening attributed to take-all; disease in long runs of cereals where TAD should have been established.	CI9126 26/7/91
1992	Yellow patches in some second and subsequent wheats. Root symptoms of take-all and also interveinal yellowing similar to Mg deficiency.	CI9210 13/3/92
	Take-all in second and subsequent wheats evident in yellow patches caused by acidity.	10/4/92

PP numbers, and later CI numbers, are codes that were used on ADAS Eastern Region Disease Intelligence Reports, indicating year and issue.

government and commercial organizations. In 1966 a conference on take-all was held at the University of Leeds with 68 delegates, 65 of whom were from the UK, and there a list of 73 British references on take-all was supplied by the Plant Pathology Laboratory (MAFF) at Harpenden. Despite all this writing about take-all, there has been previously only one book (Asher and Shipton, 1981) dedicated to the disease and this book rapidly and rightfully assumed the position of main reference work on the subject. Prior to this book, reviews by Butler (1961) and Nilsson (1969) served as general reference works and the first edition in 1972 of a book by Gair *et al.* (1983) contained useful practical information for the UK. Since Asher and Shipton (1981) there has been the First International Workshop on Take-all of Cereals, which took place in 1983 with 63 participants, representing nine countries, in Horsham, Victoria, Australia, the proceedings of which were published later (Kollmorgen, 1985). References held in the Rothamsted database (which is not claimed to be comprehensive) indicate that since Asher and Shipton (1981), publication worldwide peaked in the period 1986–1988 at about 70 publications per year. In 1991 a Home-Grown Cereals Authority Research Review, on which this book is based, was produced (Hornby and Bateman, 1991).

In farming circles take-all has always been notorious. This reputation now owes much to the lack of economic chemical controls and resistant wheat cultivars for commercial use. Yet, paradoxically, take-all has played a leading role in advancing our knowledge of diseases caused by soil-borne plant pathogens. In the UK it can be largely avoided by not growing susceptible crops consecutively, but for decades a significant proportion of UK cereals has been grown as second or subsequent cereals and has therefore been at risk from take-all. The disease is usually a problem only under these conditions in Britain, yet this is the basis for a widely promulgated, general impression of take-all as an intractable disease. Undue emphasis on certain aspects of the disease has coloured opinions and attitudes concerning the achievements of, and prospects for, take-all research and has led to the perception by some that, because take-all still exists as a problem, there has been no progress. Some of these issues and their interactions are summarized in Fig. 1.1.

It should be appreciated that many of the problems of take-all research are common to other soil-borne diseases that are difficult to control. The following observations were made in the context of a study of phymatotrichum root rot of cotton (Jeger and Lyda, 1986), but are mostly familiar ground for those conversant with take-all:

> Apart from avoidance of infested areas, management options are more strategic (i.e. available only before the start of the growing season) than tactical (i.e. available during the growing season), involving rotation, choice of early-maturing cultivars, organic and inorganic amendments of the soil and deep chisel ploughing. Each cultural practice has on occasion provided some level of control, but no one approach has consistently proved effective and economic. There are no resistant host varieties and no economic fungicide or fumigant treatments. The development of basipetally-translocated systemic fungicides may offer new opportunities for control. Despite the lack of control options, weather-based forecasting schemes would be beneficial in alerting farmers to potential losses.

Table 1.1 gives an impression of how take-all has waxed and waned in the years since it was first recorded in Britain in 1912. Garrett (1937) pointed out that the spread of the disease in wheat-growing counties was apparent rather than real, because the take-all fungus was indigenous on a number of grasses and had probably been an inhabitant of English soils for longer than wheat had been grown here. Soil type and farming methods were considered to be restricting the fungus. These views, however, pre-dated knowledge of the *Gaeumannomyces–Phialophora* complex (see 5.2), so it is possible that some of the 'take-all fungi' isolated from grasses in those days were not Ggt. Misidentifications were still coming to light in the 1970s, as in the case of a fungus isolated in a bioassay of a spring wheat soil in 1967, initially identified as *Ophiobolus graminis*, and then discovered to be *Phialophora* sp. (lobed hyphopodia) (Hornby, 1978a).

The record in Table 1.1 is very uneven. Most of the explanations for disease incidence and severity provided by the observers must be regarded as opinion, because there is no evidence of experimental verification of their assertions. Also, intelligence-gathering activities concerning take-all in Britain have been erratic and years in which take-all seems to have been quiescent could reflect the absence of records as much as absence of disease. In the last few years the regular bulletins, previously produced by ADAS, giving national information on take-all have been notably absent, although the Winter Wheat Disease Survey, which introduced observations on take-all in 1985, is still continued by the Central Science

Fig. 1.1. Five major concerns in take-all studies and how they interact.

Laboratory. Details of this survey's methods are given in Section 6.2: results up to 1988 have been published (Polley and Thomas, 1991) and there are annual reports (e.g. Polley et al., 1994, 1995, 1996). Signs of the disease above ground fluctuated in the surveyed crops and these suggest that take-all was less prevalent (i.e. fewer crops were affected) in 1994 than in 1995 or any other of the previous eight years for which there were observations on take-all (Table 1.2). This impression is supported by Table 1.3, which estimates that the percentage of wheat crops

Table 1.2. Percentage of crops of winter wheat in England and Wales affected by take-all, 1985–1995.

Take-all severity	Year										
	1985	1986	1987	1988	1989	1990	1991	1992	1993	1994	1995
Category 1	13.7	—	24.4	16.7	16.3	19.6	17.4	17.4	24.9	11.5	16.5
Category 2	7.0	—	23.4	5.7	8.5	6.8	4.7	5.1	4.3	3.1	3.1
Category 3	4.3	—	13.2	6.1	5.4	4.9	2.9	4.0	6.4	1.8	2.0
Category 4	4.7	—	9.1	0.7	3.5	0.9	0.9	1.9	1.7	1.0	0.6
% Crops with take-all	29.7	—	70.1	29.2	33.7	32.2	25.9	28.4	37.3	17.5	22.2

The numbers of crops in the survey each year are given in the final column of Table 1.4 and the severity categories (4 is worst, with > 10% of the field as patches) are based on above-ground symptoms (see Section 6.2 for details).
Source: CSL/ADAS Winter Wheat Disease Survey.

Table 1.3. Percentages of wheat fields in England and Wales growing first or consecutive wheat crops and estimates of the percentage of the national wheat crop at risk from take-all and the percentage of the national wheat harvest coming from crops not at risk from take-all.

Wheat crops	Years (19—)															
	76	77	78	79	80	81	82	85	86	87	88	89	90	91	92	93	94
First (a)	70	60	63	55	51	52	48	47	56	45	49	56	54	64	60	62	74
(b)	18	19	20	14	16	11	14	12	13	11	9	11	8	8	8	7	10
Second	23	33	27	31	31	28	28	32	20	32	31	26	32	19	24	27	16
Third	4	4	5	7	11	9	10	10	10	10	9	6	5	6	8	4	3
Fourth and subsequent	3	3	5	7	7	11	14	11	14	14	11	13	9	11	9	6	7
At risk from take-all (R)	48	59	57	59	65	59	66	65	57	66	60	56	54	44	49	44	36[a]
Harvest from crops not at risk	55	44	46	44	38	44	37	38	46	36	43	50	49	59	54	59	67

(a) All first wheats. (b) First wheats after a cereal other than wheat.
[a]This value recorded as 37% in Polley et al. (1995).
Notes
1. Crops at risk from take-all are taken here to be all wheat crops grown as second or subsequent cereals, but this is a slight overestimate because oats often behave as a non-susceptible break.
2. The percentage of the harvest of wheat grain from crops not at risk assumes first wheat crops after non-cereal breaks yield 15% more than other wheats and has been estimated as $(100-R)115/R + (100-R)1.15$.
3. Consecutive wheat crops have a regional distribution.
Sources, 1976–1988: Polley and Thomas (1991); 1989–1994: CSL/ADAS Wheat Disease Survey.

at risk from take-all was lower in 1994 than in any year since 1975. In years for which there are both surveys and disease intelligence reports, take-all tends to be given a greater significance in the latter. This is because intelligence reports have reflected the situation in second and subsequent wheats only and so could be biased by a few very severe cases. The surveys, in contrast, represent the whole wheat acreage, between one-third and two-thirds of which (as Table 1.3 shows) will not have been at risk from take-all.

Despite this unevenness in the national record of take-all, the experience of workers at Rothamsted over the last 30 years is that it has been extremely rare not to get at least moderate take-all in experiments that were managed to encourage natural occurrences of the disease. Since this has been an annual objective for most of this period, moderate to severe take-all was recorded somewhere on Rothamsted farms in most of the years in which the disease appeared to be unimportant nationally. There is also an impression that in these years the disease that did occur had a greater detrimental impact on yield than it would have done in years more favourable to disease nationally (Fig. 1.2).

Table 1.4 gives some indication of the incidence of winter wheat crops with visible patches of take-all in different regions of England and Wales in the period 1987–1995. Because there are quite large differences between the number of crops observed in the different regions, care should be taken in interpreting these figures. Wales and Cornwall/Devon each have 3 years with the greatest incidences, but the values are based on small numbers of samples. No one region consistently has a greater or lesser incidence of crops with patches than the others, and the whole table conveys an impression of a disease more widely spread than in the 1930s to 1950s, when attention was focused on a limited number of counties and soil types (Table 1.1).

The introduction of clover into England around 1620 started a shift from a deteriorating agricultural situation with regard to nitrogen to a sustainable, high-yielding, low-input agriculture based on clover. After World War II, industrially produced nitrogen fertilizers almost completely displaced clover, and nitrogen pollution (see 2.3.5) became one of the plagues of our time (Kjaergaard, 1995). Such changes need to be taken into account when trying to understand the historical record and the causes and timing of past outbreaks of take-all in the British Isles. After Lawes wrote in 1876 that 'corn, and especially barley, can be grown with profit more frequently than has hitherto been thought consistent with good farming' (Dyke, 1991), the wheat acreage decreased by about two-thirds in the ensuing agricultural depression (Smith, 1960) and significant intensification of cereals did not occur until the 1930s (Yarham, 1979). These early milestones, described in Newby (1988), are summarized in Fig. 1.3 and relevant events are shown in Table 1.5. Areas and yields of wheat, barley and oats in Great Britain are shown graphically for the period 1895–1995 in the *Cereal Variety Handbook* (National Institute of Agricultural Botany, 1996), and areas of total cereals, barley, wheat and oats and these crops as percentages of the crops and grass area in the UK are shown graphically for the years 1926–1967 in Britton (1969). In the period 1926–1935, areas of total cereals, wheat and barley were decreasing, whereas areas of oats declined and recovered. Figures 1.4 to 1.9 show certain

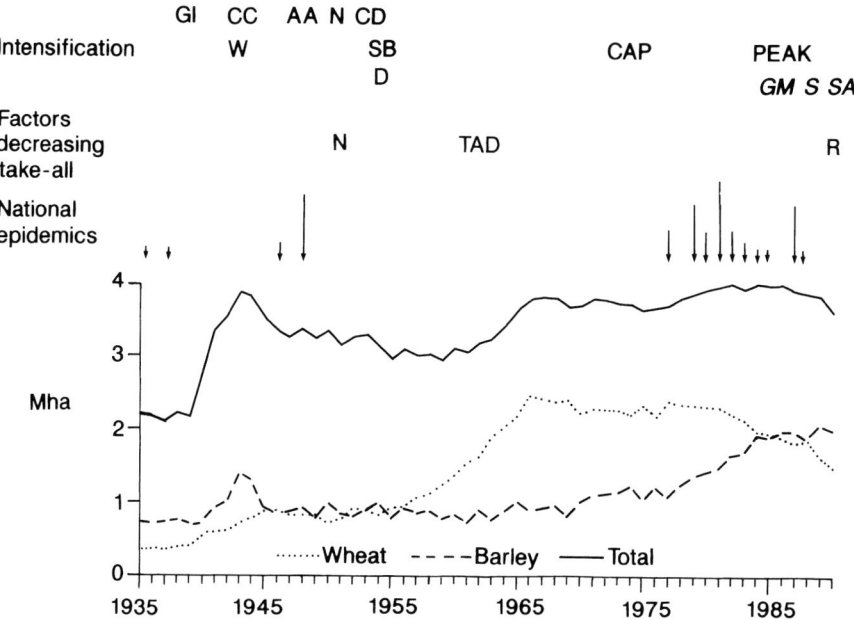

Fig. 1.2. Changes in areas of wheat, barley and total cereals in the UK during the period 1935–1990, shown in relation to events contributing directly or indirectly to intensification of cereal growing, factors likely to have decreased take-all and years when take-all was apparently widespread (national epidemics).

Key

Intensification
  Favouring: **GI**, government intervention; **CC**, compulsory cropping; **W**, weedkillers; **AA**, Agricultural Act; **N**, increased N usage; **CD**, advent of Cappelle Desprez; **SB**, continuous spring barley; **D**, derationing of animal feeds; **CAP**, CAP support; **PEAK**, peak production Opposing: **GM**, gross margins subsiding; **S**, surpluses; **SA**, set-aside. There is more information about these events in Table 1.5.

Factors decreasing take-all
  N, increased N usage; **TAD**, take-all decline; **R**, rotations (increasing proportion of first wheats).

National epidemics
  Impressions gleaned from the limited information in Table 1.1; the length of the arrows suggests the relative severity of the attacks.
  Area data only, over a slightly longer period, also appear in Fig. 1.4.

aspects of cereal production for the years 1935–1993. During World War II, the area of cereals increased from 2,154,000 ha in 1939 to a peak of 3,869,000 ha in 1943 (Fig. 1.4). At this time oats were the dominant cereal and more wheat was grown than barley. After the war, for a time, the areas of wheat and barley tended to be similar and constant and a progressive fall in oat acreages resulted in a decline to 2,942,000 ha of cereals by 1959. The late 1950s marked the beginning of a massive upsurge in spring barley production that was sustained well into the 1960s, to be followed by a period of slow decline as wheat acreages began to be increased. Another set of figures for England alone (cf. those for the UK in Fig. 1.5) shows that in 1939 wheat and barley occupied about 12% of the total area of crops

Table 1.4. Percentage of winter wheat crops with visible take-all patches, on a regional basis for England and Wales, 1987–1995.

Year	NO(N)	NO(L)	M&W	EAST	SE(W)	SE(R)	SW(B)	SW(S)	WALES	NAT
1987	28.6[b]	50.0[c]	42.3 (68)	36.2 (43)	44.0[a]	75.0[b]	69.2[a]	50.0[c]	50.0[c]	45.7 (197)
1988	13.6[a]	3.6[a]	21.4[a]	11.5 (113)	20.0[a]	3.3 (30)	7.4[a]	66.6[c]	25.0[a]	12.5 (280)
1989	11.8[b]	0.0[b]	18.2 (55)	17.0 (112)	11.1[a]	11.1[b]	40.0 (30)	20.0[c]	19.2[a]	17.5 (258)
1990	19.0[a]	3.2 (31)	15.9 (63)	11.5 (139)	10.7[a]	3.6[a]	14.8[a]	50.0[c]	3.7[a]	12.6 (326)
1991	17.4[a]	12.9 (31)	9.1 (66)	6.3 (143)	11.1[a]	0.0[a]	5.4 (37)	16.7[c]	30.8[a]	8.4 (344)
1992	7.4[a]	0.0 (52)	18.6 (70)	12.0 (159)	19.2[a]	5.3[a]	13.2 (38)	0.0[c]	39.1[a]	11.0 (373)
1993	10.7[a]	7.0 (43)	16.1 (62)	7.4 (121)	9.1[a]	24.2	19.4 (36)	16.7[a]	30.0 (30)	12.5 (345)
1994	12.9 (31)	0.0 (46)	3.0 (67)	7.3 (165)	4.4[a]	3.6[a]	9.4 (32)	25.0[a]	9.1[a]	6.0 (384)
1995	3.6[a]	0.0 (5)	1.7 (60)	5.9 (152)	31.8[a]	0.0[b]	5.6 (36)	0.0[c]	4.0[a]	5.7 (352)

NO(N), Cleveland, Cumbria, Northumberland, Durham, N. Yorks (Northallerton), Tyne & Wear.
NO(L), Humberside, N. Yorks (Harrogate), S. Yorks, W. Yorks.
M&W, Cheshire, Derbyshire, Notts, Leics, Hereford & Worcester, Gr. Manchester, Lancs, Merseyside, Shropshire, Staffordshire, Warwickshire, West Midlands.
EAST, Bedfordshire, Northants, Suffolk, Essex, Hertfordshire, Gr. London (E), Lincolnshire, Norfolk, Cambridgeshire.
SE(W), Kent, Surrey, E. Sussex, W. Sussex, Gr. London (SE).
SE(R), Berkshire, Buckinghamshire, Hampshire, Oxfordshire.
SW(B), Avon, Dorset, Gloucestershire, Somerset, Wiltshire.
SW(S), Cornwall, Devon.
WALES, all Welsh counties.
NAT, England & Wales.
() Number of samples, if > 29; [a] number of samples 20–29; [b] number of samples 10–19; [c] number of samples < 10.
Percentages are of crops in severity categories 2 + 3 + 4 (see Section 6.2).
Source: CSL/ADAS Winter Wheat Disease Survey.

and grass in England (excluding rough grazing and fruit, ornamentals and glasshouse crops). By 1959 this figure had risen to 23%, by 1979 to 37% and by 1984 to 40% (Murphy, 1989).

Wheat acreages began to increase rapidly in the late 1970s as the decline in barley acreages began to accelerate. In 1986, 2 years after cereal acreages reached a maximum, the area of wheat exceeded that of barley for the first time since 1954. Acreages fell progressively from 1984 to 1993 for total cereals and from 1988 to 1993 for barley, whilst wheat peaked at 2,083,000 ha in 1989 and has fluctuated since. Triticale was first recorded in the statistics in 1989 and achieved 11,000 ha in both 1991 and 1992. Apart from a short period around 1970, wheat plus barley increased both as a percentage of all crops and in area from about the mid-1950s until 1984 (Figs 1.5 and 1.6). However, between 1935 and 1993 there was a decrease of 1,625,000 ha in the area occupied by total cereals and grass (Fig. 1.7).

The proportions of spring to winter cereals have changed in the last quarter of a century. In 1968 the percentages of spring cereals were 14, 95 and 85% for wheat, barley and oats, respectively, but in 1994 the percentages were 5, 43 and

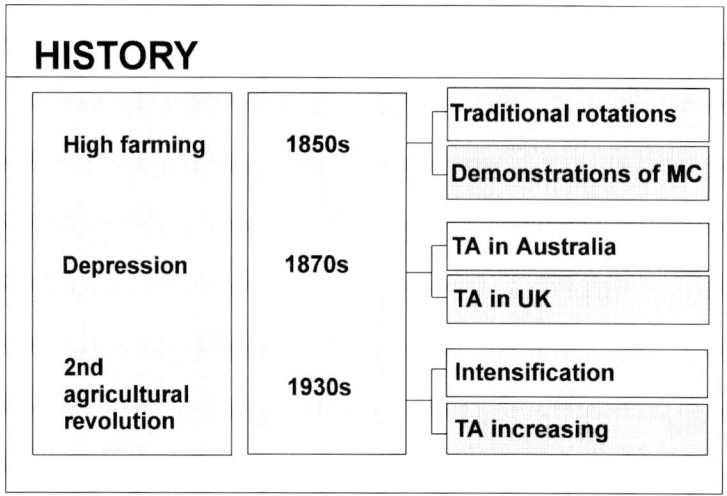

Fig. 1.3. Some historical periods in agriculture and their impact on cereal production and take-all disease. MC, monoculture; TA, take-all.

40% (Home-Grown Cereals Authority, 1995). Table 1.6 has additional information about barley in 1990–1991: crops in over 80% of winter barley fields were second or subsequent cereals, but spring barley as second or subsequent cereal crops decreased from about 60% of fields in 1990 to about 45% in 1991.

The difficulty of harmonizing relevant surveys is illustrated by considering data from the Fertiliser Practice Surveys for 1990–1991 (Table 1.6). In this table the previous cereals are not known, but winter wheat or winter barley are the most likely. There is still a trend showing an increasing proportion of first wheat crops and a decreasing proportion of wheat crops as second cereals in England and Wales, which is present, considering consecutive crops of wheat alone, for the same years in Table 1.3. Further comparison and cross-checking are difficult because the data in Table 1.6 distinguish winter and spring cereals, whilst third and subsequent consecutive crops are grouped differently to those in Table 1.3.

To put wheat production in the UK into perspective, 558.8 Mt of wheat were produced from an area of 222 Mha worldwide in 1993/94 and estimates for 1994/95 were 526.5 Mt from 215.1 Mha (Home-Grown Cereals Authority, 1995). With a production of 80.3 Mt of all types of wheat, the EU (Belgium, Luxemburg, France, Germany, Italy, The Netherlands, Denmark, Eire, the UK, Greece, Spain and Portugal) was the second biggest wheat-producing region after China (106.4 Mt) in 1993/94. After France (4.3 Mha, 28.4 Mt) and Germany (2.4 Mha, 15.7 Mt), the UK (1.8 Mha, 12.9 Mt) had the third greatest area in common wheat (*Triticum aestivum*) production in the EU (12.3 Mha, 73.9 Mt) in 1993 (Home-Grown Cereals Authority, 1995; see Table 1.24 for the 1992 areas). In that year average yield in the UK was 7.32 t ha^{-1}, the greatest recorded, although production had peaked at 14.357 Mt in 1991. In the UK, the large majority of the wheat (Table 1.7 and Fig. 1.9) and barley (Table 1.7) has been

Table 1.5. Some of the events in cereal production in Britain that may have affected take-all directly or indirectly.

Period	Event or activity	Relevance to take-all
The 1000 years before 1800	Wheat crops consisted of 'landraces'; semi-natural selection improved resistance to diseases and pests and gradually increased yield in a genetic background of very tall plants.[a]	No information, but predominant rotation in strip farming (winter cereals, spring cereals, fallow for grazing) and impoverished soils may have encouraged take-all.
19th century	Norfolk Four-course System (roots, barley, legume/grass ley, wheat).[a,b]	Would not have favoured severe take-all.
1913	Many soils deficient in P (a situation lasting until the early 1950s).[b]	Superphosphate of lime advocated for decreasing take-all.
Late 1930s	Government intervention (incentive for technical progress).[c]	
Early 1940s	Compulsory cropping.	Increased areas of susceptible crops.
	Development of selective weedkillers.[b]	Paved way to intensive cereal cropping.
1947	Agricultural Act: government subsidies.	
1950s	Beginning of increased N usage: 1951–1962, 36 to 79 kg ha^{-1} (Suffolk); 1975–1979, 93 to 135 kg ha^{-1} (England and Wales).[b]	Likely to have decreased take-all losses in Britain's progressively more intensive acreage.
	Development of continuous spring barley growing, particularly on light calcareous soils of southeast England; continuing into 1970s.[b,d]	Yield recovery and stabilization through TAD.[e]
1953	Eyespot-resistant Cappelle Desprez appears on NIAB recommended list.	A milestone on the way to intensive wheat.
1954	Derationing of animal feedstuffs.	Barley production increases.
1972	CAP support: cereal prices start rising.	
1972–1986	Wheat yields increased from 4.2 to 6.9 t ha^{-1} (from 4.9 to 6.15 t ha^{-1}: 1977–1982). Barley yields increased from 4.0 to 5.3 t ha^{-1}.	
1976	Dramatic increase in use of fungicides from about 15% of crops to a peak of around 90% in 1985 (Polley and Thomas, 1991).	
1980s	Surplus grain and set-aside.	Intensification means the proportion of the wheat crop at risk is greater than ever.
	Rise in oilseed rape production.	As a break crop, provides a profitable opportunity for less intensive cereals.
1984	Peak cereal production: > 26 Mt.	
1985	Gross margins subsiding.	
1986	Wheat area surpassed barley area for first time since early 1950s.	
1990	Increasing impact of sustainability, low input systems, environmental protection, organic farming approaches, curbs on production.	Trend to more first wheats and less of the national crop at risk.
1995	FAO Bulletin warning grain stocks below safe level – beginning of turn-around from grain surplus to shortage? (Richardson, 1995).	Possible halt or reversal of trend to more first wheats?

[a]Bingham et al. (1991); [b]Yarham (1981); [c]Britton (1969); [d]Yarham (1979); [e]Shipton (1972, 1975).

grown in England. In a winter wheat survey conducted by ADAS/CSL (which has provided information for Tables 1.2–1.4 and 1.13–1.16 and is discussed in Section 6.2 in relation to the procedures used for take-all), the percentage of wheat crops grown as first wheats in the Eastern Region, which comprised 17% of the UK wheat area in 1994 (National Institute of Agricultural Botany, 1996), tended to follow the percentages in the national crop (Fig. 1.8). Both sets of figures show that second and subsequent wheat crops exceeded 50% of wheat crops for a large part of the 1980s. The national figures were obtained using a stratified sample based on the areas of winter wheat grown in each region (the regions are listed in Table 1.4), and the numbers of crops sampled nationally are given in Table 1.4 for some of the years plotted in Fig. 1.8. For the remainder, the range was 142 crops in 1976 to 293 crops in 1986.)

Winter wheat grown in a climate of relatively even temperature and rainfall matures more slowly than elsewhere, producing higher yield and a lower nitrogen content, and is better suited for biscuit- and cake-making than for bread. The principal wheats of the world may be classified as in Table 1.8. Winter wheat comprised $c.$ 97% of total wheat in the UK in 1991. The UK exported about 3.0 Mt of wheat and 1.2 Mt of barley in the year July 1993 to June 1994 and during the same period imported almost 1.8 Mt of common wheat (soft, semi-hard, hard and

Fig. 1.4. Areas of total cereals (C), wheat (W), barley (B) and triticale (T) in the UK, 1935–1993.
Source: Central Statistical Office (1935–1995).

seed wheat) and 50,000 t of durum, *Triticum turgidum* var. *durum*, grain (Home-Grown Cereals Authority, 1995).

Bringing together the statistics on cereal areas as above and relating them to events in the evolution of cereal production helps to identify the trends in intensification in cereal production in the UK since 1933. It shows that high input–high output farming and increased frequency of cropping have been associated. Against this can be set the little that can be gleaned about national epidemics of take-all (Fig. 1.2). A useful starting point is to look at how areas of cereals have changed since the 1930s and to note three peaks in total area: the first associated with World War II, the second with increased spring barley and the third with increases in winter wheat. Despite these fluctuations this period has been essentially one of progressive intensification of cereal growing. One definition equates intensive cereals with high input–high output farming. To a large extent in the UK this has meant increased frequency of cropping and it is this aspect that is important in take-all. The proportion of wheat and barley to total crops and grass rose from below 20% in the 1950s to 30% nationally in 1966 (Fiddian, 1973). In 1971 UK farmers produced 59% of the country's needs in total cereals, but by 1981 this had grown to 103% (Yarham, 1986). By 1970, continuous spring barley was a familiar feature of many light land areas and by the mid-1970s continuous wheat

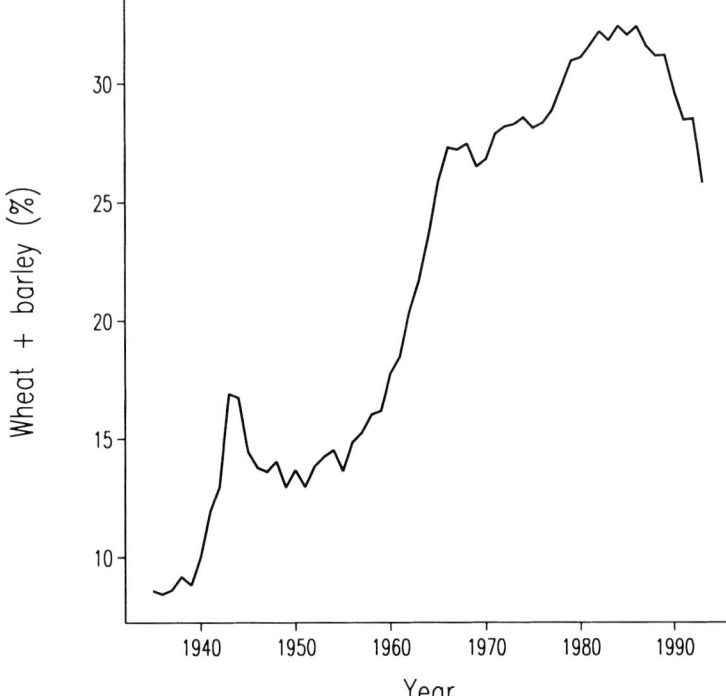

Fig. 1.5. Wheat and barley areas as a percentage of the area of all crops in the UK, 1935–1993.
Source: Central Statistical Office (1935–1995).

growing was established on well-structured chalky boulder clays of the eastern counties (Yarham, 1986). Between 1977 and 1988 fields growing wheat as a second or subsequent cereal ranged from 57% to 66% (the larger percentage occurring in both 1982 and 1987) of UK wheat fields (Table 1.3).

Massee (1912) provided the first unequivocal report of take-all in British crops, but it was increased take-all severity associated with intensification of cereal cropping in the 1930s, especially on light alkaline soils (Yarham, 1979), which was described as the first 'flare-up' of the disease in recent times (Garrett, 1950). Before the 1970s much information about outbreaks of take-all nationally tended to be anecdotal. At Rothamsted between 1930 and 1976 severe take-all was reported in 1948 (when take-all nationally was more widespread and destructive than previously), 1956, 1958, 1960, 1963, 1965, 1971, 1972 and 1974 (Hornby, 1978b). ADAS Disease Intelligence reports for the Eastern Region recorded higher take-all levels than for many years in 1977, the worst epidemic for many years in 1979 and serious yield losses attributable to take-all in 1981. Then followed a succession of years of dwindling disease, terminated by an upswing in 1987 (Tables 1.1 and 1.2; Fig. 1.2). Advisers were generally agreed that the take-all epidemics in wheat in the early 1980s were more important than those of the mid-1970s (Hornby and Henden, 1986). The generally little disease in the

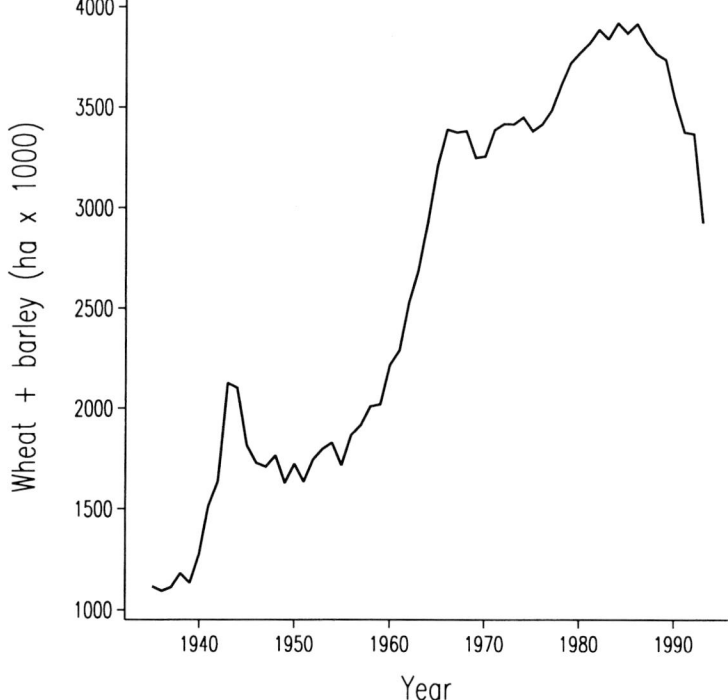

Fig. 1.6. The combined areas of wheat and barley in the UK, 1935–1993. Source: Central Statistical Office (1935–1995).

mid-1970s contributed to a premature relegation of take-all to an unimportant problem in cereal production. By 1983, a resurgence of the disease necessitated the reinstatement of take-all as a priority research objective in the AFRC. Intensification of cereals following World War II has been progressive, but the intensity (see 6.3.3) of take-all has fluctuated widely, although it seems that successive national epidemics have increased in severity.

As Fig. 1.2 indicates, some factors associated with intensive cereal growing actually work to decrease take-all, e.g. take-all decline (TAD), a natural biological control which develops after a severe outbreak of disease in consecutive susceptible cereals (see 2.4.1 and 4.3.1), and increased use of N fertilizers (see Figs 3.8 and 3.9). The interaction of weather and take-all is poorly understood. Consequently, even if it were possible to predict weather sufficiently in advance, it is still not known in any detail which seasons are likely to favour take-all. There were broad associations between weather and take-all at Rothamsted in the period 1930–1976 (Fig. 1.10 and Section 2.3.2). Colder, duller and wetter than average summers (where summer was the 3 months June–August) were associated with severe take-all. In a study of wheat monoculture in Germany, it was concluded that annual weather influenced the development of both inoculum and infection (Sturm *et al.*, 1984b).

Fig. 1.7. The combined areas of cereals and grass in the UK, 1935–1993.
Source: Central Statistical Office (1935–1995).

1.2 Take-all in the UK Today

1.2.1 *Importance*

ADAS trials and disease surveys in the late 1980s led to estimates that leaf and stem-base diseases of wheat were costing growers in England and Wales about £244 million per year, comprising £113 million in lost yields and £131 million in the cost of fungicides, without which yield losses would be much greater (R.J. Cook and R.W. Polley, 'Crop losses in wheat and barley', paper presented to the Schering Cereal Disease Symposium, 1990). These figures serve to put into context the losses caused by take-all. At a conservative estimate, losses from take-all in the same period were in the region of £16 million to £55 million per year (Table 1.11) as a result of its effects on second and subsequent wheats which, with some seasonal variation, made up about 60% of the wheat acreage (Table 1.3). Therefore, a single root disease, from which more than one-third of the acreage was not at risk, was having an average effect on farm incomes of about a sixth as much as all the other diseases put together. The 1990s have seen changes in cropping practices and fungicide usage, and surveys suggest fewer fields with above-ground signs of take-all in comparison with the 1980s (Table 1.2). It would be premature to interpret this as take-all no longer being a problem, or having much diminished in

Fig. 1.8. The percentage of wheat crops grown as first wheats in the period 1971–1991.
ADAS Eastern Region of England (———); nationally in England and Wales (- - - -).
No data available for 1983 and 1984.
Source: CSL/ADAS Winter Wheat Survey.

relative importance. Where susceptible cereals are grown consecutively, take-all is still arguably the most important and the least managed disease of wheat and barley in the UK, largely because there are no economic chemical controls or resistant cultivars. The importance of take-all extends beyond simple yield loss statements as given above. In the absence of chemical controls, the risk of serious yield losses in intensive wheat situations imposes significant additional constraints on arable farmers allied to: flexibility of cropping (relating to resistant hosts and crop rotations); soils and weather; fertilizer practice (particularly nitrogen usage); flexibility of sowing date; choice of cultivation technique; weed control practices; and the sensitivity of crop yield to climate.

Tables 1.3 and 1.9 to 1.11, incorporating data from a variety of sources, are an attempt to establish some measure of the importance of take-all in UK farming in the absence of critical data. They should not be regarded as anything more than what currently seems a reasonable starting point for discussion. Apart from the actual losses from take-all (Table 1.11), there is the question of whether crops at risk from the disease need to be grown at all. At present this seems unavoidable, bearing in mind the shortfall which would otherwise occur, the decreasing world grain stocks, an increasing UK export market and an insufficient variety of economically viable alternative crops (but see point (1), p.23).

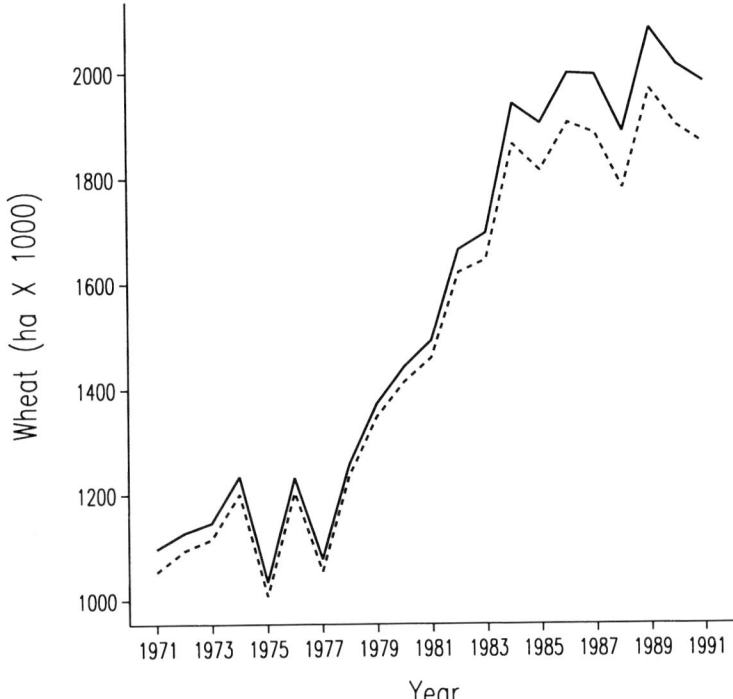

Fig. 1.9. Areas of wheat in the UK and in England and Wales for the period 1971–1991.
UK (———); England and Wales (- - - -).
Source: Central Statistical Office (1935–1995).
MAFF Agricultural Census Branch, Guildford.

At ADAS Experimental Husbandry Farms on contrasting soils, second winter wheats yielded, on average, 3% (Bridgets, 1982–1989), 6% (Boxworth, 1969–1989), 13% (High Mowthorpe, 1974–1989), 14% (Rosemaund, 1978–1989) and 16% (Arthur Rickwood, 1974–1989) less than first wheats. In the same periods third wheats yielded 8% less than first wheats at Boxworth and continuous wheats yielded 11% (Boxworth and Bridgets) and 20% (High Mowthorpe) less than first wheats. At Arthur Rickwood, in 1982–1985, third and subsequent

Table 1.6. Winter wheat, spring barley and winter barley crops in sequences of consecutive cereals in England and Wales, 1990–1991.

Year	Number of consecutive cereals	% of fields			% of estimated area		
		Winter wheat	Spring barley	Winter barley	Winter wheat	Spring barley	Winter barley
1990	1	46.5	37.1	8.6	39.4	33.0	7.1
	2	25.9	13.0	24.0	26.7	10.8	20.2
	3–4	13.9	24.1	36.0	16.6	22.4	38.3
	5+	10.4	22.0	26.5	13.1	26.5	28.9
	Unknown	3.3	3.9	4.7	4.2	7.3	5.5
	–	3443 fields	647 fields	1538 fields	1,892,298 ha	305,438 ha	849,967 ha
1991	1	50.9	48.4	11.6	45.9	42.7	8.4
	2	23.2	13.9	23.6	23.6	14.6	21.6
	3–4	13.0	19.5	38.5	14.3	20.2	40.0
	5+	9.3	11.6	19.3	11.8	15.6	22.0
	Unknown	3.6	6.6	6.9	4.4	6.9	7.9
	–	3300 fields	467 fields	1417 fields	1,961,738 ha	217,662 ha	817,918 ha

Source: Survey of Fertiliser Practice (1991), England and Wales. Personal communication from P.K. Leech: further analysis of data summarized in Chalmers *et al.* (1991).

Table 1.7. Areas (in thousands of hectares) of wheat and barley in the different countries within the UK, 1984–1993.

	England		Scotland		Wales		Northern Ireland	
Year	W	B	W	B	W	B	W	B
1984	1854.3	1445.3	71.0	438.0	10.8	49.5	3.1	45.6
1985	1805.4	1452.8	81.6	415.7	10.1	50.4	5.1	47.2
1986	1894.3	1400.0	88.8	418.2	10.5	50.6	3.8	47.9
1987	1874.4	1347.3	103.7	387.5	10.7	51.4	5.0	45.0
1988	1772.2	1395.4	98.7	389.6	9.8	51.5	5.0	42.7
1989	1958.8	1204.7	108.0	362.5	11.2	45.0	4.9	40.3
1990	1885.3	1101.8	111.0	338.3	11.3	39.2	5.8	36.6
1991	1853.1	990.3	109.7	329.1	11.8	36.2	5.9	37.3
1992	1925.1	913.4	121.5	311.3	12.6	35.6	7.4	37.0
1993	1631.3	816.3	109.1	276.1	11.8	33.5	6.7	38.6

W, total wheat; B, total barley.
Source: Anon. (1994b).

wheats yielded 16% less than first wheats. Relative yields differed considerably amongst the farms and there is no evidence of TAD in the figures. Yield differences cannot be attributed solely to take-all, and confounding factors include: different areas of crops, in the order first > second > continuous wheats; cultivars not the

Table 1.8. The classification of principal wheats by milling characteristics and baking strengths.

Milling characteristics[a] (Hard versus soft wheats)	
Extra hard	Durum (macaroni wheat), some Algerian, Indian
Hard	CWRS, American HRS, Australian Prime Hard
Medium	Plate, Russian, some Australian, American HRW, some European
Soft	Some European, some Australian, American SRW, American Soft White

Baking strengths[b] (Strong versus weak wheats)	
Strong	CWRS, American HRS, Russian Spring, some Australian
Medium	American HRW, Plate, S.E. European, Australian Prime Hard
Weak	N.W. European, American SRW, American Soft White, Australian Soft

[a]Relating to the way endosperm breaks down; [b]large loaf volume, good crumb texture and keeping properties versus small loaf, coarse open texture, low protein content.
CWRS, Canadian Western Red Spring wheat, Manitoba; HRS, Hard Red Spring wheat; HRW, Hard Red Winter wheat; SRW, Soft Red Winter wheat; Plate, Argentian wheat (HRW).
Source: Kent and Evers (1994).

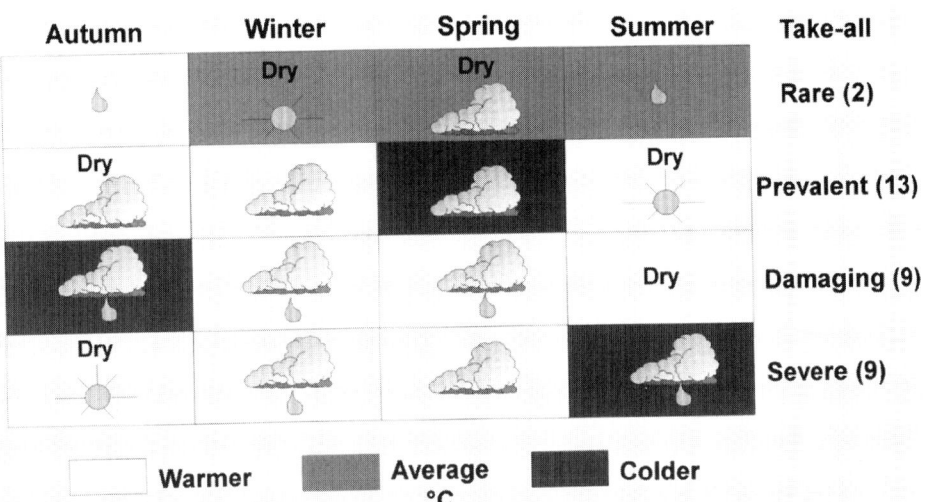

Fig. 1.10. Associations between weather and take-all, mostly in winter wheat, at Rothamsted, 1930–1976.
In this period, there were 23 growing seasons when there was some appraisal of take-all in *Rothamsted Reports* and these have been apportioned to the four disease categories given. The seasonal weather associated with each of these categories is characterized by comparison with 83-year averages. Average, or more or less than average, is shown for temperature (grey, white, black), precipitation (no symbol, rain drop symbol, Dry) and sunshine (no symbol, sun symbol, cloud symbol).
Adapted from Hornby (1978b).

same in all crop sequences; second crops probably sown later in later years; Baytan seed treatment probably used for second and continuous wheats; continuous wheats likely to have suffered grass weed problems. Further information for Boxworth is given in Table 3.5; in Table 3.11 there are some data on the effect of N fertilizer on the relative yields of first and subsequent wheats and in Table 3.18 the effects of seasonal variation are shown for Rosemaund. The expectation that first wheats would yield most did not always materialize in a long-term sequence

Table 1.9. UK cereal supplies (Mt per annum).

Cereal	1986	1987	1988	1989
Cereals				
Total home production	24.5	21.6	21.1	22.7
Total home use	20.6	20.4	19.8	20.1
Wheat				
Total home production	13.9	11.9	11.7	14.0
Total home use	12.0	11.9	11.6	11.6
Home-grown use				
milling	3.6	3.5	3.8	4.2
animal feed	6.0	5.5	5.1	5.3
industrial	0.4	0.5	0.6	0.5
seeds and sundries	0.6	0.6	0.6	0.6
total	10.7	10.1	10.1	10.7
total (% of production)	77	85	86	76

Source: MAFF.

Table 1.10. Estimated contributions to UK grain production of wheat grown after wheat or another cereal and an indication of grain shortfall if the first wheats only had been grown.

	1986[a]			1987[b]			1988[c]		
	I	II	III	I	II	III	I	II	III
Yield of wheat as 2nd and subsequent cereals[i]									
Mt per annum	6.1	7.9	7.5	6.5	8.0	7.9	6.0	7.0	6.7
Balance of national grain production[ii]									
Mt	18.4	16.6	17.0	15.1	13.6	13.7	15.1	14.1	14.4
Balance shortfall[iii]									
Mt	−1.6	−3.4	−3.0	−4.9	−6.4	−6.3	−4.9	−5.9	−5.6
%	−8.0	−17.0	−15.0	−24.5	−32.0	−31.5	−24.5	−29.5	−28.0

(i) Based on percentages of wheats not grown as a first cereal (Table 1.3). Column I, excluding all first wheats; column II, including wheat after a cereal; column III, as II, but corrected for greater yield of first wheats (15%).
(ii) Obtained by deducting the figures in the line above from the total home production in Table 1.9.
(iii) This assumes the UK needed 20 Mt of home-grown grain (Table 1.9) and does not take into account the value of exports. It is possible that any such shortfall could have been made up by using surplus barley as animal feed (J. Orson, National Cereals Specialist, personal communication).
[a] HMSO (1989a), [b] HMSO (1989b), [c] HMSO (1990).

experiment on Great Harpenden field at Rothamsted, where first wheats yielded 4.5% (1986), 12.3% (1987) and −4.4% (1988) more than subsequent wheats.

Estimates of the importance of take-all are elusive for the following reasons.

1. *It is difficult to establish whether growing wheat in rotations where there is a high risk of take-all is absolutely necessary to meet requirements.* The implications of a move away from specialization and avoiding intensive cereal production need studying. Some matters to take into consideration are: the limited number of acceptable break crops; the uncertainty over the future of global wheat production in what appears to be a period of unprecedented climatic change; the possible turn round in markets should an industrial use be found either for wheat itself, or for one of the major alternative (break) crops; area payments making second wheats an attractive proposition at the time of writing.

2. *The implications and feasibility of replacing wheat with barley (which yields less and is still less profitable than wheat) at periods of risk are not well understood (see Section 3.2.1).*

3. *Take-all survey data are inadequate and usually not comparable (see Table 2.1).* Adequate surveys would require disease data from different wheat sequences in different seasons in different locations.

4. *Estimates of loss suffer from a lack of standardization in the measures of disease that have been used.* Assessing disease is discussed in detail in Section 6.3.3.

5. *Losses have varied considerably in different years, but the characteristics of years that favour take-all are not well understood.* The interaction between disease and

Table 1.11. Estimates of losses in second and subsequent wheats due to take-all.

	1986[a]	1987[b]	1988[c]
National average yield of wheat (t ha^{-1})	6.96	5.99	6.22
Estimated loss in crops (% of all wheat)	1	4	3
National loss (Mt)	0.139	0.476	0.351
Average price (£ t^{-1})	117	115	116
Loss (M£)	16	55	41

Estimated losses based on: (i) the only national (England and Wales) figures available, which are 0.9% for 1977, 2.6% for 1978 and 2.8% for 1979[d]; and (ii) average loss for WW in East Midlands 1963–1965 of 3.8% (Rosser and Chadburn, 1968).

Monitoring at Rothamsted farms and ADAS Disease Intelligence reports suggest rankings for take-all: 1988 > 1987 > 1986. However, the Eastern Region Crop Intelligence Report for 22 July 1988 suggested that although disease levels were greater in 1988, patchiness (see Table 1.2) and yield losses were greater in 1987.

Annual loss due to *G. graminis* in the UK was given as 2.5% by Oerke *et al.* (1994) and 6% has been assumed for second and third wheat crops following a non-cereal break (Cook *et al.*, 1991). With minor infections there was no effect on yield, with moderate infections a 15% reduction and with more severe disease up to a 62% reduction (Oerke *et al.*, 1994).

Estimates of average loss are the weakest link in this kind of exercise.

£ t^{-1} is the mean of prices for three kinds of wheat (bread-making, other milling and feeding): proportions of each kind not accounted for.

References: [a] HMSO (1988), [d] Polley and Clarkson (1980); see Table 1.10 for references denoted by [b] and [c].

weather is apparent in the observations contributing to Fig. 1.10 and the interaction of these with yield is considered in Section 3.6.

6. *Dates on or growth stages at which take-all has been estimated have varied, making comparisons of disease–yield relationships difficult.* Ways in which take-all affects yield are given in Table 1.12 and experimental work on this topic is discussed in Section 6.5.2.

Meynard (1985) proposed a model identifying the different phases of plant growth during which yield components developed and, as a consequence, periods during which these components could be affected by limiting factors such as nitrogen deficiency. Based on this model, Fig. 1.12 shows how the timing of severe take-all affects different yield components, resulting in different levels of yield loss. Others have pursued this theme of the timing of severe take-all and damage to the crop. 'Late infestations', i.e. those recorded 3 weeks prior to anthesis in a winter wheat crop in which no take-all was detected in March or early April, decreased photosynthetic potential, resulting in an earlier start to leaf senescence, lower total biomass and decreased grain yield (Green and Ivins, 1984; see also 6.3.4). Hornby

Fig. 1.11. Samples of 200 winter wheat grains taken in 1991 from plots with little or no take-all (left) and from plots with severe take-all and above-ground symptoms in patches (right).
Cv. Mercia grown as a second winter wheat after oilseed rape.
The hand-threshed samples yielded the following:

	Left	Right
1000-grain weight (g, oven dry)	33.8	19.8
Yield (t ha^{-1}, 85% dry matter)	6.13	1.88
Grain size as % of total:		
1–2 mm	4.3	49.3
2–3.5 mm	94.6	49.6
Hagberg Falling Number	374	401

(1994) suggested that the timing of severe take-all could explain apparently inconsistent effects of take-all (Fig. 1.13). The ability of plants to respond to take-all by producing new roots is a consideration in any explanation of how the intensity of disease at critical stages (Seidel and Spaar, 1980) affects yields. The plant's capacity to regenerate roots depends on the growth stage, and at certain stages the rate of disease development can be masked by intensive root growth (Seidel *et al.*, 1981; see also 2.6.3). Experiments with winter wheat plants growing in hydroponic culture explored the ability of the plants to respond, by regenerating new roots, to the removal of their roots at different growth stages (Wächter *et al.*, 1979). The

Table 1.12. Evidence for effects of take-all on grain yield, some components of yield and grain quality measurements.

Yield component/ quality value	Effect	Source
Total weight per unit area[a]	Severe take-all decreased yield	Manners and Myers (1981)[b]
	In general decreased yield attributed to take-all, but N (amount and form) affected take-all	Slope and Etheridge (1971); Prew and Dyke (1979)
	Decreased, effects of rotation	Sturm *et al.* (1984a)
	Better correlation with 'whiteheads' than with root infection	Gutteridge *et al.* (1987)
	Variable effect in slight take-all category; decreased where disease moderate–severe (split-line response)	Bateman *et al.* (1990a)
	Variable; slight disease can increase yield	Herman and Dovrtěl (1991)
Number of ears	Decreased where disease most severe	Manners and Myers (1981)[b]
	Sometimes increased where disease less severe	Manners and Myers (1981)[b]
	Can be the most important effect	Seidel *et al.* (1981)
	Decreased	Sturm *et al.* (1984a)
Grains per ear	Decreased where disease most severe	Manners and Myers (1981)[b]
Grain weight (Fig. 1.11)	TGW decreased where disease most severe	Manners and Myers (1981)[b]
	TGW decreased	Sturm *et al.* (1984a)
	HLW and TGW decreased	Bateman *et al.* (1990a)
Number of ear-bearing culms	Decreased	Seidel *et al.* (1981)
Protein content	Overall decrease, but unaltered or increased (severe disease) per grain	Manners and Myers (1981)[b]; Bateman *et al.* (1990a)
α-Amylase activity	Unaltered or slight decrease	Bateman *et al.* (1990a)

[a] A selection of references is given.
[b] Includes earlier references.
TGW, 1000-grain weight.
HLW, hectolitre (specific) weight.

regenerative capacity, measured as the percentage of total root length of control plants, was 67% at Feekes growth stage 8 (flag leaf just visible) and 59% at 10.1 (first spikelets visible) and thereafter diminished to only 11.4% at 11.1 (ripening, medium milk). On the basis of this, the authors suggested that between Feekes growth stages 8 and 10.1, plants would be best able to compensate for decreases in root mass caused by take-all infection.

Examples from the field support this view. In 1991, a second winter wheat crop after oilseed rape at Rothamsted, Hertfordshire, was uneven in mid-June (GS 47), and take-all was severe in the areas of poorest growth. Hand-harvested samples taken from these areas yielded 1.88 t ha^{-1} compared with 6.13 t ha^{-1} from areas without above-ground symptoms. In a second winter wheat crop at IACR's Woburn Farm, Bedfordshire, in the same season, take-all patches did not appear until late in June (GS 69); grain yields were 3.1 t ha^{-1} within the patches and 4.7 t ha^{-1} outside the patches.

Severe disease late in the season, towards the end of grain filling, may not affect yield greatly, but may be important in inducing TAD where successive crops of susceptible cereals are grown.

7. *The effect of slight take-all on yields seems variable and may range from depression to stimulation.*

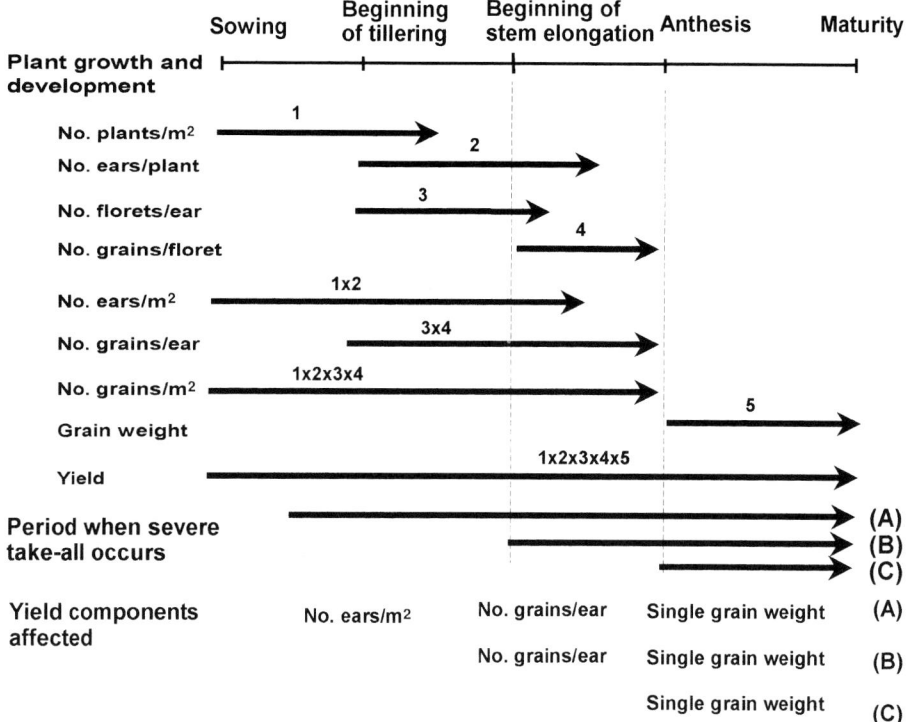

Fig. 1.12. Diagram showing growth stages of winter wheat at which different components of yield are achieved and can be affected by take-all (based on Meynard, 1985).

1.2.2 Control

The best control for take-all in the UK is not to grow consecutive crops of susceptible cereals. The drift back to more first wheats since the late 1980s has been a move in this direction. Farmers often inquire about the prospects of controlling take-all by other means such as fungicides, or resistant or tolerant cultivars, because in practice, if they grow cereals intensively, they can presently do no more than pay attention to good husbandry, assess risks (e.g. of breaking a long run of cereals, or taking a third crop) and consider options. In well-run production systems in Britain, changes in husbandry usually achieve only a modest decrease in disease. Advisers too, unable to recommend any fungicides or resistant cultivars, are limited to advice on how to avoid take-all or how to minimize its effects by correct husbandry, and to weighing the pros and cons of a limited number of options, such as shortening rotations or exploiting TAD by staying in continuous cereal cropping.

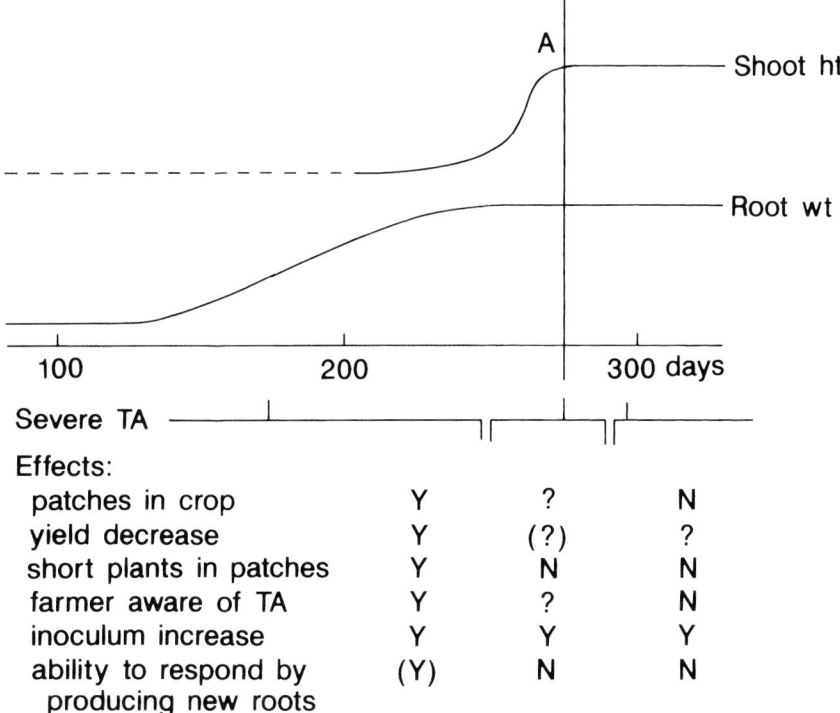

Fig. 1.13. A sketch suggesting how apparently inconsistent effects of take-all arise. The patterns of shoot heights and root weights are shown against the number of days from sowing winter wheat (x-axis). This time is divided into three approximate periods on the basis of when severe take-all first occurs, and some probable effects of the disease in each of these periods are listed.

A, anthesis; **ht**, height; **wt**, weight; **TA**, take-all; **Y**, yes; **N**, no; **(?)**, some doubt and **?**, considerable doubt that these effects will be apparent.

Modified from Hornby (1994).

An upsurge in take-all in the 1980s (see 1.2.3) renewed interest in how to control the disease. Because of little progress in other areas, biological control (see 4.3) has held centre stage for some years, although latterly less in the UK than elsewhere. There has been a change in emphasis from exploiting resident antagonistic microorganisms to the introduction of biological control agents (BCAs), a research topic that has generated much enthusiasm and optimism, but as yet no product for farmers. Generally, there is relatively little research effort going into breeding resistant or tolerant varieties (see 3.2) and apparently none in the UK, but there has been increased interest in fungicides recently (see 4.2.2).

1.2.3 Changes: evidence, causes and implications

Although take-all and the perception of it as a serious cause of yield loss in UK cereals have fluctuated widely since the 1930s, there is much anecdotal, and some scientific, evidence of changes in the incidence and severity of take-all since 1981. Nationally, disease intensity has seemed to be greatest in more recent outbreaks, especially in the first half of the 1980s (Hornby, 1992b; Fig. 1.2). Also, outbreaks of take-all in situations hitherto considered to be relatively safe are indications that take-all may be changing. In the 1980s patches of take-all in crops were more apparent than in the 1970s. They ranged from a few infected plants to extensive swathes and occurred on a variety of soils. Second and third wheat crops were particularly affected. Patches were seen, but less frequently, in winter barley crops.

Some first wheats suffered, which is uncommon, and even some continuous winter wheats (e.g. a ninth wheat on London clay in July, 1987) developed patches; usually TAD suppresses the disease in wheat monocultures. Grasses may interfere with TAD (see 3.3.3) and this could account for patches of disease in sites reported to be in decline. Grasses such as common couch [*Elytrigia repens* (L.) Desv. ex Nevski = *Elymus repens* (L.) Gould. = *Agropyron repens* (L.) P. Beauv.] have been implicated also in outbreaks of take-all in first wheats: severe infection frequently occurs on couch-infested land, irrespective of treatment with glyphosate in the previous autumn (see 3.5.5). If a dry season follows spraying, dead rhizomes, colonized by the pathogen, may not rot.

In July 1990, severe and widespread take-all was present at ADAS Gleadthorpe (formerly Gleadthorpe EHF), a typical sandland farm in Nottinghamshire. Particularly surprising were first wheats and some barley crops suffering badly from the disease, severely infected rye volunteers in a winter wheat crop, and all plants in some triticale and oat crops with similar levels of slight infection. In the absence of observations on possible causal factors such as volunteer wheat, any explanations can only be speculative. In 1994, the ADAS Plant Disease Clinic in Cambridge found 'a lot of take-all' on oats from ADAS Rosemaund in Herefordshire.

In the 1980s, occurrences of severe disease began to be associated with cereals in short rotations with oilseed rape. There were many reports of severe infections occurring in second wheats after oilseed rape (e.g. Table 3.8 and associated text), whereas in the 1970s second wheats were less often severely infected. In the 1970s, low returns on grassland enterprises resulted in more land being put into cereal production. Field beans were replaced by oilseed rape as the preferred alternative crop and the rape/winter wheat/winter wheat or winter barley rotation emerged and is widely used today. The 10 years ending in 1988 saw a tenfold increase in the acreage of rape grown in the UK. The greater yields of first wheats after oilseed rape than after winter beans (Bowerman, 1989; Table 3.8) are attributed to residual nitrogen (which is greater after rape) and the later sowing of wheat after beans. Cereal volunteers, which could aid the carry-over of Ggt, are also more common in rape than in beans because of different herbicide programmes and the use of minimal cultivation before sowing rape. The relatively large residues of nitrogen under rape may favour saprophytic survival of Ggt, which might then increase sufficiently in a following early-sown wheat to provide a level of inoculum capable of causing greater losses to take-all in a second wheat than are likely to occur in a second wheat after beans.

Apart from dramatically exposing the limitations of our generalizations, these experiences resurrect many old questions about geographical location, soil type, different cereal species, irrigation, sowing dates, weed control and previous cropping that have remained unanswered.

There are aspects of the CSL/ADAS Winter Wheat Disease Survey that add to these impressions of change and of disease in circumstances where hitherto it was expected only rarely. For instance, during 1989–1994, 12.5–31.4% of first wheat crops surveyed in England and Wales apparently had take-all (Table 1.13), with 4.0–11.7% showing above-ground symptoms (Table 1.14). Also, where there had been a break from susceptible cereals of 2 years or more, the disease was often worse than after a break of only 1 year (Tables 1.15 and 1.16). However, since there was no information about the occurrence of susceptible weeds in the breaks, or details of the crop sequences beyond the previous 4 years, this is no more than an intriguing indication. Furthermore, we have to accept that take-all may not have been the cause of the patches in the first wheats after breaks. Even in second winter wheat crops there may be pitfalls: at Woburn in July 1994 the extent

Table 1.13. Effect of non-cereal break crops in England and Wales on take-all (percentage of crops affected).

		Position in rotation			
Year	n^a	1st wheat	2nd wheat	3rd wheat	4th and subsequent
1989	246	21.9 (137)	55.6 (63)	46.7 (15)	29.0 (31)
1990	312	21.9 (169)	46.0 (100)	56.3 (16)	29.6 (27)
1991	337	19.6 (214)	32.8 (64)	57.1 (21)	31.6 (38)
1992	361	21.7 (217)	37.7 (85)	42.9 (28)	38.7 (31)
1993	328	31.4 (204)	48.9 (90)	30.8 (13)	57.1 (21)
1994	368	12.5 (273)	28.3 (60)	30.0 (10)	28.0 (25)

n = number of crops sampled; ( ) = number of crops in category.
[a]Since previous cropping data were not available for all crops sampled for take-all, these values are smaller than those in the last column of Table 1.4.
Source: CSL/ADAS Winter Wheat Disease Survey.

Table 1.14. Effect of monoculture on take-all patches in England and Wales (percentage of crops in severity categories 2 + 3 + 4).

		Position in rotation			
Year	n^a	1st wheat	2nd wheat	3rd wheat	4th and subsequent
1989	246	11.7 (137)	33.3 (63)	20.0 (15)	12.9 (31)
1990	312	5.9 (169)	23.0 (100)	25.0 (16)	7.4 (27)
1991	337	5.1 (214)	10.9 (64)	28.6 (21)	13.2 (38)
1992	361	5.1 (217)	20.0 (85)	17.9 (28)	16.1 (31)
1993	328	7.4 (204)	16.7 (90)	15.4 (13)	38.1 (21)
1994	368	4.0 (273)	8.3 (60)	10.0 (10)	16.0 (25)

n = number of crops sampled; ( ) = number of crops in category.
[a]Since previous cropping data were not available for all crops sampled for take-all, these values are smaller than those in the last column of Table 1.4.
Source: CSL/ADAS Winter Wheat Disease Survey.

Table 1.15. Effect of non-cereal break crops on take-all in England and Wales (percentage of crops affected).

	Number of years break		
Year	Nil (second and subsequent wheats)	1 year (all first wheats)	2+ years
1989	43.1	19.5	13.6
1990	43.3	18.4	22.7
1991	37.4	15.3	17.8
1992	36.3	23.6	15.9
1993	48.0	29.5	31.0
1994	25.7	10.2	14.3

2+ = 2, 3, 4 or more years (information was requested for the previous 4 years only).
Source: CSL/ADAS Winter Wheat Disease Survey.

Table 1.16. Effect of non-cereal break crops on take-all patches in England and Wales (percentage of crops in severity categories 2 + 3 + 4).

	Number of years break		
Year	Nil (second and subsequent wheats)	1 year (all first wheats)	2+ years
1989	24.3 (144)	7.3 (82)	13.6 (22)
1990	19.5 (180)	2.6 (114)	9.1 (22)
1991	14.2 (155)	1.5 (137)	8.9 (45)
1992	17.6 (171)	4.9 (144)	2.3 (44)
1993	20.7 (150)	5.8 (139)	7.1 (42)
1994	10.3 (136)	1.7 (177)	3.6 (56)

2+ = 2, 3, 4 or more years (information was requested for the previous 4 years only).
( ) Numbers of samples; because of the ways used to compute the different categories the row totals differ slightly from those in Tables 1.13 and 1.14.
Source: CSL/ADAS Winter Wheat Disease Survey.

of patches in one crop exceeded that expected on the basis of the take-all rating (TAR) (see 1.2.1) and it seems most likely that the concurrence of severe eyespot contributed to the extensive patchiness. Tables 1.13 and 1.14 also emphasize the occurrence of take-all in about a quarter to a half of fourth and subsequent wheats and the presence of patches in some of these crops. Also, in the period 1989–1994 there were 3 years in which the percentage of second wheat crops showing patches was greater than the percentage of third wheat crops showing patches, and 3 years when the opposite (i.e. the conventional expectation) occurred.

Changes in agricultural practice have influenced the disease, but none is known to correlate well with the sporadic outbreaks of severe take-all (Fig. 1.2). In fact, most seem to be of the fine-tuning kind and do not rank as major factors in explanations of national outbreaks. Weather may explain some of the variation (see 2.4.2). These sorts of association need further exploration, but it might be that the memorable 'take-all years' are the concurrence of vulnerability in the nation's rotations and certain weather conditions.

1.3 Trends in Cereal Production and Take-all Research

The UK has the greatest yields of wheat of any of the major wheat-producing countries (Brown *et al.*, 1994). Since 1984 the average has fluctuated around 6.5 t ha^{-1}, but in the preceding 45 years there had been a tripling of yields, coinciding with, amongst other things, increased fertilizer usage (see data for the period 1974–1984 in Table 3.10) and the development of wheat cultivars able to exploit this. Between 1947 and 1990 there was an increase of 2.79 t ha^{-1} in wheat yields in England and Wales (Scarisbrick and Meikle, 1993). Before 1940, yields had been remarkably stable, even as far back as 1884.

Winter wheat has been grown on the field known as Broadbalk at Rothamsted for just over 150 years, although no one part of the field has actually had 150 years of monoculture. Broadbalk is the location of one of the Rothamsted Classical Experiments (Johnston, 1994). Because similar amounts of fertilizer or farmyard manure were applied to some plots in the period 1852–1990, it is possible to gauge the effects of other factors on wheat yields (Fig. 1.14). In the 1920s, weed competition caused yields to decline, but thereafter introducing weed control, changing cultivars and introducing fungicides and crop rotation to minimize the effects of soil-borne pathogens had dramatic effects. Where wheat was grown continuously with PK fertilizer and 48 kg or 144 kg N, yields have increased by more than a half, or have more than doubled, respectively, since the 1950s, but showed a down-turn in the 1980s. Where wheat was grown in rotation, yields were still increasing up to 1990. The Broadbalk experiment has demonstrated that yields can be sustained in monocultures based on inorganic fertilizers, contrary to much popular opinion (Lewis, 1994).

The contribution of genetic improvement to the increase in wheat yields in England and Wales in the period 1947–1983 was estimated at 45% from relative yields of varieties in NIAB trials and seed sales statistics (Scarisbrick and Meikle, 1993). In other parts of Europe with similar growing conditions, e.g. France and

Germany, the rise in yields is slowing as yields approach UK levels (Brown *et al.*, 1994).

Topical issues in the UK with implications for take-all are: the question of gross margin, now that growing for maximum yield is considered outmoded; switching to more first wheats and the concomitant need for alternative crops; nitrogen set-aside (fertilizer quotas), as opposed to land set-aside, as a means of limiting production; and feasibility of wheat farming at world prices (achievable by larger farm businesses on land with productivity at least equal to that of average land in the east of England?). Issues arising out of the practice of cereal growing are summarized in Table 1.17; interestingly some are based on foreign experience reported in the British press and have little or no documentary support in Britain. They are also a subset of the many environmental impacts attributed to agricultural practices that affect air, water, soil, nature, wildlife and landscapes (Stanners and Bourdeau, 1995). The advice to plough at least 1 year in 4 or 5, have spring

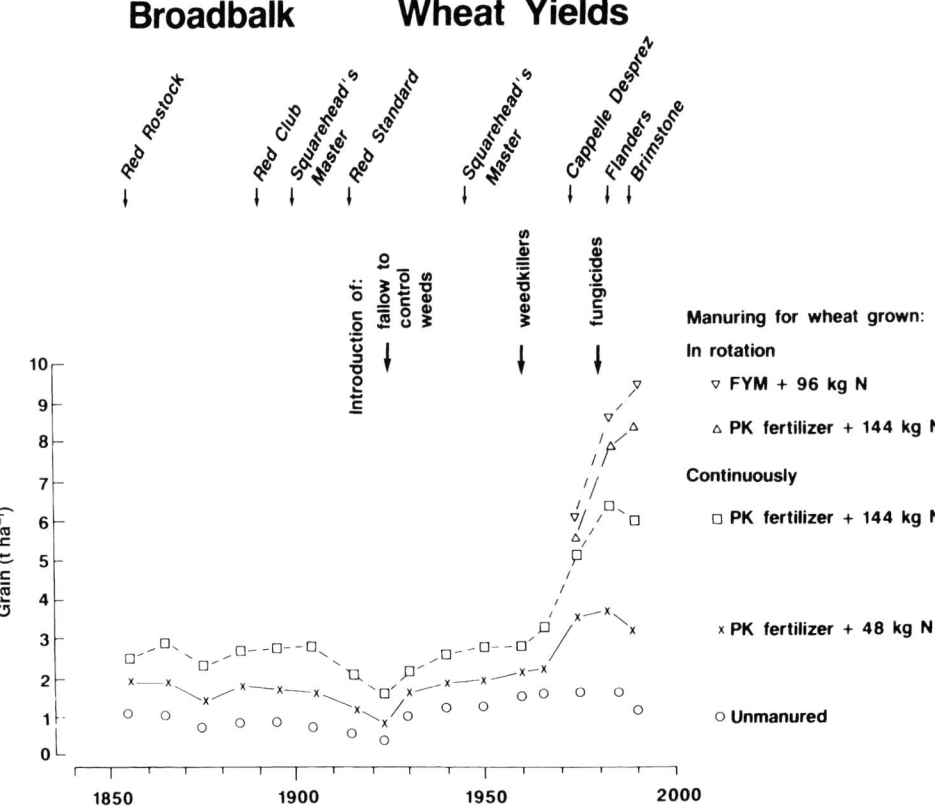

Fig. 1.14. Yields of winter wheat grown on Broadbalk, Rothamsted, from 1852 to 1990 with fertilizers and with farmyard manure, showing effects of changing cultivars and the introduction of weed control, fungicides and crop rotation to minimize effects of soil-borne pathogens.
Reproduced with permission from Johnston (1994).

Table 1.17. Summary of some practical concerns of farmers and environmentalists.

Issues	Problems	Comments and references
Manganese	Many reports from overseas indicate manganese deficiencies as an important factor in take-all severity (see 2.2).	No strong evidence to support this as a major factor in UK outbreaks (Hornby, 1985), but there has been little experimentation and manganese is the most common trace element deficiency observed in UK cereals.
Nitrate	Nitrate leaching from soil and entering water supplies. Surface or groundwater with ≥ 50 mg nitrate l^{-1} and water that is eutrophic is polluted (EC Nitrate Directive).	Leaching may be greater where plants have take-all (Mielke, 1988); conflict between minimizing leaching of nitrate (Agricultural and Food Research Council, 1988) and take-all, e.g. sowing date advice for WW.
Sulphur	Sulphur deficiencies are occurring as deposition from atmospheric pollution decreases.	Declining sulphur levels in UK wheat grain (Anon., 1995c). Adding S in summer of 1995 increased wheat and barley yields in northeast Germany by as much as 20% (MacKenzie, 1995). S applied to deficient soils can decrease take-all (Huber, 1981b).
Cold and frost	Plants with take-all are sensitive to cold and susceptible to frost.	Claimed for autumn-sown plants in Germany (Mielke, 1988).
Couch grass	Treatment of couch-infested land with glyphosate (Roundup) has been associated with increased take-all in following cereals.	See Sections 2.3.5, 3.5.5, 4.2.3, and Mielke (1988).
Straw disposal	After the 1992 burning ban in the UK, disposal problems became acute, particularly on heavy land.	One option for farmers is to change the rotation and use more break crops. If a 5-year rotation such as WW/beans/winter cereal/WW or WB/rape were adopted, there would be fewer second and no third wheats (Long, 1990).
Set-aside	Returning to normal cropping after land has been withdrawn from agricultural production may result in disease problems.	Little or no evidence to support view that ryegrass would preserve TAD (Blake, 1989).
Disease as a tool	Identifying soil problems on heavy land.	Take-all may highlight soils in need of management (ADAS, 1989). This has been discussed in the context of using root disease as a bioindicator of soil health (Hornby and Bateman, 1997).

crops in the rotation and delay autumn sowing applies where black-grass (*Alopecurus myosuroides*) is resistant to herbicides (Weed Resistance Action Group, 1991) and would be mostly in accordance with that for take-all. However, advice to use certain cover crops to minimize the leaching of nitrate might be controversial and advice to sow winter wheat early for the same purpose (Addiscott *et al.*, 1991) would conflict with advice on how to minimize take-all. New enthusiasms are likely to have conflicting effects on take-all: lower inputs, leading to less nitrogen usage and thinner crops, would most likely result in disease increase in the first case and disease decrease in the second, and reverting to older cereal cultivars that scavenge for nitrogen would probably not change take-all in British crops.

A current alternative to break crops is 'set-aside', where permitted crops such as clover or rye-grass may affect take-all in subsequent cereals. Work at Rothamsted showed that rye-grass, alone or in combination with lucerne or clover, was a better break than a ley of lucerne or clover alone. This may not always be so. In an IACR/ADAS set-aside study at Woburn, winter wheat in 1988 was followed by seven different land treatments in 1989: ploughed in the autumn (PA), winter wheat; PA, rye-grass; weeds, topped; spring-sown trefoil; straw chopped and spread in autumn, cultivated in spring; PA, cultivated in spring; PA, forage rape. In 1990 winter wheat was grown again and TARs (scale: 0–300) were 108, 83, 65, 52, 29, 22 and 20, respectively. (Information for some of these treatments over a period of three years is given in Table 3.9.)

Issues arising mostly out of ongoing research are summarized in Table 1.18 and most of them are discussed in more detail subsequently. In take-all research, it is usual to find that general areas of interest contain much interacting complexity. For instance, the rhizosphere has many involvements in take-all, such as the significance of host-specific rhizosphere effects, changes in the physicochemical environment due to uptake of NH_4^+-N, changes in the microbiota, ectotrophic growth of Ggt, and fungicides in the rhizosphere (Hornby, 1990b).

1.4 Take-all Outside the UK

This is a selective report dealing with countries where there is published evidence of the importance of take-all, especially where this is in the form of estimates of losses.

The presence of take-all was acknowledged for the first time in France on wheat by Prillieux and Delacroix (1890). Surprisingly, the disease was described as affecting the stem only and there was no mention of symptoms on the root system. The plant samples came from experimental stations near Paris, but the authors reported that similar symptoms had been observed in different parts of France and that severe damage had occurred in 1889. Perithecia were present on the first internode of wheat culms; the shape and size of the ascospores were described and the fungus was identified as *Ophiobolus graminis* Sacc.

Subsequently the disease received little attention until the work of Ponchet and colleagues (Ponchet and Coppenet, 1957, 1962a,b; Ponchet, 1962; Lemaire and Ponchet, 1963). This revival of interest was due to outbreaks of the disease on newly cropped land, previously moorland, in Brittany in the 1950s (Ponchet and Coppenet, 1957). Surveys made between 1954 and 1956 revealed that entire fields were so severely affected by take-all that harvested grain was no more than the amount used as seed. Because these fields originally had very low soil pHs, they were amended with a calcium fertilizer before being cropped with wheat. In other regions, take-all was considered as a patch disease that could cause some damage, but there was no indication of its importance.

Wheat has long been an important crop in France (Table 1.19). The area under production reached a maximum at the beginning of this century, at about 7 Mha, and remained above 6 Mha until World War I. Between the two world wars, the area of wheat fluctuated around 5.5 Mha, until it decreased with the creation of the *Office du Blé* in 1936, which imposed a reduction in the area of wheat and had as its objective a guaranteed minimum price to farmers. Since World War II, the area of wheat has remained between 3.8 and 4.7 Mha, except for a rise to 5 Mha between 1989 and 1992.

If the new interest in take-all between 1950 and 1960 is to be explained by changes in cropping which favoured disease, then the fact that the area of wheat did not increase suggests that other factors, which promoted annual grain

Table 1.18. A summary of some current research issues.

Issues	Problems	Comments
Biological control	No resistant cvs or chemical controls	A popular and attractive alternative which is proving erratic and difficult to manage
Epidemiology	Not fully understood	Incomplete knowledge makes it difficult to interpret data and compare regional findings
Factors affecting take-all	Many recorded, but relative significances unclear	Dominant factors need to be identified and interactions explored
Generalizations	All too often scientists discuss take-all without regard to regional differences	There is growing evidence and experience to show that many findings are not generally applicable throughout the world
Artificial inoculum	Severe, natural infection in soil cannot be achieved to order in field work	Artificial infestation may overcome some difficulties, whilst creating others
Yield–disease relationship	Not fully understood	One reason why it is difficult to be precise about the importance of take-all
Host damage	The way in which take-all causes root dysfunction is thought to be straightforward	This affects yield through the interaction of remaining functional root and environmental conditions
Diagnosis and identification	Traditional methods are time-consuming and laborious	Serological (e.g. ELISA) and molecular biological (e.g. DNA probes) methods are under investigation

Table 1.19. The areas and production of wheat in France since 1850.

	Year										
	1850	1880	1900	1913	1935	1950	1960	1970	1980	1990	1994
Area (Mha)	6.0	6.9	6.9	6.5	5.4	4.3	4.3	3.8	4.5	5.0	4.5
Production (Mt)	6.6	7.5	8.9	8.7	7.8	7.7	11	14	22	32	31

Sources: Anon. (1960, 1995b), Boyeldieu (1980).

production by 43% during the same period, were involved (Table 1.19). This marked the end of a long period of stagnation in production and heralded 27, 57 and 45% increases in grain yield, respectively, in each of the following decades.

This gain in productivity can be explained partly by wheat breeding programmes, which introduced new cultivars with greater yield potentials. These were exploited partly by more appropriate nitrogen fertilization (e.g. the work by Coïc, 1950, which identified nitrogen needs at different growth stages of a wheat crop and showed the advantage of applications made at the appropriate times) and partly by the introduction of foliar applications of fungicides in the 1970s. Such changes were accompanied by earlier sowing and greater seeding densities, which are known to increase the risk of take-all.

In 1994, there were 17 Mha of arable land in France, including 2.6 Mha of pastures. With some 4.5 Mha of wheat and 1.4 Mha of barley, the area of crops susceptible to take-all was about 6 Mha, i.e. about 35% of the total of arable crops and 41% of the area dedicated regularly to rotations (e.g. annual crops) (Table 1.20). This indicates that a field was likely to grow wheat or barley at least once and often twice in 3 years.

There were six major regions of cereal production in France in 1993 (Tables 1.21 and 1.22). Areas of the two cereal crops – wheat and barley – and areas of wheat crops alone are classified according to whether they were growing first, second, third, or subsequent cereal crops. Where wheat and barley are grouped, first cereals represented 56–82% of the area in the regions (mean for all France = 65.7%), while 15–37% were grown as a second cereal (mean = 27.9%) and only 2–8% as a third cereal (mean = 6.1%). When wheat alone is considered, more than 75% of any of the areas (84% as a mean for all France) was cropped as a first wheat, compared with only 8–20% (mean = 14.4%) growing a second wheat. Third and subsequent wheats are rare in France. These figures contrast with the UK, where more second and subsequent wheat crops have been grown (Table 1.3). In France, second cereals are more often barley. Of the 1.6 Mha of barley grown in 1993, 1.1 Mha (68.7%) had barley as a second cereal, 300,000 ha (18.7%) had barley as a third cereal and only 200,000 ha (12.5%) had barley as a first cereal (Table 1.23).

One reason French farmers choose barley as a second cereal is to avoid the problems of take-all. The main difference between France and the UK that perhaps explains the different proportions of first and subsequent wheats (60/40, UK; 84/16, France) is a larger choice of alternative crops for rotations in France, possibly due to more favourable climatic conditions. Wheat and barley in France currently comprise only about 73% of the total area cropped to cereals (including maize for grain, but not silage; Table 1.24), whereas they occupy about 96% in the UK. Furthermore, other crops that can be grown in rotation with wheat or barley (e.g. forage maize, oilseed rape, sunflower, grain legumes, sugar beet and fodder beet) occupy 4.44 Mha in France, but only 888,000 ha in the UK (Table 1.24). The situation in Germany, which ranks second in cereal production in the EU, is quite similar to that in France.

Table 1.20. Areas of annual crops in France.

Crops	Area (Mha)	
	1994	1988/9 (mean)
Wheat	4.592	5.022
Barley	1.404	1.801
Oats	0.166	0.219
Rye	0.045	0.065
Triticale	0.175	0.147
Maize (grains)	1.666	1.833
Sorghum	0.048	0.071
Rice	0.027	0.020
Other cereals	0.064	0.081
Total cereals	8.187	9.259
Rape/sunflower	1.757	1.820
Grain legumes	0.678	0.559
Beets	0.436	0.452
Potatoes	0.144	0.145
Maize (silage)	1.472	1.620
Set-aside	1.924	0.312
Total annual crops	14.598	14.167

Source: Anon. (1995a).

Table 1.21. Percentage of the wheat crop + barley crop areas cropped with first, second, third or subsequent cereals in the six most important regions of production in France.

Regions[a]	Number of consecutive cereal crops				Area (Mha)
	1	2	3	> 3	
Centre	56.6	35.2	8.1	0.1	0.9455
Picardie	71.1	24.6	4.3	0	0.5510
Champagne-Ardennes	72.3	23.1	3.6	1.0	0.4980
Bourgogne	56.2	37.3	6.5	0	0.4650
Poitou-Charentes	66.6	28.0	5.4	0	0.4220
Pays-de-Loire	82.4	15.6	2.0	0	0.3820
All France	65.7	27.9	6.1	0.3	6.0965

[a]Six out of 22 are listed and represent 53% of the total area cropped to cereals.
Source: Anon. (1995b).

There are no official data on yield losses caused by take-all in France and there has been no survey on the occurrence of the disease. Usually take-all is acknowledged by farmers, and even advisers, only when patches of whiteheads (see 6.2 for a discussion of this symptom) show in the field. The popular perception is simply

Table 1.22. Percentage of the wheat area cropped with first, second, third or subsequent wheat crops in the six most important regions of production in France.

	Number of consecutive cereal crops				
Regions[a]	1	2	3	> 3	Area (Mha)
Centre	75.2	22.1	2.7	0	0.725
Picardie	87.7	10.9	1.4	0	0.440
Champagne-Ardennes	90.9	8.2	0.9	0	0.381
Bourgogne	88.8	10.2	1.0	0	0.317
Poitou-Charentes	78.7	19.7	1.6	0	0.307
Pays-de-Loire	84.4	14.4	1.2	0	0.287
All France	84.0	14.3	1.7	0	4.298

[a]Representing 59% of the total area cropped to wheat.
Source: Anon. (1995b).

Table 1.23. Percentages of wheat and barley areas in France classified according to number of consecutive cereal crops grown.

Crop	1st cereals	2nd cereals	3rd cereals
Wheat	84.0	14.3	1.7
Barley	12.5	68.7	18.7

Source: Anon. (1995b).

Table 1.24. Areas (Mha) of the most important rotation crops in the three most important wheat-producing countries in the EU in 1992.

Crop	France	Germany	UK	EU
Wheat	5.080 (4.698)	2.583	2.066	16.743 (13.656)
Barley	1.800	2.408	1.297	11.502
Other cereals[a]	2.462	1.523	0.126	7.331
All cereals[a]	9.342	6.514	3.489	35.576
Forage maize	1.525	1.243	0.051	3.270[b]
Rape (+ turnip)	0.688	1.001	0.421	2.337
Sunflower	0.991	0.065		2.400[b]
Grain legumes	0.724	0.049	0.208	1.320[b]
Beets	0.512	0.580	0.208	2.250[b]
Set-aside	1.663	1.090	0.634	4.815

[a]Including maize.
[b]Figures missing for some countries in 1992; estimation based on figures from previous years.
Source: Anon. (1994c). Figures in parentheses are from Home-Grown Cereals Authority (1995).

that where there are no patches there is no take-all, where there are a few patches there is take-all but little or no damage, and where there are large patches there is a big problem which will result in significant decreases in yield. Experiments at Le Rheu (INRA, Rennes) over a period of 20 years have shown that take-all was always present on the root systems of second wheats, but it did not always cause whiteheads. These crops never yielded as much grain as first wheats and this has been attributed partly to take-all. Recent experiments using a seed-treatment fungicide from Monsanto with an efficacy against take-all resulted in an increase in yield of 2 t ha^{-1} in two crops of wheat where infestation by the take-all fungus was high and where all other diseases had been controlled by applications of foliar fungicides (Lucas *et al.*, 1994).

Data from 1981 and 1986 (Table 1.25) show that, in the northern part of France, second wheat crops yielded less than wheat grown after either beet, oilseed rape, or maize. The yield of wheat after barley was least of all (see 3.2.1 for disease effects in wheat after barley in the UK). Take-all is found almost throughout France and in 1981 it was considered to be a major obstacle to the desired intensification of wheat production. Reports in the farming press (Anon., 1981) indicated that take-all was severe in 1981 in northern France and was probably responsible for an important part of the reductions in yield in that year. On the basis of 15 years' personal experience of plant samples sent to his laboratory and of feedback from advisers and farmers, P. Lucas identified 1981, 1985, 1988, 1989, 1992, 1993 and 1994 as years in which take-all was severe and caused significant yield losses in northern France.

Experience in The Netherlands is similar to that of the reclaimed silt land of eastern England, where potato/wheat/sugar beet/wheat rotations are practised because of intensive sugar beet and potato production. Take-all problems are therefore rarely seen. Systematic annual surveys of winter wheat in The Netherlands during 1974–1986 revealed a mean annual value of 0.2% of stem bases with symptoms of take-all during ripening (incidence) and only 7% of fields with such symptoms (prevalence); roots were not examined (Daamen and Stol, 1990). This was a lower disease intensity than in England (Clarkson and Polley, 1981a) or Belgium (Lagneau *et al.*, 1986), where take-all is not considered to be particularly

Table 1.25. Mean yields of wheat after different preceding crops in the mid-north part of France.

Previous crop	Yield of wheat (t ha^{-1})	
	1981	1986
Wheat	4.6 (18.6)[a]	5.7 (18.8)
Barley	4.2 (5.2)	5.1 (4.2)
Maize	4.8 (15.6)	5.8 (14.2)
Rapeseed	5.0 (8.0)	5.8 (8.5)
Sugar beet	5.5 (13.1)	6.6 (10.7)
Potatoes	5.1 (3.7)	6.1 (3.3)

[a]Figures in parentheses are the percentages of the wheat area.
Source: Anon. (1984, 1988b).

damaging. Wheat monoculture is rarely practised and 1-year breaks normally provide effective control, except on light or acid soils where a 2-year break is recommended.) Dutch clay soils were described as very suppressive to take-all. The intensity of the disease was least in years when precipitation was high during March–June and temperatures were high in May and June (in part agreement with Hornby, 1978b). When precipitation is high, the clay soils become saturated: the high moisture content and the resulting low aeration are unlikely to favour the survival of Ggt, or host infection (Zimmermann, 1984; Heritage *et al.*, 1989). The area of winter wheat in The Netherlands was about 120,000 ha during this period and the annual national production of grain increased from 0.7 to 1 Mt (Daamen, 1990). Most of the Dutch small grains were grown in rotation with potatoes, sugar beet, peas, grass seed crops, oilseed rape or flax, which suggests that few consecutive cereal crops were grown, and this too will have contributed to the generally low level of take-all. The years in which take-all was most prevalent were 1974 > 1977 > 1975 > 1984 > 1976 (30, 15, 14, 10 and 9% of fields infected, respectively) and the years with the greatest incidence of disease were 1974 > 1984 > 1977 > 1975 = 1976 (1.0, 0.5, 0.3, 0.2 and 0.2% of tillers infected, respectively) (Daamen and Stol, 1990). There is little to suggest that these disease fluctuations were in step with those in the UK (Table 1.1).

In the former German Democratic Republic, *G. graminis* occurred locally in 1989, but had little effect on yield; it was also described as one of the principal stem-base pathogens of wheat in the former USSR (Oerke *et al.*, 1994). In the former German Federal Republic, Mielke (1983) declared that take-all may cause losses of up to 35% in wheat crops. During 1975–1985 *G. graminis* was a principal pathogen of durum wheat in Italy (Boggini, 1986).

Severe attacks of take-all were more common in Sweden in the 1950s and 1960s, but during most of the 1970s the disease was of minor importance (Yarham, 1981). This change may have been due in part to the increased usage of nitrogen fertilizer, but the main reason seems to have been a change to warmer, drier summer weather. A return to wetter conditions in 1978 and 1979 led to a resurgence of the disease.

Gaeumannomyces graminis, therefore, is widespread through central and western Europe. Zadoks and Rijsdijk (1984) published separate maps (Figs 1.15–1.17) of isodams (lines through points of equal mean annual damage) for the pathogen on barley, oats and wheat, based on data collected during 1961–1970. The exact source of the data is not given and the damage map for wheat does not coincide well with the later regional disease data for England and Wales shown in Table 1.4. (In Table 1.4, the tendency over an 8-year period is for a larger percentage of wheat crops in the west to be diseased, but generally those percentages are based on much smaller numbers of crops than those for the major wheat-growing regions to the east.) Whether or not this reflects a changing situation or inadequate data is not resolved.

The rise in wheat yields is also slowing in the USA and China, the world's two largest wheat exporters. Because wheat in these two countries is grown in areas of low rainfall, this levelling out is at about one-third to one-half the yield levels for western Europe (Brown *et al.*, 1994). A falling off in the response to increased

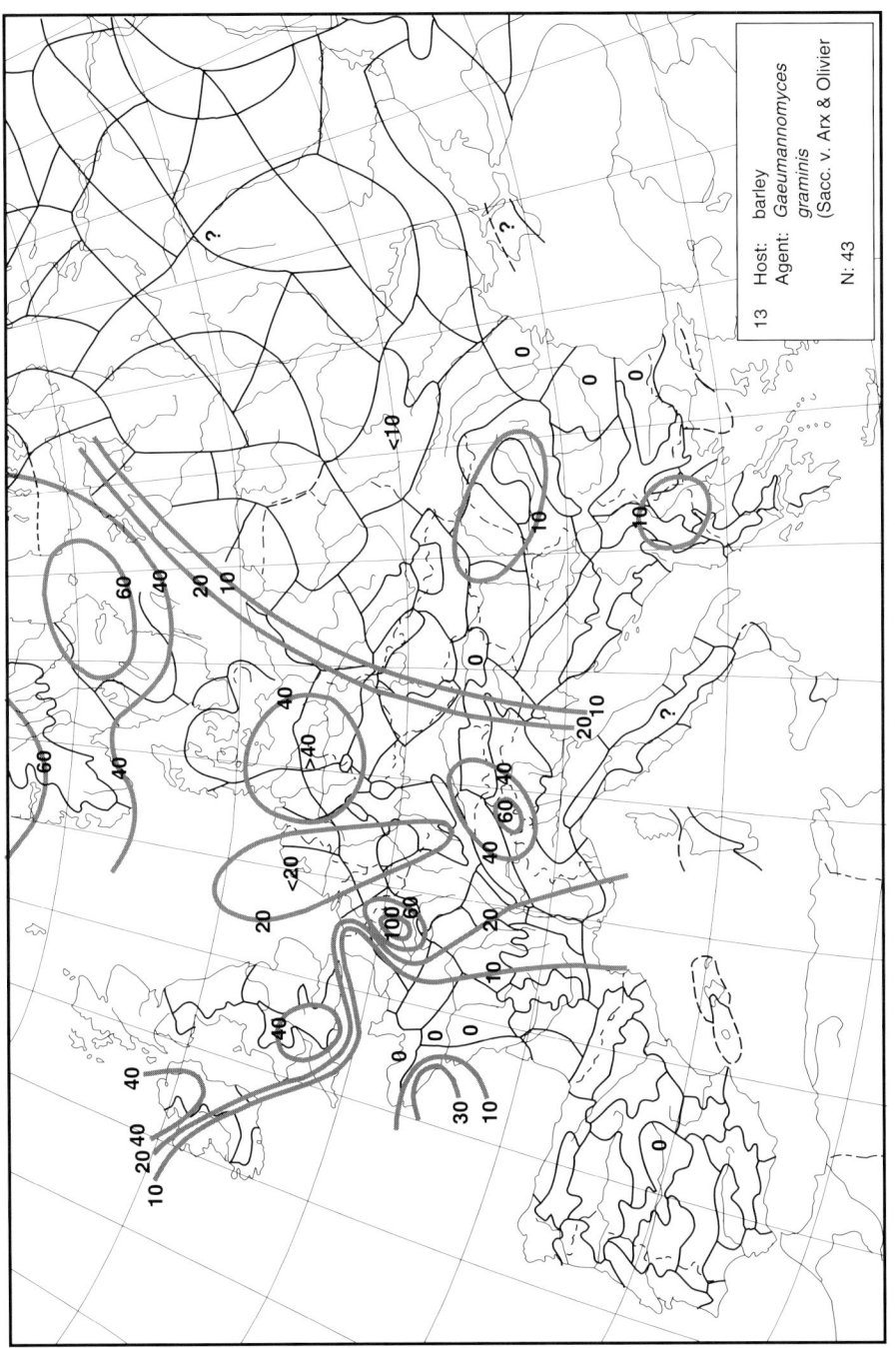

Fig. 1.15 (see p. 44).

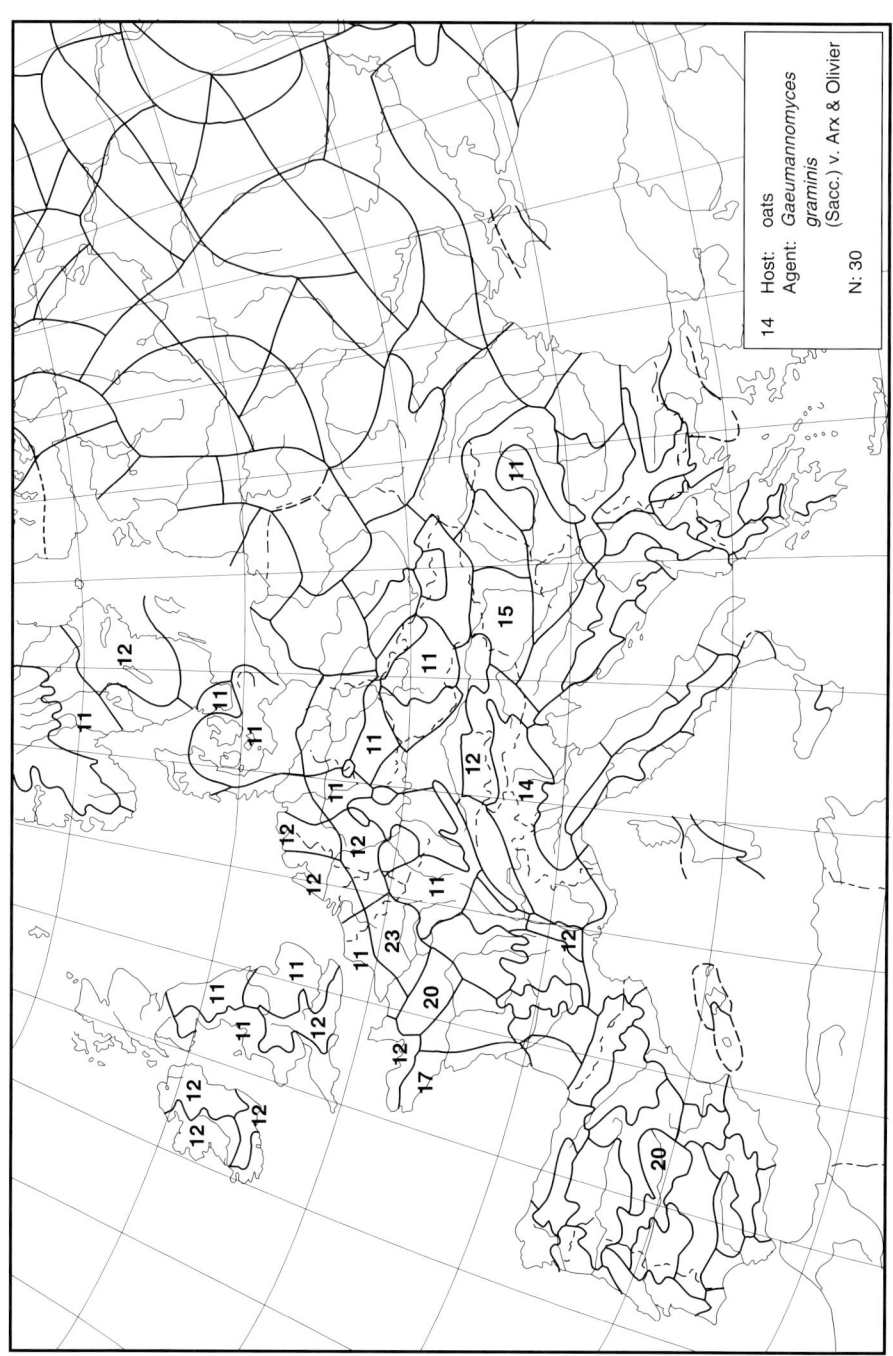

Fig. 1.16 (see p. 44).

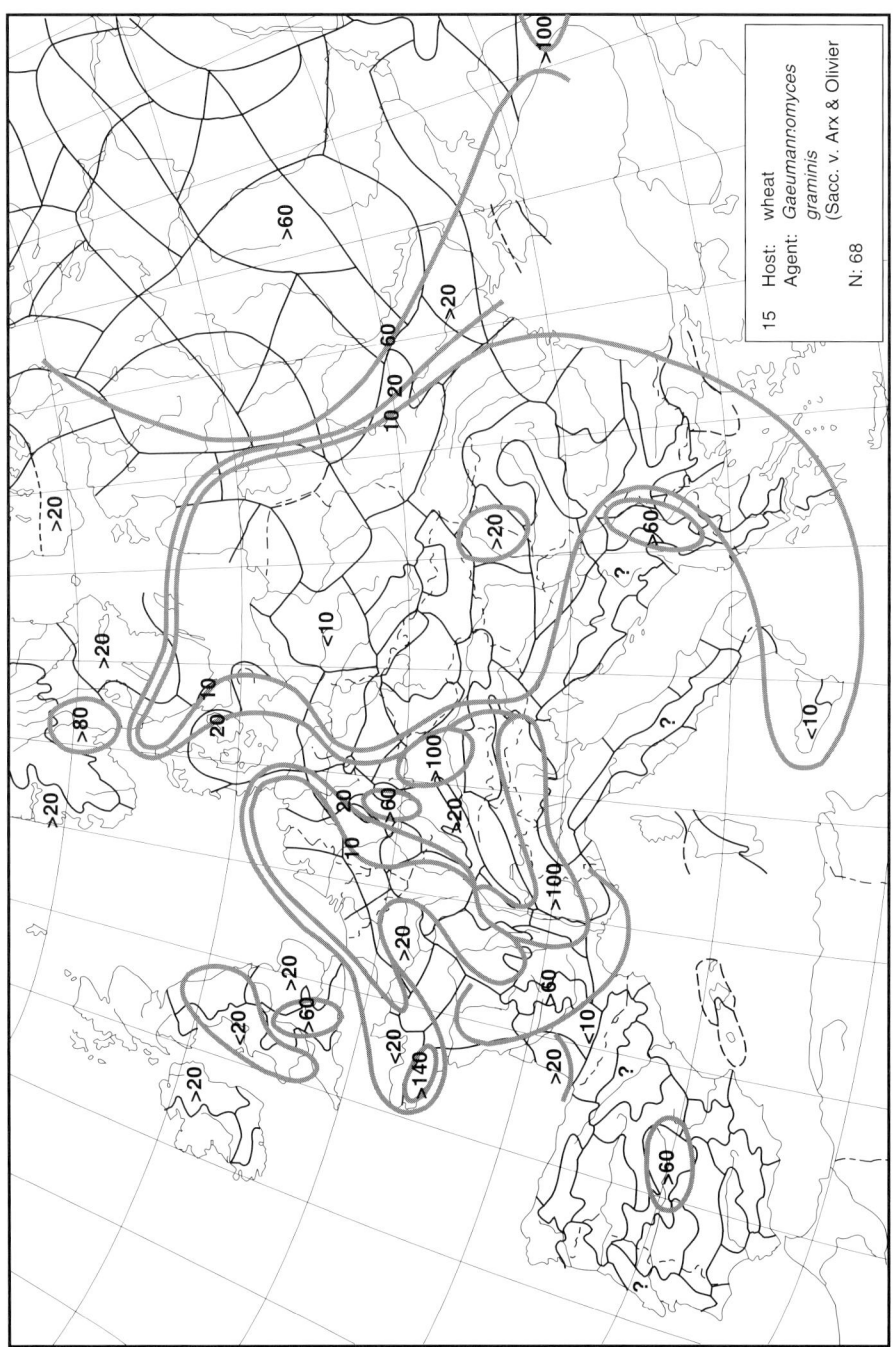

Fig. 1.17 (see over).

fertilizer application by current wheat varieties, the lack of new wheat varieties able to respond to even greater applications of fertilizer and environmental degradation (soil erosion, air pollution, waterlogging and salting) all contribute to the slowing in the rise in yields.

Take-all is the principal disease of wheat and barley in Western Australia. C.A. Parker ('Can root-rots and take-all be controlled?', paper presented at Farm Management Foundation 1976 Cropping Review, Western Australia) estimated average losses of 30%; Cotterill and Sivasithamparam (1988a) reported 25% damage to root systems on more than 80% of plants, and later (Cotterill and Sivasithamparam, 1989a) put losses at 36–40% of yield in moderate–high risk areas and 5–10% of yield overall in South Australia. In 1992 it was estimated that take-all cost South Australian growers Aus$60 million in lost yields (Mussared, 1996). Brennan (1989c) found that the fungicide flutriafol formulated as granules with superphosphate fertilizer decreased infection by 50–60% and increased yield by 25–30%. In 1986 in New South Wales losses attributed to *G. graminis* were 10–20% and 30–50% in some districts (New South Wales Department of Agriculture, 1987). Slightly earlier experiments in southern and central New South Wales had shown that liming increased the incidence of take-all: when soil pH in $CaCl_2$ solution (0.01 mol l^{-1}) was > 5, 60% of wheat and 80% of triticale were infected, but when pH was < 4.8, disease incidence was less than 5% (Murray *et al.*, 1987). [On Rothamsted farm in the UK, winter wheat died where soil had a pH (in water) of 4.5, and Cook and Veseth (1991) showed graphically that in the USA at soil pH 4, yield was about 10% of yields obtained between pH 6 and 7.5.] In these experiments, yield losses were between 26% and 77%; each increase of 1% of plants with basal stem blackening resulted in a 0.76% loss in yield. This is a somewhat greater effect of disease on yield than those given for UK cereals in section 6.3.3. Take-all was present in 32% of 56 wheat crops examined in southern New South Wales in the spring of 1993 (Lemerle *et al.*, 1996). In Victoria, triazole fungicides decreased infection by 35–50% and increased yields by 17.5–40% (Ballinger and Kollmorgen, 1988). Of Aus$303 million spent on crop protection measures in Australia

Fig. 1.15 (p.41). Isodamage contours for *Gaeumannomyces graminis* on barley in Europe for the period 1961–1970.
Mean annual damage (MAD) was characterized by a damage code (0–5) and the figures on the map are MAD × 1000.
Reproduced from Zadoks and Rijsdijk (1984).

Fig. 1.16 (p.42). Mean annual damage (MAD) caused by *Gaeumannomyces graminis* on oats in Europe for the period 1961–1970.
MAD was characterized by a damage code (0–5) and the figures on the map are MAD × 1000.
Reproduced from Zadoks and Rijsdijk (1984).

Fig. 1.17 (p.43). Isodamage contours for *Gaeumannomyces graminis* on wheat in Europe for the period 1961–1970.
Mean annual damage (MAD) was characterized by a damage code (0–5) and the figures on the map are MAD × 1000.
Reproduced from Zadoks and Rijsdijk (1984).

in 1986/87, Aus$61 million were used on the control of *G. graminis* (Brennan and Murray, 1988).

In the Pacific Northwest of the USA, the importance of take-all and other soil-borne diseases has long been recognized and reflected in the considerable amount of research into the subject. In this area, which includes especially the rain shadow areas in Washington, Idaho and Oregon, an estimated 20% of winter wheat was lost to take-all in 1983 (Folwell *et al.*, 1991). In the same year, two out of three crops were considered to be at risk from take-all, as a result of wheat being grown commonly in 2 out of every 3 years. In the Pacific Northwest it was estimated that the effective control of take-all would increase winter wheat yields by 10–50% (Heim *et al.*, 1986). Take-all has long been recognized as a problem in intensively grown cereals west of the Cascade Mountains in the Pacific Northwest. The drier areas of the Columbia basin account for over 90% of the region's wheat acreage and take-all was of little importance until the 1960s, when irrigation of wheat increased dramatically. The combination of virgin (previously desert) soil, the lack of antagonists and abundant water resulted in losses of as much as 50% of yield in recently reclaimed fields [D.J. Yarham (1984) Root rotting diseases of wheat in the Pacific Northwest of the USA. Report on a study tour to Oregon and Washington State. MAFF, unpublished, 23 pp.].

After the early 1970s, take-all was also considered to have a major influence on the productivity and variability of wheat in Indiana, USA (Huber, 1981a); growing high-yielding varieties resistant to the hessian fly (*Mayetiola destructor*) involved crop management changes that increased infection by *G. graminis*: losses were estimated at 4% in 1977 and 25% in 1973 and 1976, representing a cost to the farmers of US$6.8–50.6 million per year. In Kansas the estimated average loss in wheat attributed to *G. graminis* was 1.2% over the period 1976–1987 (the average for all diseases was 15.3%), with the greatest loss being 3% in 1976 (Sim *et al.*, 1988). Stelljes and Hardin (1995) claimed that losses from take-all and other root diseases in the USA amount to about 15% of the crop and cost the farmers US$500–1000 million annually.

In Canada, Nova Scotia was the only province in which *G. graminis* was listed amongst the principal pathogens of wheat (Oerke *et al.*, 1994). Take-all is less of a problem in the spring-sown crops of Canada than in the predominantly winter-sown crops of the mid-western states of America. It does, however, cause occasional losses, chiefly on the black soils of the prairie provinces (Yarham, 1981).

In the late 1970s, take-all began to pose serious problems for the major wheat-growing states of Brazil (Yarham, 1981). In Rio Grande do Sul in Brazil, root diseases in 1981 caused wheat losses of about 20% in wheat/soya bean double- cropping. The major disease was common root rot (*Cochliobolus sativus*), but take-all was important where farmers had limed fields heavily for the soya bean crops (Hornby, 1982). For the same region, Tomasini *et al.* (1983) gave average yields for the four commonest varieties of wheat in the range 1.53–1.66 t ha^{-1} and reported that *G. graminis* was found in only 19 of 430 fields, where it was responsible for yield losses of 0.37 t ha^{-1}. There was a survey of root diseases of wheat in the state of Paraná in 1981 (Diehl *et al.*, 1984): common root rot and take-all occurred in western and central-southern regions, and common root rot alone in the northern region. In Chile (Hornby, 1982) take-all was a major problem that seemed to be not so well controlled by rotation as it was in Europe. *G. graminis* decreased yield by 20% in 1983 (Cortazar *et al.*, 1987), but at La Platina Experimental Station, Santiago, it was particularly notable in only two seasons in the period 1965–1986 (Cortazar, 1989).

In 1981, Trolldenier (1981) suggested that take-all was the second most important cause of loss in wheat worldwide, after black stem rust. There are few cereal-growing regions from which it has not been reported. By 1984, *G. graminis* had been recorded in Africa (eight countries), Asia (nine countries), Australasia and Oceania, Europe (24 countries), North America and South America (Fig. 1.18). New occurrences relating to the species continue to be recorded, as for example the first report of the variety *graminis* (Ggg) from wheat roots in new agricultural land in Egypt (Fouly *et al.*, 1996a). Oerke *et al.* (1994) listed *G. graminis* as one of eight categories of major fungal pathogens of wheat on a worldwide scale, but rated it as amongst the most important pathogens only in Oceania. In eastern and western Europe, rusts and *Erysiphe graminis* were considered more important, and in Western Europe also *Septoria* spp., *Pseudocercosporella herpotrichoides* and barley yellow dwarf virus (BYDV). In Asia, bunts, rusts and smuts were more important; in North America, rusts, smuts and *Septoria* spp., and in Latin America, rusts, *Cochliobolus sativus* and BYDV. It was stated that diseases like take-all and common root rot increased in importance as cultivation became more intensive.

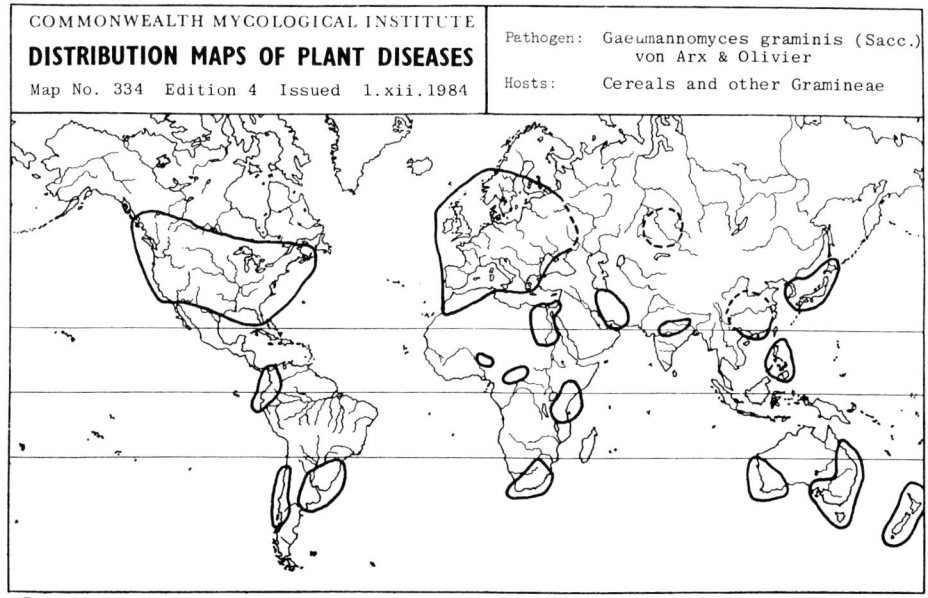

Fig. 1.18. Distribution map for *Gaeumannomyces graminis*.
The dashed lines indicate unconfirmed or uncertain identifications.

Disease and Epidemiology

2.1 Preamble

2.1.1 Concerning the host

Whereas Chapter 5 is devoted to the pathogen, information about the host is introduced where appropriate throughout the book, e.g. cereal types and production history (Chapter 1), resistance (see 3.2) and nutrition (see 3.4). Here, it is necessary to consider aspects of the cereal host's root system in order to understand the sections that follow.

Cereals have a fibrous root system, with two distinct parts: seminal and nodal (crown) roots, which differ in time of appearance and potential depth of rooting (Klepper and Rickman, 1990). The root axes initially grow at successively increasing angles from the vertical before turning to make up a series of downward-growing axes. This allows successive axes to explore wider and wider cylinders of soil as the season progresses. In a mechanistic computer simulation of the paths of growth of the main axes of seminal roots of wheat exposed to inoculum of Ggt (Gilligan, 1985a; Gilligan *et al.*, 1994), the first seminal root was assumed to grow vertically and the rest were approximated by equations for three-dimensional parabolas with variable angles of rotation at the crown. The secondary and tertiary branches on root axes are thinner and shorter than the axes and are presumed to be primarily absorptive (Klepper and Rickman, 1990). The period of branching of wheat roots may be measured in degree-days (GDD) and is about 250 GDD (base temperature 0°C). WHTROOT, a model of the cereal root system, is degree-day driven (Klepper and Rickman, 1990).

In the UK, the top several centimetres of first seminal root axes of winter wheat may have several anucleate cortical cell layers by late February, and by mid-April nearly five of a total of six layers may be dead (Deacon and Henry, 1981). Subsequent pairs of seminal roots may show similar or smaller amounts of root cortex death. Since root cortex death occurs as rapidly in sterile conditions, it is thought to be endogenously controlled and to be associated with the transition from nutrient absorption of younger regions to a predominantly conducting function of older regions of the root. Although cortical cell death does not seem to differ significantly amongst crops within an established sequence of winter wheat crops, the crops preceding a run of cereals can influence the amount of cell death, with grass being associated with more dead cells than swedes followed by potatoes. This may explain why *Phialophora graminicola* frequently persists better in wheat after grass than in wheat after other crops, because this fungus seems to colonize root cells just before or just after their death. Laboratory experiments (Brown and Hornby, 1987) have shown that cortices of proximal and middle regions of roots on 5- to 18-day-old seedlings grown on agar have fewer nucleate cells when nitrogen is supplied in the ammonium form, rather than as nitrate or a mixture of the two forms. Nitrate prolonged the survival of nucleate cells in the outer layers of the cortex. In gnotobiotic cultures take-all was greatest where nucleate cells were fewest and root exudation greatest, and the effect was strongest with ammonium nitrogen. (Other effects of forms of nitrogen are discussed in Section 3.4.1.)

2.1.2 Concerning interactions

The 'disease triangle' (pathogen–host–environment interaction) and the 'disease pyramid' (pathogen–host–environment–time interaction) help in visualizing the complex of factors that relate to disease causation (Bateman, 1978). There is a hypothesis applicable to all the relationships embodied in the disease pyramid. The so-called multiple component hypothesis of pathogenesis and parasitism 'focuses attention on the nature of the interacting components of host . . . and parasite . . . as determining factors in parasitic or pathogenic relationships' (Bateman, 1978). This assumes that host and parasite each have factors that are potentially favourable and factors that are potentially unfavourable to the other. Thus, four 'environments' (host environments favourable or unfavourable to the pathogen, and vice versa) are discerned and each of these has multiple components. The sum of the components of the four environments and their interactions determines the type of relationship. Huber and McCay-Buis (1993) used multiple component analysis to

- facilitate the organization of additional information that leads to more effective management;
- provide the mechanism for conceptually dissecting the favourable and unfavourable environments that each host and pathogen bring to the relationship over time.

Huber and McCay-Buis's (1993) multiple component evaluation of take-all indicated several approaches to biological control other than classical antagonism (see 4.3). There has been an inadvertent universal selection of manganese-reducing organisms for plant growth promotion and biological control. It is possible, therefore, that there is a generalized mechanism involving manganese (see 3.4.3), whereby biological control is achieved by inhibition of virulence, rather than by antagonism, and/or increased availability of manganese to the plant.

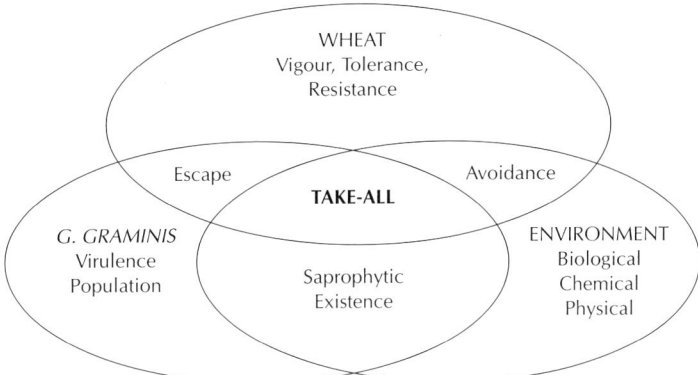

Fig. 2.1. The *Gaeumannomyces graminis*–wheat–environment interaction: take-all occurs only when favourable conditions of the three factors of plant, pathogen and environment overlap; disease severity is determined by the 'relative favourableness' of individual components of the whole system.
Figure reproduced from Huber and McCay-Buis (1993), caption modified.

Take-all is the result of the Ggt–wheat–environment interaction, in which the sum, balance and interaction of all potential components determine the nature of the interaction (Fig. 2.1). Each part of the disease pyramid can have an effect on the 'relative favourableness' of a component of the system in determining severe disease.

2.2 Disease Cycle

Detailed information on the pathogen and the disease is available in Asher and Shipton (1981); what follows is a general outline and a recapitulation of some of the salient points of the disease cycle in new diagrams and photographs. Take-all disease cycles published in texts for students (Parry, 1990), or for farmers and advisers (Weston, 1944), show the main stages of the disease, but are generalized and simplified. The schemes given here (Figs 2.2 to 2.5) are generalized for Hertfordshire and Bedfordshire, and are probably generally applicable to the UK. They include a higher level of epidemiological realism than previously published schemes. The circular presentation is a convenient device for showing the cyclical nature of the processes under consideration, but leads to a succession of concentric annuli with increasing diameters (Figs 2.4 and 2.5). The relative sizes of these annuli have no significance in these schemes. Also, for the sake of clarity, not all variations that are likely to be encountered are shown. In Fig. 2.2, for instance, in the period marked 'primary infection of seminal roots' there is, very occasionally, exceptional disease development leading to a seedling blight, where plants are severely diseased or killed (Plate 2.6). In the USA, take-all may cause symptoms

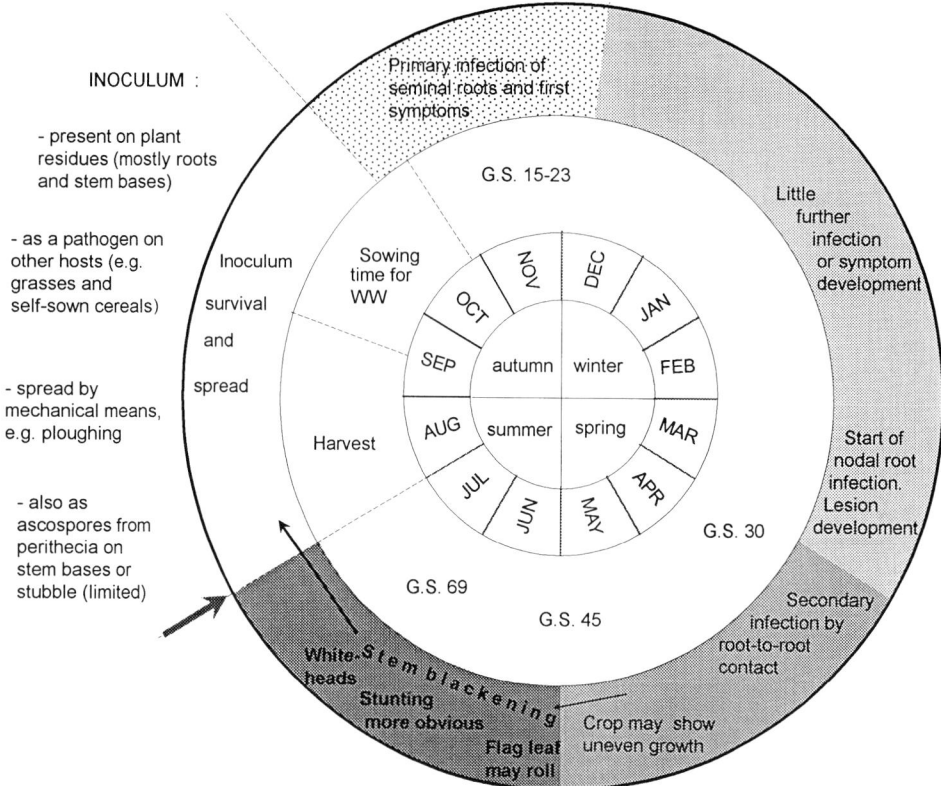

Fig. 2.2. Generalized annual disease cycle for take-all in winter wheat in the UK.
The outer annulus marks stages in symptom development, with shading indicating an increasing effect of disease on the host. Symptoms of early root infection have been described in detail in Hornby and Fitt (1981); amongst the more obvious are lignitubers, or papillae, around hyphae attempting to penetrate living cells and root lesions. The stages of the disease can be related to season (centre circle), month of the year (inner annulus) and growth stage of the host (middle annulus). The TAR (described in 6.3.3) of the crop may often double between GS 45 and GS 69–71.
Aspects of inoculum survival and spread between successive crops are listed outside the outer circle.
The diagram should be read clockwise, starting at the arrow (harvest of a previous crop).

characteristic of nitrogen deficiency or magnesium deficiency in young plants and is consequently often misdiagnosed (Huber and McCay-Buis, 1993). There are reports from Australia of Ggt growing through soil (see Sections 2.7 and 4.7), but in the UK growth in soil seems to be very limited, or is confined to colonized organic residues and therefore is not featured in Fig. 2.3.

Garrett (1938) suggested that the alternation between an active parasitic phase and a declining resting phase was characteristic of a highly specialized root-infecting fungus, confined to existence upon its hosts because of microbial competition in the soil. In later works he initially called such an organism a 'semi-obligate parasite', but soon introduced the phrase 'ecologically obligate parasite' (Garrett, 1956). This idea was developed in Hornby (1994) to show some general areas of autecological interest in relation to Ggt (Fig. 2.6). Away from

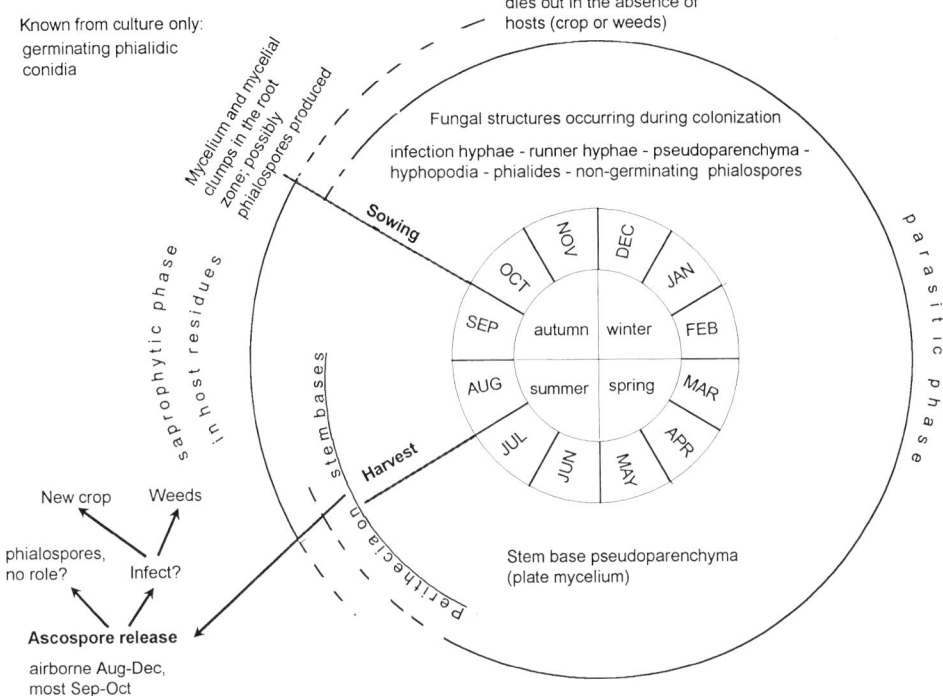

Fig. 2.3. Generalized life cycle of *Gaeumannomyces graminis* var. *tritici* associated with winter wheat in the UK, showing stages associated with the growing plant (parasitic phase) and soil-borne and air-borne stages (saprophytic and dispersal phases).

Individually, signs of the fungus may not be good diagnostics; for example, runner hyphae are not produced just by Ggt, and runner hyphae of Ggt may grow epiphytically over the roots of rye and oats without disease (Huber and McCay-Buis, 1993).

living hosts, therefore, Ggt resides mostly in host residues in the soil in a declining, saprophytic state. Such material constitutes the bulk of inoculum in established arable fields and it has been a conventional focus for autecological studies. Consequently, there is a large body of research on the nature, sources, amounts, behaviour, potential, survival and relationships to disease of this inoculum (references throughout this book; for earlier reviews and further literature see Asher and Shipton, 1981). On and inside hosts, when Ggt is parasitic and/or pathogenic, its biomass increases. The internal environment is important in the study of autecological relationships, for it is inside the host that interactions operate which influence the extent of disease and the effect that the disease has on yield. Adjacent to the interface between soil and host are special niches: to the soil side, the rhizosphere (Hornby, 1990b), and to the host side, the outer cells of the root, where in cereals natural, non-pathogenic death occurs in the cortex (Deacon and Henry, 1981). In these niches the study of the autecology of Ggt leads into current research topics on biological control mediated by interactions with other microorganisms (see 4.3).

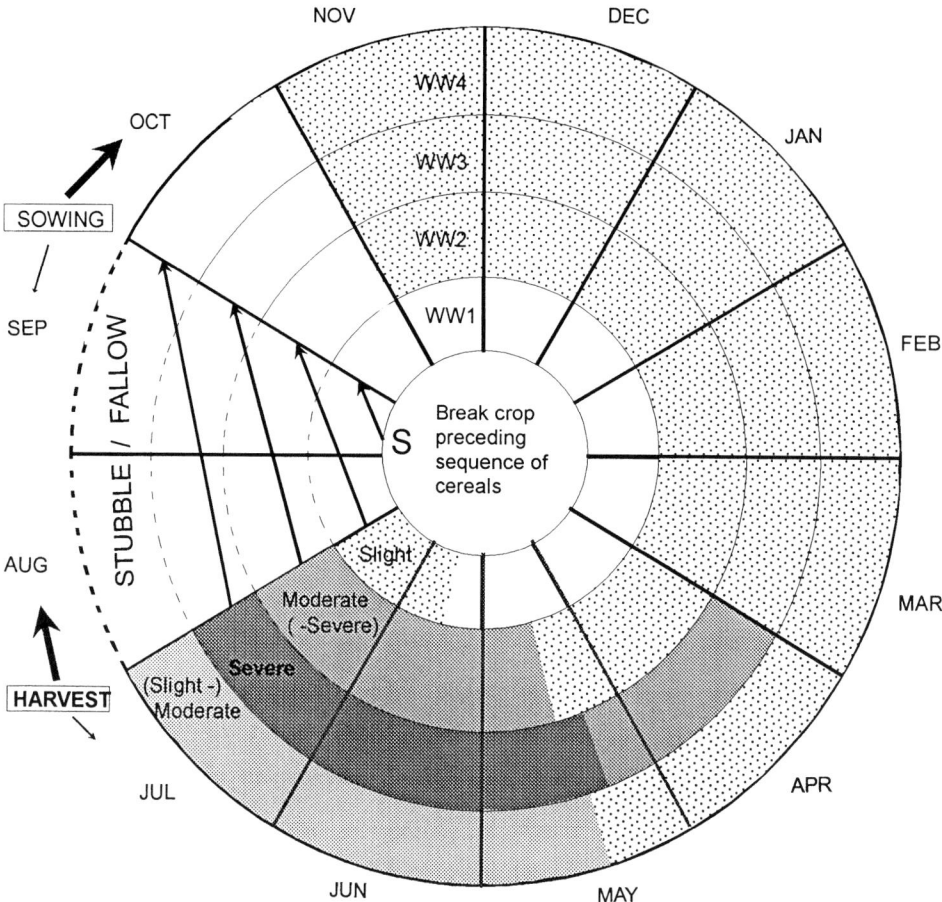

Fig. 2.4. Annual cycles of take-all intensity in four successive winter wheat crops grown after an effective break crop: an impression based on many years' experience at Rothamsted Experimental Station and not an accurate representation of any particular epidemic.

Read this diagram by starting in the middle at **S** and move clockwise, following the arrows. Sowing and harvest are not fixed dates because of weather and for other reasons: the thickness of arrows associated with these labels on the diagram indicates the relative likelihood of the operation being earlier or later than the date shown. It is considered advisable to delay sowing when the risk of take-all is great (see 3.5.1). WW1 . . . WW4 = consecutive winter wheat crops.

The disease intensity levels indicated by increasing intensity of shading are none[a], slight, moderate and severe (see 6.3.3). This classification is not sensitive enough to detect the early differences, referred to in Fig. 2.16, that may occur amongst sequences.

[a]An effective (clean) break crop usually decreases inoculum of Ggt to the extent that by autumn very little infectivity can be detected by conventional soil bioassay. Grass weeds, volunteer cereals and other hosts in break crops may seriously offset this effect and in some cases actually increase inoculum (see 3.5.5).

Some of the stages referred to in Figs 2.2 and 2.3 are illustrated. The fungus spreads along cereal roots by means of dark runner hyphae (Fig. 2.7). From these, hyaline branches invade the cortical cells radially and finally enter the stele, rapidly colonizing and destroying the phloem, thereby cutting off the supply of

Disease and Epidemiology

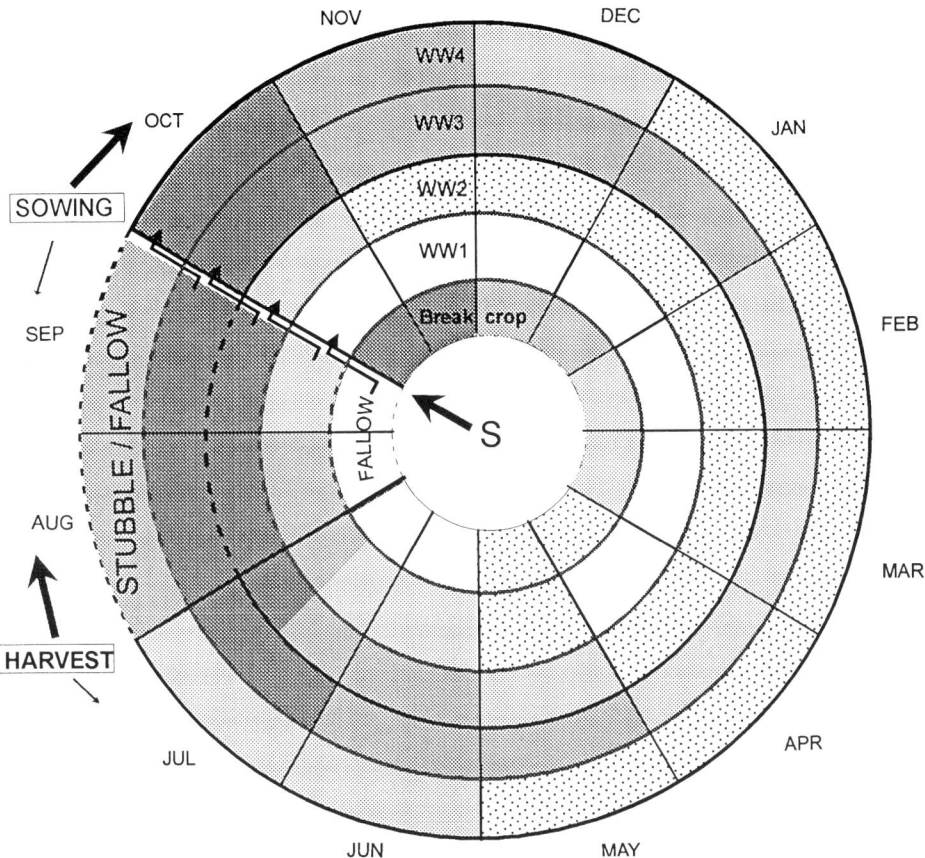

Fig. 2.5. Annual fluctuations in soil infectivity during a sequence of a break crop and four successive wheat crops; based on many years' experience at Rothamsted Experimental Station.
Read this diagram by starting in the middle at S, move clockwise and then follow the arrows. The intensity of the five levels of shading indicates the relative amount of disease on bioassay plants, from less that 2% (unshaded), through 2–10%, 11–25%, 26–50%, to more than 50% (darkest) of roots infected.
WW1 . . . WW4 = consecutive winter wheat crops.
Between consecutive crops of spring barley, inoculum estimates based on soil infectivity tests tend to show a distinct mid-winter minimum (Hornby, 1975; Hornby and Henden, 1986). There are fewer data for winter cereals, but the trend tends to be weaker (see Fig. 4.3, which implies a decrease in soil infectivity between autumn and spring) or absent.

photosynthate to the portion of the root below the damaged tissue; subsequent colonization of the xylem, causing decreased uptake of calcium and water, is often manifested as dark root lesions (Fig. 2.8a,b). These are the symptoms usually used in assessing diseased plants. Non-germinating phialospores (Fig. 2.9) may be produced in soil or in culture, and about harvest time the sexual stage may become apparent on stem bases as perithecia (Fig. 2.10; Plate 1.5), which produce asci and ascospores (Fig. 2.11).

Campbell and Benson (1994) classified *G. graminis* as a fungus exhibiting a polycyclic life cycle (i.e. more than one life/disease cycle during a growing season), indirect initial infection (i.e. where an intermediate spore type is produced between

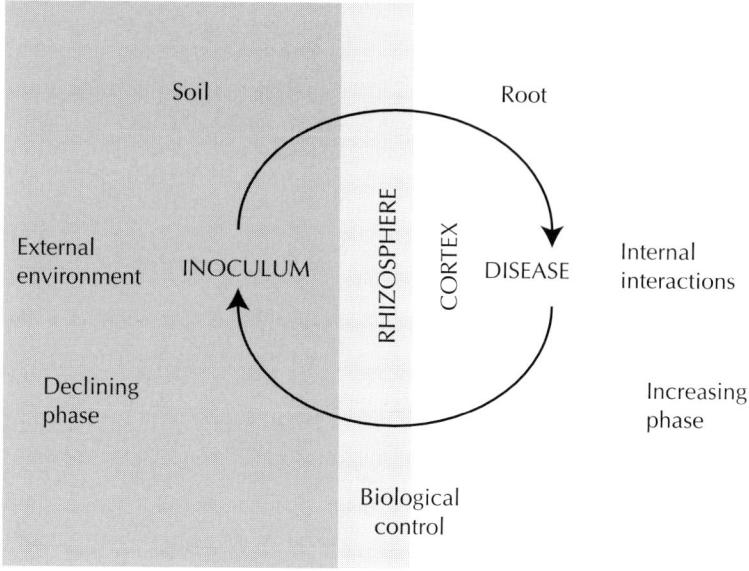

Fig. 2.6. Some general aspects of autecological research on the take-all fungus. From Hornby (1994).

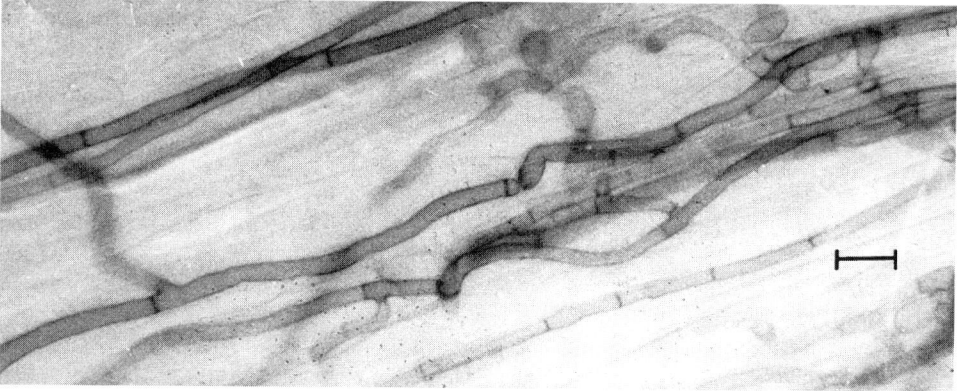

Fig. 2.7. *Gaeumannomyces graminis* var. *tritici*: runner hyphae on roots of a wheat seedling from a pathogenicity test. Bar = 10 µm.

the resting/survival propagule and infection) and a dispersal mechanism based on movement of spores prior to initial infection. They referred to early work (discussed in Asher and Shipton, 1981) which showed that ascospores could infect the proximal parts of seminal roots of seedlings that had germinated on the soil surface. These observations led to the conclusion that volunteer (self-sown) wheat seedlings growing from seed shed at harvest were at risk from infection by ascospores. Ascospores, however, still do not have a fully substantiated role as the initiators of take-all epidemics in the UK (Hornby, 1981) and phialospores and germinating

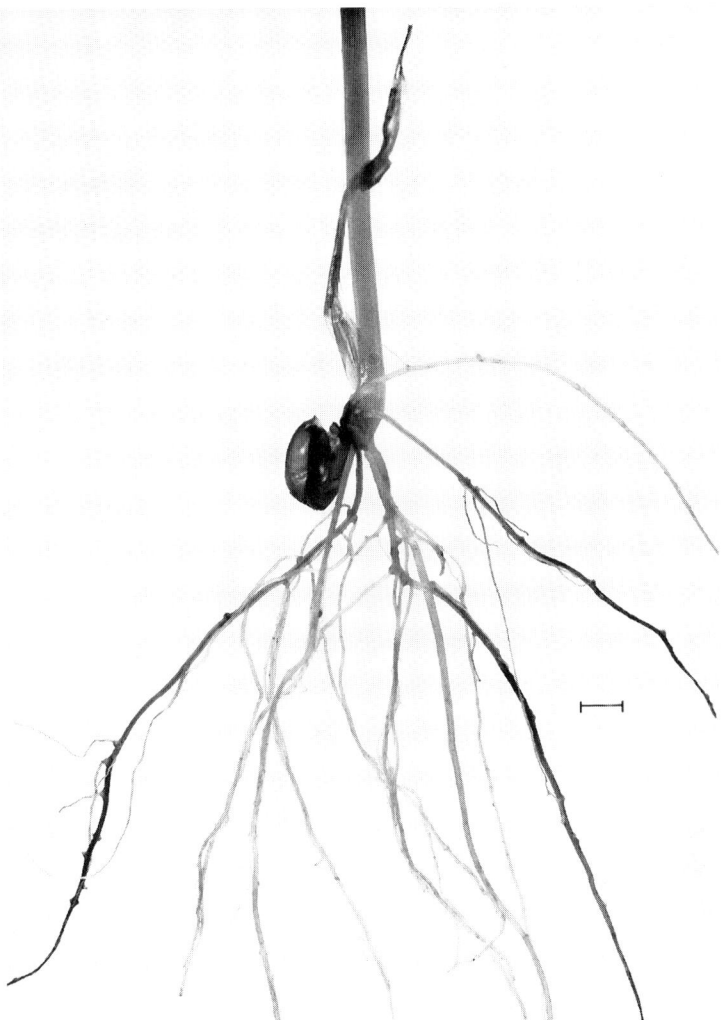

Fig. 2.8a. *Gaeumannomyces graminis* var. *tritici*: take-all lesions on roots. Winter infection of cv. Cappelle Desprez in the field, GS 14. Bar = 0.5 cm.

phialidic conidia seem to have no roles in epidemics. The Campbell and Benson (1994) classification, therefore, would seem to relate only to special cases, such as the introduction and establishment of *G. graminis* in the virgin soils of new polders in The Netherlands (discussed in Asher and Shipton, 1981). In fact, experience in the UK would favour putting Ggt into another of Campbell and Benson's categories, namely polycyclic, direct initial infection (i.e. infection occurring primarily by mycelial growth from a resting or survival spore) and dispersal by growth of mycelium. In the case of Ggt, there is no 'resting or survival spore', although plant residues colonized by mycelium serve a similar role in this context. Secondary infection by mycelial growth may be autoinfection or alloinfection (Gilligan,

Fig. 2.8b. *Gaeumannomyces graminis* var. *tritici*: take-all lesions on roots. Summer infection of cv. Cappelle Desprez; severely rotted roots and blackened culm bases, GS 71. Bar = 0.5 cm.

1985a). How frequently these types of infection occur within a single season is not known, although alloinfections are probably relatively few. If new infections occurring on neighbouring plants are taken to indicate spread, then the disease may be better described as oligocyclic (term in Gilligan, 1985a) within a season, rather than polycyclic, though simulation studies suggest otherwise (see 2.6.4).

2.3 Disease–Environment Interactions

2.3.1 Introduction

Take-all is able to develop where the soil pH is between 5.5 and 8.5, where soil temperatures are between 5° and 30°C [although severe infection is restricted to a narrower range, given as 5–15°C (Cook, 1981) and 12–20°C (Hornby, 1981)], near to field capacity and sufficiently well structured for aeration to be good. In many regions these conditions either occur naturally or are created artificially by liming, irrigation and cultivation for the benefit of the crop (Cook, 1981). As a

Fig. 2.9. *Gaeumannomyces graminis* var. *tritici*: phialides and phialospores produced on PDA. Bar = 10 μm.

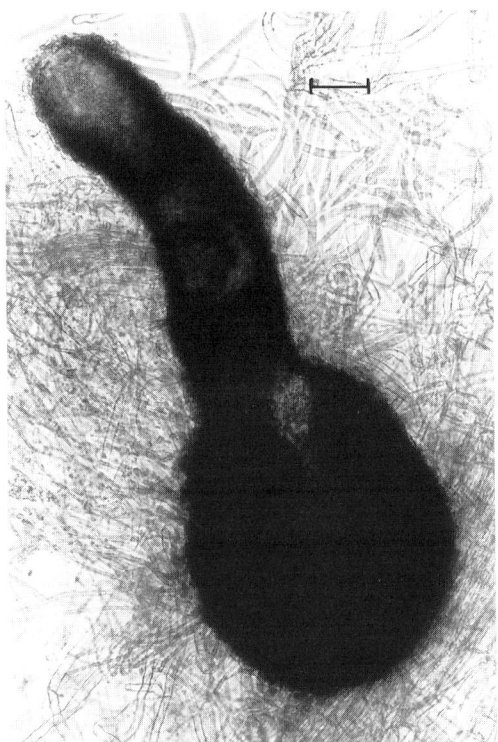

Fig. 2.10. *Gaeumannomyces graminis* var. *tritici*: perithecium produced after 6 weeks on excised wheat roots in a rotting test[a]. The perithecium has been ruptured and many asci can be seen at the top right. Bar = 50 μm.
[a]Details of this test are in Holden and Hornby (1981).

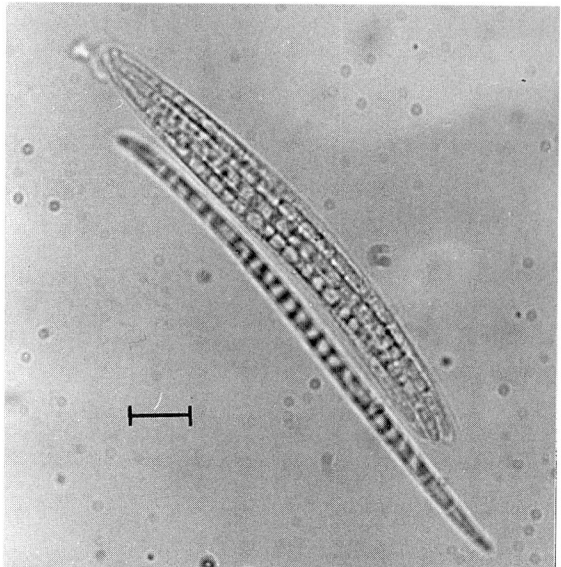

Fig. 2.11. *Gaeumannomyces graminis* var. *tritici*: ascus and one free ascospore from the perithecium in Fig. 2.10. Bar = 10 μm.

result, the disease is widely distributed throughout the temperate wheat-growing regions of the world and has been recorded on wheat at high altitude in subtropical or even tropical areas (Garrett, 1981; more detail in Section 1.4).

The effect of take-all on crop yield depends on the balance between disease development and host growth. Certain seasonal weather patterns, described in Section 1.1 and Fig. 1.10, are often associated with severe 'take-all years'. These weather patterns exert their effect by encouraging pathogen growth and/or by reducing the ability of plants to tolerate root loss. This general example of a disease–environment interaction illustrates the need to consider the environment as experienced by the pathogen (predominantly below-ground factors such as soil temperature, pH, texture, structure, nutritional status and water status) and the environment as experienced by the host (below-ground factors, and above-ground factors such as evaporative demand and sunshine). Some of these factors can be manipulated to some extent as part of normal crop husbandry and can therefore offer a method of disease management (see Chapters 3 and 4). Underlying these efforts, however, will be the natural variations in soil type and climate that occur among different areas, whether they be part of the UK, different countries, or larger geographical regions. There will also be natural variation in the weather amongst seasons.

2.3.2 Weather and seasonal effects (British Isles)

Take-all years (details in Section 1.1), when visible effects of take-all are widespread, represent the most severe end of a spectrum of disease intensity that is weather dependent. Because weather patterns determine, to a large extent, the national loss suffered through take-all in a given year, an understanding of the weather–disease relationship is desirable. Some earlier findings on the effects of weather on take-all were summarized by Clarkson and Polley (1981b).

In 1948 take-all was widespread and destructive in England (Table 1.1). Moore (1948) attributed the severity of the epidemic to early planting in the autumn of 1947 (the soil being dry) whilst the fungus was still active in the stubble residues, and a following mild winter that permitted active growth of the fungus along the seminal roots. Exceptionally heavy rain in January 1948 caused much soil nitrogen to be leached and root growth was retarded. This was interpreted as favouring the disease, but dry weather in early spring was believed to have checked fungal growth along the roots and to have helped in preventing a general occurrence of the 'true take-all symptom' (plants stunted and yellow, failing to produce ears, or killed outright). The onset of wet conditions was apparently responsible for renewed fungal activity and whiteheads were prevalent well before harvest in most parts of the country.

In 1948, many crops that followed a 1-year break without cereals were affected by take-all. Moore (1948) traced the origin of this to the wet summer of 1946 and the severe winter that followed. Stubbles were not ploughed under until late in the spring of 1947 or, if they had been ploughed, the stubble residues remained frozen and unrotted in the ground. The cold winter was followed by a hot, dry summer with very dry soils and in many districts the stubble was still unrotted when autumn sowings began. This could have almost nullified the effect of a 1-year break.

Thus it seems that weather conditions well before sowing can influence how take-all develops subsequently. Today we are little further ahead in relating weather to take-all development – due primarily to a lack of quantitative information on take-all severity, collected on a national basis. As in Moore's time, much of our understanding is based on qualitative information, or field experience.

The inter-relationship between seasonal weather and take-all has been studied using descriptive information on take-all severity from 33 years of the *Report of Rothamsted Experimental Station* (Hornby, 1978b). Take-all levels were described as rare, prevalent, damaging or severe. After dividing the weather data into autumn, winter, spring and summer and ascribing values above or below the long-term mean for temperature, rainfall and sunshine, certain trends were apparent (Fig. 1.10). Both 'prevalent' and 'damaging' take-all years occurred when summers were warmer and drier than average, but 'prevalence' was associated with cold, bright springs and 'damage' with warm, dull springs. Hornby (1978b) suggested that it was a host response to weather that determined the difference between the two categories of years. The picture is confused by those years when take-all was categorized as severe, as these were generally associated with below-average temperatures and rainfall during the summer. Clearly, a finer breakdown of the cropping cycle is desirable.

During the spring and summer, winter wheat goes through a number of developmental stages within a few weeks; as a result, short-term weather effects could be critical. For example, 1987 was generally wet during the spring and summer, but a short, hot, dry period in late May/early June was sufficient to put crops under moisture stress at a critical point in their development. Even winter wheat with modest levels of take-all had patches of disease visible above ground.

Hornby and Henden (1986) reported on percentage plants and percentage roots infected with take-all and on soil infectivity during 16 years of continuous spring barley grown on sandy loam at Woburn, Bedfordshire. Changes in cultivar, nitrogen rates and sowing dates may have accounted for some of the changes in disease that marked an upsurge in take-all which began in the late 1970s and entered its terminal phase in 1983. This upsurge was unexpected because the site had apparently developed TAD following a peak of disease in 1969. Unusually dry years in 1975 and 1976 may have decreased inoculum and wholly or partially eliminated TAD, so that a build-up of disease followed, similar to those after breaks from cereals. On balance, the upsurge was considered to have been mostly weather-determined and to have been a reflection of a general long-term fluctuation of the disease in the British Isles. There is support for widespread, long-term changes in the intensity of take-all in monocultures of spring barley from two Irish sites 50 miles apart at Clonroche and Oak Park (Cunningham, 1985; also discussed in 3.3.3). Take-all peaked in third and fourth crops at Oak Park in 1967 and 1968, with an index (see 6.3.3) of just over 40%, and then decreased erratically almost to extinction in 1979. In 1981 there was a recovery to about 20%. At Clonroche, where the incidence of take-all was generally greater, there was a pronounced peak of disease in a second crop in 1966 and minor peaks in 1969, 1972, 1976, 1978 and 1982 (highest level since 1972), at levels that were considered likely to have caused losses in wheat.

The ADAS Cereal Disease Survey (see 6.2) has collected data on take-all severity from a random sample of winter wheat fields (stratified by region to reflect the area of wheat grown). In 1977–1979, root systems were assessed and the percentages of plants with slight, moderate or severe take-all recorded. In 1985, 1987, 1988 and 1989 the disease was assessed on a field scale by crop growth and premature ripening. Roots were checked to ensure that take-all was associated with the symptoms. Table 2.1 shows the survey results for these years (Polley and Thomas, 1991; 1989 data unpublished).

Table 2.1. ADAS Cereal Disease Survey: measures of take-all nationally at growth stages 73–75.

Severity rating on roots	Percentage plants			Patch rating in crops[a]	Percentage fields			
	1977	1978	1979		1985	1987	1988	1989
Slight	8.9	29.8	24.9	1	13.7	24.9	16.8	16.3
Moderate	1.6	8.1	7.1	2	7.0	22.8	5.4	8.5
Severe	1.3	2.4	3.1	3	4.3	13.2	6.1	5.4
				4	4.7	9.1	0.7	3.5
% crops affected[b]	58	91	93		29.7	70.0	29.0	33.7

[a]The higher the rating, the greater the area of disease plants (see Section 6.2 for details).
[b]1977–1979 data are not comparable with 1985–1989 data.

There are a number of problems in interpreting these data:

- They were collected for a limited number of years (10 years is probably a minimum period for a valid analysis of weather effects).
- The two methods of assessment cannot readily be related.
- Take-all severity was not compared with visible crop effects or yield.

Nevertheless, the severity of take-all in 1987 is readily apparent (as discussed in Section 3.6, where more detailed effects of weather at a single site are also discussed).

2.3.3 Temperature and water potential

Soil and plant tissue water-potential data may offer a useful means of interpreting the effects of weather on take-all development and yield loss, because they are a direct measure of water availability. Unfortunately such data are both scarce and difficult to interpret. Soil water potential varies with depth down the profile; and the water potential of plant tissue fluctuates diurnally and depends on (amongst other factors) the crop canopy, sunshine, temperature and wind speed. Furthermore, calculation of the soil water potential from more commonly recorded rainfall or soil moisture deficit data is likely to be inaccurate.

According to work from the USA, Ggt cannot grow in wheat tissues at a water potential below about -4.5 MPa and its growth rate is halved below -2 MPa. Wheat plants in dryland conditions in the USA are reported to be commonly at -2.5 to -3.5 MPa between the tillering and heading stages. Wheat at -5 MPa was recorded during an exceptionally dry season in the Pacific Northwest of the USA (Cook *et al.*, 1972; Papendick and Cook, 1974). Under dry conditions, therefore, growth of the pathogen within the root tissue can be halted. If this occurs at a time when further root development can still occur, the effects of the disease may be reduced. If, however, the development of the crop is too advanced for substantial root growth to occur, the additional stress of dry conditions may combine with a restricted, diseased root system to exacerbate crop loss. Although little or no firm information is available on the relationship between soil water potential and take-all development in the UK (but see 3.6. and 4.5), field experience suggests strongly that take-all development is arrested in dry soil conditions.

The growth of Ggt is also decreased if the soil becomes too wet. In predicting the extent of fungal growth towards host roots in soil, Heritage *et al.* (1989) took into account the effect of different soil water contents. Percentage water-filled pores (% WFP) was used as the measure of soil saturation (40% WFP corresponding to approximately -0.1 MPa on the sandy soil used). In glasshouse experiments, growth towards roots was not affected in the range 40–70% WFP, but was less at 80% WFP. This reduction in growth of Ggt as the soil approached saturation was attributed to an increase in anaerobic activities (such as denitrification) and a decrease in aerobic activity due to decreased soil-gas exchange. An indirect effect of an antagonistic microflora, reduced susceptibility of roots to infection, or the production by the roots of a fungal inhibitor under diminished soil oxygen conditions may also have been involved. It was calculated that all growth of the fungus would cease at 86% WFP. Whole-body computed tomography X-ray (X-ray CT) scans of soil around single roots in experimental systems (Grose *et al.*, 1996) showed that volumetric water content around such roots is spatially heterogeneous, suggesting the existence of sites in the rhizosphere that might favour or restrict fungal growth. In soil surrounding wheat roots, the most favourable regions for the growth of Ggt were found to be nearest the roots, as distinct from the outer rhizosphere, or the bulk soil.

The saprophytic survival of Ggt has been much studied (Shipton, 1981). Temperature and water potential are important in determining microbial activity, and hence inoculum survival. Shipton (1981) cited Australian work in which survival of Ggt was tested in naturally infested soil subjected to a range of temperature and moisture regimes. Viability remained almost unaltered for 45 weeks if the soil was maintained either dry (−25 to −98 MPa soil matric potential) and cool (15°C), or moist (−0.4 to −0.7 MPa) and cool. Much viable inoculum still remained after soil was kept very dry (−98 MPa or less) and hot (35°C), or wet (−0.01 to −0.02 MPa) and cool. Only in hot, wet soil was the fungus eliminated within 4 weeks. Thermal inactivation of inoculum during the 3-month summer fallow between harvest and sowing winter wheat is an important limiting factor in take-all development in Kansas (Bockus *et al.*, 1994a). Consequently, soil shading, arising from volunteer wheat, double-cropping or no tillage, tends to prevent high soil temperatures and thereby favours inoculum survival.

In other Australian work, Wong (1984) studied the saprophytic survival of Ggt and three avirulent fungi in soil under controlled temperature and moisture regimes. All of the fungi survived longest in cool (15°C), dry (<−10 MPa) soil, with warm (30°C), dry soil giving the second longest survival period. All the fungi were eliminated from warm, moist (−0.3 MPa) soil within 3 months. Survival was intermediate under cool, moist conditions, which favoured Ggg more than the other fungi tested. There were also differences in survival between different isolates of Ggg and of *Phialophora graminicola*. These findings have implications for the maintenance of TAD as well as for the selection of potential BCAs.

2.3.4 *Soil type: physicochemical aspects*

The texture, structural stability, depth, available water capacity, pH and nutritional status of the soil can all affect take-all severity.

Nutritional status and pH are manipulated routinely by humans and are discussed in Sections 3.4 and 3.5.4. An example of soil pH interacting with soil structural stability occurs on the chalky boulder clays of East Anglia. These are soils of a texture (clay or heavy clay loam) that would normally be prone to structural problems. However, in some series (notably the Hanslope series) the high pH maintained by the soil's natural calcium carbonate content encourages stable aggregates to form. The resultant good soil structure does not inhibit rooting, available water is adequate and take-all problems are rare. In contrast, the boulder clays of the Ragdale series are often deeply decalcified and can suffer from poor structure and waterlogging in winter; take-all problems are more frequent on these soils (Catt *et al.*, 1986). This effect of pH contrasts with the normal situation in which high pH is associated with severe take-all.

Moore (1948), drawing on earlier work of Garrett at Rothamsted, stated:

> the fungus travels along the roots more quickly when the soil is of a light texture, alkaline, moist and warm . . . Thus the light textured, alkaline soils of the Yorkshire and Lincolnshire wolds, the East Anglian ridge, the Chilterns, the chalk downs of Hampshire and Wiltshire, and the Cotswolds are the chief danger areas for take-all in this country.

This statement covers several aspects of the soil environment and its effect on take-all. 'Light texture' can have two effects: first, by tending to produce a loose seed bed which favours spread of the fungus; and second, by having poor water availability which subjects plants with damaged root systems to water stress late in the season. Furthermore, such soils are often shallow due to erosion by water and wind, thus restricting rooting depth.

The relationship of texture to 'moist and warm' soil conditions is more complex. In 'normal' seasons even light soils can be assumed to be near field capacity in the early spring. Lack of water is therefore unlikely to limit growth of the pathogen until late spring or early summer. Even if its growth is halted early, sufficient root damage may already have occurred for the ensuing moisture stress on the plant to have a severe effect. Ggt is likely to be more active in light soils than in heavy soils in the spring because they warm more rapidly. The damage caused to the crop depends on the balance between the growth of the root system and the growth of the fungus. Crop growth increases as the temperature rises in the spring and the ability of increased plant growth to compensate for increased pathogen activity depends partly on the effect of temperature on partitioning between root and shoot growth. This subject has been discussed in detail by MacDuff (1989) and is complicated by interrelationships between soil temperature and soil moisture effects and by transient and acclimatization responses by the plant. Nevertheless, there is some evidence that increasing soil temperature towards the optimum for root growth results in an increase in the shoot : root dry matter ratio. The plant may therefore be diverting more of its resources into shoot growth just at the time when root growth is required to overcome the effects of increased pathogen activity (see implications for disease assessment in 6.3.3 and discussion on the timing of severe disease in 1.2).

2.3.5 *Nitrate leaching*

Although the effects of form, amount or timing of nitrogen fertilization on take-all are quite well documented (see 3.4), the effects of root disease on nitrogen uptake by plants have received less attention. Nitrogen uptake by a winter wheat crop grown after oats and by a winter wheat crop grown after barley, which had more take-all, was similar until anthesis, but thereafter uptake by the infected crop was much less than that by the crop with negligible take-all (Prew *et al.*, 1986). Recently attention has turned to the possibility that diseased plants leave more nitrogen in the soil to be leached as nitrate (Table 1.17). Leaching is a concern because too much nitrate in drinking water has been linked to blue baby syndrome and stomach cancer in humans, and in natural bodies of water it leads to eutrophication, which destroys fish and vegetation. Arguments for decreasing nitrate in drinking water and the limit set have been challenged (Monckton, 1996). Nevertheless, a voluntary Nitrate Sensitive Area scheme already exists in the UK and Nitrate Vulnerable Zones that will be subject to mandatory measures are being introduced.

Information on the impact of crops with take-all on the leaching of nitrate is limited. The increased yield of wheat after soil fumigation may not be due to the flush of nitrogen released by killing the microbial biomass, but to the control of root pathogens, which results in improved root health and consequently improved nutrient absorptive capacity of the roots (Cook, 1992b). The finding that more nitrate (22 kg ha^{-1}) was left in the top 180 cm of soil in a plot that was not tilled, where take-all was significantly more severe, than in a similar depth of soil in a tilled plot (Cook *et al.*, 1992) apparently supports this contention. However, differences in nitrogen mineralization between untilled and tilled soil could have explained the differences in nitrogen content between the two profiles.

Almost all of the nitrate at risk of being leached over the winter period in the UK from soils where winter wheat is growing comes from mineralization of organic nitrogen (Macdonald *et al.*, 1989). This mineralization is largely a function of past inputs of nitrogen, which were determined by the long-term cropping history. Consequently, practices such as minimizing areas sown to spring crops, decreasing periods when soil is left bare, avoiding autumn applications of nitrogen fertilizers or organic manures, as well as restricting ploughing up old grassland and the clearing of old woodland, will tend to minimize nitrate accumulation in autumn from the mineralization of soil organic nitrogen. Further experimentation at Rothamsted (Macdonald *et al.*, 1991), in which ^{15}N-labelled fertilizer was applied to wheat, led to recoveries of 37–55% of this nitrogen in the crops, with the low percentages associated with crops where yield was limited by take-all (39%) or black grass, *Alopecurus myosuroides* (37%). In general terms the Rothamsted work suggests that if nitrogen were applied to suit the crop, about 50% would be recovered in the crop, 25% would be lost as gas and 25% would be in organic form in the soil. The major factors determining residual nitrate in soil at harvest are choice of crop, soil total nitrogen and clay content.

Disease may prevent crops meeting their potential, so that unused applied nitrogen is left at risk of being leached from the soil. For example, potatoes left much more residual nitrate than wheat, except at a sandy loam site where take-all severely limited wheat growth (Macdonald *et al.*, 1997). More recently, in a preliminary study at Rothamsted, total nitrogen in the crop, soil mineral nitrogen (NO_3^--N and NH_4^+-N) and take-all were measured inside and outside patches of take-all in a fifth winter wheat crop that had been sown in September. Total nitrogen was least and soil mineral nitrogen most where take-all was severe (Fig. 2.12). In these conditions it seems that the average amount of mineral nitrogen left in the soil after a first cereal crop is 50 kg N ha^{-1}, but after severe take-all this can be as much as 150 kg N ha^{-1}, hence leaving much more nitrogen available for leaching. If this applies generally where severe take-all occurs then the impact on the environment could be considerable, particularly in Nitrogen Sensitive

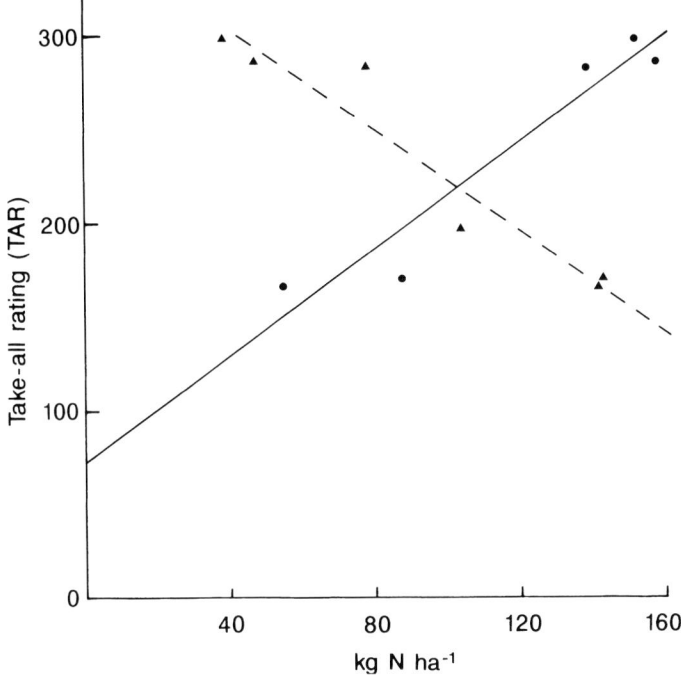

Fig. 2.12. The relationships between take-all and total uptake of nitrogen by a winter wheat crop at Rothamsted (dotted line, $y = 352.9 - 1.312x$, 90.3% of the variance accounted for) and between take-all and total mineral nitrogen (NO_3^--N and NH_4^+-N) in the top 50 cm of soil (continuous line, $y = 72.8 + 1.428x$, 89.9% of the variance accounted for).

Areas. In an ADAS trial in a Nitrogen Sensitive Area in Lincolnshire, it was found that take-all tended to reduce the off-take of nitrogen in the grain and therefore to increase the risk of nitrogen leaching (N. Paveley, personal communication). Where nitrogen loss was calculated by multiplying drainage volume by its inorganic N content, the loss was greater under consecutive wheat crops where take-all decreased yield and therefore nitrogen uptake by the crop (Webster and Goulding, 1995).

In France, observations on nitrogen uptake by the wheat plant were made in an experiment to measure the efficacy of an experimental seed treatment fungicide, MON41100, for take-all (Lucas *et al.*, 1994). In a field previously cropped with wheat, plots treated at three rates of MON41100 and control plots showed different take-all disease indices at growth stages 31 and 69 (Table 2.2). All plots were fertilized similarly and were protected from foliar and stem base diseases by fungicide sprays. Yields of 2.37, 2.85, 3.65 and 4.35 t ha^{-1} were obtained, relating to the descending scales of disease shown in Table 2.2. The amount of NO_3^--N and NH_4^+-N left in the soil profile to a depth of 75 cm was assessed at harvest and revealed no significant differences between treatments. Based on the nitrogen content of the harvested grain, an estimate was made of nitrogen taken up by the crop. (The assumption was that 3 kg of nitrogen were taken up for every 100 kg of grain harvested.) The amount of nitrogen taken up plus the amount of nitrogen left in the profile indicate that 63, 43 and 13 kg N ha^{-1} were leached to below 75 cm at the three highest levels of disease, respectively (Table 2.2).

These results show that take-all may increase leaching of nitrogen. Moreover, when the canopy shows some unevenness or some weakness at tillering, or at the

Table 2.2. Effect of a fungicide applied to wheat seed on incidence of take-all, nitrogen uptake by plants, nitrogen left in the soil profile (0–75 cm) and nitrogen leaching below 75 cm, at harvest.

Treatments	Disease index (0–4)[a]		Nitrogen (kg ha^{-1})			
	GS 31	GS 69	In soil at harvest (1)	Taken up by the crop (2)	(1) + (2)	Lost from the profile[b]
Control	2.46	3.39	60	71	131	63
Fungicide (MON41100, g a.i. kg^{-1} seed)						
0.5	1.51	3.23	65	86	151	43
1.0	1.29	3.19	71	110	181	13
1.5	1.03	2.64	63	131	194	–
SE (20 df)	0.159	0.216				

[a]See Section 6.3.3.
[b]Comparison with best protected plots (i.e. 1.5 g a.i. MON41100 kg^{-1} seed).

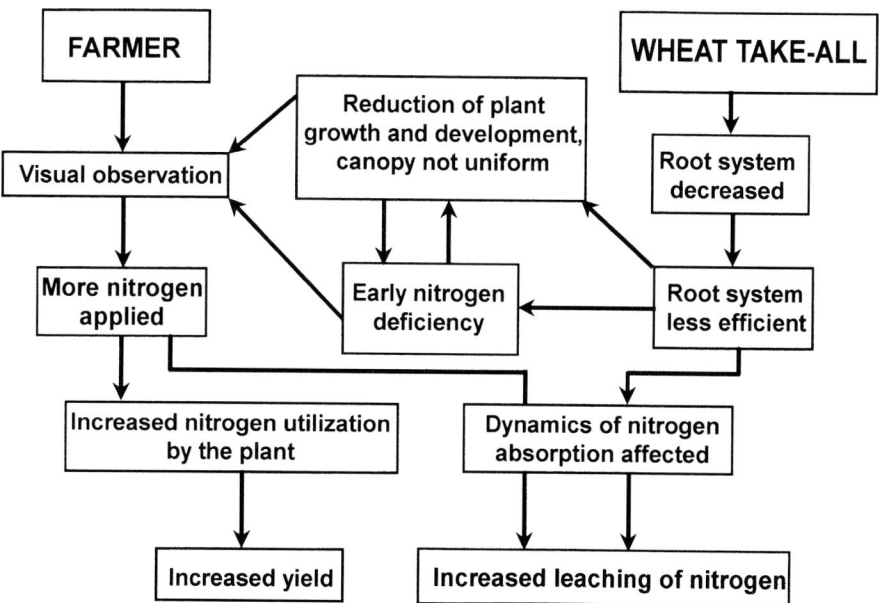

Fig. 2.13. A scheme showing how nitrogen utilization and leaching may be influenced by take-all.

beginning of stem elongation, the farmer may increase nitrogen fertilization in the expectation of some beneficial effect on plant growth. Such practice probably contributes to more nitrogen loss and more ground water pollution. These effects and hypotheses are summarized in the scheme in Fig. 2.13. They suggest points for further study:

- What is the real impact of take-all on nitrogen uptake by plants, taking into account the incidence and severity of the disease and the growth stage of the plants?
- Can different approaches to nitrogen fertilization be devised for wheat crops at risk from take-all to reduce yield loss without increasing the leaching of nitrogen?

2.4 Suppressive and Conducive Soils

Soils that suppress plant disease, or the pathogenic fungi within them, are known as suppressive soils. The natural biological control of take-all (see 4.3.1) provides examples of such soils. Soils in which infection occurs unimpeded are often referred to as conducive soils. Soil suppressiveness is a widespread and well known phenomenon, though usually complex and not easily understood. Types of suppressiveness and their possible mechanisms have been reviewed with particular reference to take-all (Hornby, 1983). A list (Hornby, 1983, adapted from a review by Walker, 1975) of types of take-all suppression and suggestions as to the categories of suppressive soils they represent is reproduced in Table 2.3.

The properties of suppressive soils, the causes of suppression and its role in disease management have been the subjects of many investigations by take-all researchers. Rovira and Wildermuth's (1981) review of soils suppressive to take-all contained three lists of properties of different kinds of suppressiveness, which included Walker's (Table 2.3), and a synthesis based on earlier versions and their own experience. Much of the research on soils suppressive to take-all published since 1981 has attempted to identify mechanisms and has often concerned the transferability of suppressive factors from one soil to another. Authors have referred to suppressiveness as 'specific' or 'general' – terms used in Rovira and Wildermuth's (1981) version of types of suppressiveness (Table 2.4).

Soil suppressiveness can be an important factor in the epidemiology of take-all, occurring either as a consequence of repeated cropping of the host plant (as in TAD), or as a natural property of the soil. Suppressive soils have also been a popular source of potential BCAs for many take-all researchers, who have reintroduced these organisms into the same soil or introduced them into different soils with the intention of suppressing Ggt or the disease. Lines of research that have predominated in different countries in recent years are discussed below.

2.4.1 Recent research in the UK: take-all decline (TAD)

TAD, defined as suppression type I in Table 2.3 and discussed in Sections 3.3.3 and 4.3.1, is of particular significance to cropping systems in the UK, where the first

Table 2.3. Suggested relationship between types of take-all suppression and types of suppressive soil.

Suppression type	Name or characteristic	Development requirements[a] H	P	D	Development period	Suppressive soil (SS) type[b]
I	Take-all decline (TAD)	+	+	S+	Few seasons	Induced (by monoculture)
II	Long-term	+	+	+	Over 200 years of permanent cultivation	Long-standing
III	Long-term, without severe disease	+	+	–	Many years of wheat	Long-standing
IV	Non-host (crop-dependent)	–	–	–	One to several seasons	Induced a) by rotations, b) by grass or grass/legume leys
V	Added pathogen	–	+	–	Short	Induced
VI	General antagonism	–	–	–	'Immediate'	Not usually considered within SS range
VII	Cross-protection	+	–[c]	–	'Immediate'	Introduced[d]

[a]H, host; P, pathogen; D, disease; s, severe; +, required; –, not required.
[b]This refers to the induction procedure (cropping) and the length of time during which it was induced; see Hornby (1983) for more detail.
[c]Avirulent pathogen may be added to the soil.
[d]Where organisms are added to the soil rather than on the seed.
Source: Hornby (1983).

Table 2.4. Causes and the main characteristics of two types of suppressiveness to take-all in soils.

Transferable (specific) suppression	Non-transferable (general) suppression
Causes	
Continuous wheat with take-all (TAD soil)	Increased microbial activity with increased organic amendments and fertility build-up
Additions of Ggt mycelia to soil	NH_4^+-N uptake by roots
Addition of certain other fungi (antagonists/competitors?) to soil	Non-host plants
	High soil temperatures
Characteristics	
Eliminated by moist heat (60°C for 30 min)	Not eliminated by moist heat
Eliminated by chemical fumigants	Reduced but not eliminated by fumigation
Operates below 20°C and masked by non-transferable suppression above 25°C	Operates at all soil temperatures but increases above 5°C
Operates primarily in the rhizosphere	Operates primarily in bulk soil

After Rovira and Wildermuth (1981).

convincing evidence of the phenomenon was obtained more than 30 years ago at Rothamsted in soils that had grown successive crops of wheat. Some farmers have specialized in cereal monoculture (recent examples discussed in Section 3.3.2), and for three decades the research at Rothamsted was principally concerned with natural suppression in such systems. The development of hypotheses on the mechanisms of TAD has been the subject of a number of reviews (summarized in Rovira and Wildermuth, 1981; Hornby, 1983) and here the discussion will be limited to describing new research and ideas.

The robustness of TAD and its ability to survive through break crops that are not hosts, or in the presence of different host species for different periods of time, require further investigation. The loss of TAD, once it has been built up by continuous cropping of wheat, could represent a loss of investment, particularly if shorter sequences of cereals with their high risk of severe take-all are introduced subsequently. The results of a long-term crop sequence experiment at Rothamsted, involving wheat, barley and triticale, have far-reaching implications for our understanding of the epidemiology of take-all and mechanisms of suppression (some of the rotational implications are discussed in Chapter 3). They provided a demonstration of TAD for the first time in monocultures of winter barley or of winter triticale (Hornby and Gutteridge, 1995); its occurrence in spring barley, now less commonly grown than winter barley, had already been demonstrated. Substituting barley or triticale (which usually suffer less from take-all) for winter wheat when severe take-all was expected did not achieve a TAD that protected subsequent wheat: wheat grown after the barley suffered from severe take-all (Hornby

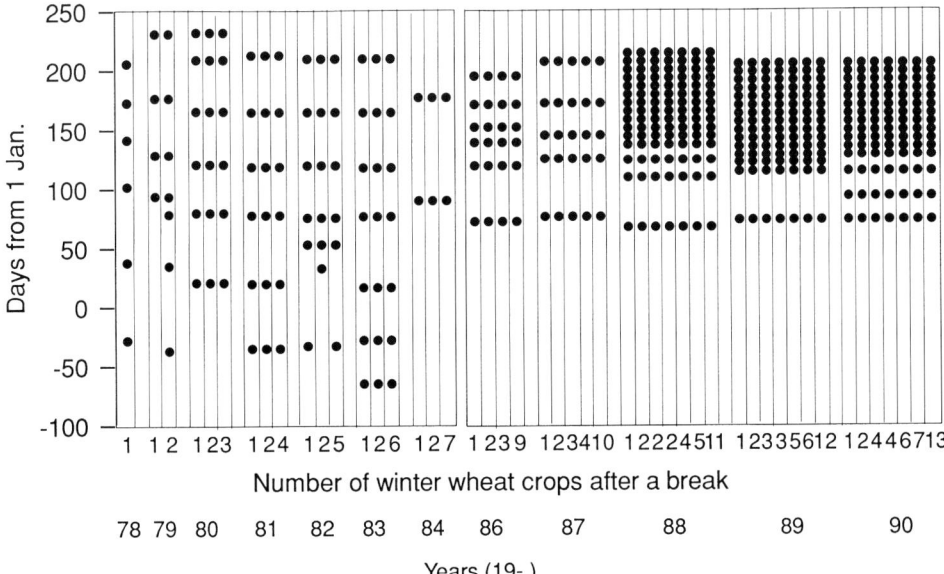

Fig. 2.14. Scheme showing wheat sequences available for sampling for take-all in each year of a phased sequence experiment (CS212) on winter wheat at Rothamsted and the occasions (dots) during the growing season when each sequence was sampled.

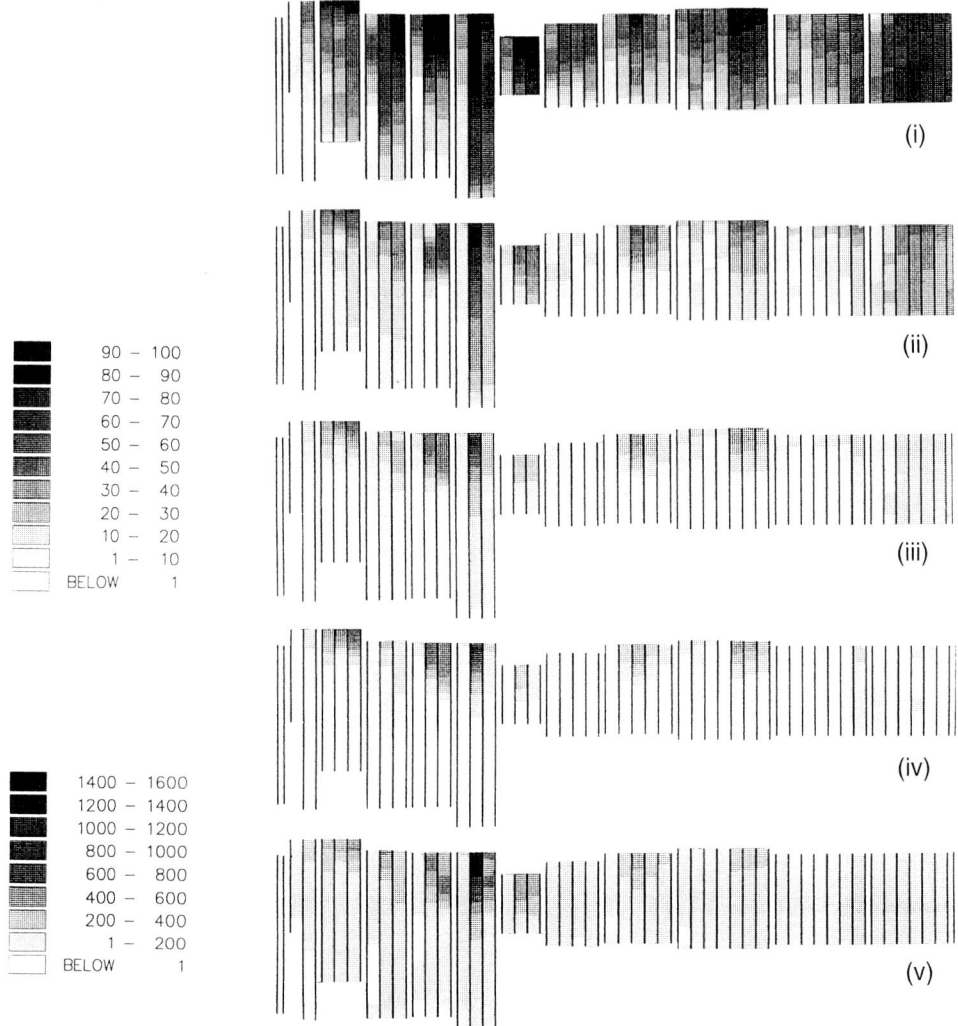

Fig. 2.15. Two-dimensional representations comparing five different measures of take-all in a phased sequence experiment (CS212) on winter wheat at Rothamsted.
Measures of disease: (i) % plants infected; (ii) % seminal roots infected; (iii) % nodal roots infected; (iv) % all roots infected (also shown in 3-D in Plate 7.18); (v) number of infected roots per plot.
Years, sequences, parts of the growing seasons considered and sampling dates are explained in Fig. 2.14. The mapping options chosen resulted in the outer columns being narrower than the others; this should be ignored.
Disease variation is on a grey scale from white (no disease) to the darkest shading (most disease). For ease of comparison, all percentage data are presented on the same scale with eight class limits: < 1, 10, 20, 30, 40, 60, 80 and 100%. The numbers scale (representation (v)) has seven classes with limits of < 1, 200, 400, 600, 1000, 1200 and 1600.

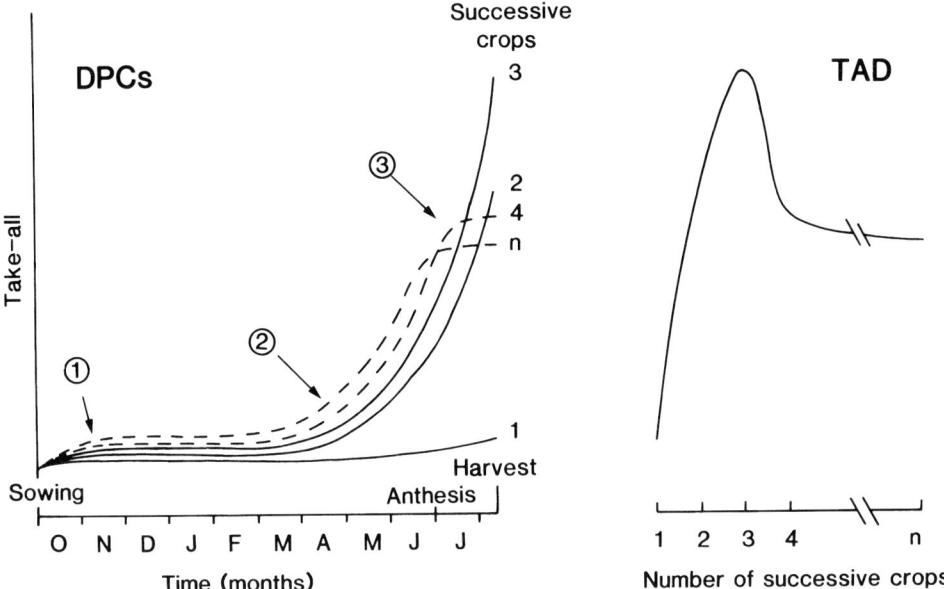

Fig. 2.16. The sketch to the left illustrates some concepts of TAD in years with moderate to severe disease, based on studies of DPCs in phased sequence experiments. Arrow 1: initial disease levels are positively related to the number of previous crops. Arrow 2: until early summer, disease increases are greater in sequences with the most previous winter wheat crops. Arrow 3: disease ceases to increase at about anthesis in plots that subsequently exhibit TAD. The 'classical' TAD relationship amongst the sequences in July is on the right.
Reproduced from Hornby (1992a).

and Gutteridge, 1995). Wheat and barley have been shown to be associated with genetically different forms of Ggt (see Chapter 5) and they may also develop their own, distinct antagonistic microfloras.

In the experiment illustrated in Figs 2.14, 2.15 and Plate 7.18, crops of winter wheat in phased sequences were assessed at intervals during each growing season and provided new evidence for the expression of TAD late in the season (Fig. 2.16). This is an interesting contrast to studies in very different soils in Western Australia, where suppression of the pathogen in its saprophytic phase seems to be important (see 2.4.3). Soil sampled early in a season or the use of seedlings or young plants may not be the most appropriate experimental materials for investigating the mechanisms of suppression in TAD in the UK, or for testing BCAs obtained from TAD soils.

Most hypotheses to explain the mechanisms of TAD implicate changes in the antagonistic microflora (i.e. changes occurring outside the pathogen). Table 5.1 summarizes other work that has associated changes in the pathogen with stages in epidemics and assesses the strengths of the evidence. It includes recent research using a DNA probe to investigate pathogen populations (described in Chapter 5), which raises the possibilities that changes in the pathogen may be involved in the progress of take-all epidemics, or may be instigated by the changes in the microflora.

2.4.2 Recent research in France: nitrogen form and bacteria

In France, cereal monoculture has not been such a feature of arable agriculture as it has in the UK and so there has been less emphasis on research on field-based studies of TAD. However, observations on effects of different fertilizers and an interest in biological control have prompted detailed investigations into the natures of soil conduciveness and soil suppressiveness, following leads established in other parts of the world (see 2.4.3 and 2.4.4).

The form of nitrogen fertilizer applied has been reported to affect the severity of take-all in some parts of the world. For example, NH_4^+-N has often been reported to decrease disease compared with NO_3^--N (see 3.4.1). Researchers in France investigated the roles of form of fertilizer nitrogen and fluorescent pseudomonads (bacteria already implicated in suppression of take-all and the most favoured organisms for use as BCAs; see 4.3.2. and 4.3.3) in soil suppressiveness. NH_4^+-N was shown by bioassay to make the soil less conducive to take-all than NO_3^--N, especially when the amount of inoculum was great (Sarniguet *et al.*, 1992a). The effect was not correlated with amounts of aerobic bacteria or fluorescent pseudomonads, but seemed to involve qualitative changes in bacterial populations (changes in proportions of different species or types) in the soil and rhizosphere as the different root systems (seminal and, later, nodal) developed. NH_4^+-N led to a predominance of antagonistic fluorescent pseudomonads whilst NO_3^--N encouraged 'deleterious' bacteria that encouraged disease (Sarniguet *et al.*, 1992b). In 'crops' of wheat grown successively in pots, in which suppressiveness was induced where Ggt was present, a dramatic increase in populations of fluorescent pseudomonads occurred in the rhizosphere of a first healthy crop, with a further increase in subsequent crops that was greater on plants that were infected with Ggt (Sarniguet *et al.*, 1992b). Qualitative changes in the bacterial population were associated with root necrosis (Sarniguet and Lucas, 1993). Figure 2.17 shows a simple scheme of the supposed events.

In a parallel study of take-all patch in turf grass, the ratio of number of fluorescent pseudomonads to the number of all bacteria increased from outside the patch towards its centre where recolonization by grass (*Festuca* sp.) had begun and, presumably, disease suppression was occurring (Sarniguet and Lucas, 1991, 1992). A large proportion of the fluorescent pseudomonads from the centre of the patch were shown to be antagonistic *in vitro* to the pathogen, Gga.

2.4.3 Research in Australia: pathogen suppression during saprophytic growth

In contrast to soils in the UK, a significant amount of saprophytic growth of Ggt has been reported in some soils in Australia (see 2.7). It appears to be associated with a low level of organic matter in these soils. Such extended saprophytic growth in the pathogen's life cycle could be more sensitive to antagonism than mere survival on crop residues, or colonization of roots (Simon and Sivasithamparam, 1988a, 1989). A soil sandwich technique (Grose *et al.*, 1984; Glenn *et al.*, 1988) for measuring hyphal growth through soil has been used in most of the Australian studies of suppression.

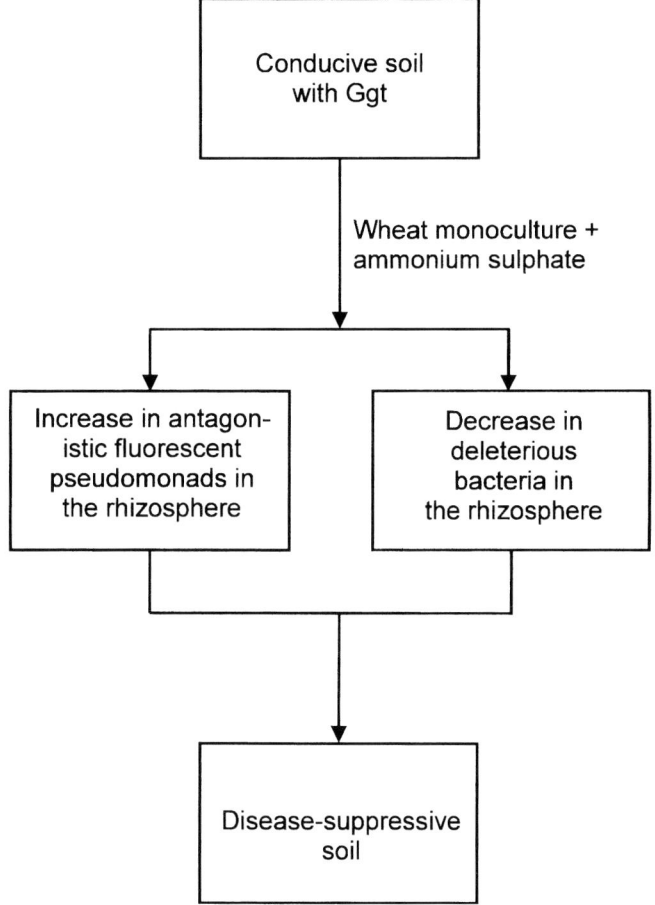

Fig. 2.17. Proposed sequence of events following application of ammonium sulphate to soil continuously cropped with wheat and infested with Ggt. Based on Sarniguet *et al.* (1992a, 1992b).

Suppression and the effects of form of nitrogen in Western Australian soils were subjects of a series of publications (Simon and Sivasithamparam, 1988a–d; Simon *et al.*, 1987a, 1988b), the main elements of which are covered in a short review (Simon and Sivasithamparam, 1989). Pathogen suppression and disease suppression appeared to be related and were found in soils with cereals in rotations with pasture or legumes. (Pathogen-suppressive and disease-suppressive are terms frequently used interchangeably, but here they distinguish suppression of the pathogen growing saprophytically, or surviving in the soil, from suppression of the pathogen growing parasitically; see Hornby, 1983.) Two out of three continuous wheat soils were pathogen-suppressive and all three were disease-suppressive. Transferable suppression was demonstrated only where 1% of a suppressive soil was added to the same soil that had been γ-irradiated. The term transferable suppression was preferred to specific suppression, because it was not clear if the suppression transferred was specific to the virulent Ggt isolates used or if it was due to specific organisms. Adding nitrogen as ammonium sulphate to soil growing continuous wheat increased suppression, which was associated with an increase in *Trichoderma* spp. (well known as BCAs that are effective against various pathogenic fungi). The activity of *Trichoderma* spp. was lower when the soil was also treated with lime. Acidification with ammonium sulphate, without liming, was postulated to decrease microbial activity and increase suppression by *Trichoderma* spp. Antibiotics produced by one species, *T. koningii*, were implicated in this suppression (also reported in Dunlop *et al.*, 1989). These hypotheses are summarized in Fig. 2.18, which is adapted from Simon and Sivasithamparam (1988d).

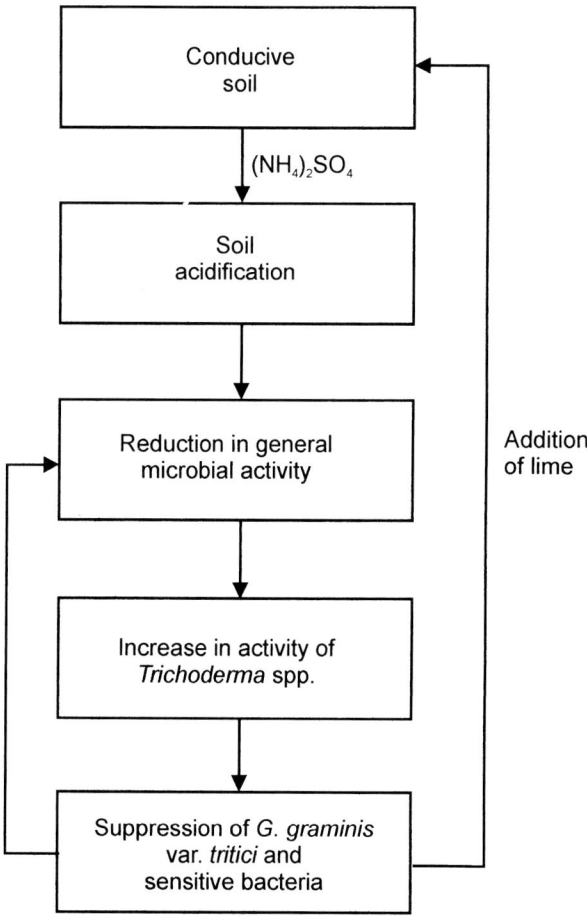

Fig. 2.18. Possible sequence of events following repeated application of ammonium sulphate to a Western Australian (Newdegate) soil resulting in suppression of the saprophytic growth of Ggt by *Trichoderma* spp.
Redrawn from Simon and Sivasithamparam (1988d).

Mycophagous amoebae recovered from take-all-suppressive soils from long-term wheat monoculture and permanent pasture in South Australia have been shown to attack and lyse hyphae of Ggt, suggesting that they may have a role in pathogen suppression in those soils (Chakraborty and Old, 1982; Chakraborty *et al.*, 1983; Chakraborty and Warcup, 1983, 1984). Amoebae from soils in the Pacific Northwest, USA, have also been shown to attack and lyse hyphae of Ggt (Rovira and Wildermuth, 1981).

2.4.4 Recent research in the USA

Much of this research is based on wheat-farming systems in the Pacific Northwest (Washington), inland regions of the northwest (Montana) and the climatically very different south-central region (Georgia); cropping systems in some of these regions are described in Sections 2.7 and 3.7.

An influential research programme on biological control of take-all at Pullman in Washington State, begun in 1968, has been described in several review articles (Cook, 1992a, 1993, 1994) and is referred to in Section 4.3.3. That work identified fluorescent pseudomonads as principal agents of transferable suppressiveness and led to their investigation as BCAs. Some clarification of the processes involved emerged from experiments in which highly suppressive soil from long-term wheat monoculture and uncropped soil were compared after adding inoculum of Ggt (Cook *et al.*, 1986). Similar effects, including increased lesion numbers, in both soils after pasteurization suggested that general suppression was operating. Specific suppression may have resulted mainly from the inhibitory effects of secondary colonists of developing lesions, and general suppression may have resulted mainly from the inhibitory effects of secondary colonists of the organic food base carrying the inoculum, i.e. during the pathogen's saprophytic phase.

A comparison of eight Montana soils showed that two were highly suppressive, but that different mechanisms were probably operating in them (Andrade *et al.*, 1994a). Mycoparasitism was thought to be responsible for suppression in one soil, from which two sterile fungi with exceptional antagonism towards Ggt were isolated, and antagonism by actinomycetes and possibly *Pseudomonas* spp. in the other. Also in Montana, a soil from a rotational cropping system and conducive to take-all was shown to have slight general suppression, and two soils that were suppressive after more than 10 years of wheat monoculture had transferable suppression that differed in effectiveness in tests using different soils (Andrade *et al.*, 1994b). Further, it was found that a suppressive factor that was not activated in its own soil could become active when transferred to a different soil – an observation that adds to the complication of interactions among chemical, physical and biological causes of suppression.

Most soils that were double-cropped continuously, usually with wheat and soya bean, in Georgia did not become suppressive, whilst specific suppression was observed in wheat monoculture (Rothrock and Cunfer, 1986). However, suppressiveness did occur with wheat/sorghum double cropping, a system which resulted in a lower soil pH than the wheat/soya bean system (Rothrock, 1989). These findings are among many reports that implicate soil acidity as being a factor in suppression.

2.4.5 *Synthesis: mechanisms and regional differences*

Since 1981 the definition and characterization of some types of suppressive soil have been in need of revision. Nonetheless, progress has been made in identifying mechanisms of suppression that are likely to be operating in particular soils. These may be operating alone, among other mechanisms, or as interactions. However, many of the suppressive soils that have been the subject of experimentation cannot be regarded as examples of TAD as defined in Table 2.3. This is because they have not arisen in sequences of wheat in which disease build-up followed by disease decline have been demonstrated.

In the search for BCAs, the microorganisms tested have often been those recovered from suppressive soils. The assumption is, therefore, that mechanisms of action of the BCAs will be the same as those that operate in natural disease suppression. Such mechanisms were listed, with representative examples, by Thomashow and Weller (1996) as competition, parasitism/predation, induced resistance and antibiosis. These mechanisms, as they relate to BCAs, are discussed further in Section 4.3.3. Control of take-all in wheat by *Pseudomonas* spp. was listed among the examples of antibiosis in Thomashow and Weller (1996), but all the other mechanisms, except induced resistance, have been reported as occurring in naturally suppressive soils.

Simon and Sivasithamparam (1989), referring to Western Australian soils, listed four phases of saprophytism of the pathogen: (i) survival in crop residues; (ii) growth in soil; (iii) colonization of organic residues; and (iv) ectotrophic growth preceding infection. The hypotheses recounted here suggest that biological suppression occurs in all of these phases, as well as in the subsequent pathogenic phase. The mechanism in Fig. 2.17, based on research in France, relates to disease suppression and events in the rhizosphere; and that in Fig. 2.18, from Australian work, to suppression of the pathogen in its saprophytic growth stage. Both may be special cases that do not operate generally and also they may not be the only mechanisms in the soils for which they are described.

The regional research described above seems to suggest regional differences in mechanisms, but the mechanisms researched may to some extent reflect the interests and techniques of the researchers, as well as differences in soils, climates and cropping systems.

2.5 The Disease Progress Curve

An epidemic is defined as a change in disease intensity in a host population over space and time (Kranz, 1974). In the case of root disease, much of the temporal dynamics remains hidden beneath the surface, rather like an iceberg (Gilligan, 1994a). Studies of plant pathogens in populations of plants may be qualitative (e.g. where do epidemics of take-all occur?), quantitative (e.g. estimating amounts of disease and changes in disease over time and/or space), or analytical. An example of the last category would be the description of relationships among disease, time and/or environmental and biological variables. Such descriptions are frequently empirical, but can also be mechanistic, using a mathematical equation (the 'model') to predict disease or some component of the disease cycle (there is a more general treatment of modelling in Section 2.6). The disease progress curve (DPC) is central to the study of plant disease epidemiology, presenting a picture of disease dynamics and a summary of the interactions among the host population, pathogen population and environment.

2.5.1 Description

DPCs may be plotted as disease intensity against a measure of time or host growth, or described in terms of a mathematical function of time. To determine whether certain aspects in the 'shape' of the DPC are significant usually requires fitting the disease progress data to mathematical functions such as a simple polynomial regression equation (involving powers of the explanatory variable). Except for some controlled environment experiments (e.g. Asher, 1972; Wildermuth and Rovira, 1977) and detailed studies of the dynamics of inoculum in soil over several years (Hornby, 1975), the take-all progress curve was rather neglected before the 1980s and it received no mention in the index of Asher and Shipton's (1981) book. Subsequently, it has received attention in studies of: (i) sequences of winter wheat crops at Rothamsted (Hornby and Gutteridge, 1988; Werker *et al.*, 1991); (ii) analysis of selected agronomic variables in multi-factorial experiments (Bateman, 1986; Christensen *et al.*, 1987; Werker and Gilligan, 1990); (iii) comparison of different models to describe take-all epidemics in first and subsequent wheat crops (Brassett and Gilligan, 1989).

Models to describe DPCs range from a simple straight line relationship to a series of linked equations incorporating host growth and inoculum decay (see 2.6.2). Linear models (i.e. models that are linear in the parameters; the response functions may be curved) are easily calculated by regression and are very flexible in terms of shape (Gilligan, 1985a). Non-linear models require iterative optimization algorithms, and optimization is easiest with few parameters, approximately quadratic functions, small correlations between parameters and good initial parameter estimates (Payne *et al.*, 1993). Non-linear models are less flexible, but appeal intuitively because of biological-like properties with respect to shape and the parameter values that describe these properties, such as the upper or lower asymptote. There are, however, very few examples where disease progress data for take-all have been fitted to either linear models or non-linear models. The cost of obtaining disease progress data for take-all and the variability in the shapes of disease progress curves may have contributed to this. Linear models may be useful in smoothing out oscillations in the data that are not real statistically, and in separating out long-term trends in inoculum or disease from shorter-term fluctuations within growing seasons (Werker *et al.*, 1991).

2.5.2 Interpretation

Despite much flexibility in how DPCs can be described, the models frequently reveal little of the underlying biological mechanisms responsible for the shapes of the curves. Quantitative relationships between the various components of the infection chain, such as inoculum density, inoculum decay, the rate of new primary infections and subsequent spread of the fungus, are often such that shifts in the balance may cause significant changes in the shape of the DPC (Brassett and Gilligan, 1988). For example, including root growth and inoculum density as variables in a model of a monocyclic root disease often leads to a sigmoidal curve (Jeger, 1987), so if certain data are well described by a logistic function, it cannot be concluded that the disease is polycyclic, or that the intrinsic (underlying) rate of disease increase is constant. Biological properties under consideration must have an experimental basis from which models may be developed and tested.

It is known that the amount of initial inoculum of the take-all fungus is important and that this is determined primarily by cropping history, the severity of disease in the previous crop and the rate of inoculum decay. It is also known that the amount of inoculum at sowing is not always correlated with the disease at harvest and that certain environmental and biological properties of the soil, such as moisture and suppressiveness, significantly influence the rate of disease progress. Werker *et al.* (1991) ascribed the differences in shape observed amongst take-all epidemics in first, second and continuous wheats (Fig. 2.19) to differences in the initial inoculum densities and rates of disease spread (secondary infections). It was proposed that the primary mechanism of TAD (see 4.3.1), manifested by a reduction in disease severity between the fourth and sixth years of continuous cereal growing, was a reduction in the rate of disease spread caused by unknown biological factors. Certain groups of bacteria are popular explanations that have been proposed, particularly by workers outside Britain (see 4.3.2.) This decrease

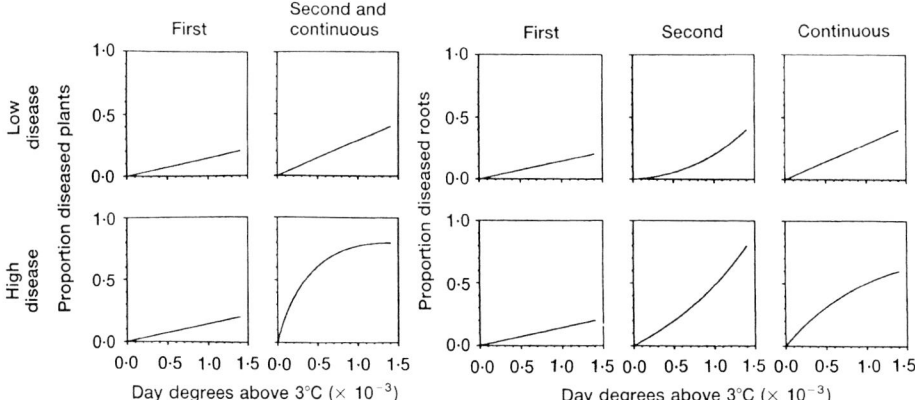

Fig. 2.19. Shapes of disease progress curves for proportions of diseased plants (left) and proportions of diseased roots (right) in first, second and continuous wheats in years of low disease (top) and years of high disease (bottom).
Reproduced from Werker *et al.* (1991).

in the rate of secondary infection may occur relatively early in a sequence of consecutive cereals, but may be camouflaged by the presence of high inoculum densities following second and third wheats (Werker *et al.*, 1991). In some instances, continuous wheats may show more disease towards the end of the growing season than do second wheats. This may be a seasonal effect where a predominance of primary infections rendered secondary infections significantly less important. The absence of TAD, therefore, is not necessarily the explanation, but this cannot be discounted. Brassett and Gilligan (1989) fitted data for the progress of take-all (more explanation of what is involved is given in Section 2.6) in first wheats and in second and subsequent wheats to: (i) a linear model (third order polynomial); (ii) standard non-linear models (the logistic, and logistic with allowance for host population growth); and (iii) custom-built non-linear models (incorporating parameters for primary and secondary infection, and with allowance for host-population growth and decay of inoculum).

The last group (with three parameters) yielded a description of the data that was consistent with biological constraints and fitted the data equally as well as the polynomial.

2.5.3 *Analysis*

There are sophisticated methods available for comparing DPCs (e.g. Madden, 1986; Gilligan, 1990a), but these have rarely been applied to take-all, where separate analyses of variance for each sampling occasion have been usual. The effect of treatments on the dynamics of the disease have not therefore been subjected to statistical rigour. Werker and Gilligan (1990) used the AUDPC and the linear and quadratic contrasts (equivalent to the slope and amount of curvature below or above the straight line) to distinguish between categories of effects of treatments and interactions on disease progress. In such analyses, later sowing typically showed an initial decrease in disease, but the subsequent progress of disease was faster, such that by harvest the differences in disease among sowing date treatments were small or insignificant (cf. Section 3.5.1). In one year, however, the DPCs diverged. This was attributed to a decrease in primary infections due to high soil moisture causing a decay of inoculum and a greater dependence on existing infections for subsequent disease development.

2.5.4 *Some practical problems*

It is not easy to explain how relatively small amounts of infection in one year can give rise to severe disease in the following year. Slope *et al.* (1979) showed that soil infectivity, measured by wheat-seedling bioassay, differed after the harvest of different crop sequences and that this difference was reflected in take-all in the following crops. In first and second wheats grown after 'clean' breaks, soil infectivity increased from May to harvest; and the more infective the soil in April, the more rapid the increase in soil infectivity. If 20% of roots or more are infected in a soil bioassay in September, that field should be regarded as having a high risk of take-all for a following susceptible cereal. Severe take-all in winter wheat in summer may be related in some cases to the amount of autumn infection and, in general, the earlier severe take-all occurs, the greater the yield losses will be.

The insurance value of using artificial inoculum in field experiments on take-all and the possibility of generating less patchy infection in the crop has attracted some researchers (see 6.4). Disease from artificially applied inoculum may not simulate natural disease realistically, and comparing DPCs has shown the need for caution. Where disease progress in continuous winter wheat within and between seasons was compared in naturally infested plots and in plots receiving artificially applied inoculum (on colonized, killed oat grains) (see 6.4.2), natural infection required more time (several seasons) to reach its peak than did the disease in the artificially created epidemic, which peaked in the season in which the inoculum was applied (Fig. 6.10). Weather may have been implicated: the beginning of the experiment coincided with the first of two successive warm, dry summers in 1989 and 1990 that adversely affected survival of natural inoculum. On the other hand, sowing the seed and putting artificial inoculum into warm, dry soil in the first autumn (1989) may also have caused poor initial performance of that inoculum. In another experiment in 1992 (Bateman and Hornby, unpublished data), inoculum applied in autumn in more favourable conditions resulted in severe disease and yield losses of from 35% to 67%. In the experiment shown in Fig. 6.10, TAD followed a peak of disease in both naturally and artificially infested plots and the disease progress curves within seasons were not greatly different at similar stages of the epidemics. The peak of disease in the natural epidemic was associated with greater patchiness (determined by above-ground symptoms and by root blackening) and more severe disease within the patches than occurred in the artificially infested plots. The extent of enlightenment derived from such experiments depends on which variable is used for measuring disease progress (see 6.3.3).

The importance of primary and secondary inoculum in models has been emphasized (see 2.6), but measuring these in practice presents a challenge. The contribution of primary infection has been estimated mathematically and, hence, by elimination the contribution of secondary infection (D. Bailey, personal communication). Disease progress from a low density of particulate inoculum was analysed with respect to the seminal and nodal root systems of winter wheat separately. Because of the rapid decay of particulate inoculum, primary infection was largely restricted to the seminal roots, whilst secondary infection affected both seminal and nodal roots.

In the 25 years prior to 1995, large sets of disease data were collected from long-running field experiments at Rothamsted and Woburn. These and computer graphics allowed epidemics to be pictured in ways hitherto impossible. Graphical displays provide the best summaries of data, simplify the aspect of the data by appealing to our natural ability to absorb visual images, and (it is to be hoped) provide a global view of the information, thereby stimulating possible explanations (Greenacre, 1988). Plate 7.18 and Fig. 2.15 illustrate two ways of presenting aspects of a take-all epidemic in an experimental site. Figure 2.14 gives some idea of the amount of time and sampling that was necessary to provide the data for such representations. Early in the experiment, emphasis was on sampling to give a wide coverage of the growing season, but after 1983 similar numbers of samplings were confined to the period from spring to harvest and then, after 1987, the number of samplings in this shorter period was approximately doubled. Werker *et al.* (1991) used weighted linear functions to overcome the statistical problems of repeated observations within plots, variable sampling frequencies and intervals within and amongst the first 9 years of this experiment. Plate 7.18, a three-dimensional representation, dwells on the interaction of cropping, season and one measure of disease (i.e. percentage of all roots infected), whereas Fig. 2.15 compares a two-dimensional represenatation of that data with disease measured as percentages of plants infected, of seminal roots infected and of nodal roots infected, and numbers of infected roots per plot. These presentations show a complexity that is all too often ignored. Perhaps one of their most obvious features is the fluctuation in the amount of disease through the years, which to some extent reflects what happened nationally (see 1.1 and Fig. 1.2). Another is that the different measures of disease show different degrees of sensitivity at different stages of the epidemic (Fig. 2.15). For instance, disease differences amongst sequences were picked up earlier in the season using seminal root infection (Fig. 2.15 ii) compared with nodal root infection (Fig. 2.15 iii), and sequences differed more obviously where disease was measured as percentage of plants infected (Fig. 2.15 i) than where numbers of infected roots per plot were counted (Fig. 2.15 v).

Such studies of the 'anatomy' of epidemics provide new insights. For example, they have suggested that the rate of disease increase increases annually in the first few consecutive crops, but then, usually in a fourth crop, it is dramatically capped in June (Fig. 2.16). This may be a major manifestation of the TAD phenomenon that has previously been overlooked (Hornby, 1992a). Normally TAD is visualized by comparing disease in different crop sequences once in the summer. Much of the research into the mechanisms of TAD, as well as on suppression by introduced BCAs, has concentrated on seedlings and young plants, partly for reasons of ease and convenience. Results from such work may not explain effects occurring later in the season.

2.5.5 *Conclusions and future research*

No one model describes the progress of take-all in all or even the majority of situations. Exponential, monomolecular ('simple interest' increase in disease) or logistic models (illustrations of representative standard non-linear curves are given in Payne *et al.*, 1993) contribute little towards explaining the biological mechanisms that generate the DPC. The long duration of a take-all epidemic, infections arising from two sources of inoculum (a dwindling reservoir of infested plant remains and an increasing reservoir of infected roots) and the sensitivity of the pathogen to changes in the soil environment (expressed as a variable infection rate) do not lend themselves to conventional modelling. However, epidemics in which some of the variables predominate may be satisfactorily described by, for example, models that incorporate components for primary and secondary infection (Brassett and Gilligan, 1988), additionally allow for inoculum decay and host growth (Brassett and Gilligan, 1989), or alternatively allow for variable infection rates as a function of seasonal factors such as temperature and moisture

(Waggoner, 1986). Whilst such models increase in sophistication, validation of the individual components with experimental data lags behind, because of problems such as the infrequent attainment of an asymptotic level of disease prior to harvest. The appeal of many models lies in their equilibrium behaviour, but the stability properties of epidemics of plant pathogens in general have received little attention and it is not apparent that these should dictate the structure of models for disease progress (Gilligan, 1985a). Interpretation of parameter estimates of DPCs relies on the fact that these apply throughout the sampling unit and that the associated errors amongst them are independent (C.A. Gilligan, personal communication). It is widely recognized that diseases are not distributed randomly throughout the crop, and indeed that the spatial distribution of disease is an integral part of the dynamics of host and pathogen in which initially small and well defined disease foci grow and coalesce (e.g. Hornby et al., 1989). When dispersal of a pathogen is localized, as in soil-borne diseases, heterogeneity in inoculum density and in the disease conduciveness of the soil environment may obscure the interpretation of epidemiological mechanisms where disease progress is derived from a process of averaging.

> . . . despite the undisputed influence of mathematics and the computer on the thinking approach to research problems of epidemiologists, epidemiology remains an experimental discipline. Experiments provide the data bases for models and in turn help to test them. It is from this mutual interplay of theory and empirics that epidemiology derives its scientific thrust and charm.
>
> (Kranz and Rotem, 1988)

2.6 Modelling

The word model has been much used in Section 2.5 in the specific sense of a mechanistic description using a mathematical equation. In this section, modelling and its application to take-all are considered in more detail. 'Model' in a wider sense means a simplified representation of a system or subsystem made for the purpose of demonstrating or calculating a function of the system (Zadoks and Schein, 1979). There are, therefore, many types of model, which may be classified hierarchically (Campbell and Madden, 1990) as:

mental or tangible
        physical or abstract
                qualitative or quantitative
                        mathematical
                        statistical.

What now follows concerns abstract models, because they are more relevant to epidemiology. These may be classified also by their purpose or use as descriptive, predictive or conceptual. They may be empirical or mechanistic, linear or non-linear and may be regarded as working analogies. If a policy or a strategy creates the same result as with a real system, the model is a valid one (French, 1985). Models may also be 'parsimonious', in that they trade 'off a slight reduction in fit or explanatory power for reduction in the complexity of the model' (Philippi, 1993).

Analytical mathematical models and simulation models are types of mechanistic model. Simulation models are used where mathematical analysis is too troublesome or impossible, but they are often costly and last-resort techniques. Simulation models may be analogue simulation (for systems that change continuously with time and may be expressed in terms of sets of differential equations), or digital simulation (for systems that undergo discrete changes at various points in time) (French, 1985). The conceptual take-all models discussed below refer to idealized systems, whereas field epidemics are characterized by systematic and stochastic variation, which is expressed by patchy occurrence (Plates 5.12, 5.13, 6.16 and 6.17) and uneven rates of disease development (Gilligan, 1990b).

Knowledge of environmental parameters is far less developed for root diseases than for foliar/shoot diseases. Consequently, few simulation models have been attempted for root diseases and there is a scarcity of even simple descriptive models (MacDonald, 1994). It is not surprising, therefore, to find that models arrived late in the study of take-all. However, since Asher and Shipton (1981), the disease has featured in research on the following models.

- For the behaviour of soil-borne pathogens in the rhizosphere: Gilligan (1985a), Wilkinson *et al.* (1985a), Gilligan and Simons (1987).
- For the generation of epidemics from primary and secondary inoculum: Gilligan (1990b, 1994a,b), Gilligan *et al.* (1994).
- Extended to allow for host growth: Gilligan (1990b, 1994a).
- Extended to allow for biological control: Gilligan (1994a,b).
- Extended to link rhizosphere behaviour to epidemic behaviour: C.A. Gilligan (personal communication).
- Extended to allow for spatial variability: Gilligan (1994b).
- Extended to allow for the dynamics of lesion growth: Gilligan *et al.* (1994).
- For the simulation of patterns of disease in the crop: Hornby (1994).
- For the influence of cropping systems on disease: Colbach (1995).

MacDonald (1994) argued that the conceptualization, testing and refinement of models is necessary for the identification and delineation of the key environmental parameters that affect individual diseases. To date, perhaps the most useful contribution of modelling take-all has been in identifying areas of epidemiological or biological ignorance.

2.6.1 Qualitative models

Throughout the take-all literature there are schematic representations, usually based on experimental findings, that offer explanations of subprocesses of the disease. Since these schemes purport to represent the essential qualities or attributes of the original (Campbell and Madden, 1990), they may be regarded as qualitative models. Amongst examples in this book are Figs 2.2–2.5, 2.17 and 2.18. Earlier examples are a hypothetical mechanism for TAD based on nutritional changes in the rhizosphere caused by the interaction of disease and microflora

(Brown et al., 1973) and the effect of inoculum age and bioassay temperature on estimates of soil-borne inoculum (Hornby, 1975). Sometimes, such qualitative models are upgraded, as in the case of the explanation of changing patterns of take-all in successive wheat crops (Fig. 6.2), which led to the cellular automaton described in 2.6.7.

2.6.2 Mathematical models: temporal

This section is based on a review by Gilligan (1994a) and is about summarizing the progress of epidemics in 'disease progress' (see 2.5) and 'inoculum dynamic' curves, showing cumulative amounts of disease or inoculum (per unit area or volume) plotted against time. A model for the dynamics of Ggt would need to allow for primary and secondary infection sources, increase in the density of roots and loss of infectivity of inoculum. Conceptually this is achieved by linking the following three differential equations.

- Rate of change of infected roots per unit time:

$$\frac{dN_i}{dt} = (r_p P + r_s N_i)(N - N_i); \quad N_i(t = 0) = N_{i0} \quad (1)$$

- Rate of change of total roots per unit time:

$$\frac{dN}{dt} = r_N N \left(1 - \frac{N}{\kappa_N}\right); \quad N(t = 0) = N_0 \quad (2)$$

- Rate of change of inoculum per unit time:

$$\frac{dP}{dt} = r_d P; \quad P(t = 0) = P_0 \quad (3)$$

The derivation of equation 1 is discussed below in general terms to give some idea of the kinds of equations used in this approach to modelling. The symbols used in the equations are listed in Table 2.5. In nearly all the equations listed and in Table 2.5, host units (N) are taken to be roots, but plants may be considered as host units as in equation 4.

Published data on survival of inoculum of Ggt in soil usually produce exponential curves or curves with a short lag period before an accelerating followed by a decelerating loss of inoculum. However, these curves show no cyclical behaviour.

If new infections arise only by primary infection from a reservoir of inoculum in the soil, the rate of increase in infected plants per unit of time may be described by the following equation:

$$\frac{dN_i}{dt} = r_p P(N - N_i) \quad (4)$$

If P and N are constant, this equation may be solved analytically to yield a monomolecular model for cumulative increase in the density of infected plants, but such a model is highly simplified and does not take into account the growth of roots and the decay of inoculum.

Table 2.5. Symbols used in equations 1–12.

Symbol	Definition
K	Compound parameter equivalent to φS
N_i	Density of infected roots
N	Density of host roots; specifically in equation (7) the number of seminal roots per pot
N_0	Initial density of roots
N_{i0}	Initial density of infected roots
P_0	Initial density of inoculum (propagules per unit volume)
P	Density of propagules of inoculum per unit area or volume of soil
r_d	Relative rate of decay of inoculum (relative to P)
r_N	Intrinsic rate of production of new roots
r_p	Net rate of new primary infections per unit inoculum
r_s	Intrinsic rate of secondary infection
S	The proportion of inoculum still infective at the time of sowing 'crop' plants
t	Time; specifically in equation 8 the time in days since incorporation of inoculum
U	Expected number of uninfected seminal roots per pot
δ	Relates to time of onset of epidemic
κ	Carrying capacity (upper limit to density of infected tissue = N in equations 5 and 6)
κN	Upper limit to root density of a plant in equation (2)
λ'_1	Mean number of infections per root derived from soil inoculum that has been exposed to 'volunteer' seedlings
λ'_2	Mean number of infections per root derived from inoculum comprising infected volunteer roots
λ_2	Mean number of infections per root derived from the dual sources of inoculum
ρ	Daily rate of removal of inoculum
φ	Probability that a root is exposed to and infected by a propagule

Infection by root-to-root spread seems to be the principal mechanism by which plants become infected by Ggt. The logistic model and its close variants dominate the description of this form of epidemic, in which secondary infection predominates. Thus, for the logistic model,

$$\frac{dN_i}{dt} = r_s N_i \left(\frac{\kappa - N_i}{\kappa} \right) \qquad (5)$$

integration gives,

$$N_i = \frac{\kappa}{1 + \exp[-r_s(t - \delta)]} \qquad (6)$$

The model has wide application, but it takes no explicit account of inoculum and relegates primary infection to a subservient role. The production and growth of roots and the effects of disease on the root population are ignored.

The following example of modelling processes in take-all epidemiology is taken from the work of Gilligan and Brassett (1990). It concerns the development of primary infection models to include successive cycles of secondary infection by the use of a recurrence relation. Probability models for the escape of infection courts from disease were initially concerned with primary infection and related to a single time of observation (Gilligan, 1990b). They incorporated the concept of the pathozone – the volume of soil surrounding a subterranean plant organ within which the centre of a propagule must occur if it is to have any chance of infecting the organ (Gilligan, 1985b). Data were obtained using a model system in pots, where roots of wheat seedlings were exposed to infection from 'propagules of Ggt' (colonized millet seeds) in soil and from infected roots of 'volunteer' seedlings grown prior to sowing 'crop' seedlings (Brassett and Gilligan, 1990; also referred to in 4.4.1). There were five factors: 'volunteer' density (three levels, including 0), inoculum level (two levels), sowing date (late and early), soil texture (two levels) and control of 'volunteers' (two levels).

The probability of a root escaping infection is given as $(1 - \varphi)^P$ (Gilligan, 1985b) and consequently the expected number of uninfected roots may be approximated by a Poisson model when φ is small and P is large:

$$U = N \exp(-\varphi P_0) \tag{7}$$

To allow for the decay of inoculum in the soil, a sigmoid model was selected as a general descriptor of the change in S.

$$S = \exp[1 - \exp(\rho t)] \tag{8}$$

Equation 7 now becomes:

$$U = N \exp(-\varphi PS) \tag{9}$$

Using the compound parameter λ to represent φPS and assuming the two sources of inoculum are independent, simultaneous exposure of roots to soil and to root inoculum can be incorporated:

$$U = N \exp(-\lambda'_1) \exp(-\lambda'_2) = N \exp(-\lambda_2) \tag{10}$$

In fitting models to the data, use was made of the fact that the parameters φ and S occurred throughout as a product, allowing equation 9 to become:

$$U = N \exp(-KP) \tag{11}$$

From these equations it was possible to estimate the parameter K using data for disease in the absence of volunteers. N and P were known, U was observed and separate values for K could be estimated for different combinations of sowing date and soil condition (as affected by the addition of sand). Subsequently, λ'_1 was calculated as the product of KP for each treatment combination, furnishing a measure of the effectiveness of initial inoculum in causing disease. Estimation of λ_2 was from data for pots with 'crop' plants in which there was infestation by 'volunteer' seedlings. Finally the ratio λ'_1/λ'_2 – which summarizes the relative contribution of the combined (soil and root) and single (soil) sources of inoculum to the final level of infection – could be computed for each combination of sowing date and soil condition.

In this simple probability model to estimate and compare the relative efficiencies of the single and dual sources of inoculum, the final amount of disease represents a balance between removal and multiplication of inoculum. It does not take explicit account of secondary infection between roots within crop plants. However, the presence of 'volunteer' seedlings was found to increase the effectiveness of inoculum to cause disease, and sowing date and soil condition affected the efficiency of both soil and root (from infected 'volunteer' plants) inoculum. These findings are in accordance with the results of field experiments. The way in which volunteers can offset the benefits of late sowing in the field is discussed in Section 3.5.1.

Other models attempt to link primary and secondary infection with allowance for root and inoculum dynamics. These may take the form of growth equations, or computer simulations, and pay more or less attention to death of roots, latent and infectious periods and anatagonistic interactions between pathogens and other microorganisms or soil fauna. Brassett and Gilligan (1988) formalized a model for primary and secondary infection in equation 12. This equation for the rate of change in the number of infected roots per unit area combines monomolecular and logistic models for primary and secondary infection, respectively.

$$\frac{dN_i}{dt} = r_p P(N - N_I) + r_s N(N - N_i) \qquad (12)$$

Fitting non-linear models of the progress of disease caused by soil-borne pathogens (such as those related to the equations above) is rarely possible because adequate data are scarce. One reason for this is the expense of collecting them. Take-all is a good example and also illustrates other problems, such as harvest interrupting progress towards an asymptote and DPCs that differ markedly in shape (Werker *et al.*, 1991). In Section 2.6.5 the application of a linear model approach is described for one of the biggest sets of data ever collected from a take-all experiment.

2.6.3 *Selecting disease variables*

Take-all illustrates an 'unsettling "twist in the tale" about modelling and analysis of temporal progress of disease' (Gilligan, 1994a). There are occasions when assessments of the proportion of diseased plants indicate an apparently static epidemic, but the disease is actively spreading throughout the root system. Also the rate of spread of infection and disease may be greater within diseased plants than through the entire plant population (Werker and Gilligan, 1990).

Under these circumstances no one measure of disease is adequate (see for example Fig. 2.15). The danger of relying on single measures of disease is also discussed under disease assessment in Section 6.3.3, particularly the situation where the amount of disease does not decrease, but an increase in the total root population density results in a decreased proportion of diseased roots. Gilligan (1994a) selected five variables to measure disease and also estimated the density of roots per plant. Their significance in relation to experimental findings was as follows:

- The ratio of diseased to total plants is *disease incidence*. It increased sigmoidally during the season and since the plant population was static an increase in incidence reflected the dynamics of root growth.
- The ratio of diseased to total roots measures *average severity of disease* per plant throughout the entire population.
- The ratio of diseased to total roots averaged over diseased plants measures *local severity of disease*. It omits zero values and may be used to assess rate of spread of disease amongst roots on diseased plants.

2.6.4 *Mathematical models: spatial*

Host growth and spatial pattern recur in considerations of models for both overall disease progress of soil-borne plant pathogens and for components of the infection chain (a valuable concept for use in constructing models for disease progress of soil-borne plant pathogens) (Gilligan, 1985a). Computer simulation is very flexible and logically suited to the synthesis of models for the components of the infection chain. One stochastic model (Gilligan, 1994b; Gilligan *et al.*, 1994) simulates spatial and temporal spread of Ggt on seminal roots in small populations of plants grown in pots in a controlled environment. This pot approach contrasts with field-work approaches to modelling (see 2.6.5), which use the data from designed experiments or sample surveys, with concomitant, comparatively crude disease measurement, low sampling frequencies and poor control (if any) over environmental variables. Protagonists of detailed experimentation in controlled environments on components of the epidemic argue that it permits detailed testing of mechanistic hypotheses. If these hypotheses are meant to apply to the field, it is to be hoped that the situations created in pots represent what happens in the field, but as yet that is far from proven.

The pot model was designed to use theory of disease spread between roots and information from experiments concerning the probability of infection from initial inoculum in the soil and the rate of spread of disease on seminal roots. Its objective was to synthesize information on the dynamics and orientation of the growth of seminal roots and the dynamics of primary and secondary infection of Ggt. The model was written as a FORTRAN simulation program and stochastic variation was introduced using Monte Carlo methods (i.e. methods making use of random numbers).

Three kinds of secondary infection from runner hyphae were distinguished: re-infection of the same root; infection of another root on the same plant via the crown; and cross-infection between roots on different plants. Pathogen parameters concerned: estimates of size and density of inoculum; rate of growth of the fungus on roots; and probabilities of primary and secondary infection for given distances between inoculum source and host units. Fourteen host parameters concerned orientation, density, emergence, rates of growth and size of roots. The main output variables were total and infected root length, numbers of infections, proportion of infected roots and numbers of primary and cross-infections. The model was tested against independent data sets and was found to simulate them satisfactorily.

The model enabled a sensitivity analysis to be made on the effects of changing inoculum density, rate of fungal growth and critical distances for primary and cross-infection on the dynamics of infection by Ggt. This revealed that the proportion of infected roots was most influenced by the inoculum density and the maximum distance for primary infection; distance for secondary infection had little effect. Fungal growth rate principally controlled the density and length of infections on roots. Following the first wave of primary infections, average length of separate infections first increased and then decreased (because of dilution caused by the initiation of the first cross-infections), before increasing again rapidly (as the new infections grew and overlapped). The first cross-infection occurred 7 days after primary infection started.

In earlier work (Gilligan, 1985a), the relationship between inoculum density and number of cross-infections on the seminal roots of five seedlings had been simulated, using different minimum distances at which cross-infection could occur. The greater the minimum distance, the more the maximum number of cross-infections occurred at low inoculum densities (e.g. about 60 occurred with a minimum distance of 1 cm and an inoculum density of 0.4 propagules cm^{-3}, but only about 12 with 0.2 cm and 1.6 propagules cm^{-3}). Distances over which inoculum in the soil is effective depend on several factors, including the suppressiveness of the soil (Hornby, 1981, and Section 2.4). Although Wilkinson et al. (1985a,b) found differences between a suppressive and a conducive soil in this respect, these appeared not to be an effect of TAD. The estimated mean distance for infection (EDI), i.e. the distance that an inoculum source can be from the root and still cause infection, was affected by associated microbiota, particularly when the food base of Ggt inoculum was near the minimum threshold size. However, antagonists responsible for TAD were not implicated when inoculum consisted of particles of colonized oat grains, nor did TAD affect minimum effective particle size. This work suggests that the suppressiveness of TAD limits lesion development after infection.

From pot to field plot, spatial aspects of diseases and pathogens have been analysed using stratification and block-quadrat anaylsis. The approach was used by Werker (1988) to describe variation in Ggt in large plots of wheat in a multifactorial field experiment. A block-quadrat approach, both empirical and statistical, combining temporal and spatial analysis and giving details of data handling and subsequent deductions (Hornby et al., 1989), is also discussed in Sections 6.3.1 and 6.3.4.

Time series techniques are alternative methods of detecting spatial pattern. Conventional designs based on random sampling do not deal directly with the relationship between variance and distance. Spatial heterogeneity depends on absolute location, whereas spatial dependence (autocorrelation) depends on surrounding values. The recognition of spatial pattern is, therefore, dependent on the scale of observation as well as on the scales of the process under study. Consequently, the description and analysis of patchiness is limited by the techniques for observation, sampling and analysis.

2.6.5 Empirical statistical models

There are examples in take-all research of linear and non-linear models in this category. One statistical model was used for estimating the EDI (see 2.6.4) for Ggt (Wilkinson *et al.*, 1985a). Another used probability theory to estimate mean distances for infection to occur (Gilligan, 1985b, also referred to in 2.6.2).

Orthogonal polynomial contrasts and analyses of data for early and late sampling times were used in a model (Gilligan, 1990b; Werker and Gilligan, 1990) that provided insight into the dynamics of disease development in a large factorial experiment over three seasons. In August, the percentages of diseased roots per diseased plant were significantly different, with less disease in a late-sown crop. In early spring other variables (percentage of diseased plants, number of diseased roots per plant, percentage of diseased roots, number of diseased roots per diseased plant) were significant and all indicated less disease in late-sown plots. However, the percentage of diseased roots per diseased plant was greater in late-sown plots, because there were fewer roots per plant. The linear trends between the two sampling dates showed that disease in the overall plant population accelerated more in the late-sown plots, catching up with that in the early-sown plots. Nevertheless, diseased roots per plant actually decelerated more in the late-sown plots because of the more rapid increase in root density in this treatment. This approach promoted detailed dissection of potential mechanisms for disease proliferation (Gilligan, 1990b).

A similar example, using a linear model approach, was the extraction of certain dynamic trends from a parsimonious dataset relating to nine seasons of a winter wheat experiment with take-all in different crop sequences (Werker *et al.*, 1991). Curves were separated into linear, quadratic and cubic components, where: (i) the mean amount of disease was related to the AUDPC; (ii) linear contrasts gave a measure of average rate of change of disease over time; (iii) quadratic contrasts gave a measure of the average degree of curvature – positive values implying acceleration, negative values deceleration, in the rate of disease increase; and (iv) cubic contrasts were sensitive to inflexion points, e.g. when an accelerating epidemic changed to a decelerating one.

Because treatment differences may be hidden by simple averaging over seasons, Werker *et al.* (1991) analysed the way that the components for trends within seasons varied with the trend amongst seasons. Mean levels of disease were relatively low in 1979–1980 and 1986–1987 and relatively high in 1981–1984 (Fig. 2.15 and Plate 7.18). The rate of disease progress in second wheats was greater than in continuous wheats and the rate in continuous wheats was greater than in first wheats. Disease progress in second wheats was associated with a positive quadratic contrast (i.e. an accelerating epidemic), whilst a negative quadratic contrast (i.e. a decelerating epidemic) was common in continuous wheats (Fig. 2.19). The average rate of increase in the diseased root population in continuous wheats (i.e. four or more consecutive crops) was approximately half that in second wheats. This suppression of disease under wheat monoculture was more evident in the root than plant population and in years of high rather than low levels of disease. By contrast, first wheats differed little between years. The contrast coefficients were used to make allowance for variable sampling intervals, sampling frequency and period of observation.

This work suggested two hypotheses about the epidemiological mechanisms for suppression of disease in monoculture:

1. Changes occur in the carrying capacities of populations of roots for Ggt caused by prior colonization of the roots by antagonistic microorganisms.
2. A change in the balance between primary and secondary infection is involved in the dynamics of TAD. Primary infections assume a greater importance in continuous wheats due to the build-up of high inoculum densities; secondary infections play a greater role in the disease dynamics of second wheats because of the lower inoculum densities that follow the slow progress of disease in first wheats.

Relating the view expressed in (2) above with equation 12 (the model for primary and secondary infection in Section 2.6.2), progress of disease in second wheats is initially slow (low P, although r_p may be high), but increases as secondary spread assumes more importance (high r_s) as secondary infections build up and plant roots grow. In continuous wheats, inoculum densities are higher in the early part of the growing season (high P) but the rate of disease progress is slow (low r_p and r_s), and this becomes apparent later in the season when there is a decline in primary infections due to shortage of susceptible tissue and loss of infectivity of soil inoculum.

2.6.6 Host growth and yield

Some of the models discussed above take some account of the growth of wheat roots. There are wheat crop simulation models that have been developed for a variety of purposes: analysing experimental results, predicting crop responses in novel conditions, aiding decision management and assessing how changes in environmental conditions may affect crop productivity (Porter *et al.*, 1993). In comparisons of three such models, one (AFRCWHEAT2), which included subroutines that described the movement of water and nitrogen within the soil profile (Porter, 1993), was the more accurate predictor of development, whereas another (CERES-Wheat) gave a better prediction of final yield. However, none of the models considered the effects of other nutrients, weeds, pests or diseases (Porter *et al.*, 1993). CROPSIM-Wheat, another model describing the growth and development of wheat, was designed to facilitate, amongst other things, addition of computer routines or procedures concerned with the effects of disease (Hunt and Pararajasingham, 1995). This was considered an aspect necessary for models to be useful in the general field situation, but there was no disease example in this paper.

Modelling higher plant processes has been claimed to focus clear attention on deficiencies in our knowledge of root growth. It is recognized that information on numbers and locations of root meristems, elongating zones, ageing zones, associated rates of water and nutrient uptake and exudates will be needed to interface with future models describing rhizosphere microbial dynamics and activities of specific root pathogens (Klepper and Rickman, 1990). The converse, that future work on modelling of subterranean epidemics must turn ever more frequently to models for host growth, has been acknowledged (Gilligan, 1985a; Huisman, 1982).

2.6.7 *Simulation of patch development: cellular automaton*

In computer science, simulation is the mimicking of real-life situations (French, 1985). A classic computer simulation is 'Life', in which a colony of cells grows according to a few simple rules. Cells are placed on a grid and in each generation the succeeding generation is determined by the number of neighbouring grid squares containing a cell (e.g. 2 or 3 means survival; > 3, death through overpopulation; < 2, death through isolation; birth occurs in an empty square with three neighbouring squares occupied). The initial pattern of cells determines how the population develops or dies out, and in the early 1980s one of the aims of games based on 'Life', which had become popular for home computers, was to devise an initial pattern that would last the greatest number of generations. This approach has not been much used in plant pathology, but the program PatchMaker, written in Pascal by D. Hornby, J. Antoniw and D.D. Hornby and referred to in Hornby (1994), applies it to simulating take-all disease patterns, using simple rules for disease spread and severity in a population of wheat plants. Such a model is a cellular automaton and although this one was initially written with take-all and TAD in mind, many of its features are applicable to other diseases that occur in patches.

A cellular automaton is a dynamical system that treats space and time in a discrete fashion and in which each point (or cell) can have any one of a finite number of states (Silvertown *et al.*, 1992; Adamatzky, 1994; Lyman Hurd, 1996: What are Cellular Automata (CA)? [On-line]. Available: http://alife.santafe.edu/alife/topics/cas/ca-faq/general/general.html). As in 'Life', the states are updated by local rules, i.e. the state depends on the cell's own state one time-step previously and also the states of its nearby neighbours at the previous time-step. The updating of cells is synchronous. Cellular automata are 'bottom-up models' that generate global behaviour from local rules. Such automata have been used to model clonal growth, competition, colonization, succession, the influence of fire, regeneration and weed spread in higher plant studies. An automaton with a random initial starting arrangement was used to simulate competitive interaction of five grass species by Silvertown *et al.* (1992). Stochastic cellular automata can be sound models for real epidemics and have shown that the trajectory of an epidemic depends strongly on the initial distribution of infections in space (Adamatzky, 1994).

Some basic ideas that raise the question of the nature of take-all patches (Fig. 2.20) were used as a basis for PatchMaker, which attempts to simulate patterns of diseased plants within a field of 50,000 plants in 250 rows. The whole field or sections of it may be viewed on the screen, where individual plants are represented by colour-coded pixels or symbols, indicating the plant's disease status. An initial selection of foci, i.e. plants with slight take-all, occurring at random in an array of healthy plants (the 'field'), starts the process. After this the current model is

Disease status

	Build-up				Decline	

Cycle:
0   1   2   3   4   5   6

```
                                                    1
                                          1       1 2
                                  1     1 2     1 2 3
                          1     1 2   1 2 3   1 2 3 2
                  1     1 2 1   1 2 3   1 2 3 2   1 2 3 2 1
0   1   1 2 1   1 2 3 2 1   1 2 3 2   1 2 3 2 1   1 2 3 2 1 0
        1       1 2 1   1 2 3   1 2 3 2   1 2 3 2 1
                1       1 2     1 2 3     1 2 3 2
                        1       1 2       1 2 3
                                1         1 2
                                          1
```

Fig. 2.20. How disease might develop from a focus of one plant with slight infection in an array of healthy plants, where plant and row spacings are similar. At each step, an infected plant changes its disease severity (initially increasing and then decreasing after a peak severity in a crude simulation of take-all decline, which in nature usually requires several consecutive crops to develop) and neighbouring uninfected plants become infected.
Disease severity: 0, healthy; 1, slight; 2, moderate; 3, severe.
Bold numbers show the centre of the developing patch; after cycle 3, part of the right-hand side of the patch is omitted to save space.
Wider row spacing produces elongated patches.
From Hornby (1994).

deterministic (i.e. having the same outcome for that particular starting point and the same subsequent choices of options). There are simple rules for disease spread from plant to plant and for changes in the disease severity of infected plants. Four disease levels are recognized: none (healthy), slight, moderate and severe. The combination of these rules creates concentric zones of disease severity, which spread until, finally, all plants in the field are severely infected (model default).

If the disease decline option is chosen (Fig. 2.20), then there is recovery back to a lesser level of disease following severe infection until, finally, all the plants in the field are at that level. When trying to use this decline option realistically, there are two important considerations. Firstly, TAD (Hornby, 1979) rarely, if ever, eliminates the disease, so the decline back to healthy shown in Fig. 2.20 would not be a realistic outcome. Secondly, the manifestation of TAD requires successive crops, because the plants with severe disease inducing the phenomenon rarely, if ever, recover within a crop. Patterns become more complicated as spreading patches coalesce and, when cultivation between successive cereal crops is simulated, causing displacement and dilution of inoculum. Such disruption creates distributions of infected plants from which disease would develop in the following crop that are quite different from the initial starting point of a few slightly infected plants occurring at random.

Applying the rules for spread and changing disease severity in following crops extends the range of patterns obtained. Patterns of severely infected plants emerge in the model that in some cases superficially resemble patterns of patches seen in the field. However, since the model is based on points representing regular plant spacings within and among drilled rows, the patterns that develop are initially conspicuously angular (as in Fig. 2.20) and unlike the more rounded patches of diseased plant foliage often seen in fields, as in Plates 5.12–5.15, 6.16 and 6.17. Patchiness in the field is probably partly affected by the orientation of the leaves and stems of a plant that 'produces its mainstem leaves to form an overall shape similar to that of a hand-held fan' (Rickman and Klepper, 1991) and the different positions in the drill furrow of seed that gives rise to such plants. Diseased plants may also have a smaller canopy and a paler appearance and may be overgrown by healthier plants. The representation of plants by symbols for disease severity in a two-dimensional array with regular spacing is unlikely to result in a realistic picture of patchiness, but it may come close to simulating the zones of disease severities exhibited by the plants that underlie the visible patchiness in the field.

One iteration of the rules is termed a cycle, and the user may modify the model's settings after each cycle. Between each cycle a history of the options that were chosen and counts of plants in each of the disease categories is displayed. Model settings which will affect the disease patterns obtained are: (i) row spacing; (ii) the endpoint chosen for the disease severity scale (e.g. severe disease, or disease decline); and (iii) cultivation up, down, right or left in the field.

At present, with many cycles followed by cultivation, patterns can be created that begin to resemble real disease patterns in crops at, or after, peak disease in a polyetic epidemic (see book cover). (A polyetic epidemic is one that increases in intensity over many years, but which may be obscured if the amplitude of the annual cycle is large because of factors such as weather; Zadoks and Schein, 1979.) Attempts to create realistic patterns in a biologically more meaningful way involved choosing model settings which: (i) did not allow recovery after severe infection within crops; (ii) accepted the induction of decline principles by severe infection and subsequent carry-over of these; (iii) regarded TAD as a limit to disease build-up at that location; and (iv) applied cultivations between crops that displaced and diluted inoculum. However, these choices brought about stabilization by the third crop, with a few diseased plant positions expressing TAD. Apart from identifying obvious deficiencies in the current model, this highlights also our lack of understanding of the actual processes and consequently fuels the interplay between data and ideas.

Trying to modify PatchMaker so that it acts like the real system, prompts ideas and generates insights about that system. Modifications currently under consideration to improve the model are as follows:

- Severely infected plants inducing the onset of TAD should not partially recover. Infected plants occupying the same positions in the following crop would be the first to show disease suppression.

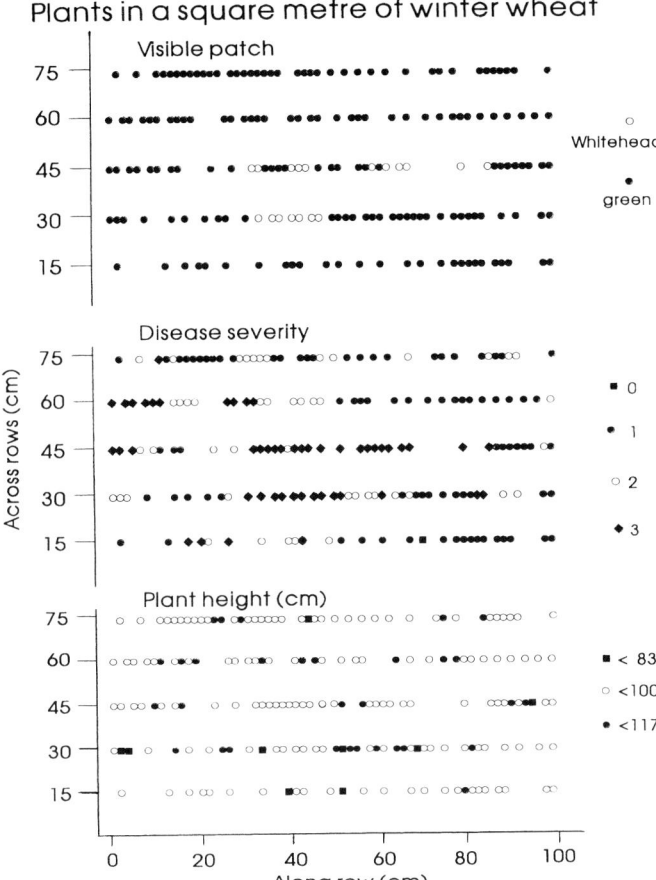

Fig. 2.21. Investigation of plants in a 1 m × 1 m square of a first winter wheat crop at St Gelven, near Rostrenen, Cotes d'Armor, France, on 2 July 1992. Top: plan showing a small cluster of whiteheads. Middle: plan showing disease severity scores for each of the plants. Bottom: plan showing the height categories of the plants.
From Hornby (1994).

- Plants growing where TAD is operative should develop take-all as do plants in other areas, until the set level at which disease is suppressed is reached. Such plants should retain the capability of infecting neighbouring plants.
- A stochastic component should operate after the initial selection of infected plants. As can be seen in Fig. 2.21, real patterns of diseased plants in the field are not concentric zones each containing plants of one disease severity level. Also, field crops tend to be gappy and plants are not in exactly the same positions each year.

These and other improvements will be hard work in detail, but at least the aims can be specified clearly. The prospects for increasingly valid simulations of TAD in a cellular automaton remain good.

2.6.8 A study of the effects of crop succession and cultural practices: combining linear and non-linear models

At a site near Toulouse in France, 11 different crop successions were established over a period of 12 years (Colbach *et al.*, 1994). In the 13th year (1981/82) winter wheat was grown and managed identically throughout the experiment and take-all was assessed at flowering. Previous crops were classified as either hosts (i.e. they increased the disease risk for subsequent winter wheat), non-hosts (decreased the disease risk), or 'amplifiers' (they increased disease risk when associated with host crops in the crop sequence). Maize and ryegrass behaved as 'amplifier' crops at Toulouse. A standard linear regression model was developed to explain take-all severity, measured as percentage of the root system infected, in terms of crop succession and irrigation (Table 2.6).

Table 2.6. Estimates of effects in a linear model[a] explaining take-all severity on winter wheat at Toulouse in terms of crop succession and irrigation.

Constant		13.36*	
Irrigation	+		−
	6.94*		0
		Precedent crop	
	Host	'Amplifier'[b]	Non-host
Ante-precedent crop			
Host	17.29*	16.37*	4.22
'Amplifier'[b]		0.29	
Non-host	7.98*		0

[a] take-all severity = constant + irrigation effect + precedent crop effect + (precedent crop effect)/(ante-precedent crop effect) + error. (13)
In this equation, / is the nesting operator.
The estimates are coefficients of regression on dummy variates. For example x_1 = 'Constant'; x_2 = 0 for no irrigation and 1 for irrigation; x_3 = 0 for non-host, non-host and 1 for host, host, x_4 = 0 for non-host, non-host and 1 for host, 'amplifier', etc.
The estimates show mean differences and those marked * are significantly different from 0 at $P = 0.05$.
[b] See text for explanation.

Another 15 crop sequences, including wheat monoculture, were compared at Grignon, France, in the period 1980–1993; some details of this experiment are summarized in Colbach *et al.* (1996). The level of take-all (expressed as percentage of diseased plants, or as total length of necrotic root per plant at flowering) was described as a function of the number of consecutive cereals (Colbach and Huet, 1995). It was assumed that where successions of wheat were interrupted with spring barley, disease development would be similar to that in a wheat monoculture (this view is not supported by work reported in 3.2.1). There were three to five different runs of consecutive cereals present every year and most of these occurred in more than one year. The initial model tested was:

$$disease = constant + year\ effect + f(x) + block\ effect + error \qquad (14)$$

where $f(x)$ represents disease level as a function of the number of consecutive susceptible cereal crops, x. This function has to take into account disease increasing to a maximum, then decreasing (TAD), before stabilizing at an asymptotic value. The non-linear model

$$f(x) = a\,(x^2 + bx)\,(x^2 + c)^{-1} \qquad (15)$$

would do this, but it requires a large number of points and more high values of x than the data provided, in order to give a correct estimation of the asymptote value a. Accordingly, a simpler model in the form of the following second-degree equation was adopted:

$$f(x) = a.\ln(x) + b.[\ln(x)]^2 \qquad (16)$$

where $a > 0$ and $b < 0$.

The level of disease was already high in the first year of wheat (disease incidence = 64%) but peaked (84% of plants infected) for $x = 3.9$ (i.e. fourth consecutive cereal crop), before decreasing to 74% of plants infected in the tenth year of monoculture. The trend for disease severity (length of necrotic root per plant) was similar, but more pronounced than for disease incidence: the mean length of necrotic root was 2.3 cm in the first year ($x = 1$), 4.9 cm at $x = 3.5$, and 3.1 cm at $x = 10$. The decline after the peak of disease was faster for disease severity than for disease incidence, which suggests that TAD is explained better by less extensive spread of necrosis than by a decrease in rate of primary infection. A similar conclusion has been reached by others (Cook *et al*., 1986, and Section 2.6.4).

The percentage variance in disease explained by the model was low (35% for disease incidence, 36% for disease severity). The patchy nature of the disease may be a reason for this: there was significant variability in disease incidence between blocks.

The high level of disease in the first wheat was attributed by the authors to grasses in the preceding lucerne crop. The subsequent increase in disease in successive crops was small, irrespective of how disease was measured. Local conditions may not have been favourable to take-all between 1986 and 1989 and the disease maximum achieved may have been insufficient to stimulate a build-up of antagonistic microflora necessary for a strong decline in disease. There may also have been a year × monoculture duration interaction, because 1988, 1989, 1992 and 1993 were years in which take-all was particularly severe and caused significant yield losses in northern France (see 1.4). The unbalanced nature of the dataset may also have confused the picture: fifth and ninth wheats were represented only in 2 years and sixth to eighth and tenth wheats only in 1 year, whereas second to fourth wheats were represented in 4 years and first wheats in 8 years.

Take-all, eyespot, sharp eyespot and brown foot rot were four foot-and-root diseases of wheat considered in multifactorial experiments in France, in which the effects of cropping sequence, soil structure, crop residues and the management of wheat were examined (Colbach, 1995). The details for take-all are also published separately (Colbach *et al.*, 1997). For each site and each growth stage, linear models (analysis of covariance) were used to express disease levels as a sum of the effects of the experimental factors and of covariables of the host population (described as a 'static' approach). Using all the data, a series of synthetic models was produced to explain the levels of each disease at heading in terms of site, elements of the cropping system and covariables of the host population. Non-linear models, with a parameter for primary infection cycles and one for secondary infection cycles, were fitted to the data for each treatment, so that the two parameters could be studied as a function of elements of the cropping system and covariables of the host population (described as a 'dynamic' approach). This resulted in a second series of synthetic models to explain the parameter values (the process is explained in more detail below). On the basis of this work, it was suggested that for a particular set of environmental and economic constraints a cropping system with the least risk of disease can be chosen (Colbach, 1995). In relation to effects on take-all, there existed a hierarchy and interactions amongst cultural practices and it was recommended that in disease management such practices should not be considered individually (Colbach *et al.*, 1997).

Non-linear models proposed by Brasset and Gilligan (1989) and by Colbach (1995) for fitting field disease progress data for Ggt are given in Table 2.7. Those

Table 2.7. Different uses of a model to explain the progress of take-all in winter wheat.

Equations		Characteristics	Authors
$dI/dt = (k_1 p + k_2 I)(N - I)$ p = initial amount of inoculum N = no. of plants I = no. of infected plants t = degree-days	(17)	Two infection processes	Colbach (1995)
$dI/dt = (r_1 P_0 + r_2 I)(N - I)$ P_0 = initial amount of inoculum N = number of roots I = no. of infected roots t = degree-days	(18)	Two infection processes; takes into account growth of the host	Brasset and Gilligan (1989)
$dI/dt = (r_1 P + r_2 I)(N - I)$ as above with $P = P_0 \exp(-r_p t)$ r_p = rate of decay of inoculum per 1000 day-degrees	(19)	Two infection processes; takes into account growth of the host and decay of inoculum	Brasset and Gilligan (1989)

Equation 18 uses the symbols I, r_1 and r_2 instead of N_i, r_p and r_s, respectively, but is otherwise the same as equation 1 and a simplification of equation 12.

of Brassett and Gilligan (1989) use number of infected roots as the output variable and integrate root development and inoculum decay. That of Colbach (1995) uses percentage of infected plants as the output variable. All include the idea of primary infections arising from inoculum in host residues and secondary infections arising from inoculum on the living host. Consequently, each has its associated infection rates: r_1 and r_2 (Brassett and Gilligan); k_1 and k_2 (Colbach). The distinction lies in how the models are used, rather than in the models themselves.

Whereas Brassett and Gilligan's equations aim to describe and interpret curves for the progress of disease caused by soil-borne pathogens, Colbach's equation attempts to classify cropping systems according to general disease development and assumes that a simple, two-parameter equation (without taking into account inoculum decay or root development) is sufficient. Colbach's initial equation (Table 2.7), which is monomolecular and logistic, was modified to model the percentage of diseased plants (Y), by dividing by N, which is the number of plants after total plant emergence (assumed to be constant, i.e. no losses during winter, or due to take-all):

$$dY/dt = (k_1 p + k_2 Y N)(1 - Y) \qquad (20)$$

The following constants were defined: $c_1 = k_1 p$, associated with the primary cycle, and $c_2 = k_2 N$ associated with the secondary cycle. After integration, the percentage of infected plants, as a function of time expressed as cumulative day-degrees (above 0°C after sowing), is:

$$Y = \frac{1 - \exp^{-(c_1 + c_2)t}}{1 + \frac{c_2}{c_1} \exp^{-(c_1 + c_2)t}} \qquad (21)$$

This equation was successfully fitted ($r^2 = 0.99$) to data from a plot on a site known to favour take-all, which was assessed every 2 weeks after GS 30. It was then fitted to take-all data obtained from different experimental treatments on three sites (at Le Rheu in 1992 and 1993 and at La Verrière and Chartres in 1992). The treatments were sowing date (early vs. late), sowing density (high vs. low), amount of nitrogen applied (high vs. low), form of nitrogen applied (low ammonium content vs. high ammonium content) and previous crop residue (buried vs. taken away). Take-all was assessed at GS 15, 30, 50 and 80. Fitting these data provided estimates for the parameters c_1 and c_2 for each combination of experimental treatments at each site. For each set of estimated values for the two parameters, a linear model was used to explain the values as a function of various factors analysed and covariables measured:

$$\text{parameter} = \text{constant} + S + M + D + F + R + SM + SD + SF + SR + \text{error} \qquad (22)$$

In this model the main effects are: S, sowing date; M, plants m^{-2}; D, nitrogen dose; F, nitrogen form; R, crop residue. Two-factor interactions are shown as SM, SD, SF and SR. The results of this work are discussed in Section 3.5.1.

It should be pointed out that, once beyond the complexities of the mathematics and statistics, much of what emerges from these modelling exercises often seems only a description or confirmation of what is already well known empirically (e.g. in this case, the health value of certain rotations). It then becomes easy to criticize the models as being too simple, general or selective and the experimental data as coming from unsuitable sites, or being inadequate for the job in hand. Consequently, the explanatory models generated seem restrictive, valid only for the experimental locations used and factors tested. Comparing cropping systems for their disease risks has limited value also if there is no link to the effect of the disease on yield. Many of these points have arisen elsewhere in this chapter. The fact that there are as yet no take-all models in general use to assist farmers is because the disease system is neither simple nor well understood. Modelling is helping to identify the lacunae in our knowledge.

2.7 Comparisons with Other Regions

The severity of take-all in different countries is determined by cropping patterns, which may have been more or less determined by a take-all problem or threat, as well as by climate and soil type. Here, brief summaries of situations described in Yarham (1981) are used to introduce more recent information.

In Australia the disease is reported to be worse on calcareous sands and sandy loams than on clay loams and clays. Severe attacks can also occur on acid sandy soils in Western Australia (WA) (cf. acid old pasture sites in England, Section 3.5.4). It is worse in southern Australia than in the northern wheat belts: north of latitude 32°S it is not a major problem for wheat growers. Severe attacks of take-all are usually associated with above-average rainfall in winter and spring, and in the wetter parts of Australia (annual rainfall in excess of 450 mm) the prevalence of take-all and *Septoria* results in little wheat being grown. It seems that inoculum of Ggt is present at a greater density and is more infectious in WA than in the UK, because of differences in climate (Mediterranean-type vs. cool temperate), soil type and agricultural practice (Cotterill and Sivasithamparam, 1989a). It was suggested that hot, dry summers promote the survival of the fungus and that the sandy and nutritionally poor soils favour the disease. Wheat is commonly rotated with pasture or legumes in order to build up soil nitrogen. Unfortunately, the pasture grasses in WA are predominantly of a type which carries Ggt and so they effectively maintain the disease between crops. Low microbial activity in the soil may also favour survival of inoculum. Observations of weather patterns have led to a take-all prediction model for southern Australia: research attributed to D. Roget of CSIRO concluded that take-all severity can be predicted from measurement in the preceding season of how much take-all inoculum was in the soil and how much rain fell in the spring (Mussared, 1996). Two dry years will apparently decrease the disease to negligible levels and 2 good rainfall years would bring the disease to a peak.

Research on the saprophytic survival and the behaviour of Ggt in soil has continued in Australia (see 4.7), using, for example, inoculum and plant chambers, in which inocula were kept separate from roots by a metallic gauze of pore size 40 µm, through which hyphae could grow (Wildermuth *et al.*, 1984), or a soil sandwich technique (Glenn and Sivasithamparam, 1991). The former apparatus has demonstrated that in soil fumigated with methyl bromide, Ggt from ground oat inoculum grew 6–9 mm in 15 days and 6–14 mm in 28 days to colonize seminal roots and cause lesions. In natural soil, hyphae colonized roots 7–14 mm away in 15 days. Lesions formed only from inocula up to 8 mm from roots. In the absence of roots, hyphae grew 5 mm in fumigated soil and 2 mm in natural soil (Wildermuth et al., 1984). These distances are little different from those in earlier reports from the UK and elsewhere (Hornby, 1981). They are considerably different from the 4 cm of growth from infected straw pieces at low temperatures in a soil sandwich (Grose et al., 1984) and the 6 cm in 8 weeks through a mixture of Gabalong soil and yellow sand (Glenn and Parker, 1988).

In Indiana, USA, TAD occurs in wheat, although its effect is less apparent on light sandy soils than on silt loams or clays. Occasional, severe attacks of take-all have been recorded in first wheat crops after pasture, lucerne or, more especially, soya beans. There is a controversial report (mentioned in Walker, 1981) of isolates of Ggg from soya beans that are highly virulent to wheat. Farmers in the irrigated areas of the Pacific Northwest of the USA now practice rotation, or utilize TAD in continuous wheat. Where annual rainfall is 200–400 mm, wheat is alternated with fallow and take-all seldom, if ever, occurs, although brown foot rot (*Fusarium*) can cause severe damage (D.J. Yarham, personal communication). Take-all was first reported in Alabama in 1983 (Gudauskas et al., 1984). There, wheat is double-cropped with cotton, soya beans or peanuts; double cropping with soya beans became very popular in the late 1970s and early 1980s. The soya beans may help to maintain inoculum of Ggt; and high moisture during the winter, characteristic of the southeast, favours the fungus. In 1990, 113,314 ha of winter wheat were grown in Alabama and the first detailed survey of root pathogens of winter wheat in the southeastern USA (Chen et al., 1996) revealed that Ggt was the pathogen most frequently isolated (26% of the fungi identified).

TAD is now believed to be present in some fields in Montana, USA, especially in the east in long sequences of consecutive wheat crops. The peak of disease occurred in fifth or sixth wheats in these fields, when most of the yield was lost. However, without alternative crops, some farmers continued to grow wheat and by the tenth crop, or later, yields had returned to near normal. This pattern is reminiscent of TAD as originally described at Rothamsted, except for the longer time scale, possibly imposed by the long cold winters and short growing seasons slowing down the polyetic epidemic. The complexities of take-all suppression were demonstrated by research on Montana soils suggesting that different biological suppressive factors were operating in soils from different wheat fields in Montana (Andrade et al., 1994a, discussed in 2.4.4). Further east, in North Dakota, a take-all epidemic in two out of four plots of irrigated continuous spring durum wheat proceeded more like those in Britain, reaching a peak of disease in the third season (Stack and Thompson, 1988). Disease in another plot levelled off in the fifth season. Generally the disease levels seemed low, with a reported maximum of 21% take-all; yields were decreased by 15% for each 10% of disease. It was not explained what '% take-all' represents.

In Brazil, wheat is sown in the autumn and harvested in the spring to be followed by soya bean as a summer crop harvested in the autumn. In general, this double-cropping is repeated year after year and, combined with high soil moisture levels, temperatures of 12–20°C and heavy dressings of lime (to benefit the soya bean crop), it results in severe take-all. There is no evidence of TAD in this system, perhaps because the soya bean crops prevent the establishment of an antagonistic microflora.

The absence of take-all in fields cropped to wheat for 25 years or more in Hokkaido, Japan, was attributed to sterile fungi that suppressed take-all by aggressively colonizing wheat roots (Narita and Suzui, 1991).

In Europe, TAD has been reported to occur in France (Lemaire and Coppenet, 1968). Switzerland has some of the longest runs of intensive cereal cropping in the world, with some high altitude fields cropped with spring barley for over 200 years. Take-all in these fields is negligible, yet it does occur in barley grown in rotation at lower altitudes in the same area. The soil from monocropped high alpine fields of barley has been shown to be extremely antagonistic to take-all, although this is not necessarily the usual TAD effect. Not a single example of TAD had been recorded in monocropped wheat in Switzerland. In Germany take-all occurs on good, aerated soils (e.g. loamy sands, sandy loams and peaty soils) where wheat follows wheat (Mielke, 1988). More than usual take-all and whiteheads were observed in wheat in 1992–1994 on loess and black earth soils in several German Länder, and this was attributed to high rainfall in late spring and unsuitable crop rotations (Mielke, 1995).

In South Africa, the severity of take-all is related to soil type and climate; it is most serious in areas of winter rainfall in the Cape Province and in the eastern Free State where monoculture of wheat is practised on light-textured soils. The disease has been reported to be very severe in wheat following lucerne or lupins and in rotation with soya beans or groundnuts.

Factors influencing take-all worldwide are discussed further in the next chapter.

Take-all and Cereal Production Systems 3

3.1 Factors that Affect Take-all

A useful table of factors that affect take-all (Table 3.1) was compiled by Huber and McCay-Buis (1993). This list tends to be general and contains some factors that currently appear to be more important in some regions than in others (cf. Fig. I.1). The view that take-all is an expression of specific interacting favourable components has already been introduced (see 2.1) and in this chapter some of these components are examined in more detail. Much take-all research has been concerned with identifying the factors that have the greatest effect on the disease and applying the knowledge gained to disease management (e.g. Hornby *et al.*, 1990a). That there may be some key factor that would unify all the disparate disease effects reported in the literature is a possibility that has recently emerged through the championing of manganese (Huber and McCay-Buis, 1993). The view expressed is that all the conditions influencing take-all are correlated with their effect on manganese availability, and that practices that are recommended to decrease take-all specifically increase manganese availability. These ideas are explored later in Section 3.4.3.

3.2 Host Resistance

Limitations in the flexibility of cropping that involves cereals are determined partly by the susceptibility of the crops to take-all. The usual way of avoiding take-all, or disrupting the build-up of disease, is to grow non-host crops, usually non-cereals, as discussed in 3.3.1. Here the use of cereals that are host to Ggt, but have reduced susceptibility, is discussed. Much recent research towards finding resistance in cereal hosts has taken place outside the UK and Europe.

In the USA, older cultivars of hard red winter wheat (Table 1.8 contains a classification of principal wheats) suffered less yield decrease than high-yielding soft white winter wheat cultivars with the same amount of root necrosis (Huber and McCay-Buis, 1993). The latter have a greater nutrient requirement and lower

Table 3.1. Components of the wheat plant, the pathogen (*Gaeumannomyces graminis*), the biotic environment, and the abiotic environment that are favourable and unfavourable for take-all.

Favourable for disease	Unfavourable for disease
Wheat plant	
Nutrient deficiency	Nutrient sufficiency
Nutrient-inefficient (especially manganese)	Nutrient-efficient (especially manganese)
Susceptible	Tolerant or resistant
Low seed manganese	Manganese accumulation in seed
Slow root growth	Fast (extensive) root growth
High yield potential	Low yield potential
Limited siderophore production	High siderophore production
Stimulation of manganese oxidation	Suppression of manganese oxidation
Limited root exudation	High root exudation
No production of toxic root exudates (especially avenacin)	Production of avenacin (and other inhibitors)
High plant population	Low plant population
Stress-sensitive	Stress-tolerant
Nitrate nutrition	Ammonium nutrition
G. graminis	
Highly virulent	Non-pathogenic
Large population	Absent or small population
Fast growth rate	Slow growth rate
Oxidizes manganese	Unable to oxidize manganese
Produces avenacinase (for oats)	Unable to produce avenacinase
Insensitive to soil antagonists	Sensitive to soil antagonists
Insensitive to temperature	Sensitive to temperature
Nitrate nutrition	Ammonium nutrition
Biotic environment	
Large population of manganese oxidizers	Small population of manganese oxidizers
Low population of manganese reducers	High population of manganese reducers
Fast rate of nitrification	Slow rate of nitrification
Low general microbial activity	High general microbial activity
Abiotic environment	
Nitrate nitrogen	Ammonium nitrogen
Alkaline soil pH	Acid soil pH
Organic fertilization (animal manure)	Inhibition of nitrification
Imbalanced nutrients	Balanced nutrients
Optimum temperature for manganese oxidation	Temperature for manganese reduction
High soil moisture	Low soil moisture
Conducive soil	Suppressive soil
Soya bean or lucerne prior crop	Oat or lupin prior crop
Loose seedbed	Firm seedbed
High plant population	Low plant population
Plant stress	Manganese, choride fertilization
Short monocropping	Long monocropping (TAD)
Early seeding	Late seeding
Application of lime (calcium carbonate)	Application of 'acid' fertilizers
	Paddy rice preceding wheat in rotation[a]

[a]This additional entry has recently come to light as a result of cooperative research in China, where newly cultivated, alkaline, desert land in Ningxia and Gansu Provinces of North Central China, used for wheat, frequently suffered 30–50%, and sometimes 60–80%, whiteheads in the first year of irrigated cultivation. Winter wheat monocropped in dry-land, rain-fed areas for over 200 years also suffered from 20–30% whiteheads (D.M. Huber, personal communication).
After Huber and McCay-Buis (1993).

efficiency of partitioning vegetatively stored nutrients; they are also more susceptible to take-all. Soft red winter wheat cultivars also differed in susceptibility, reflecting increased nutrient efficiency in early growth, a more prolific root system and/or the ability to accumulate high levels of manganese in the seed (see 3.4.3). Such differences amongst wheats in the duration of nutrient uptake necessary to fulfil the needs of the developing grains also need to be considered when trying to assess the effects of severe outbreaks of take-all at different times during crop growth (Figs 1.12 and 1.13).

3.2.1 Exploiting different cereal species

Long practical experience has resulted in an empirical understanding of what is and what is not possible in relation to take-all. Whilst barley remains the most popular alternative to wheat as a second cereal on soils where take-all risk is high, trials on the peat soils of the ADAS Arthur Rickwood Experimental Centre have shown the value of triticale (a wheat × rye hybrid) in this situation. On the light soils of ADAS Gleadthorpe Experimental Centre, rye may be cropped with impunity as a second cereal but, despite its relative resistance to the disease, it cannot be relied on as a break crop in a cereal sequence. Consequently its use as a catch crop to reduce nitrogen leaching in the autumn and winter is limited because of the increased risk of take-all in any subsequent cereal. In the absence of Gga, oats provide an excellent break crop, but at Gleadthorpe they have been found to be infected to a limited extent by Ggt.

Different cereal species, therefore, have different levels of tolerance to take-all. Apart from oats, which are almost immune to infection by Ggt, wheat is the most susceptible cereal, rye the most resistant and barley is intermediate (Scott, 1981; Mielke, 1992). However, reports from eastern Germany suggest that take-all was an important contributor to the 20% losses in yield that occurred in rye monoculture (Roth *et al.*, 1984; Wendland, 1986). Also, on red duplex and calcareous Mallee soils in Victoria, Australia, annual losses due to take-all have been the same (7%) for wheat and barley (Anon., 1994c). The production of a fungitoxic glucosidic compound, avenacin, is considered largely responsible for the resistance of oats to take-all, although less cortical cell senescence, making oats less susceptible than other cereals to root invasion, was also suggested as a contributory factor (Yeates and Parker, 1986b).

The relative susceptibilities of wheat, barley, triticale, rye and oats were investigated in plots inoculated with Ggt in Georgia, USA (Rothrock, 1988b). All cereals except oats were susceptible. There were significant decreases in the yields of wheat, barley and triticale, as well as in tillers per metre, 1000-kernel weight and plant height of these cereals. The susceptibilities were wheat > triticale > barley > rye. Wheat showed the greatest yield reductions (except in comparison with triticale in 1987) and the greatest reductions in height, tillers per metre and 1000-kernel weight. Oats and rye did not suffer significant decreases in yield. The differences in take-all amongst these cereals were taken to be true differences in susceptibility, not differences in disease expression, or rate of compensatory root growth. There were only small differences in susceptibility amongst cultivars within a species (see next section). When, in the same localities, wheat (cv. Stacey) was grown after 2 years of growing the same cultivars of triticale, barley or rye, take-all was severe (Rothrock and Cunfer, 1991). In comparable sequences, wheat following wheat had significantly less take-all (for which explanations are offered as this was perhaps unexpected) and wheat following oats had little or no take-all. All the susceptible cereals seemed, therefore, to support sufficient inoculum potential to cause significant levels of disease and decreased yields in following wheat crops. Wheat following oats often had detectable levels of disease. It was concluded that differences in the susceptibilities of cereals cannot be used for the management of take-all in a subsequent cereal.

More than half the world's production of triticale has been from Europe (Table 3.2). Rye has a high level of resistance to take-all and its chromosomes can be transferred readily into wheat (Wallwork, 1989). The expectation for this cereal was that it would contain the natural resistance to take-all of rye whilst retaining some of the valuable characteristics of wheat. However, triticale was closer to wheat than rye in its susceptibility to take-all at the seedling stage (Linde-Laursen *et al.*, 1973; Mielke, 1974), but mature plants were intermediate in susceptibility between wheat and rye, octoploids being more susceptible than hexaploids (Hollins *et al.*, 1986). In East Germany, triticale was at first considered just as susceptible as wheat to take-all and consequently the same cultural practices were recommended for both cereals (Mögling, 1984). Later, it was suggested that winter triticale, grown with a range of fertilizer, growth regulator and fungicide treatments, was not very susceptible to take-all since yield losses did not occur (Mögling *et al.*, 1988). When wheat, barley, triticale and rye were compared in the UK with either low or high fertilizer inputs, wheat was the most susceptible, rye the most resistant and triticale and barley intermediate, the last two with approximately equal susceptibility (Gutteridge *et al.*, 1993). There appeared to be no advantage in changing to triticale from either wheat or barley except where wheat was grown on less fertile soils with low inputs, in which situation grain yield was increased.

Table 3.2. Production of triticale in 1989.

Country	World area		World production		Mean yield ($q\ ha^{-1}$)
	Mha	%	Mt	%	
Australia		6.8		4.0	
China		37.5		23.6	15–17
France		8.8		13.9	42
Germany FR					52
Italy					24–25
Poland		37.5		47.0	33
Portugal					24–25
Spain				4.0	24–25
Switzerland					42
Total	1.6		4.2		26.5

From Kent and Evers (1994).

On light organic soils, triticale outyielded wheat where wheat, but not triticale, showed visible take-all (Cleal, 1993). Compared with winter wheat cv. Galahad, superior yield and specific weights in excess of 72 kg hl^{-1} made the triticale cv. Purdy potentially more profitable and possibly more readily marketed when grown as a second cereal in situations with a high risk of take-all.

In Australian work (Wallwork, 1989), the levels of resistance varied between triticale lines, not all of which contained the full set of seven rye chromosomes. Although there was a trend for fewer rye chromosomes to be associated with less resistance, the resistance of triticale cultivars could not be predicted on the basis of rye chromosomes alone. Consequently, recommendations for the use of triticale in take-all-infested areas should be made only after careful experimentation and field testing. In the USA, Huber and McCay-Buis (1993) reported that triticale lines with the rye chromosome conditioning nutrient efficiency were as resistant as rye; other lines were as susceptible as wheat.

Some degree of disease escape in barley may be aided by its more vigorous production of secondary roots compared with wheat. Also, the effect of the disease may differ on winter barley and on winter wheat because of different rates of crop development. At Rothamsted, under wheat, take-all often increases rapidly between May and July and is most strongly related to yield at host GS 69–71. Barley usually reaches this stage in early June, at which time wheat is usually in boot (GS 45), and most of the grain filling in winter barley is completed before take-all reaches its peak. There is approximately a 3-week period between barley reaching

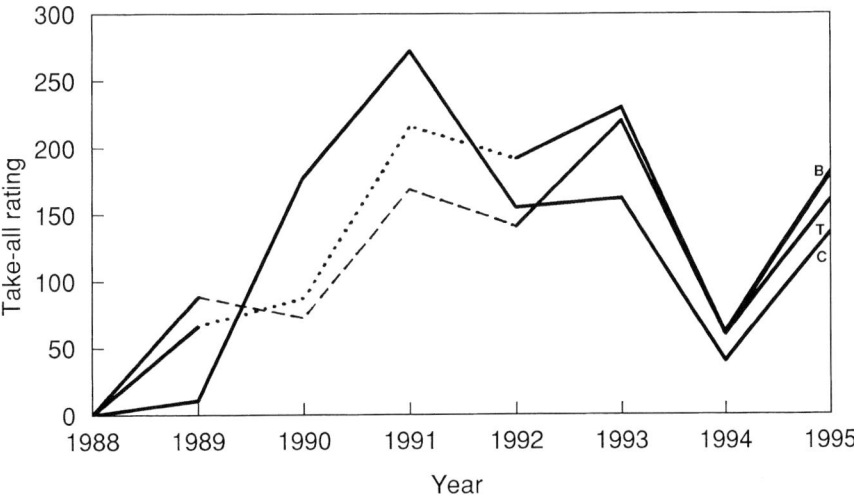

Fig. 3.1. The effect of replacing winter wheat by winter barley or winter triticale for the 3 consecutive years in which continuous winter wheat crops were considered to be at high risk of take-all.
Dotted line, winter barley; dashed line, winter triticale; solid line, winter wheat.
Where points on solid lines coincide, B, T and C distinguish between sequences containing barley, triticale or continuous wheat, respectively.
Data are from a large field experiment on Rothamsted Farm described in outline in Hornby and Gutteridge (1995).

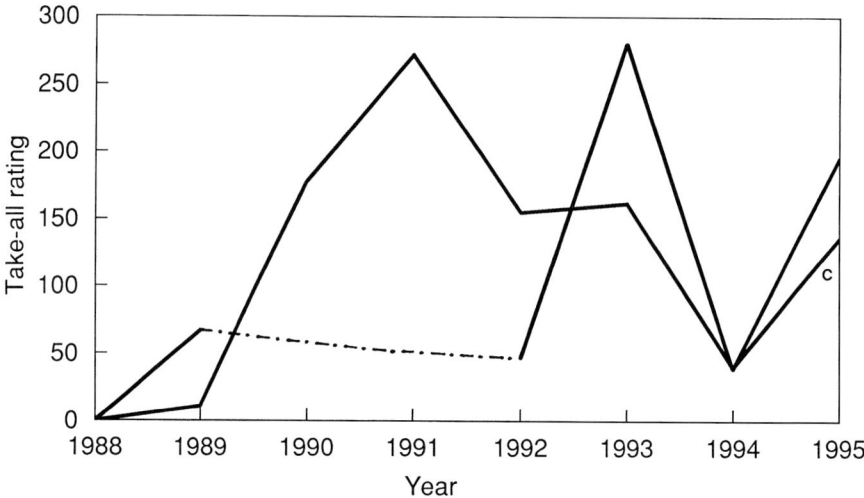

Fig. 3.2. The effect of replacing winter wheat by spring barley for the 3 consecutive years when continuous winter wheat crops were considered to be at high risk of take-all.
Dash–dot line, spring barley; solid line, winter wheat. Where points on solid lines coincide, C identifies the continuous winter wheat sequence.
Data are from a large field experiment on Rothamsted Farm described in outline in Hornby and Gutteridge (1995).

GS 69 and wheat reaching the same stage, during which time the take-all severity rating on wheat may double. This was one reason for proposing barley as a 'bridge' over high risk periods in sequences of wheat. This hypothesis, extended to include other less susceptible cereals, was tested as part of a large, long-term field experiment at Rothamsted, some of the findings of which were reported in Hornby and Gutteridge (1995). Take-all was slight in crops of triticale, winter barley or spring barley that were substituted for wheat after two wheat crops (Figs 3.1 and 3.2). In the second year of triticale or winter barley as bridge crops, disease became more severe, but it declined in a third consecutive year. With spring barley as a bridge, disease lessened progressively during the 3 consecutive years it was grown. However, none of the 3-year bridges prevented severe disease on the resumption of winter wheat and increase in take-all was greatest after the spring barley bridge. Also, despite the assumed development of TAD in spring barley grown annually since 1975 (and in most years since 1902) on the 'Exhaustion Land' Classical experiment at Rothamsted, there was severe take-all when a wheat crop was grown in 1992 (Gutteridge *et al.*, 1996). These observations fit the impressions of Northumbrian farmers in the 1960s, reported to D.J. Yarham, that wheat after spring barley suffered more from take-all than did wheat after wheat. In the USA, neither rye nor triticale provided protection against take-all in a following wheat (Huber and McCay-Buis, 1993).

The resistance of oats to take-all in the USA was said (Huber and McCay-Buis, 1993) to be correlated with resistance to grey speck (manganese deficiency; see

3.4.3). Some oat lines susceptible to grey speck were susceptible to take-all and did not protect a following wheat crop from the disease. Take-all of oats has been rare since grey speck resistant cultivars have been available. (Interestingly, in earlier work – Butler and Jones, 1961 – it was mentioned that grey speck was popularly called 'roadside take-all' in South Australia, but the disorder was not considered to have any connection with the disease, take-all.) Consequently, oat take-all occurs only where soil manganese levels are well below the level for physiological sufficiency. The susceptibility of oats to take-all growing in soils characteristically low in manganese in Western Australia may be due to a requirement for manganese in the production of avenacin, or to insufficent manganese for general plant requirements.

3.2.2 Cultivar resistance and possible mechanisms

Hollins *et al.* (1986) found little difference in susceptibility to take-all amongst currently available winter wheat cultivars and expectations for finding useful resistance in winter wheat are small in Britain (Scott *et al.*, 1989), although resistance from other cereals is feasible (Fig. 3.3; see 3.2.3). One UK report suggested that cultivars may differ in the extent to which they aid increases in inoculum from small populations of the take-all fungus (Widdowson *et al.*, 1985): at Saxmundham in Suffolk there was twice as much inoculum after cv. Avalon (a bread quality wheat) as after cv. Norman (a feed quality wheat). Elsewhere in the world different levels of resistance have been associated with an ability to utilize fertilizers, or to become immunized by avirulent fungi (see below). A graded

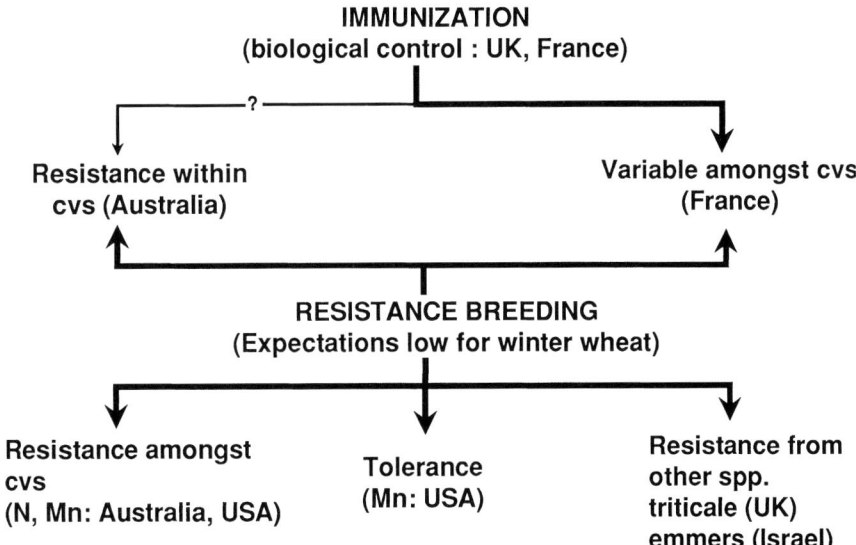

Fig. 3.3. Scheme based on reports of within-cultivar and between-cultivar differences in immunization against, or resistance to, take-all in winter wheat. Regional, mineral and other cereal associations are shown (further explanation in text, particularly Section 3.2.3).

response to nitrogen by plants with severe take-all has been recorded in the world wheat collection. In Australia and the USA, resistant cultivars are efficient users of manganese; in the USA, greater concentrations of manganese in the tissues of take-all-tolerant wheat are correlated with smaller populations of manganese-oxidizing bacteria in the rhizosphere.

Comparisons of the yields of different wheat cultivars have indicated that some are more tolerant than others of take-all. In 6 years up to 1983, in tests in naturally infested small plots and tests in inoculated small plots, Wächter (1984) found that cv. Fakta developed least take-all out of nine cultivars grown. It was suggested that a cultivar's tolerance could be assessed on the basis of yield and the number of fertile shoots. However, attempts to grow mixtures of cultivars to exploit different tolerances of take-all produced no beneficial effects (Mögling, 1987).

Recent searches for cultivar resistance have also been made in Australia. Penrose and Neate (1994) tested wheat genotypes grown in the field and in sand culture. Root cortex browning and the extent of root colonization in sand culture were associated with resistance correlated with root disease in the field, mostly in wheats that matured early. These results were considered to be sufficiently encouraging to make further evaluation of resistance characters worthwhile. Earlier, Penrose *et al.* (1986) suggested that wheat had two levels of resistance which, although low, could be useful. Differences in the thickening of cortical cell walls in the seminal roots of wheat seedlings, providing a mechanical barrier to infection, seemed to be responsible for differences in the susceptibility of cultivars to take-all (Penrose, 1987). Nilsson (1973c) had already established differences in the susceptibility of cultivars by inoculating seminal roots in a laboratory assay.

Differences between the infection characteristics of different cereal species have been proposed as the basis for seeking differences in susceptibility between cultivars (Solel *et al.*, 1990). For instance, manganese nutrition is among the characteristics of plants associated with susceptibility to take-all. Manganese might increase the biosynthesis of defence-related phenolics and lignin and thus resistance to take-all (Graham and Rovira, 1984). Rengel *et al.* (1993) analysed the effects of manganese fertilization and isolates of Ggt differing in virulence on wheat genotypes that differed in their efficiency of manganese utilization. Added manganese was most effective in increasing resistance in a wheat genotype that was least efficient in its utilization of manganese. Subsequent work suggested that a manganese-efficient genotype had a more efficient mechanism for conversion of phenolics to lignin than did a manganese-inefficient one, and that this accounted for a greater resistance to take-all (Rengel *et al.*, 1994).

Attention has also been drawn to early maturity amongst barley cultivars (Rothrock, 1988b), where earliness was related to smaller decreases in yield. Such minimizing of yield losses due to take-all may arise because the host develops more before the pathogen's activity increases as soil temperatures rise in the spring (cf. the earliness argument used in relation to barley vs. wheat in Section 3.2.1). The author pointed out that earliness had been associated with decreased damage due to take-all on wheat as long ago as 1919 in Australia.

There has been an extensive search for varietal differences in wheat and barley in Japan (Oyanagi *et al.*, 1990). A comparison of 244 wheat cultivars, 56 six-rowed barleys, 50 naked barleys and 34 two-rowed barleys was made using four different fertilizer treatments over 3 years. Whiteheads occurred more with composts, or composts plus compound fertilizer, than with only compound fertilizer or no fertilizer (increasing nitrogen fertilizer usually decreases take-all in wheat in the UK: see 3.4.1). The order of susceptibility was: wheat > naked barley > six-rowed barley > two-rowed barley. Whitehead symptoms were less in later-heading wheats and six-rowed barleys. Stem-base blackening, a symptom of severe disease, was closely correlated with whiteheads in all but seven of the wheat cultivars.

3.2.3 Novel breeding programmes

The exploitation of known resistance in other species, referred to in Fig. 3.3, would require the collaboration of plant breeders and molecular biologists. An example is the possibility of transferring the ability to produce avenacin, thought to be a cause of resistance to Ggt, from oats to wheat, or to other root-inhabiting microorganisms. This would require much research into biosynthetic pathways and the creation of transgenic wheat plants. The subsequent behaviour of Gga, which currently appears rarely to attack wheat in Britain, could be a major factor in determining the eventual use of such engineered organisms. Australian reports of oat-attacking strains of Ggt are a cause of concern in this context. Ggt has also occasionally been found causing take-all lesions on the roots of oats in England (an example was referred to in 3.2.1).

Linde-Laursen *et al.* (1973), evaluating resistance in *Triticale*, *Aegilops* (goat-grass) and *Haynaldia*, concluded that the high levels of resistance occasionally found within the *Triticeae* were unlikely to be exploited successfully in breeding wheat because of their dilution when combined with *Triticum* germplasm. Subsequent efforts to incorporate alien amphiploid and chromosome substitution lines into wheat have produced variable results (Conner *et al.*, 1988). Among the species in which resistance has been sought is the diploid wild wheat, *Triticum tauschii* (Eastwood *et al.*, 1993). (The origins, classification and names of wheats are given in Cook and Veseth, 1991.) Moderate resistance was found in two out of 50 accessions of a tetraploid wild wheat, emmer (*T. turgidum* var. *dicoccoides*), in preliminary screening in Israel. Subject to verification, emmer was considered as source material for transferring resistance to common wheat, *T. aestivum* (Solel and Anikster, 1988). Mielke (1992) proposed *Hordeum bulbosum*, *H. bogdani* and *H. brevisubulatum*, which had very low susceptibility to take-all, as possible sources of resistance for barley.

Large-scale testing of somaclonal variants of wheat derived from callus cultures showed some differences in the early generations, but the most promising lines appeared more susceptible in later generations (Eastwood *et al.*, 1994). This was an interesting approach to identifying sources of resistance which ultimately proved not to be of practical use in breeding programmes.

Immunization or induced resistance by avirulent root parasites apparently varies amongst cultivars, according to French work (Lemaire *et al.*, 1982). Root colonization by a take-all-suppressive strain of *Pseudomonas fluorescens* also varied amongst cultivars (Weller, 1986). Immunization by avirulent fungi is a biological control phenomenon originally researched in France and Britain. It may have been responsible for the better plants seen in the field in Australia and selected as 'resistant lines', although few (five out of 100) survived further selection in controlled conditions (C.A. Parker, in Kollmorgen, 1985). Because the amount of variation between wheat cultivars in resistance to take-all is so small, attempts have been made in France to exploit the variability observed in the level of cross-protection induced by a hypovirulent isolate of Ggt. The initial aim was to enhance the efficacy of the biological control investigated by Lemaire *et al.* (1979a,b; see 4.3.4). In a programme started in 1973, plants that produced most weight of grain after being grown from seed coated with a hypovirulent isolate of Ggt in the presence of much inoculum of virulent Ggt were selected for breeding. Pedigree breeding was based on agronomic characteristics and laboratory tests for assessing the capacity for cross-protection (Lemaire *et al.*, 1982). Field tests have still not shown any benefit from these cultivars as part of a biological control package, because other factors limit the efficacy of the hypovirulent isolate of Ggt in biological control (see 4.3.4).

Current investigations in France include an exploration of the variability in disease tolerance, or escape; the ability of the plant to produce new roots when attacked by Ggt. Differences between cultivars have already been found, mainly in the number of nodal roots produced at equivalent growth stages (P. Lucas, unpublished results). From observations on these cultivars in heavily infested soil, their ability to compensate for disease-induced root loss was not, on its own, sufficient to avoid serious yield losses in conditions of severe take-all. Even so, some interest remains in combining different techniques that separately have only small effects on the disease, but act at different stages of the epidemic and so may have a cumulative effect in protecting the plant.

3.3 Rotations

Since take-all in the UK is usually a problem only where cereals are grown frequently, it can be avoided quite easily. One view is that if the proportion of cereals is not more than 40% of the total acreage in the rotation, losses from take-all, eyespot and cereal root eelworm could almost be prevented (Slope, 1966). In Germany, Mielke (1988) recommended that susceptible cereals should not exceed 60% of the arable acreage if rotation were to control take-all. Rotations with 80% and 100% cereals generally suffered increased incidence of eyespot, take-all and cereal cyst nematode and decreased yields in a crop rotation trial in East Germany (Steinbrenner *et al.*, 1984). The results also showed lower yields of spring barley following winter wheat in rotations with 80% or more cereals, but a significant increase in the incidence of take-all was recorded in a rotation with 60% of cereals.

3.3.1 Changes in rotational practice

This section amplifies the relevant parts of the historical outline of cereal production in Section 1.1. The traditional rotations such as the Norfolk four-course (wheat, roots, barley, ley) (Weston, 1944) were well-balanced rotations which limited soil-borne diseases, including take-all. Such rotations formed the basis of UK agriculture until the early 1930s. However, the minutes of the Agricultural Improvement Council for England and Wales, Crop Husbandry Group, meeting in November 1948, record S.D. Garrett as reporting that, although rotations had previously kept it in check, take-all had been favoured by 'the intensive growing of cereals' during the agricultural depression of the 1930s and during World War II. Yield losses of over 50% were recorded in some fields in the south. In the post-war period, cereal growing became a very attractive option, especially in the drier parts of the country, as increasing costs of production and a decline in the availability of labour coincided with increases in the availability of modern machinery, artificial fertilizers and effective agrochemicals. Moreover, the decline in the population of working horses reduced the market for oats (which are resistant to Ggt) and thus increased still further the intensification of wheat and barley growing.

In the eastern counties, where very little grass was grown, the proportion of wheat and barley in the rotations was considerably greater than in the country as a whole (see 1.1), with many farms adopting a policy of very intensive, often continuous, cereal growing. The record harvest of 1984 highlighted the problems of over-production and gave impetus to a move back towards the introduction of an increasing proportion of break crops in arable rotations. This move was assisted by the rapid expansion of oilseed rape following the introduction of low erucic acid cultivars in the late 1970s. Pulses, too, increased in popularity – the English acreage rising from 37,300 ha in 1959 to 139,600 ha in 1986. There has also been sporadic interest in other crops such as lupins and sunflowers but, as yet, these are still not fully established as UK crops. For many years, therefore, arable agriculture on the heavier soils of eastern England has been dominated by cereals (especially wheat), oilseed rape and pulses, with cereals frequently occupying up to two-thirds of the total acreage in such rotations as wheat, wheat, rape, wheat, wheat, beans. More recently the increasing profitability of linseed has provided yet another break crop which fits well into a system of agriculture in which the combine harvester provides the only necessary piece of harvesting machinery.

On lighter mineral soils and on fen peats the risks of take-all are greater, but the ability to grow root crops such as potatoes, sugar beet, and (in some areas) carrots and other field vegetables has always presented the farmer with a wider range of available break crops. On such soils barley provides a safer alternative to wheat as a second cereal. Take-all was considered largely responsible for the variable yield of second wheats that often occurs on the organic soil (loamy-peat, 30–66 cm over sand and gravel, organic matter 28–34%) of the black fens (Madge, 1987). On the very lightest land rye, with its ability to withstand drought and its moderate resistance to take-all, has always been a popular crop, though in recent years it has met competition from triticale as a crop for such situations.

The trend towards increasing further the proportion of non-cereal crops in arable rotations has, however, been limited both by economic and pathological considerations. Over-cropping with pulses can lead to problems with a foot rot complex of fungi (predominantly *Fusarium* and *Phoma* spp.); rape, spring beans and linseed are all susceptible to *Sclerotinia sclerotiorum*, and winter beans, if cropped too frequently, can suffer severe attacks of *Sclerotinia trifoliorum*.

Fig. 3.4. Wheat cv. Cappelle Desprez after an oat break (left) and after wheat and the development of severe take-all (right).

Short sequences of cereals can thus be valuable in reducing disease risks for other crops in the rotation but, of course, such sequences carry their own disease problems, particularly in the form of take-all. Oats grown in place of wheat or barley in a cereal sequence obviate the risks of take-all (Fig. 3.4) over most of the UK and the introduction of naked oat cultivars has contributed to some revival of interest in this crop. However, the market for oats remains limited and for the foreseeable future cereals susceptible to take-all are likely to remain the basis of arable agriculture in the UK.

Table 3.3 shows estimates for England, at the beginning of the 1994 and 1997 seasons, of the yields, costs and gross margins of combinable arable crops of average yield; the crops listed are suitable for farms on heavy land in East Anglia. Table 3.4 illustrates some of the rotational options available to the arable farmer and highlights the fact that take-all is not the only disease that has to be considered in planning a rotation. It is suggested that take-all was responsible for some of the differences in this table, even though the figures are based on data produced at Boxworth EHF, which has a Chalky Boulder Clay soil (Hanslope series) that is considered relatively safe from severe take-all. Consequently, were this exercise to be done for soils and sites at greater risk from take-all, greater differences would be expected. Given that continuous wheat on the Boxworth farm yields 7.2 t ha^{-1} and knowing the yields, prices and marginal costs of other crops grown on the farm, a similar exercise can be carried out to estimate the costs of a range of alternative

Table 3.3. Predicted average yields and gross margins of combinable crops in England for the years 1994 and 1997.

Crop	Year	Average yield (t ha^{-1})	Sale value (£)	Area payment[b] (£)	Variable costs[b] (£)	Gross margin[b] (£)
1st winter wheat[a]	1994	7.7	730	185	215	700
	1997	8.0	1120[c]	–	250	870
2nd winter wheat[a]	1994	6.7	635	185	235	585
	1997	7.0	1015[c]	–	275	740
3rd winter wheat[a]	1994	c. 10–15% below 2nd wheats	–	Same as 2nd wheats	Similar to 2nd wheats	–
	1997		–			–
Winter barley[a]	1994	6.0	550	185	200	535
	1997	6.3	625	265	220	670
Winter oats[a]	1994	5.5	505	185	155	535
	1997	6.5	650	265	180	735
Oilseed rape (winter)	1994	3.0	390	420	215	595
	1997	3.0	595	425	230	720
Winter beans	1994	3.6	340	345	125	560
	1997	3.4	390	385	165	610
Spring beans	1994	3.4	325	345	120	550
	1997	3.2	370	385	160	595
Peas (dried)	1994	3.5	350	345	205	490
	1997	3.8	430	385	245	570
Linseed	1994	1.8	190	430	195	425
	1997	1.5	225	515	170	570

[a]Cereals costed at feed, rather than milling or malting, prices.
[b]See Table 3.7 for explanation.
[c]Output = sale value + area payment (because of decisions pending the area payment is taken to be the same as for 1996).
– No estimate given.
From Nix (1993, 1996).

rotations for use at Boxworth (Tables 3.5 and 3.6). A glossary of the farm management terms used in these tables, and elsewhere in the text, is provided in Table 3.7.

The relative profitability of these rotations will obviously be determined by the relative prices received for their various components. To maintain his overall gross margins a farmer needs to retain such flexibility of cropping as will enable him to adjust his rotation to take account of price changes. Losses from take-all can greatly limit this flexibility. However, the current system of area payments (Table 3.7) in the EU countries to some extent buffers the effect on gross margins of falling yields in the second and third years of a cereal sequence.

Of the rotations listed in Table 3.4, the most profitable is so short that, in the longer term, it increases the risks of *Sclerotinia sclerotiorum* and clubroot (*Plasmodiophora brassicae*) in the rape, and of *S. trifoliorum* and the fusarium–phoma foot rot complex in the beans. While fungicidal control of *S. sclerotiorum* is possible in rape, it increases the costs of production and once the pathogen is present in a field it puts at risk most other non-cereal break crops. Although *S. trifoliorum* attacks

Table 3.4. Approximate gross margins for rotations based on combinable crops in England and Wales for 1993/94.

	Total gross margin (£ p.a.) on a 300 ha farm	
Rotation	'Actual'	'Theoretical'[a]
1. OSR[sc]/WW/WW/WW/WBa	175,080	183,240
2. WBe[tf]/WW/WW/Li[s]/WW/WW/WBa/OSR[sc]/WW	179,499	182,666
3. Pe[sf]/WW/WBe[tf]/WW/WW/WBa/OSR[sc]/WW	182,438	184,219
4. WBe[tf]/WW/WW/WBa/OSR[sc]/WW	183,750	186,125
5. WBe[tf]/WW/SBe[sf]/WW/Oa/WW/WBa/OSR[sc]/WW	185,833	185,833
6. WBe[tf]/WW/WW/OSR[sc]/WW/WW	186,250	191,000
7. WBe[tf]/WW/WW/OSR[sc]/WW	188,400	191,250
8. WBe[tf]/WW/OSR[sc]/WW	191,625	191,625

[a]The 'theoretical' gross margins show how the 'actual' gross margins would change if control of take-all reduced by 50% the differential between the yields of first wheats and the yields of second and third wheats.
WW, winter wheat; WBa, winter barley; WBe, winter beans; SBe, spring beans; OSR, oilseed rape; Oa, oats; Li, linseed.
Crops susceptible to take-all are underlined.
[c]Susceptible to club root; [f]susceptible to fusarium/phoma foot rot; [s]susceptible to *Sclerotinia sclerotiorum*; [t]susceptible to *S. trifoliorum*.

Table 3.5. Yields and gross margins of crops in arable rotations at ADAS Boxworth (10-year average yields at 1994 prices).

Crop	Yield (t ha^{-1})	Sale value (£)	Area payment (£)	Variable costs (£)	Gross margin (£ ha^{-1})	Theoretical gross margin[a]
OSR	3.21	417	420	215	622	–
WW	8.42	800	185	215	770	770
WW	7.34	698	185	235	648	699
WW	7.27	691	185	235	641	696
WBE	3.34	315	345	125	535	–
WW	7.85	746	185	215	716	716
WW	7.70	731	185	235	681	689
WW	7.13	677	185	235	627	662

[a]Assuming that 50% of the difference between first wheats and second plus third wheats has been recouped by the control of take-all.
OSR, oilseed rape; WW, winter wheat; WBE, winter beans.

few crops other than winter beans and clover, its presence limits the choice of non-cereal break crops to species susceptible to *S. sclerotiorum*. Moreover, if fusarium foot rot complex occurs in beans grown in too short a rotation, it limits the opportunity of introducing spring beans or, more importantly, peas into the system. Thus, even when a range of break crops is available, it is often judicious (to decrease disease risks to those crops) to lengthen the rotation by the introduction of short sequences of wheat or barley. Take-all reduces the profitability of introducing such measures.

Table 3.6. Actual gross margins of arable rotations and theoretical gross margins at ADAS Boxworth, assuming that take-all control had prevented 50% of the yield loss in second and third wheats.

Rotation	Mean gross margins (£ ha^{-1})	
	Actual	Theoretical
Rape/wheat/beans/wheat	660	660
Rape/wheat/wheat/beans/wheat/wheat	658	671
Rape/wheat/wheat/wheat/beans/wheat/wheat/wheat	656	673

Table 3.7. Explanation of farm management terms.

Term	Explanation
Area payment	A production support payment made by Government to compensate for the withdrawal of price support for agricultural commodities. The payment is made on the basis of the area of a crop grown irrespective of the yield of that crop. The payment may be made on uncropped land taken out of production under the 'set-aside' scheme.
Gross margin	The total value of the crop produced (including any area payment) less the variable costs.
Intervention	The creation of an artificial alternative outlet for grain (other than durum wheat, feed barley and rye) of specified minimum quality where purchase is made from public funds at pre-determined prices.
Intervention standards	Quality standard which must be met by grain sold into intervention; based, for wheat, on specific weight and milling quality.
Variable costs	Those costs which can be readily allocated to a specific enterprise and which will vary in direct proportion to the scale of that enterprise – e.g. fertilizers, agrochemicals, seeds, contract costs and casual labour (permanent labour, machinery, etc., are classed as fixed costs).

In October 1992, more far-reaching changes in rotational practice were ushered in by a revision of the EU Common Agricultural Policy (CAP) which reduced the support for market prices by reducing the price at which grain was brought into intervention and by imposing tighter intervention standards. At the same time a requirement was introduced that 15% of land on which area payments could be claimed (primarily land destined to be cropped with cereals, oilseed rape and pulses) should be set aside and either left uncropped (but with green cover) or used for growing, under contract, crops for uses other than human or animal feed. The set-aside land could either be rotated round the farm or left in set-aside for 5 years – but in the latter case a higher percentage (18% for the 1994 harvest) had to be entered for the scheme. There have been subsequent modifications of the regulations, which remain subject to change from time to time. From the point of view of take-all, the most important change in the set-aside scheme has been the ruling that, whilst the rotational set-aside period remains from 15 January to 31

August, the land need not be left uncultivated for the whole of the set-aside period. The cover crop can be sprayed off with non-selective herbicides from 15 April and ploughing is now permitted in May.

A survey of 800 farmers carried out for Schering Agriculture (Anon., 1992a) illustrated the way in which these changes influenced cropping patterns in the first year of the set-aside scheme. Acreage reductions of 9% in winter wheat, 43% in spring wheat, 21% in winter barley, 35% in spring barley and 25% in oilseed rape were reported, but there had been increases of 14% in beans and of 3% in linseed.

In theory, set-aside offers a range of rotational options to the farmer. Set-aside land can be used for growing crops (including cereals) for industrial purposes, or planted with a cover crop such as mustard, which is later ploughed in as green manure. Most frequently, however, the land is simply left uncultivated and unsown, the natural regeneration of volunteer plants and weeds providing the necessary green cover. The influence that take-all has on some of these options is discussed in Section 3.3.4.

The favourable experience with spring barley monoculture (Table 1.5) notwithstanding, it took longer for continuous wheat growing to become an accepted farm practice, even though continuous wheat had been grown successfully in Hertfordshire as long ago as the 1860s (see 1.1). Yield losses during the peak take-all years, before TAD (see 2.4.1 and 4.3.1) was established, constituted an unacceptable risk for many wheat growers. The following section describes how differences in soil and weather can influence the extent to which continuous cereal growing, or other cropping options, may be adopted with impunity.

3.3.2 *Effects of soil type and climate on rotational options*

Just how great the risks of severe take-all are in the years of peak disease depends on a number of factors, especially soil type, temperature and rainfall (see 2.3). Advisory experience and take-all data from ADAS Experimental Centres (discussed in more detail in 3.6) allow the following generalizations to be made.

On the very light Bunter Sandstone soils of Nottinghamshire, for example, and the peat soils of the East Anglian fenland, the risk of take-all is so great (except in very dry seasons) that two wheat crops are very rarely grown in succession. The well-structured Chalky Boulder Clay (Hanslope series) soils of Cambridgeshire, north Essex and west Suffolk, on the other hand, are less conducive to the development of the disease. Equally importantly, root growth is usually so good in these Hanslope soils (as long as drainage is not impeded) that plants continue to perform well even when a high proportion of those roots bear take-all lesions. On such soils it is thus possible to grow consecutive wheat crops through the peak take-all years without suffering catastrophic losses. Farmers on these soils were, therefore, at the forefront of the move towards continuous wheat growing. On soils with a high silt content (e.g. those in the north fens or the Old Red Sandstone soils of Herefordshire) take-all can be very severe in wet seasons. The same applies to the very poorly structured clays of the London Basin and parts of the West Midlands, where conditions not only favour infection but also so restrict rooting as to exacerbate the root loss caused by the disease. Surprisingly, given the traditional association of the disease with light, alkaline soils, severe attacks of take-all appear to be relatively infrequent on the chalk soils of, for example, the Yorkshire Wolds.

The complex interaction between soil and climate and its influence on rotational options is further illustrated by reference to observations made over the years at ADAS Gleadthorpe on the normally high-risk Bunter Sandstone soils of Nottinghamshire. In a series of spring barley trials carried out in the 1930s, very little disease developed at this site, which was later interpreted as resulting from the soils being too dry (see 2.3.3) during the spring and summer for the pathogen to develop on the roots (Yarham, 1981). In years when rainfall is well below average in both winter and summer, even winter wheat can escape severe attack if the crop is not irrigated.

By the 1970s wheat was being grown continuously by many farmers on a wide range of soils in eastern and southeastern England. There are anecdotal accounts that on the higher risk soils barley was often used to get through the take-all peak; this conflicts with more recent findings (see 2.4.1 and 3.2.1). On the fen peats and the lightest of the mineral soils, however, the risks were generally considered too great to embark on cropping wheat continuously (though continuous barley might be grown on the light soils).

Examples of where continuous winter wheat has been a viable proposition are occasionally reported in the farming press. At one farm in Wiltshire on thin, chalky soil some fields had been in continuous wheat for more than 10 years (Abel, 1994). Because of the exposure, the crops were difficult to establish and subsequently required careful husbandry, involving sprays seven times a year, including doses of magnesium, manganese and copper. On a farm in Northamptonshire on medium to heavy soil (Allan, 1996) some fields had been in wheat for 30 years and the shortest run of wheat was 10 years. Although yields declined during the first four consecutive wheat crops, they subsequently levelled off at about 10–15% less than those of the farm's first wheats, giving 9.0–9.8 t ha^{-1} during 1993–1995. This was achieved by inversion ploughing, subsoiling only if essential, avoiding early drilling, achieving good tilth and firm seedbeds, controlling couch grass 'to cut take-all', keeping the soil above pH 6.5, applying nitrogen as three split applications of urea at the top end of the recommended rate (no details were given, but Ministry of Agriculture, Fisheries and Food (1994) recommendations would suggest a total nitrogen application in the region of 230 kg ha^{-1}) to compensate for 'residual take-all', mapping weeds and the rigorous control of grass weeds by herbicides and cultural methods (including hand roguing wild oats).

3.3.3 Exploiting TAD

Whilst the natural biological control of take-all known as TAD is effective, its full implementation would require extensive continuous cropping, which has been impossible, mostly for reasons of economics. Many of the relatively small number of farmers who do grow wheat continuously are probably helped by TAD. However, the need for wheat, the lack of suitable, profitable alternative crops and the unsuitability of some wheat-growing areas for alternative crops mean that much wheat is still grown as part of short runs of cereals, in which the take-all problem persists and TAD is unlikely to develop.

The effectiveness of exploiting TAD as a method of controlling the disease is illustrated by photographs taken by the ADAS Aerial Photography Unit (Figs 3.5 and 3.6). Figure 3.5 shows a contrast of healthy and uniform first wheat with second wheat in which take-all was restricting rooting and making the plants very sensitive to variations in soil conditions. In Fig. 3.6 a second wheat with severe take-all and a fourth wheat that is not visibly suffering from the disease are shown in the same field. Given the link between severity of take-all and TAD (see 2.4.1 and 4.3.1) and the fact that in the first year of its appearance the disease often exhibits a very patchy distribution, it is not surprising that it can take some years for TAD to become established over a whole field. This is well illustrated by the sequence of aerial photographs of a field at Bradwell-on-Sea, Essex (Fig. 3.7) showing take-all peaking in different areas, until by the fifth wheat the whole field was in TAD. The next year, 1987, was such a bad one for take-all (Table 1.1) that the disease showed up again. A decreased yield in that sixth year (Fig. 3.7), associated with a moderate attack of take-all, seems to argue against the reliability of TAD as a method of disease control. Even when the peak disease years are passed, Ggt still remains an active pathogen on cereals grown in the field and if conditions are particularly conducive to its development (as they were in 1987) it can still cause appreciable losses, despite TAD. The only consolation is that in the absence of TAD the losses are usually even greater. This point is emphasized by the fact that the yield in 1987 was still considerably greater than yields in the peak disease years of 1984 and 1985. In the wet autumn of 1987, soil conditions were so poor at the time of sowing that the winter wheat crop failed and the field was re-drilled with spring wheat for the 1988 harvest. The damage caused by the autumn and winter cultivations was so great, however, that the spring crop suffered from the effects of poor soil structure, once again exacerbated by take-all.

At the Clonroche and Oak Park sites in the Republic of Ireland (Cunningham 1985; see also 2.3.2) take-all peaks were reached in the early years of spring barley monoculture, but the disease was much more severe at Clonroche than at Oak Park. Thereafter the disease declined at both sites, but subsequently there was considerable year-to-year variation in disease severity. Assuming that the decline in disease was TAD, then at Clonroche secondary disease peaks during TAD were almost as high as the primary peak that preceded TAD at Oak Park. That TAD followed a much lower 'peak' at Oak Park than at Clonroche is of interest, because it appears to conflict with the view that TAD is triggered by a severe attack of the disease and that a safe rotational position for a wheat crop would be after a preceding wheat that had suffered a severe attack of take-all. Hornby and Henden (1986) observed similar secondary disease peaks after a major peak early in a spring barley monoculture, and cautioned that decreases in disease after peaks of take-all may have a variety of causes and are therefore not an infallible guide to the onset of the disease-suppressive condition known as TAD. There are other instances of take-all incidence increasing and then decreasing without serious losses ever being experienced.

Grass weeds may interfere with the TAD phenomenon and this may explain patches of the disease at sites reported to be in decline (see 3.5.5). At Rothamsted, after 2 years of *Holcus lanatus*, a carrier of Ggt, severe take-all developed in a subsequent first wheat and continued in the next two wheat crops (see 4.3.1). Presumably the antagonistic microfloras associated with TAD in wheat do not operate where different, conducive microfloras associated with the different grass hosts predominate (see 2.4.1 and 5.5). It is assumed that a 1-year break in a sequence of cereals showing TAD will not totally destroy the decline factor(s), but the frequency or length of the breaks that would eliminate TAD are not known.

3.3.4 *Effects of different break crops*

The value of different cereals as break crops in relation to take-all is discussed in 3.2.1.

There was a dramatic increase in the area of oilseed rape in the UK in the mid-1980s from approximately 100,000 to 400,000 ha (Anon., 1987, 1988a), at a time when reports of severe take-all in second wheat crops began to increase. It appeared that take-all is more severe in second wheats after rape than after other break crops. Table 3.8 shows the yields of wheat following oilseed rape or beans at ADAS Boxworth. The marked difference between the yields of first wheats following oilseed rape and those following winter beans is due to a number of factors including the effects of residual nitrogen (which is greater after rape than after beans) and the fact that wheat after beans will tend to be later sown and, therefore, will often fail to achieve its full potential. The differences (though very small) in the yields of the second wheat crops may reflect, at least in part, differences in levels of root disease in the two rotations. Because of the different herbicide programmes used in the two crops, and because rape is often sown after minimal cultivation rather than after ploughing, cereal volunteers (potential carriers of Ggt) are more likely to survive in rape than in beans. (Plate 8.20 shows Ggt growing on a rape root in a pot test and, because of the possibility that rape could itself be a carrier, controlling cereal volunteers after sowing rape has been recommended in Germany; see 4.6.) Moreover, the greater amount of nitrogen under the rape will favour the saprophytic survival of the fungus, which will then build up rapidly in the following early-sown wheat. Taken together, these factors are likely to lead to more inoculum following a first wheat after rape than following a first wheat after beans. In neither case will the first wheat itself suffer greatly from the disease, but the greater amount of inoculum present at the end of the first season could lead to

Table 3.8. Field yields of wheat at ADAS Boxworth.

Wheat crop	Yield as % of continuous wheat (continuous wheat yielded 7.2 t ha^{-1})	
	After oilseed rape	After winter beans
First	117	109
Second	102	107
Third	101	99

Source: Bowerman (1989).

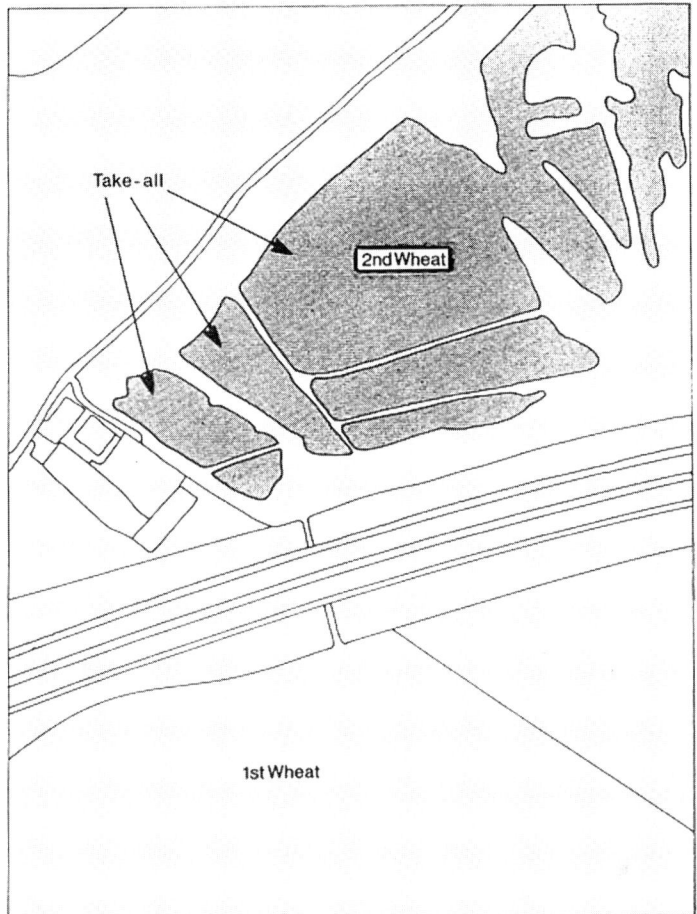

Fig. 3.5. An aerial photograph (opposite) and explanatory drawing (above) of fields of first and second winter wheat crops at Toddington, Bedfordshire.
In this area, take-all is favoured by a light soil, conducive to the spread of Ggt over the roots, and a clay subsoil that impedes drainage. Infection was severe in the second wheat, but first wheats south of the main road were uniform.
Photograph: ADAS Aerial Photography Unit.

greater losses in the second wheat after rape than in a second wheat crop after beans.

The effects of root exudates of oilseed rape, or products of its decomposition, on the control of pests and diseases have been the subject of some speculation and research. Rotations including *Brassica* spp. with high glucosinolate content might contribute to natural pest control systems, particularly through effects on weed seed germination (Brown and Morra, 1995). Ggt was shown to be inhibited by roots of seedlings of canola (*B. napus*) and mustard (*B. juncea*) rotation crops grown in the wheat belt of southern Australia (Angus *et al.*, 1994). Mature tissues, which contained low levels of isothiocyanates (volatile, antifungal breakdown products

of glucosinolates), had much reduced activity against the fungus. Therefore sowing fodder brassicas directly into pastures (which may support Ggt) is one possible means of suppressing this and other root pathogens (Kirkegaard *et al.*, 1996). In a survey of fields in southern New South Wales, take-all occurred in wheat after pasture, lupin or a cereal crop, but not after rape (Lemerle *et al.*, 1996). Although apparently contradicting the inoculum argument of the previous paragraph, a similar effect of increased take-all in a second wheat crop following rape might arise if this natural biofumigation decreased the microflora antagonistic to Ggt, as did commercial fumigants (see 4.2.1).

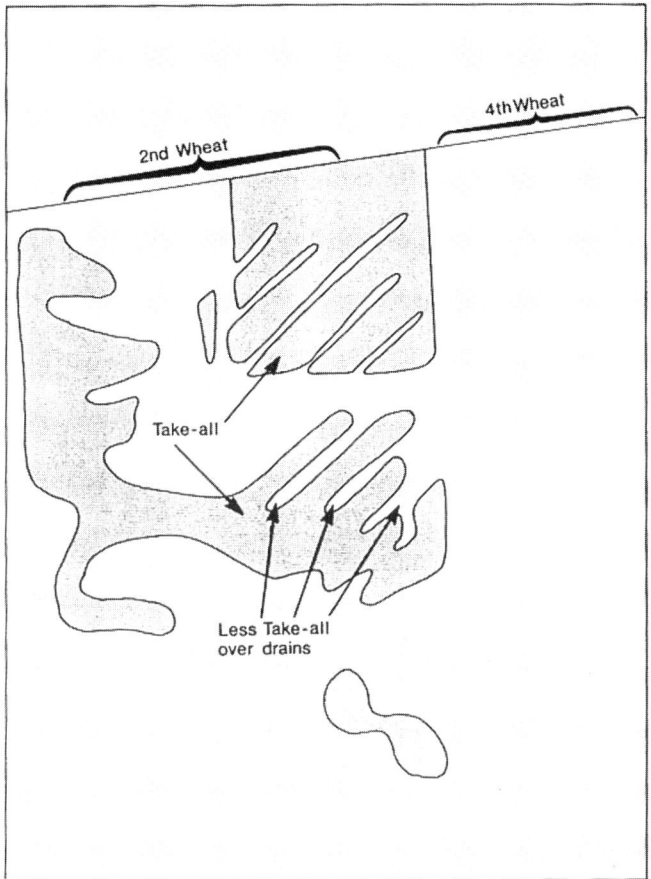

Fig. 3.6. An aerial photograph (opposite) and explanatory drawing (above) of second and fourth wheats at Boughton, Norfolk.
Take-all is visible in the second wheat, but not in the fourth. This appears to be a straightforward case of TAD, but the difference between the crops is probably accentuated because the second wheat was cv. Brock, which seemed to be particularly prone to the symptom of premature ripening.
Photograph: ADAS Aerial Photography Unit.

Catch crops of non-susceptible species could reduce the severity of the disease in the following crop by using up in the autumn the nitrogen necessary for the saprophytic survival of the fungus and by subsequently releasing that nitrogen to the benefit of the crop during the following season (cf. Garrett and Buddin, 1947).

The effect on take-all of set-aside as a component of arable rotation has been investigated in a number of experiments on ADAS (Yarham and Symonds, 1992; Yarham and Gladders, 1993) and Rothamsted sites. The first experiments were started before the management regime that was to be permitted on set-aside land had been agreed. The treatments therefore included a bare fallow, which proved to be most useful for evaluating the effects of the other treatments on the incidence of take-all. Set-aside and fallowing treatments were compared with a treatment

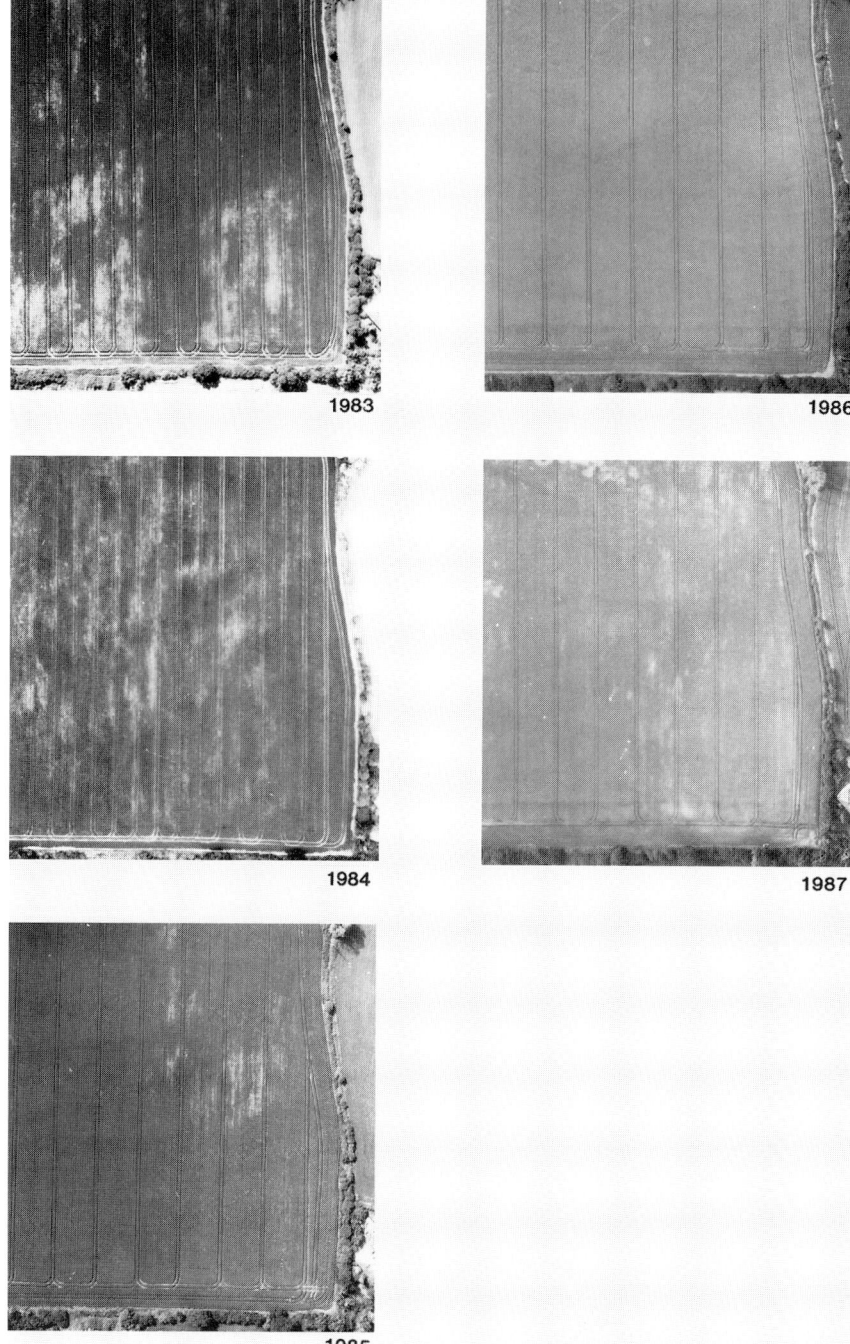

Take-all and Cereal Production Systems 127

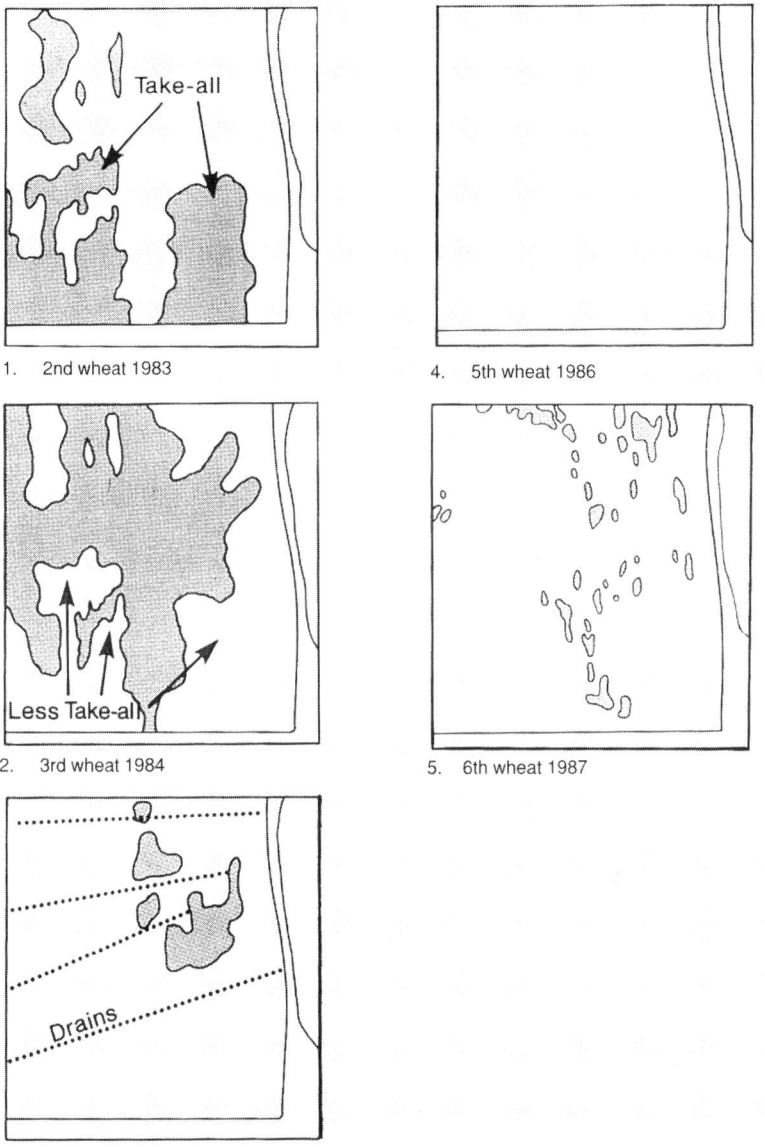

1. 2nd wheat 1983
2. 3rd wheat 1984
3. 4th wheat 1985
4. 5th wheat 1986
5. 6th wheat 1987

Fig. 3.7. Aerial photographs (opposite) and explanatory drawings (above) showing the development of take-all in five successive crops (second to sixth) of winter wheat at Bradwell, Essex.

In the sixth wheat (1987) shown in the last photograph, conditions so favoured take-all that patches of severe disease developed even though TAD was apparently operating.
Yields for the winter wheat crops in the sequence were:

            1982 (first wheat)   10.18 t ha^{-1}
            1983   8.50 t ha^{-1}
            1984   6.86 t ha^{-1}
            1985   6.86 t ha^{-1}
            1986   8.51 t ha^{-1}
            1987   7.75 t ha^{-1}

Photograph: ADAS Aerial Photography Unit.

Table 3.9. Effect of set-aside and fallow treatments on take-all in subsequent crops of winter wheat[a].

Number of wheat crops after treatment	Site	Take-all index (0–100)[b] in July (GS 69–75) following the treatments			
		Continuous wheat	Ryegrass	Natural regeneration	Bare fallow
First	Boxworth	23	20	20	9
	Drayton	36	29	22	21
	Woburn	55	12	17	4
Second	Boxworth	32	28	25	14
	Drayton	48	39	48	43
	Woburn	65	35	65	67

[a]The take-all indices are means from 3 years in three experiments at each site. The Woburn results were re-calculated from TARs (i.e. TAR ÷ 3) and include those referred to in Section 1.3, where only 1 year's data are quoted. In that particular experiment, the wheat crop grown in the set-aside year (1989) had a very high TAR (251, or 84 on the 0–100 scale) and so the TAR was low in the first wheat after the set-aside year, because of TAD, compared with the three-experiment mean given here.
[b]See 6.3.3.

maintained in continuous wheat throughout. Results that are relevant to current set-aside rules from the two ADAS sites (Boxworth and Drayton) where take-all was most in evidence and from a Rothamsted site (Woburn) are summarized in Table 3.9.

Differences between the sites and between years makes interpretation of these results difficult, but overall mean disease severity indices show that take-all in the first year after the set-aside treatments was less severe than in the continuous wheat crops and it was least severe after bare fallow. This intermediate severity suggests that volunteer wheat plants in the natural regeneration and ryegrass leys acted as hosts to Ggt. Ryegrass itself can be host to the pathogen, but it is not known what contribution this made to its survival in these experiments. *Phialophora graminicola*, an antagonist of Ggt (see 4.3.1), was present only where ryegrass was grown. Its incidence varied between years and between sites, but it was generally more in evidence at Woburn than at Drayton, and at Drayton than at Boxworth. At Woburn and Drayton, disease increased less between the first and second wheat crops after ryegrass than after natural regeneration or bare fallow, whilst at Boxworth there was little difference between treatments in this respect. This suggests that *P. graminicola* may have been having an effect on disease severity at the first two sites.

Site-to-site and year-to-year variation in the effects of set-aside on take-all is likely in large measure to reflect the number of volunteer cereal plants on the field in the set-aside year. Experience shows this to vary considerably not only between sites and years but also from place to place within the same field. During the winter of 1993/94, for example, set-aside fields were seen in which large areas were virtually devoid of any volunteers whilst others were completely covered. This could obviously contribute to the patchiness of take-all in a subsequent crop. More severe take-all could develop in areas where there were most volunteers, but if TAD had developed previously, such areas of volunteers could also lead to its preservation through the set-aside year. In the first year after set-aside, take-all is likely to be most severe in areas where there were most volunteers, although in subsequent years the reverse would be the case. Patchiness in the distribution of volunteers makes it more difficult to obtain a clear picture of the effect of set-aside treatments on take-all.

In the series of trials discussed above, the set-aside land was not ploughed until August and the effects of the treatments on take-all may therefore be different from those that would have been obtained with ploughing in the spring, as was allowed under revised regulations from 1993 to 1996. Results of experiments at Rothamsted and ADAS Drayton where natural regeneration set-aside was cultivated at different times during spring and summer, were variable. Decreases in take-all in subsequent crops, resulting from early cultivation of the set-aside land, were not as great as expected. It was concluded that in years when weather and soil conditions favour the disease, the consequences of delaying cultivation are likely to be greater than in hot, dry years that are less favourable to the disease (Jenkyn *et al.*, 1996). On organic farms only, it is permissible to sow a nitrogen-fixing legume in the set-aside year. Work at Rothamsted (Slope *et al.*, 1982) has shown that ryegrass alone or in combination with lucerne or clover is a better break in terms of its effects on subsequent take-all than a ley of lucerne or clover alone.

3.4 Crop Nutrition

The influence of crop nutrition on take-all has been reviewed by Huber (1981b) and Hornby (1985). The elements known to be essential for all higher plants may be classified on the basis of relative amounts required as major nutrients: carbon (C), hydrogen (H), oxygen (O), nitrogen (N), phosphorus (P), sulphur (S), potassium (K), calcium (Ca) and magnesium (Mg); or minor nutrients: iron (Fe), manganese (Mn), copper (Cu), zinc (Zn), molybdenum (Mo), boron (B) and chlorine (Cl). Deficiencies in both major and minor nutrients can increase infection and exacerbate yield losses caused by the disease. The central role of many minerals in disease response means that it is doubtful that a completely satisfactory understanding of most host–parasite interactions will be reached until the associated nutrient relationships are elucidated (Huber and McCay-Buis, 1993).

Reporting on pot experiments studying the effect of soil factors and fertilization on take-all, Trolldenier (1985) summarized much of what is known. Deficiences of the three major elements, nitrogen, phosphorus or potassium, lead to greater yield depressions in diseased plants than in healthy plants. A favourable effect of NH_4^+-N, as opposed to NO_3^--N, may be linked to decreased pH of the rhizosphere as well as increased numbers of rhizosphere organisms and increased root respiration. Higher rates of NO_3^--N have an ambivalent effect which both increases the disease and helps the plant to escape disease by producing more roots. The causes of the favourable effects of phosphorus and potassium are not well understood, but those of potassium are most obvious at moderate levels of disease. The favourable effect of optimal potassium as KCl may be through the ability of both ions to lower the water potential of the plant. These points are elaborated below.

3.4.1 Nitrogen

Reporting in 1948 to the Crop Husbandry Group of the Agricultural Development Council for England and Wales, S.D. Garrett noted that, whilst in Australia the crops which succumbed most readily to take-all were those on phosphate-deficient soils, in the UK the most susceptible crops were those on nitrogen-deficient soils. Since 1948 there has been a considerable increase in nitrogen usage on cereal crops. The increasing availability of low-cost nitrogen fertilizers, and the excellent responses obtained from their application to modern cereal cultivars, encouraged this increase. The trend was particularly marked in the late 1970s and early 1980s (Table 3.10). In recent years, however, the recognition that nitrates in drinking water might lead to health problems has led to demands for a tax on nitrogen fertilizers and to the establishment by EU governments of Nitrogen Vulnerable Zones and by the UK government of Nitrogen Sensitive Areas, within which farmers are compensated for strict limits being set on nitrogen usage. The problem in relation to take-all is that if restricting nitrogen increases take-all, then loss of roots will make it even more likely that more of the nitrogen applied will leach into ground waters (see 2.3.5).

In Table 3.1 there are opposite effects of nitrogen: high nitrogen levels favour the saprophytic survival of the pathogen in the absence of a host crop, but an infected cereal crop generally benefits from additional nitrogen, which helps plants to cope with root loss caused by the disease. The practical significance of this will be immediately apparent – the higher the disease risk, the higher will be the cost of nitrogen fertilizer to offset its effects. Whilst additional nitrogen will usually decrease the losses caused by take-all, the effects of this nutrient on the disease are complex and variable. In trials carried out by ADAS soil scientists between 1981 and 1986, yields fell in both the 'no N' treatment and the 'optimal N' treatments and then recovered as the fields passed through the years of high risk from take-all. However, the yield decreases were not so great where optimal levels of nitrogen were given (Table 3.11). The amount of N necessary to achieve optimal yields was greatest during the high risk years. At Saxmundham, Suffolk, where various rates of nitrogen were applied to both a first and a second wheat crop (Widdowson *et al.*,

Table 3.10. Nitrogen usage (kg ha^{-1}) on winter cereals in England and Wales, 1974–1984.

Crop	1974	1978	1981	1984
Winter wheat	90	125	162	187
Winter barley	91	106	143	150

Data from Church (1979), Church and Leech (1982), Leech and Chalmers (1985): cited in Yarham (1986).

1985) take-all severity decreased in the second wheat as the rate of nitrogen increased (Fig. 3.8).

Nitrogen will not eliminate take-all but can reduce its severity, as Fig. 3.8 clearly shows. However, this relationship is not always present or clear, possibly because disease severity varies considerably between years and locations. The greatest effect of nitrogen on take-all severity is seen when comparisons are made between treatments receiving no nitrogen and those receiving up to 100 kg N ha^{-1}. In the example in Fig. 3.8, take-all was severe and nitrogen had a dramatic effect. Unpublished data from the ADAS set-aside trials described by Clarke and Cooper (1992) show that where take-all is less severe a much smaller response may be expected (Table 3.12). In these trials differential nitrogen treatments were applied to wheat in the first year after set-aside, but only the nil nitrogen treatment was carried on into the second wheat. There was no effect of nitrogen rate in the first wheat crops, when take-all intensity was slight. Take-all was a little more severe in the second wheats. In 1990 there was no effect of nitrogen on take-all at any site. In 1991, however, at both Boxworth and Gleadthorpe

Table 3.11. Yields of successive wheat crops in relation to N fertilizer.

Crop	No. of crops in sample	Yield when no N applied (as % of first wheats)	Yield with optimal N (as % of first wheats)	Optimal rate of N (kg ha^{-1})
1st wheat	57	100 (6.43 t ha^{-1})	100 (9.40 t ha^{-1})	156
2nd wheat	34	70	87	183
3rd wheat	8	62	86	220
4th wheat	2	57	78	196
> 20 wheats	2	79	90	188

Compilation of mean data from ADAS trials, 1981–1986.

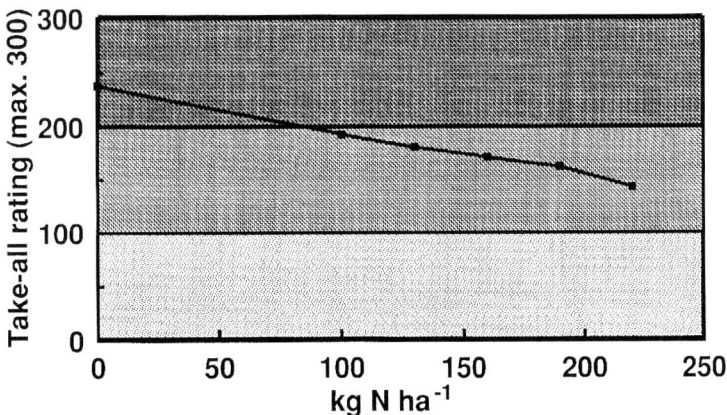

Fig. 3.8. Effect of amount of nitrogen on take-all severity in summer in winter wheat at Saxmundham, Suffolk.
Light shading, slight take-all; moderate shading, moderate take-all; dark shading, severe take-all. Data from Widdowson et al. (1985).

take-all tended to be negatively correlated with the amount of nitrogen applied in the previous year. The averages from all sites and years are shown in Fig. 3.9 to allow comparison with the Saxmundham data (Fig. 3.8). An effect of residual

Table 3.12. Effect of nitrogen rate on take-all in ADAS 'set-aside' trials.

Site	Wheat crop	Test year	Take-all index at the following N rates (kg ha^{-1})						
			0	80	120	160	200	240	280
BR	1st	1991	4.5	5.3	4.0	3.8	4.6	3.3	3.4
BW	1st	1989	25.0	24.1	22.0	22.6	22.3	22.6	21.5
	2nd	1990	27.4	26.0	23.2	22.1	22.1	23.1	20.6
	1st	1990	15.3	14.8	14.4	13.4	11.1	13.9	15.5
	2nd	1991	40.0	32.7	28.4	22.7	29.7	22.2	22.5
	1st	1991	18.9	14.2	19.7	17.7	22.8	16.6	15.9
DT	1st	1989	29.4	27.5	31.7	32.1	25.8	27.1	35.6
	2nd	1990	41.2	38.3	47.2	43.9	43.6	41.2	45.9
	1st	1990	23.6	21.3	23.1	25.4	24.9	21.3	20.6
	2nd	1991	29.1	30.9	31.3	28.6	28.6	30.8	32.0
	1st	1991	26.4	24.9	26.9	27.4	23.9	25.3	23.8
GT	1st	1989	13.8	10.0	9.8	15.5	17.0	17.5	16.7
	2nd	1990	31.8	26.3	29.2	24.9	24.7	26.3	30.3
	1st	1990	22.6	19.6	21.3	17.5	19.4	18.6	20.6
	2nd	1991	42.2	29.3	30.3	26.0	27.3	28.1	26.1
	1st	1991	26.4	24.9	26.9	27.4	23.9	25.3	23.8
HM	1st	1991	14.6	17.9	14.1	14.1	15.4	15.5	14.5

BR, ADAS Bridgets (Hampshire); BW, ADAS Boxworth (Cambridgeshire); DT, ADAS Drayton (Warwickshire); GT, ADAS Gleadthorpe (Nottinghamshire); HM, ADAS High Mowthorpe (North Yorkshire).

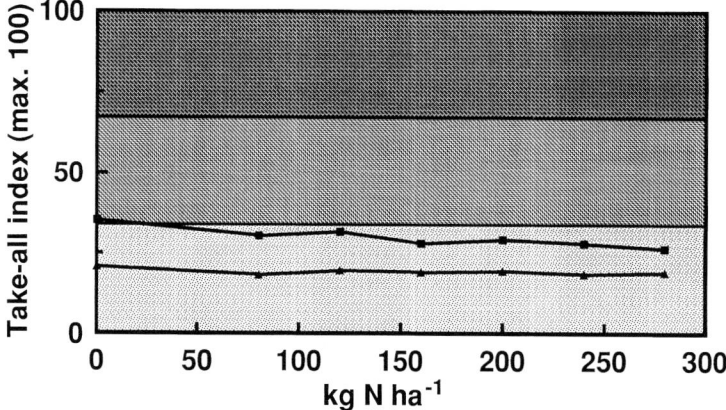

Fig. 3.9. Effect of amount of nitrogen on take-all severity in summer in winter wheat crops using data from ADAS set-aside experiments.
▲, first wheats; ■, second wheats.
Light shading, slight take-all; moderate shading, moderate take-all; dark shading, severe take-all. The data plotted are the mean data from Table 3.12.

nitrogen on take-all is also suggested by data from another trial at ADAS Boxworth, where take-all was usually least where large amounts of nitrogen had been applied to the previous crop as well as to the current crop (Table 3.13). The effect of nitrogen (at least of NO_3^--N) on the ability of a crop to withstand an attack of take-all may be due more to its increasing the numbers of healthy roots than to its decreasing the numbers of diseased ones (Table 3.14). Apparent decreases in disease in spring, when it is measured as percentage of roots infected, are discussed in Sections 2.6.3 and 6.3.3.

Table 3.13. Effects of residual nitrogen on take-all severity in winter wheat cv. Mercia in July 1990 at ADAS Boxworth.

Treatment (kg N ha^{-1})		Take-all index (0–100)
1989	1990	
210	160	44.0
210	240	49.5
210	0	50.5
210	80	53.2
120	160	53.3
0	240	56.1
120	240	58.2
0	160	62.7
120	80	63.7
0	120	65.4
0	0	65.4
0	120	66.1
0	80	66.6
0	280	67.1
120	0	69.2
180	0	69.6
150	0	69.6
90	0	70.5
60	0	74.3
0	40	77.8
	F	1.32
	P	0.2
	SED	11.87
	c.v. (%)	21.9

Data from L. Vaidyanathan (personal communication).

Table 3.14. Effect of nitrogen on take-all and root production in ADAS Soil Science Department trials carried out in the Eastern Region in 1983.

Nitrogen rate (kg ha^{-1})	% Roots with take-all	Roots per shoot	Shoots m^{-2}	Roots m^{-2}	Healthy roots m^{-2}
0	43	13	461	5802	3290
120	33	15	702	10252	6910

Mean data from L. Vaidyanathan (personal communication).

The form of nitrogen can also influence disease development. There is evidence from many countries that use of NH_4^+-N, rather than NO_3^--N, will decrease the severity of take-all (e.g. Cook and Reis, 1981; MacNish and Speijers, 1982), although in pot experiments it actually increased the disease (Darbyshire *et al.*, 1979) and NH_4^+ in the rhizosphere was associated with more extensive lesion development on axenically grown wheat roots (Brown and Hornby, 1987). Hornby and Goring (1972) suggested that there is an optimum ratio between the NH_4^+-N and NO_3^--N for minimizing the effects of take-all.

The use of NH_4^+-N, as opposed to NO_3^--N, is less important in the UK than in Australia and the USA. It has been argued (L.V. Vaidyanathan, personal communication) that in many UK soils the rate of nitrification is likely to be so rapid as to convert the NH_4^+ ions to NO_3^- before they can have much effect on the disease. In practice most farmers continue to use ammonium nitrate as their nitrogen source (though in recent years increasing use has also been made of urea). Ammonium sulphate represented less than 0.5% of nitrogen used on cereals in 1984 (Hornby, 1985), but it is now becoming more widely used, because manufacturers producing it as a by-product are beginning to exploit the agricultural market and extraction of SO_2 from power station emissions is leading to an increased need for sulphur-containing fertilizers (Table 1.17 and Section 3.4.2). Early advice in Europe for controlling take-all recommended the use of ammonium sulphate, unless soils were too acid (Hornby, 1985). It may have affected the disease by causing a decrease in rhizosphere pH, although the evidence for this occurring in UK soils is weak and variable.

In two experiments comparing the effects of nitrogen fertilizers applied in spring to winter wheat at Rothamsted, take-all was most severe with calcium ammonium nitrate ('Nitro-Chalk') and least with ammonium sulphate, but the differences were not significant (Gutteridge *et al.*, 1987). Ammonium chloride, urea and 'Nitro-Chalk' with potassium chloride were also included in these comparisons. Trials in the late 1980s at six ADAS Experimental Centres and the Morley Research Centre showed no consistent differences in effects on take-all in second wheats of: (i) calcium nitrate; (ii) a liquid fertilizer containing ammonium nitrate + urea; and (iii) the same liquid fertilizer + 'Didin' (a nitrification inhibitor). Only at the Morley Research Centre did treatment (iii) result in take-all being significantly less than where calcium nitrate had been used. In a further series of trials there was less take-all at two of three ADAS sites where ammonium sulphate (with or without 'Didin') was used than where ammonium nitrate was applied (D.J. Yarham, unpublished data).

3.4.2 Other major nutrients

Apart from nitrogen, there has been relatively little research on the effects of the major nutrient elements on take-all since that described by Huber (1981b). At that time, N, P, K, S and Mg had been shown to decrease and Ca and K shown to increase take-all severity. Also, environmental and cultural conditions were known to modify the individual and collective effects of these elements.

Phosphorus
Phosphate-deficient soils tend to favour Ggt and take-all, and application of phosphate fertilizer to such soils is usually beneficial within limits. Severe take-all patches are sometimes found to be associated with areas of phosphate deficiency. Soils with an adequate supply of phosphate (above index 2; Ministry of Agriculture, Fisheries and Food, 1980) are less prone to take-all than those showing a deficiency (index 0–1). When P-deficient soils of index 0 received applications of 120 kg P_2O_5 ha^{-1} as superphosphate in experiments at Rothamsted (Slope *et al.*, 1984), TARs were approximately halved, making the treated soils similar to soils with adequate reserves of P, i.e. of index 3. On Chalky Boulder Clay in Cambridgeshire there was a negative association between take-all and extractable P in the range 14–33 µg g^{-1} and a positive association between the disease and exchangeable K in the range 155–282 µg g^{-1} (J.A. Catt, personal communication). Take-all was severe when wheat was grown after many years of spring barley, especially where levels of soluble P were less than 10–12 µg g^{-1} (Fig. 3.10). If a wheat crop is grown after a break-crop in a phosphate-deficient soil, inoculum builds up rapidly. Addition of phosphate to a second crop to rectify the deficiency will not reduce the risk factor as the inoculum will already be present and is not affected by the nutrient. Severe disease occurs earlier in deficient soils than in soils rich in phosphate.

In an Australian glasshouse experiment using a slightly acid grey sand and six levels of P, the incidence and severity of take-all were related to P supply and to P status of the wheat test plants (Brennan, 1988). The percentage of roots infected where P was severely deficient and where it was adequate were 35% and 13.7%, and 24% and 2.3% for seminal and nodal roots, respectively. In pots containing

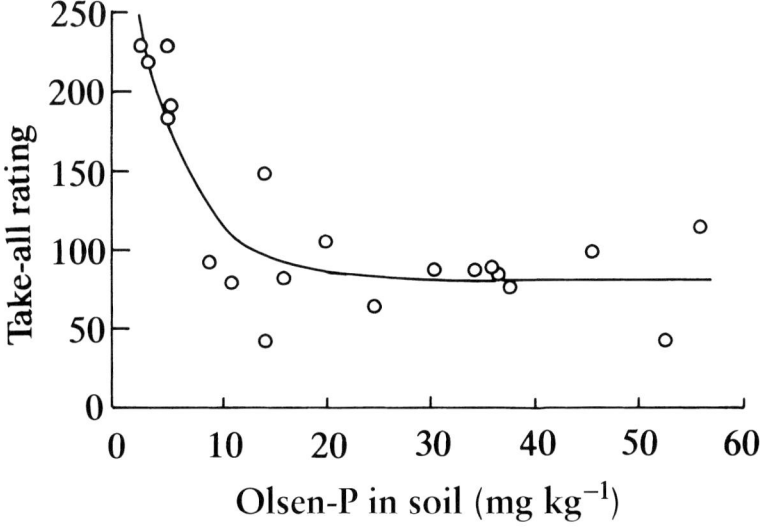

Fig. 3.10. Relationship between take-all intensity (take-all rating in summer 1992) and bicarbonate-soluble phosphorus (Olsen-P, measured in September 1993). Reproduced from Gutteridge *et al.* (1996).

3.1 kg of the same soil, adequate amounts of nitrogen and phosphorus to overcome take-all for 46 days were 400 mg and 100 mg, respectively (Brennan, 1989b). Increasing amounts of P and N at concentrations below these levels had the effect of decreasing the length of proximal lesions and increasing the length of root between the crown and the proximal lesion. The main beneficial effect of P fertilizer appeared to be the stimulation of root development, thereby allowing some roots to escape infection.

In the field in Australia applications of P (as superphosphate) in the range 0–40 kg P ha^{-1} were made annually to a soil initially severely deficient in P (Brennan, 1995). Where take-all was of moderate–low severity in the nil P treatment, adding P fertilizer controlled the disease. With more than 65% of roots infected in the nil P treatment, the effectiveness of P in decreasing the disease declined. Even with applications of 40 kg P ha^{-1}, take-all still affected plants and caused large yield decreases, underlining the need to keep inoculum densities low by suitable rotations. Greater applications of P would be uneconomic and inappropriate for trying to control Ggt by maximizing the plant's resistance and facilitating disease escape.

A vesicular-arbuscular mycorrhizal fungus (*Glomus fasciculatus*) appeared to have the same effect on take-all as applications of P (50 µg g^{-1}) to a P-deficient sandy soil containing 0.5 µg P g^{-1} soil (Graham and Menge, 1982). Each treatment increased the P status of the host and induced decreases in root exudation which were correlated with subsequent decreases in disease severity. If the treatments were applied together, the P application severely inhibited colonization of roots by *G. fasciculatus*.

Potassium
Observations on the effect of potassium on take-all, which is more complex than that of phosphate and depends on the availability of other elements (Huber, 1981b), have been sparse (Trolldenier, 1982). Mattingley *et al.* (1980) showed that a large dressing of potassium (120 kg ha^{-1}) increased take-all in phosphate-deficient soils and this effect was not prevented by the application of a small dressing of phosphate (15 kg ha^{-1}). In pot experiments investigating the effects of two levels of potassium nutrition (0.75 g and 3.0 g K$_2$O per pot as KCl) and five levels of Ggt (2, 6, 15, 30 and 60 colonized cereal grains per pot) on spring wheat in Germany, grain yields were always more affected than straw yields (Trolldenier, 1982). The level of K had no effect at very high inoculum densities, but at high densities the high K level increased straw yield and ear number per pot in relation to healthy plants (there was a grain yield response to this level of K, but it was less than in healthy plants). At low and moderate inoculum densities, straw yield and numbers of ears were not affected by take-all. At moderate densities grain yield and 1000-grain weight were depressed more at the lower level of K. Further work in pots (Prade and Trolldenier, 1990) confirmed that K deficiency also resulted in increased denitrification in the rhizosphere. Denitrification, measured by the acetylene inhibition technique, was also increased by infection by Ggt. Infection where K was deficient resulted in most denitrification. Trolldenier's (1982) measurements of K in grain and straw, perhaps because they were made using plants

grown in pots, did not support the hypothesis of earlier workers that yield losses are due to restricted nutrient uptake.

Sulphur
Despite low sulphur requirements (10–30 kg S ha^{-1}) cereal crops are susceptible to S deficiency (Withers and Sinclair, 1994). In cereals in Britain, however, S deficiency is a recent development and a qualitative model (McGrath *et al.*, 1995) predicted that 11% of British land (southeast Scotland, the Scottish borders, East Anglia, the Welsh borders and southwest England) was at a high risk of S deficiency and that a further 22% was at medium risk. The model also predicted that should the goal of 40% of the 1980 level of SO$_2$ emissions be achieved, then nearly all the cereal growing regions of Britain would be at risk of S deficiency. Any effects of power station emissions on take-all may not be equatable with applications of S to deficient soils (which decrease take-all, Table 1.17). It was suggested (Huber, 1981b) that limited observations on S and take-all probably reflected the sufficient availability of S in most soils. Should more British crops be grown in S-deficient soils, take-all is likely to increase, but to what extent is difficult to predict.

3.4.3 Minor nutrients

In general, at the time of Huber's (1981b) review, there was little information about the effects of most micronutrients on take-all. Cl had been shown to decrease take-all severity, and a combination of Zn, Mn, Cu, Mo and B had increased take-all without nitrogen, but had decreased it when applied with ammonium sulphate.

Chlorine
Chloride-containing fertilizers were found to decrease take-all in the USA and this was initially explained by their helping to maintain the optimum ratio between the NH$_4^+$-N and NO$_3^-$-N mentioned in Section 3.4.1 by slowing down nitrification, i.e. the conversion of NH$_4^+$ to NO$_3^-$ in the soil (Christensen and Brett, 1985; Powelson *et al.*, 1985). Since NH$_4$Cl fertilizer appeared to enhance the ability to yield despite take-all, this hypothesis was later rejected (Christensen *et al.*, 1990) in favour of the view that the chloride effect was more likely to be due to the fact that Cl$^-$ uptake lowers the osmotic potential of the leaf tissue, thereby enabling the plants to retain turgor despite root loss due to the disease. Chloride-containing fertilizers had no more than slight, transitory effects on take-all in ADAS trials in the UK (Werker and Gilligan, 1990).

Manganese
Since Huber's (1981b) observations, the effects of manganese on take-all have received considerable attention, especially in the USA and Australia (e.g. Rovira *et al.*, 1985; Wilhelm *et al.*, 1988; Marschner *et al.*, 1991). Large areas of southern Australia are marginal or deficient in manganese (Neate, 1994). Huber and McCay-Buis (1993) suggested that this element has a central involvement in take-all development and severity, exerting its greatest influence by delaying disease development and restricting lesion size, as well as increasing host root

growth and influencing ammonium metabolism. Put simply, the argument is that manganese availability is inversely related to take-all. Disease occurs only when the interaction of pathogen, environment and microbes induces a deficiency of manganese in the plant that is sufficient to decrease resistance. The severity of the disease reflects the extent of influence of various factors on manganese availability and its interaction with the physiology and resistance of the host. (These ideas are expressed generally in Fig. 2.1.) In the traditional view of take-all, the disease was associated with alkaline soils of low fertility and it is pointed out that these are also the conditions characteristic of immobilized manganese and manganese deficiency. Conditions that increase nitrification tend to decrease manganese availability and vice versa (Table 4.16); thus manganese is linked with other major nutrient factors in take-all, as for example the observation that the ammonium form of nitrogen decreases the disease (see 3.4.1).

The following paragraphs summarize a review of manganese and take-all by Huber and McCay-Buis (1993) and one or two subsequent references. In the USA, manganese deficiency symptoms were picked up in soya beans growing where take-all had been severe in a previous wheat crop. There were no differences in concentrations of manganese in the soil, but populations of manganese oxidizing organisms were ten times greater in the rhizospheres of deficient soya beans than in the rhizospheres of plants that appeared normal. These observations were later confirmed with winter wheat plants in commercial fields in Indiana and were interpreted as indicating a dynamic role for biological components in influencing manganese availability.

Manganese is not only an integral component of enzymes; it is also an activator for many enzyme-catalysed reactions in plants. In this role it interacts with nitrogen metabolism and is involved in respiration, photosynthesis (which affects root exudates, which in turn affect rhizosphere microorganisms), hormone metabolism, and the synthesis of secondary metabolites (e.g. phenols and lignins) involved in the plant's defence against pathogens. Seeds of cultivars of soft red winter wheat containing 'high' levels of manganese (in the range $53.1–68.3$ $\mu g\ g^{-1}$) produced, overall, plants with increased seedling vigour, more extensive root systems, 11% less whiteheads and 165 kg ha^{-1} more yield than seeds of the same cultivars with 'low' levels of manganese (in the range $37.8–55.7$ $\mu g\ g^{-1}$), when grown in soils naturally infested by Ggt (McCay-Buis *et al.*, 1995). Treating seed with manganese decreased take-all and increased yield, but not as consistently as seeds with high manganese content. Applying manganese directly to soil or as a seed treatment produced variable results in controlling take-all. Applying manganese via leaves is ineffective because it is immobilized in the phloem.

In terms of the pathogen, virulence is correlated with manganese oxidation. Combinations of isolates that were virulent at 25°C or 30°C with isolates that were virulent at 8°C or 15°C resulted in increased disease severity or severity at the level of the most virulent isolate. Furthermore, a synergistic effect resulting in severe take-all occurred in inoculated, non-sterile, conducive soil, which contained a large population of manganese-oxidizing microorganisms.

Mn^{4+} (MnO_2) is reduced by microorganisms both anaerobically and aerobically and by plant roots, which absorb Mn mostly as Mn^{2+}. Birnessite, a common Mn oxide mineral in soils, was formed in a black fungal precipitate obtained by growing Ggt *in vitro* in a nutrient broth containing $MnSO_4$ (Schulze *et al.*, 1995). A direct test of the hypothesis that Ggt decreases the host's defences prior to invasion, by catalysing the oxidation of soluble Mn^{2+} to insoluble Mn^{4+} on and around roots, was achieved using micro-XANES spectroscopy (Schulze *et al.*, 1995). (XANES stands for 'X-ray absorption near edge structure' and this particular spectroscopic technique has the potential for measuring Mn oxidation states directly within plants and soil without pretreatment of the sample.) In tests on agar containing Mn^{2+}, the Mn around roots infected by Ggt was predominantly Mn^{4+} and precipitates containing Mn^{4+} occurred in the roots. Whereas Ggt oxidizes manganese at some distance from its mycelium and may thus compromise a host's resistance mechanisms, Ggg, which is weakly pathogenic or non-pathogenic, oxidizes manganese only around lobed hyphopodia and related structures. More bacteria capable of oxidizing manganese in the rhizosphere have been found associated with Ggg than with Ggt.

The link between NH_4^+-N and manganese availability is explained here in general terms. Wheat can use NO_3^--N and NH_4^+-N equally well. In soil the process of nitrification (biological oxidation involving the bacterial genera *Nitrosomonas* and *Nitrobacter*) converts NH_4^+ to NO_3^- within about 3 weeks. Consequently there is the need to use nitrification inhibitors – for example, to stabilize NH_4^+-N in organic manures – in order to maintain the beneficial effects of the ammonium form, such as decreasing take-all. NO_3^- in the plant is translocated to the leaves and other above-ground tissues to be reduced to amino nitrogen and metabolized to amino acids with little modification of the rhizosphere microflora. However, as NH_4^+-N is absorbed by roots the pH of the rhizosphere decreases. The NH_4^+-N is metabolized to amino acids in the roots, and root exudates rich in organic compounds are produced. The reducing environment so created stimulates manganese-reducing organisms (Sarniguet, 1990), suppresses manganese-oxidizing organisms and increases manganese, phosphate and zinc uptake by the plant, thus limiting take-all.

Application of manganese on the seed often stimulated oxidizing organisms in the rhizosphere, limiting the availability of the element to a very short period. Applying manganese-reducing bacteria at the same time as manganese was more effective. This increased manganese uptake in the plant. In every case, however, the effect was a short-term one and manganese-reducing bacteria have failed to persist throughout the long period during which take-all can develop on wheat.

Many components of the soil environment correlate with nitrification, but they also correlate more consistently with manganese availability (see Table 4.16). Lime changes soil pH and redox potential, stimulates nitrification and increases populations of manganese-oxidizing microorganisms; all these changes also decrease manganese availability. Loose seedbeds increase nitrification, aeration, the oxidation of manganese and take-all. Organisms that oxidize manganese increase in the rhizosphere during runs of two to five consecutive crops of wheat, but decrease to levels similar to those of a 'climax ecosystem' under long-term monoculture.

In the Australian experiments, foliar application of Mn had little effect on disease but application of $MnSO_4$ to the seed or, more particularly, to the soil effectively reduced take-all. In Australian soils showing manganese deficiency, fewer and smaller take-all lesions occurred on roots when manganese was supplied to the soils as $MnSO_4$ (Graham and Rovira, 1984). Different explanations were advanced, such as a direct toxic effect on Ggt, or an indirect effect through the physiology of the plant, resulting in modification of root exudates and, in consequence, the rhizosphere microflora suppressing ectotrophic growth of Ggt, or in stimulation of the host's defence reaction through increased lignin production. Wheat genotypes were also found to react differently to take-all in manganese-deficient soils. One genotype efficient at taking up manganese was resistant, whilst two others that were inefficient were very susceptible to take-all and a fourth was intermediate for both characteristics (see also 3.2.2). Where manganese was not deficient, all four genotypes became infected similarly with take-all (Wilhelm *et al.*, 1990). Recent results indicate that in certain wheat genotypes bacterial microfloras may play a role in the expression of manganese efficiency and tolerance of take-all (Rengel *et al.*, 1996).

Observations that the virulence of Ggt is correlated with its manganese oxidizing ability led to the hypothesis that physiological immobilization of manganese around the infection court in advance of Ggt hyphae could compromise plant resistance mechanisms such as lignituber development and secondary metabolism (Graham, 1991). Differences in manganese oxidizing ability were shown on PDA containing different concentrations of manganese, as Mn^{2+}, by two isolates of Ggt from Rothamsted Farm (Z. Rengel, personal communication). One produced a strong, even deposition of a brown oxide (Mn^{3+} and/or Mn^{4+}) to the full extent of the colonies at concentrations ranging from 10 to 200 mg Mn l^{-1}, whilst with the other the colouration was weak and localized. The radial extension of both isolates was much decreased at concentrations of 400 mg l^{-1}.

In European wheat growing, manganese deficiency is not a major problem and severe take-all can develop in soils with sufficient available manganese. Applying manganese fertilizer has been only partially successful in controlling take-all. Therefore, the question of whether manganese is a requisite for achieving good control of the disease, or only an indicator of important activities such as those of manganese-reducing microorganisms antagonistic to Ggt, is still not clear.

Relationships between manganese concentration in soils (as Mn^{2+}) and the conduciveness of these soils to take-all were described in France (Lucas and Sarniguet, 1990). Soils that were fertilized with ammonium sulphate and became less conducive to the disease showed also a higher level of Mn^{2+} (80 ppm vs. 57 ppm). This may have been due to decreased pH (from 6.6 to 5.7), but Dommergues and Mangenot (1970) suggested that reduction of manganese in soils between pH 5.5 and pH 8 was mainly due to microbial activity. Furthermore, when adding inoculum of Ggt to soils collected from a maize monoculture (pH 5.8), or a wheat monoculture (pH 6.4), in which wheat seedlings were grown for 40 days, the Mn^{2+} content of the soils increased respectively from 39.1 to 47.0 ppm in the soil conducive to take-all (maize monoculture) and from 38.5 up to 59.7 ppm in the suppressive soil (wheat monoculture). Because Ggt is an oxidizer of manganese (Schulze *et al.*, 1995), the reduction of manganese was probably due to other microorganisms that were more numerous or more efficient in the suppressive soil (Lucas and Sarniguet, 1990).

Manganese deficiency is the most common minor element deficiency observed in UK cereals (McGrath and Hellon, 1986), but there is little data to support or to reject the view that manganese plays a significant role in outbreaks of take-all. A seven-point scale (0–6) has been used for estimating the susceptibility of wheat cultivars to manganese deficiency; cereals with 6–16 ppm Mn in leaf dry matter would normally be deficient and those with 30+ ppm non-deficient (Batey, 1971).

In England and Wales, 61,000–101,000 ha were estimated to be susceptible to manganese deficiency in 1971 (Batey, 1971). Only on certain soils are large areas likely to be affected; most commonly these tend to be peaty soils, cultivated podzols or mineral soils with high organic matter content with pH in excess of 6.0. The problem is worst in wet seasons and cold, wet weather, and on limed soils that were previously acid. Impeded drainage, puffy soils and poorly compacted soils all tend to increase manganese deficiency. In the 1990s attention was drawn to deficiency problems in cereal crops on light soils and some fen soils. Sometimes these were attributed to previous frost damage and farmers were advised to apply a corrective spray of manganese (ADAS, 1992a). Manganese deficiency was described as widespread and a little earlier than usual in cereals in spring 1996, even in fields not usually associated with the problem (Abel, 1996). Deficiency symptoms can occur also on severely deficient fields in the UK in the autumn: early-sown winter barley on high pH (> 6.5) sandy or organic soils is most likely to show general paling of foliage, usually in areas, developing to yellowing of younger leaves and death of older leaves, and the deficient crops are more sensitive to frost damage and winter kill. Spraying deficient areas with manganese sulphate (4.5 kg ha^{-1}) between October and early December, which includes a risk of scorch in frosty weather, has been recommended (ADAS, 1992b). Crops likely to benefit from autumn applications, especially oats and barley on light sandy soils, sandy organic land and in fields with a history of manganese deficiency, comprise about 5% of the country's arable area (Anon., 1992b), which suggests a bigger deficiency problem than that identified in 1971.

Other minor nutrients
Deficiencies of certain other minor nutrients can also exacerbate the disease. In growth chamber and field experiments in the USA, zinc and copper suppressed take-all (Reis *et al.*, 1982) and were most effective at low disease intensity. It was also suggested that iron, as well as manganese, might be important in conditions where they were deficient, although they were not effective in the growth chamber experiments. More recent work on copper and zinc has centred on Canada and Australia. A 'special fertilizer' containing these and other minor and major elements has been patented in China for use against *G. graminis* (Table 4.6).

That copper deficiency exacerbates take-all (Wood and Robson, 1984) was confirmed in Australia; at higher rates of copper the lengths of proximal lesions on seminal roots were decreased, and the length of uninfected root between the crown and the proximal lesions on nodal roots was increased. Because the intensity of the lesions on nodal roots was not affected, it was suggested that a kind of disease escape was operating. In copper-deficient soils in parts of the Wimmera, in the Victorian wheat-growing belt of Australia, the effect of applying copper to correct the deficiency was improved by additional application of gypsum (Gardner and Flynn, 1988). One explanation for this is that calcium ions compete with copper ions for binding sites on the soil's organic matter, thus increasing the availability of copper. Copper-deficient plants produce less lignin, which provides a barrier to infection by Ggt (cf. discussion in manganese section above); copper, like boron and manganese (Graham, 1983), is involved in the formation in the plant of phenolics that have antifungal properties. The incidence and severity of naturally occurring take-all in wheat were decreased by applying copper fertilizer to deficient soil in Western Australia (Brennan, 1989a). There, higher levels of fertilizer than were needed to achieve maximum grain yields did not further decrease the plants' susceptibility to take-all, and copper at five times the normal fertilizer rates was apparently not toxic to inoculum of Ggt in the soil. Copper is a key element for disease control in wheat and barley in Alberta, Canada, where 10% of the crop land is deficient in available copper, and take-all was controlled by additions of 11 kg of $CuSO_4$ ha^{-1} to soils with < 0.6 ppm available Cu (Evans *et al.*, 1995). Applications of copper to deficient soil increased grain yields and decreased the percentage of roots with take-all from 15– > 80% to 5–30% in a trial on eight wheat cultivars (Evans *et al.*, 1992).

As well as suppressing take-all, lowering the soil pH increases the availability of iron, which may have a role similar to that of manganese, and the availability of one may affect the availability of the other (Graham, 1983). Zinc deficiency has been shown to increase the severity of take-all in Western Australian soil (Brennan, 1992). In pot experiments in the UK, Zn deficiency acted synergistically with take-all to increase the number of fluorescent pseudomonads in the rhizosphere, suggesting that, as with manganese (see previous section), bacterial microflora play a role in the expression of micronutrient efficiency and tolerance of take-all in some wheat genotypes (Rengel *et al.*, 1996). Little information is available on other trace elements and take-all.

3.5 Husbandry

3.5.1 *Sowing date*

Whilst early sowing can be an advantage in first and long-term wheats, it can have a negative effect in those years in sequences of wheat crops when take-all is usually at its peak (Table 3.15); this can sometimes result in substantial losses in yield (Table 3.16). The demands on farm labour can be great in autumn, especially on an arable farm on heavy land, and it is important to sow autumn crops before the weather deteriorates. Many UK farmers, therefore, like to sow as much as possible of their cereal acreage in September. On poorly structured, heavy soils there is not only the risk of early sowing increasing the severity of take-all, but also the risk

Table 3.15. Effects on wheat yields of sowing in September compared with sowing in the first half of October in East Anglia.

When sown	Yields in t ha^{-1} (number of crops in the sample)			
	1st wheats	2nd wheats	3rd wheats	4th wheats
September	7.64	7.07	6.96	7.33
	(108)	(83)	(42)	(94)
October	7.44	7.32	7.09	7.17
	(103)	(115)	(81)	(140)

Survey data from a Suffolk-based agricultural consultants' group, via G. Cousins, cited in Yarham (1986).

Table 3.16. Effect of sowing date on the severity of take-all and the yield of infected wheat at ADAS Rosemaund.

Crops and harvest years	Take-all index (0–100)				Grain yield (t ha^{-1})	
	November–January		Harvest			
	Sept. sowing	Oct. sowing	Sept. sowing	Oct. sowing	Sept. sowing	Oct. sowing
Second wheats						
1983	70	22	93	72	2.07	5.05
1984	19	1	22	17	10.38	10.51
Third wheats						
1983	33	20	77	53	8.08	8.57
1984	18	5	93	59	7.35	9.41

Source: R.W. Clare, I. Ap Dewi and D.J. Yarham (unpublished data).

that, by opting for later sowing, a wet October may force such a delay in drilling that yields suffer substantially, despite take-all being less severe. A recent trend towards sowing spring cultivars of quality wheat in the autumn, because they can be sown much later in the year than winter cultivars, might also benefit from the fact that later sowing decreases take-all.

Many trials have shown that early sowing can increase the severity of take-all (e.g. Powelson *et al.*, 1985; Clare *et al.*, 1986; Prew *et al.*, 1986). When winter barley was grown after a sequence of spring barley crops, earlier sowing resulted in more severe take-all (Jenkyn *et al.*, 1992b). The sowing dates of the preceding crop had a continuing residual effect and in subsequent winter barley crops TAD was more evident in sequences of early-sown crops than in sequences of later-sown crops. These effects probably reflected differences in the amount of inoculum left after the preceding crop and the rate of development of TAD.

Although differences were not significant at the conventional 5% level in a crop sequence experiment at Rothamsted, soil bioassays after harvest in August

Fig. 3.11. Soil infectivity, determined by bioassay, in soil after early-sown (E) and late-sown (L) crops of winter wheat at Rothamsted.
▫, plants; ▨, roots.
Bar = SED, shown only where $P \leq 0.05$.
D. Hornby and R.J. Gutteridge (unpublished data).

suggested that a first winter wheat crop sown early (in mid-September of the previous year) left more inoculum in the soil than one sown a month later (Fig. 3.11). It is likely that, with delayed sowing, conditions for infection of subsequent wheats were less favourable and less inoculum of Ggt survived to infect than where crops were sown early. The incidence and severity of disease in April and July in subsequent wheat crops that were sown either early or late are given in Fig. 3.12. The differences between early and late sowings persisted through to July in both the second and third wheats. Soils were much more infectious after the second wheats than after the first wheats (Fig. 3.11) and therefore after second wheats comparisons were made between soil infectivity after harvest (August) and infectivity prior to a late sowing in October. The two bioassays of the early-sown crop resulted in similar infectivities, but infectivity decreased between the two bioassays where there was no crop. This suggests that the early sowing maintained soil infectivity through root infection. In support of this view, infection in the third wheat crop sown on 15 September 1992 already had 40% of plants infected with take-all at GS 12 (second leaf expanded) at the time the later crop was sown. This crop later developed severe take-all, whilst take-all in the third wheat crop sown late was slightly less severe.

In UK conditions, therefore, provided the soil is warm (above 10°C) and moist when seed of susceptible hosts is sown in fields infested with Ggt, then germination is quick (5–10 days) and take-all is frequently well established 4 weeks after sowing. With later sowings, however, lower soil temperatures delay germination and create conditions less favourable for infection by Ggt. There is also less inoculum of the take-all fungus available to cause infection. Consequently, such crops are usually infected much less than early-sown crops, and because the disease usually develops little over the winter period, differences in take-all, relating to sowing dates, can be detected in the spring.

At harvest, some grain will inevitably be spilled. The amount lost by spillage depends on the conditions at harvest and how the combine harvester is set. At this

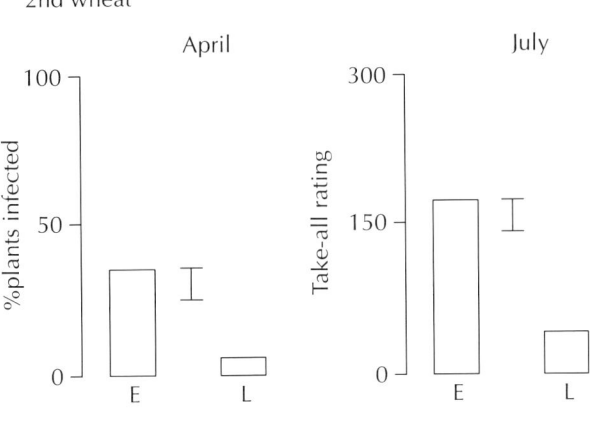
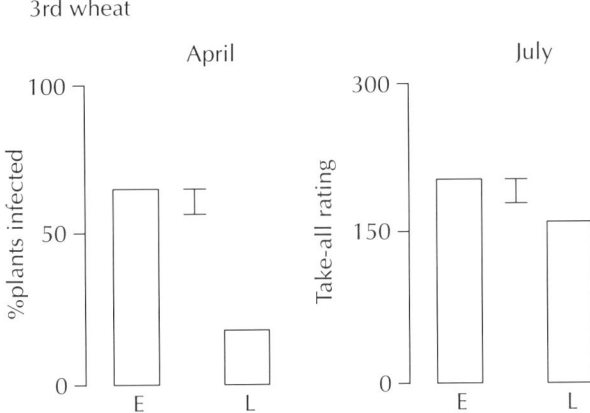

Fig. 3.12. Incidence and intensity of take-all in April and July in second and third wheat crops that were sown either early (E) or late (L). Bars = SEDs.

time the soil is warm and if it is also moist the lost grain soon germinates. The resulting volunteer (self-seeded) plants act as hosts to many pathogens, including Ggt. At Rothamsted, field plots were ploughed after harvest of a first winter wheat crop in preparation for sowing a second wheat crop early (September) or later (October). On some of the later-sown plots, volunteers were strictly controlled, whilst on others wheat seeds were broadcast at 50 kg ha^{-1} to simulate an average volunteer population. Later, the 'volunteers' on some plots were destroyed by power-harrowing prior to late sowing in October. Take-all was assessed in the following July and TARs were found to be greatest in the early-sown crop and least in the later-sown crop in which volunteers had been controlled (Fig. 3.13). Where volunteers had been encouraged, there was an intermediate amount of disease. Volunteers can therefore partially negate the beneficial effect (in terms of decreased take-all) of delayed sowing (volunteers may have a role in infection by ascospores, Section 2.2, and are a factor in a model discussed in 2.6.2).

In the data behind the modelling work (Colbach, 1995) referred to in 2.6.8, sowing date always affected parameter c_1 (i.e. primary infections) whereas parameter c_2 (i.e. secondary infections) was influenced only on sites that were the most favourable to take-all (La Verrière and Le Rheu in France in 1992, which had the highest infection rates). Early sowing always increased c_1, which is consistent with previous results (Hornby *et al.*, 1990a) and the fact that early sowing allows a longer period favourable to infection before winter. The effect on c_2 was variable: positive at Le Rheu in 1992 and negative at La Verrière. There was a positive correlation between plants m^{-2} and parameter c_1 at the sites most favourable to the disease, but the influence of this covariable on c_2 was variable. It is likely that high plant density early in the crop, when roots are still few and short, increases the probability of contact between inoculum in the soil and living roots, while the effect would be less consistent when the root system is well developed.

Parameter c_1 was increased and c_2 was decreased by high nitrogen rates and both were reduced by ammonium fertilizer. Nitrogen can stimulate both the

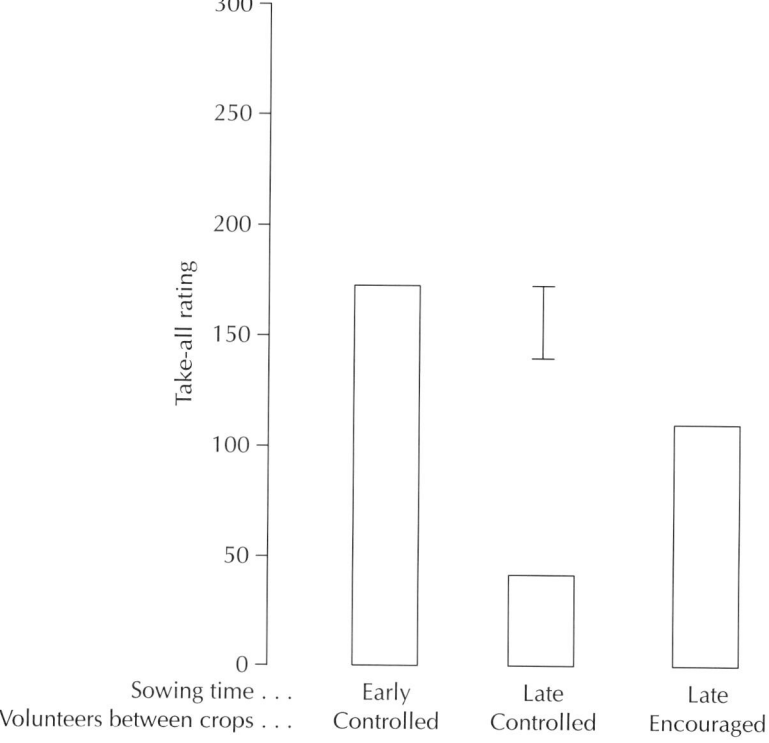

Fig. 3.13. Intensity of take-all in July in second wheat crops sown early or late following summer fallow in which wheat volunteers were either controlled or encouraged. Bar = SED.

pathogen and the antagonistic microflora (Sarniguet et al., 1992a,b). Early infection of seminal roots may allow fluorescent pseudomonads to develop in lesions to the extent that they interfere with the later spread of the pathogen, especially where ammonium nitrogen has been applied. Straw treatment did not affect the disease and treatments other than sowing date were usually significant only when sowing date was significant. Sowing date therefore seemed to be the dominant factor and its interactions with the other treatments were the strongest of all interactions. Several treatments had their greatest influence, or were significant, only on sites favourable to the disease. Thus, each factor seemed to increase risk due to the other effects, and factors with small effects could only influence disease if those having large effects were also favourable to its expression.

3.5.2 Method of cultivation

There are several reviews of the effect of cultivation method on take-all (e.g. Yarham, 1981; Neate, 1988). The effects recorded have been very variable both between and within the countries where trials have been carried out. In the Pacific Northwest of the USA, for example, direct drilling was found to increase take-all (Moore and Cook, 1984), whilst in Czechoslovakia it was said to decrease the severity of the disease (Novotný and Herman, 1981). In the UK, trials in the 1970s (see below) failed to show any strongly consistent effect of cultivation technique on take-all (Yarham and Norton, 1981), but in more recent work the disease was often observed to be more severe after tine cultivations than after ploughing, although the effect was generally small (Jenkyn et al., 1988). As yet there have been very few cases where a severe attack of take-all could be attributed to the incorporation of straw. It is possible, however, that the widespread and long-term adoption of 'non-plough' straw incorporation techniques following the ban on burning (Table 1.17) could lead to more problems on certain soil types.

Direct drilling produces a firmer seedbed than traditional cultivation techniques and this may explain why, since the first UK trials in the 1960s (Brooks and Dawson, 1968), there have been reports of less take-all after direct drilling than after mould-board ploughing. However, in ADAS trials in the 1970s (mainly on heavy soils and on sites where straw had been burned), final levels of disease (percentage of plants infected), averaged over 29 comparisons, were: plough, 35%; reduced cultivation, 37%; direct drill, 36% (Yarham, 1981). In investigations by ADAS and IACR-Rothamsted the effects of straw incorporation on take-all have also been rather inconsistent. Mould-board ploughing, as compared with other methods of cultivation, has consistently decreased take-all less in trials where straw has been incorporated than in trials where it has been burned. One possible explanation for this may be the effects of straw disposal and cultivation practices on volunteer plants, which play an important role in the survival of Ggt in the post-harvest period. Populations of volunteer plants are likely to be greater where straw has not been burned and ploughing will control such volunteers more effectively than other cultivation methods. The effect of burning on volunteers may also explain how, at some sites and in certain seasons, take-all in tine-cultivated plots has been decreased by straw burning. Other factors, such as the effect of cultivation and straw disposal practices on the time of crop emergence, may also influence the severity of the disease in such comparisons.

3.5.3 *Quality of cultivation*

There is general recognition amongst farmers that good drainage, good soil management and the preparation of a good, firm seedbed are important if take-all losses are to be minimized. In this respect the method of cultivation has generally been perceived as being of less importance than excellence of cultivation.

Wheat in the first year after a break will often produce a good crop, despite faulty husbandry. In the second or a subsequent year, however, with take-all destroying part of its root system, a much higher standard of soil management is required to produce a satisfactory yield. On Chalky Boulder Clay, where drainage had been neglected and the mole drain channels had become blocked, a first wheat crop appeared satisfactory. However, an aerial photograph taken in 1987 (Fig. 3.14) shows that a second wheat performed very badly, except in the immediate vicinity of the underlying tile drains. Examination of the root systems showed them to be severely infected with take-all (Yarham *et al.*, 1989). Another outbreak of

Fig. 3.14. An aerial photograph of a winter wheat crop on poorly draining Chalky Boulder Clay at Nuthampstead, Hertfordshire, on 14 July 1987.
Take-all is showing as pale areas where mole-draining had been neglected for the previous 10 years. The dark lines show better growth over the tile drains.
Photograph: ADAS Aerial Photography Unit, previously published in Yarham *et al.* (1989).

(a) May

(b) late June

(c) 4 years earlier – drainage system installed

Fig. 3.15. Aerial photographs of a field at Hevingham, Suffolk.
On 9 May (a) and in late June (b), 1988, severe take-all in winter wheat shows as pale areas where the soil structure was damaged by lifting potatoes in the previous, wet July. In (c), the installation of a tile drainage system 4 years earlier is shown; the system had not been extended into the area where the potatoes were grown and so the upper edge of the area of severe take-all is coincident with the boundary of the undrained part of the field.
Photographs: ADAS Aerial Photography Unit.

take-all following poor husbandry is shown in Fig. 3.15. Two years before the aerial photographs (a) and (b) were taken, the field had carried peas at the top end and potatoes at the bottom. The potatoes had been irrigated, after which it had rained. When potato lifting began, the soil was still too wet to work easily and lifting was abandoned until the soil had dried out. As in the previous case, the first wheat presented no problems, but the second wheat showed a large area of severe take-all where the potatoes had been harvested under wet conditions and where soil structure had, in consequence, been damaged. The soil varied from well structured boulder clay at one end of the field to sandy loam over gravel at the other. Not surprisingly, take-all was worse at the lighter end of the field, but straight-line boundaries revealed by aerial photographs in early May (Fig. 3.15a), and late June (Fig. 3.15b), were obviously of man-made origin rather than soil-type boundaries.

These examples underline the need for good soil management to minimize losses from take-all. In the past few years, however, ADAS has encountered a number of severe attacks in situations where the farmer has been justly proud of his cultivations and seedbed preparation. It seems that it is possible to prepare a fine, friable seedbed which is ideal for sowing with a modern air-drill, but which is also ideal for the spread of Ggt on roots. The use of the para-plough to improve soil structure has also been associated occasionally with increased severity of take-all. The effect of a loose seedbed in aggravating the severity of take-all is well documented and was recognized by Australian farmers even before the true cause of the disease was known (Griffiths, 1933; Garrett, 1981). Subsequently, similiar observations were made elsewhere, including in the UK, although it must still be emphasized that compaction and bad drainage on heavy land may also result in serious losses from take-all (Yarham, 1981).

There have been little good experimental data on this subject, perhaps because of the difficulty of achieving different levels of soil compaction experimentally, and so the types of observation described above have been the basis of recommendations to farmers. An opportunity arose to record take-all at Wrest Park, Silsoe, Bedfordshire, where a site was prepared in 1986 to compare compaction with and without tractor traffic. The soil was clay (typical calcareous pelosol of the Evesham Series) and the treatments are described in Table 3.17. Details of some of the treatments and of a 12 m gantry used in the zero traffic regime are in Chamen *et al.* (1992). In the first year, differences in compaction of the top 20 cm of the soil were seen, but the effect at lower depths was minimal. Maximum differences in compaction occurred after 3 years (W.C.T. Chamen, personal communication). In the early years of this experiment, the crop rotation was chosen to avoid the damaging effects of take-all, but in autumn 1991 winter wheat was sown after spring barley and patches of stunted, prematurely ripened plants were observed in July 1992 and take-all confirmed. Where the husbandry technique had produced a light, fluffy seedbed, as in the gantry ploughed or 'gantry speculative' plots, take-all was significantly more severe than in the other treatments where the seedbed had been consolidated more (TARs 189 vs. 81 and replicates 6 and 12, respectively, SED 34).

Table 3.17. Effects of conventional and zero traffic on seedbeds and soil consolidation in an experiment at Silsoe Research Institute, Bedfordshire, begun in 1986.

Treatment	Description	Result
1. Tractor, ploughed	Ploughed in the autumn and normal traffic over the plot. Tractor tyre positions varied such that the whole plot was covered by wheelings at some time during the season. Tram lines were used at later growth stages.	Consolidated soil, fine seedbed
2. Tractor, cultivated	Tine cultivated in the autumn, then as in 1.	Most consolidated soil, firm seedbed
3. Gantry, ploughed	Ploughed in the autumn, but all operations were from a gantry, i.e. there were no tractor wheelings on the plot.	Least consolidated soil, fluffy seedbed
4. Gantry, cultivated	Tine cultivated from a gantry in the autumn, then as in treatment 3.	Consolidated soil, firm seedbed
5. Partial gantry	Tractor ploughed in the autumn, then all subsequent operations until harvest were from a gantry.	Soil more consolidated than in treatment 3, coarse seedbed
6. Gantry speculative	Use of a modified commercial gantry. Cultivations were by a powered rotary machine from the gantry, to a depth equivalent to the ploughing.	Similar to treatment 3

3.5.4 Soil pH

Soil management practices include the maintenance of suitable pH as well as cultivations. Severe attacks of take-all are sometimes associated with acid patches in fields and one crop consultant in the UK is known regularly to advocate liming for the control of the disease. Since take-all is normally considered a disease of alkaline soils (Reis *et al.*, 1983) this appears at first sight to be anomalous. 'Acid patch take-all' has, however, been reported from Australia and Sivasithamparam and Parker (1981) pointed out the great variability amongst strains of *G. graminis* in respect of their tolerance of pH ranges. Careful assessment by ADAS of a sandy clay loam site ploughed out of old pasture in East Suffolk showed a negative correlation between pH and take-all (log transformation of percentage of roots with take-all) at pH values above 5.4 (below pH 5.4 the pathogen is likely to be inhibited). The correlation accounted for 35% of the variance. It was considered likely that a third factor (e.g. soil organic matter) was influencing both pH and disease.

3.5.5 Herbicides

The effects of herbicides on root diseases have been reviewed by Altman (1985). Yarham (1981) briefly reviewed literature more specifically concerned with the effects of weeds and herbicide use on take-all and this is mentioned further in Section 4.2.3. Perennial grass weeds such as couch grass (*Elytrigia repens*) play an important role in carrying infection through non-susceptible break crops and, by competing for nitrogen and other nutrients, weeds can increase the susceptibility of the crop to the disease. However, the application of certain herbicides (e.g. mecoprop and benzoic acid derivatives) have occasionally been shown to exacerbate the effects of the disease (Nilsson, 1973a,b; Tottman and Thompson, 1978). Trifluralin, diquat + paraquat, and (more particularly) glyphosate increased take-all on unsterilized (but not on sterilized) soil, possibly because the herbicides caused 'a shift in soil microbial populations away from those antagonistic to the pathogen' (Mekwatanakarn and Sivasithamparam, 1987). Glyphosate, a translocated non-residual phosphonic acid herbicide, may indirectly affect take-all by modifying the soil microflora, killing plant tissues that stimulate manganese-oxidizing microorganisms, complexing with soil manganese to decrease its availability to the host, or increasing the inoculum potential of the pathogen on weakened grasses (Huber and McCay-Buis, 1993). Certain soil-borne fungi act as glyphosate synergists and enhance its herbicidal effect. *Pythium* spp. are examples and so may be *Gaeumannomyces graminis*, *Rhizoctonia* spp. and *Fusarium culmorum* (Descalzo et al., 1996).

Cases of severe take-all in first wheats on couch-infested fields occur frequently and farmers often fail to recognize the importance of linking their weed control and rotational practices so that action to control couch is taken before, and not after, a break crop. In the spring of 1989 a wheat crop in Kent was so severely affected with take-all that the farmer ploughed up part of it and sowed maize instead. The wheat was the first crop after beans which had been heavily infested with couch grass and the field had been sprayed with glyphosate before the wheat was sown. The wheat in the aerial photograph of a field in Bedfordshire (Fig. 3.16) was grown after a 2-year ley that followed a long sequence of cereal crops. Because the ley had been infested with couch, it was sprayed with glyphosate before the field was cultivated in preparation for the wheat. Ggt was apparently carried through the leys on the couch and built up further on the dead rhizomes after spraying. The take-all patches towards the bottom of the photograph coincide with the worst area of couch infestation (Yarham et al., 1989).

Carrying out ADAS advice to control couch before a break crop, rather than before a cereal, may not always solve the problem. In March 1990 severe take-all patches occurred in a first wheat after oilseed rape in the Weald of Kent. The crop preceding the rape had been infested with couch, but the field had been sprayed with glyphosate before the rape crop was taken. The take-all patches in 1990 were coincident with the couch patches sprayed off in 1988, and the remains of the old, dead rhizomes could still be found in them (J. Batchelor, personal communication). Test plants grown in compost amended with fragments of the dead rhizomes became severely infected with take-all.

Couch grass not only serves to carry the pathogen through break crops, but also sometimes appears to prevent or delay the development of TAD. Figure 3.17 shows a number of fields on boulder clay in west Cambridgeshire. The farmer had experienced take-all in the fields in the previous 2 years in third and fourth wheat crops. He had continued to crop wheat on the land, assuming that by the fifth crop TAD would decrease the severity of the disease. The fields were, however, infested with couch and part of the area had been treated with glyphosate in the autumn before the fifth crop was taken. The worst of the take-all-affected areas coincided

3.6 Effects of Take-all on the Sensitivity of Crop Yield to Climate

In considering the effect of soil type on take-all and why some rotational options are better in some areas than others (see 3.3.2), reference was necessarily made to climate. The profitability of arable farming depends on seasonal variations in the weather which affect not only crop growth but also the incidence and severity of diseases. Because take-all is not at present controlled by fungicides, it acts as a very significant factor in increasing the vulnerability of crop yields to the vagaries of the climate. Assuming that much of the difference in yield between first and second wheats at ADAS Rosemaund was due to take-all, the data in Tables 3.18 and 3.19 serve to illustrate this point. The greatest difference between first and second wheats at Rosemaund was recorded in 1983, a year characterized by weather which was: (i) mild and wet in late autumn, which would favour infection of the

Table 3.18. Seasonal variation in yields of first and second wheats at ADAS Rosemaund.

	Yield (t ha^{-1})		
Wheat crops	1982	1983	1984
First	8.2	8.8	10.8
Second	6.6	5.8	9.4
Second as % of first	80%	66%	87%

Table 3.19. Weather data for ADAS Rosemaund.

Variable	Harvest year	Oct.	Nov.	Dec.	Jan.	Feb.	March	April	May	June	July
Temperature (mean °C)	1982	8.1	7.1	−0.6	2.0	4.8	5.7	8.1	11.0	15.2	16.0
	1983	9.9	7.6	4.1	6.7	1.5	6.3	6.5	10.1	14.1	18.9
	1984	9.9	6.9	5.3	3.6	3.2	4.4	7.9	14.3	16.5	17.5
	30-year mean	9.9	6.1	4.4	3.2	3.5	5.3	7.7	10.8	13.8	15.5
Rainfall (mm)	1982	67	30	100	57	34	78	24	24	175	28
	1983	53	78	62	43	18	43	100	86	16	81
	1984	77	25	91	87	35	52	6	63	47	6
	30-year mean	54	67	65	62	47	50	43	55	49	53
Sunshine (h day^{-1})	1982	3.5	1.9	1.2	1.5	1.2	4.4	5.5	6.9	5.1	5.0
	1983	2.4	2.1	1.0	1.6	2.4	2.8	5.0	3.1	5.5	7.2
	1984	3.8	1.1	2.0	2.0	1.7	1.6	7.4	5.6	7.4	8.5
	30-year mean	3.0	2.1	1.5	1.6	2.3	3.5	5.0	6.1	6.6	6.1

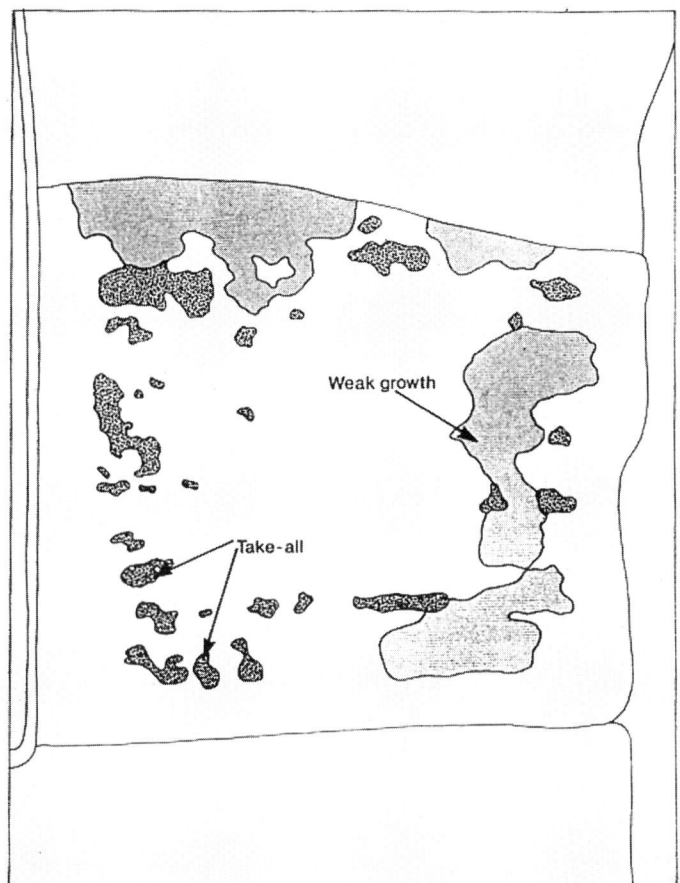

Fig. 3.16. An aerial photograph (opposite) and an explanatory drawing (above) of a winter wheat crop at Billington, Bedfordshire, on 16 April 1983.
Take-all shows as pale areas that coincide with areas that had been infested with couch grass in a preceding 2-year ryegrass ley, which had followed a long run of cereal crops. The field had been treated with glyphosate before it was ploughed and drilled with wheat.
Photograph: ADAS Aerial Photography Unit, previously published in Yarham et al. (1989).

seminal roots by Ggt; (ii) wet and dull in April and May, which would favour infection of the developing nodal roots; and (iii) hot and dry in June, which would exacerbate the effects of earlier root loss. Thus, whilst 1983 was potentially a higher-yielding year than 1982, the yields of second wheats in that year were substantially less than the 1982 yields. (Elsewhere in the UK, take-all was severe in 1983: Plate 7.18 and Table 1.1.) Note the comparison with 1984 when the very dry April would have checked disease development, and (although July was very dry) rainfall in June was not much less than average.

In East Anglia in 1987 take-all caused considerable damage, but in 1989, although there was much take-all in late winter/early spring, comparatively few crops suffered severely from the disease later in the season. A possible explanation

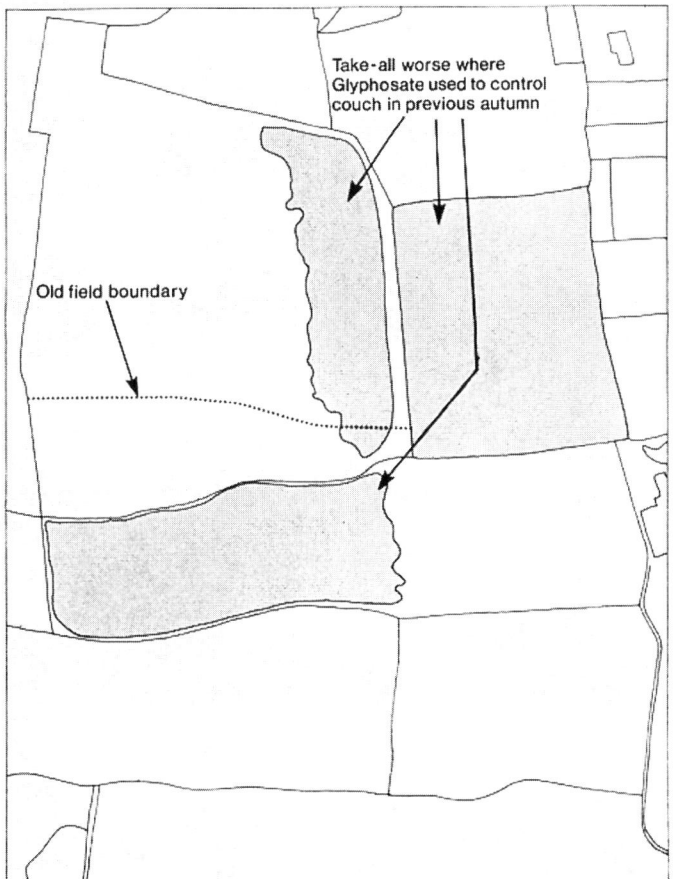

Fig. 3.17. An aerial photograph (opposite) and an explanatory drawing (above) of a fifth wheat crop, on a site previously infested with couch grass, at Comberton, Cambridgeshire, on 16 July 1988.
Severe take-all shows as pale areas where glyphosate was used to control couch in the previous season.
Photograph: ADAS Aerial Photography Unit.

is suggested by the soil moisture deficit (SMD), which in April was actually rather less (and hence more conducive to take-all development) in 1989 than in 1987 (Fig. 3.18). In 1989 the SMD rose very rapidly in May, when nodal root production would still be very active. Because the SDM would have been greatest in the upper soil, Ggt would have been checked there, allowing crops to grow away from the disease. A few crops that had been so badly damaged earlier in the season that they were unable to produce enough replacement roots could account for the few severely diseased crops in 1989. In 1987, the soil remained moist throughout much of May, but crops were then subjected to severe moisture stress in June, when it was too late for them to replace roots destroyed by the disease.

Such seasonal variations in take-all make it impossible to predict the severity of the disease in any particular season, or to find any consistent relationship between disease levels in late winter and yield. Take-all and the severity of its effects on yield vary widely from site to site even when cropping sequences at the various sites are comparable. Some of the site-to-site variation will be due to climatic differences, but these are difficult to separate from the effects of soil type. Data from three ADAS Centres (Table 3.20) show how take-all never built up to serious

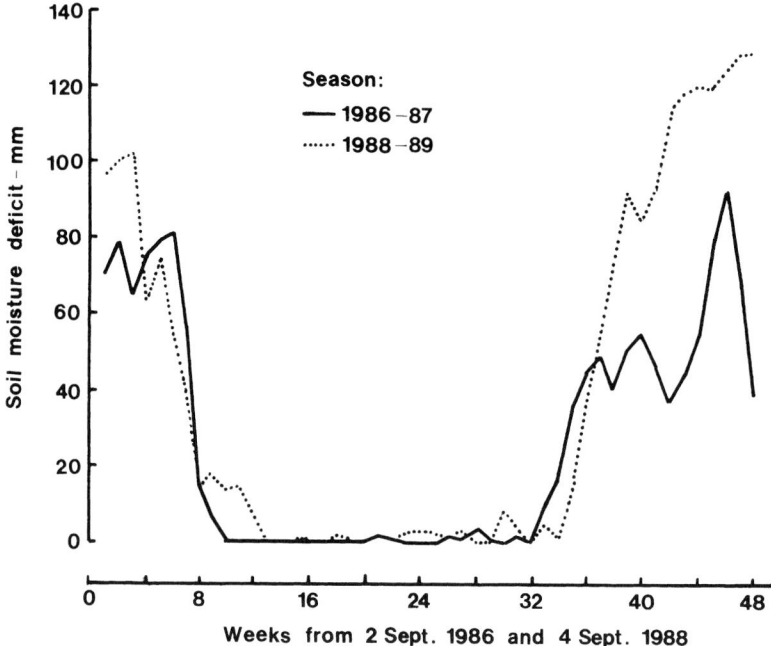

Fig. 3.18. Soil moisture deficits recorded in 2 years at Honnington, Suffolk. Meteorological Office Excess Calculation System (MORECS) square 141 (Mid Suffolk).

Table 3.20. Soils, weather and take-all severity in third wheats on ADAS farms, 1966–70.

Characteristics	High Mowthorpe, North Yorkshire	Boxworth, Cambridgeshire	Rosemaund, Herefordshire
Soil type	Silty clay loam	Well-structured calcareous clay	Silty clay loam
Soil series	Andover	Hanslope	Bromyard
Parent material	Chalk	Chalky Boulder Clay	Old Red Sandstone
Mean annual rainfall (mm)	763	593	669
Mean annual temperature (°C)	8.0	9.4	9.0
Mean autumn (Oct.–Dec.) temperature (°C)	6.0	7.1	6.8
% plants with > 25% roots infected (Zadoks growth stage 59)	0.3–5.2 (mean = 2.3)	0.5–56.9 (mean = 25.0)	33.1–80.9 (mean = 49.0)
Yield of third wheat crops as % of yield of first wheats (1964–69)	89.3	78.1	67.6

proportions at High Mowthorpe, even after 8 years of wheat cropping, whereas at Boxworth and Rosemaund disease levels tended to peak in the third years of the sequences. The subsequent decline was much more gradual at Rosemaund than at Boxworth, though yields at Boxworth continued to fall as a result of increased weed competition and other factors. Cropping tests on the different soil types together at one or more locations, possibly as soil monoliths (blocks of soil in single pieces with their structures preserved), might remove variability due to the weather factor. Such experiments have been discussed in the past but remain to be carried out.

Rotation cannot, therefore, be considered in isolation from the edaphic and climatic factors operating on the farm. The fact that take-all never became severe at High Mowthorpe is surprising, given the accepted wisdom that the disease is favoured by alkaline soils. It might be argued that the lower autumn temperatures on this High Wolds farm delayed the establishment of Ggt on the seminal roots of young plants. However, there is only $0.8°C$ difference in average autumn temperatures (and only $1°C$ difference in mean annual temperature) between High Mowthorpe and Rosemaund and the greater rainfall at the former site might be expected to increase the risk of take-all. This suggests that the primary operative factor in this comparison is edaphic rather than climatic. In contrast, to account for the greater levels of take-all at ADAS Bridgets on the chalk downs of Hampshire than on the very similar soil type at High Mowthorpe, it seems necessary to invoke climatic factors. Autumn (October–December) temperatures are $1.3°C$ higher, and mean annual temperatures $1.6°C$ higher, at Bridgets than at High Mowthorpe. The difference in disease severity between Rosemaund and Boxworth could partly be due to climatic differences (Rosemaund being wetter than Boxworth) and partly to edaphic factors, since Bromyard soils have a high silt fraction and seem to be particularly prone to the disease, whilst well-structured Chalky Boulder Clay soils (e.g. Hanslope series) tend to be relatively safe from severe attacks, as long as they are well managed. Because plants on Chalky Boulder Clays are often so well rooted, they can still perform quite well even when a large proportion of their roots is infected. The aerial photograph (Fig. 3.19a) of a wheat field affected by take-all reveals a very distinct pattern of poor growth, and a later photograph (Fig. 3.19b) of the same field taken when the soil was bare shows a soil pattern which exactly matched the earlier pattern of disease. The effects of the disease had been least severe on the calcareous clay loam (Hanslope series) that occupied most of the field, and most severe on a siltier and more poorly draining colluvial modification of the Hanslope soil, where temporary waterlogging had restricted root growth in the winter.

3.7. Comparisons with Cereal Production Systems in Other Regions

Large parts of the preceding discussions, especially those concerning crop rotations and nutrition, are based on experience and evidence from ADAS, IACR-Rothamsted and INRA. As cereal production systems vary between climatic zones

and continents, so does the impact of take-all on those systems. In considering wheat production systems of some specific regions which contrast with the UK and Europe, this section expands on Section 2.7, where factors that affect take-all in these regions are discussed. Some of the different locations representing the USA, Australia, the UK and Brazil are illustrated in Figs 3.20 and 3.21 and Plates 6.16 and 6.17. Typical UK locations are also well illustrated in the aerial photographs used for case histories in this chapter.

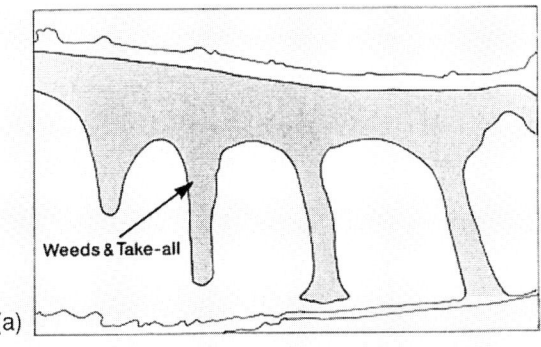

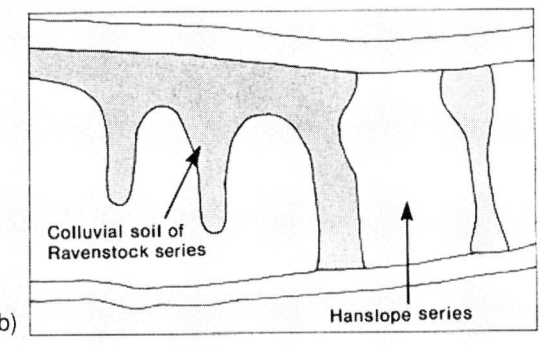

Fig. 3.19. Aerial photographs (opposite) and explanatory drawings (above) of a field at Childerley, Cambridgeshire.
(a) Wheat affected by take-all showing a distinct pattern of poor growth. (b) A later photograph of the same field when the soil was bare showing a soil pattern that matched exactly the earlier pattern of disease. The soil types are described in the text.
Photographs: ADAS Aerial Photography Unit.

In the USA, the types of wheat grown and wheat-growing practices differ from region to region (Cook and Veseth, 1991). In the Pacific Northwest, particularly eastern Washington and Idaho, most of the wheat grown is soft white winter wheat. The soil there is ideal for small grains and economic incentives have meant that wheat (or barley) is grown at least every second year in many fields, either with irrigation or without ('dryland' wheat) (Cook, 1992a). Where wheat has been grown more intensively (for example, in the irrigated areas of the Columbia Basin), take-all has caused considerable losses in second and third wheats, with TAD becoming well established by the time a sixth wheat crop was grown. A break of 1 year from a susceptible crop is usually sufficient for effective take-all control, although a small but significant amount of disease can be carried over in irrigated crops and in areas of high rainfall (Rovira *et al.*, 1990). The soils there are particularly prone to erosion by water and wind, encouraging the use of minimum tillage. Whilst tillage provides no advantage to wheat growth (Cook, 1992a),

Fig. 3.20. Irrigated spring wheat in Montana, USA, June 1991.
Take-all problems occur in this cropping system. The irrigation system, a single boom rotating on a central pivot, is similar to those now used on some East Anglian farms, mostly to irrigate non-cereal crops. Photograph: G.L. Bateman.

minimum tillage, especially where crop debris is left on the surface, allows retention of soil moisture and so encourages take-all (Cook and Veseth, 1991).

In the northwestern state of Montana, both winter wheat and spring wheat are grown, mostly in continuous cultivation, and other crops grown include alfalfa (lucerne). The winter wheat is almost always grown as dryland wheat. In spite of the long cold winter its early establishment in autumn allows ripening during the dry summer. As in the Pacific Northwest, dryland wheat does not usually suffer from take-all. Spring wheat is grown in fields irrigated by sprinklers (Fig. 3.20) so that the plants can make full use of the 120 growing days. In 1991, about 100,000 ha of wheat were grown in this way. In irrigated wheat, which suffers from take-all, a year of fallow is usually enough to reduce the disease greatly, but this practice is seldom followed because of the loss of income from a field in fallow. In the absence of suitable and profitable alternative crops, continuous wheat growing has apparently resulted in TAD in some fields. The rapid development of take-all in seedlings and the need for relatively short-term protection perhaps explain how significant take-all control can often be achieved by seed treatments containing the fungicide triadimenol (see 4.2.2). Typically, the sub-crown internodes become very blackened and this seems to account for much of the early damage.

As elsewhere, rotations are important in controlling take-all in Australia (Rovira *et al.*, 1990). The rotation is often wheat with fallow or grain legumes, but

Fig. 3.21. Seedling spring wheat in the Mallee area of northern Victoria, Australia, in August 1983.
Wheat in this and other parts of Australia is liable to suffer from take-all and cereal cyst nematodes. The mobile sand dune illustrates the proneness of this light soil to erosion where the native Mallee scrub has been cleared and, especially, where the vegetation has been damaged by rabbits.
Photograph: G.L. Bateman.

a rotation of cereals and pasture has presented particular problems where cereal growing is combined with sheep farming. This is because of the importance of grasses in maintaining or increasing populations of Ggt; at a site in Western Australia, increasing the proportion of grass in pasture increased the amount of take-all in the following wheat crop (MacNish and Nicholas, 1987). Rotations are important for controlling both cereal cyst nematodes and take-all in the Wimmera and Mallee areas of Victoria. The introduction of new crops, especially legumes such as faba beans, lupins and lentils, has decreased the need for long periods of fallow, but rotations have increased the incidence of rhizoctonia root rot (Anon., 1994c).

Wheat is grown throughout South Africa in regions ranging from those with a Mediterranean-type climate and winter rainfall in the Western Cape to those in the north with summer rainfall and colder, drier winters (D.B. Scott, 1990). The greater part of the wheat production comes from these northern areas and much of the crop is irrigated. Both spring wheat and winter wheat are grown, but winter wheat only where the winters are sufficiently cool for vernalization. Take-all occurs commonly in the Western Cape Province, the Orange Free State and in all areas where the crop is irrigated; this situation is comparable with irrigated wheat growing in the USA. In the Western Cape Province, legumes are often grown in rotation with wheat or barley. Grasses such as *Hordeum murinum* and *Lolium perenne* occurring in these crops offer a means of carry-over of the disease between host crops.

The history of the take-all problem in Brazil is an interesting example of interactions occurring adversely between climate, soils and husbandry methods. Take-all became a major constraint to wheat production in southern Brazil, a region with high rainfall, on soils that were once acid but had been limed to prevent aluminium toxicity (Reis *et al.*, 1982). Effects of take-all in Rio Grande do Sul can be seen in Plate 6.17.

Strategies for Management

4.1 Agronomic Practices

The previous chapter shows how factors that affect take-all are interlinked (e.g. Fig. 3.3) and there is often no strong reason for regarding any one factor as more significant than any other. The search should be for the key, or keys, to this interlinked structure and manganese has been considered in this context (see 3.1 and 3.4.3). In high input systems, as in the UK, changing any of these factors within the normal range of husbandry practices achieves only a fine tuning of disease control (Hornby, 1985). Even so, at present the best control strategy available for take-all is 'damage limitation', i.e. the manipulation of agronomic practices to decrease the effects of take-all, and the hope that the weather will favour crop growth more than pathogen development. The narrowness of this strategy may be a significant constraint to profitable wheat production and sensible cropping patterns (see 3.3.1).

4.1.1 Host nutrition

There is much work spanning many years to show that nutrient deficiency usually exacerbates take-all and that, in appropriate situations, applying NH_4^+-N, KCl or other mineral fertilizers may be beneficial in decreasing take-all or its effects on the host (see 3.4). Despite this knowledge, there are no generally practised crop nutrition strategies in the UK that are specifically aimed at controlling or minimizing take-all. Fortunately fertilizing crops to achieve maximum or optimum yields tends to decrease take-all in comparison with poorly fertilized crops. Where take-all is taken into consideration, nitrogen applications may be arranged with the intention of compensating for the effects of the disease (see 3.3.2). Contrary nutritional strategies aimed at ensuring sufficient take-all to provide an adequate test of treatments are often used in field trials and experiments. For instance, in their work on continuous spring barley Hornby and Henden (1986) used less than the recommended rate of nitrogen specifically to favour take-all.

A major obstacle to the adoption of a general crop nutrition strategy on the grounds that it minimizes take-all is finding one that is generally reliable. For example, dividing the spring dressing of nitrogen to Rothamsted fields, so that 40 kg ha^{-1} were applied in late February or early March and 160 kg ha^{-1} in April, resulted in less take-all, especially after wet winters (Prew *et al.*, 1986), but the effects could be variable (Hornby *et al.*, 1990a). Werker and Gilligan (1990) found that timing and amounts of spring nitrogen were irregular and temporary in their effects on disease and concluded that decisions involving them would be regarded as neutral in a management strategy for control of the disease; the application of 40 kg N ha^{-1} in autumn led to a transitory lessening of disease. In other experiments autumn nitrogen dressings have had no advantage over spring applications in terms of take-all control (Hornby *et al.*, 1990a).

4.2 Chemical Control

Chemical control of take-all has been attempted experimentally by soil treatments (including fumigants), sprays and seed treatments. The progress of research in these areas up to 1988 was reviewed by Bateman (1989). A few years later, Ggt was regarded as a pathogen of wheat whose chemical control status was 'unsatisfactory', with a 'need for improvement', according to Becker and Schwinn (1993). This account summarizes Bateman's review and describes subsequent developments: prospects for controlling take-all by chemical treatments are summarized in Table 4.1.

Table 4.1. Chemical control treatments and their prospects for controlling take-all.

Treatment	Prospects	Region	Reasons
Soil fumigation	None commercially.	An important experimental tool, Pacific Northwest, USA (Cook, 1994).	Expensive; toxic; polluting; problem of take-all resurgence.
Fungicide spray	None unless downwardly mobile fungicides found.		Poorly targeted; no chemicals with appropriate properties.
Soil-applied fungicide	Useful as in-furrow treatment in simple soils. Poor prospects otherwise, unless chemicals with novel combinations of physicochemical properties found.	Australia, especially on fertilizer granules (e.g. Ballinger and Kollmorgen, 1988).	Carrier available; little inactivation in local soils. Insufficient activity in complex soils.
Seed treatment fungicide	Promising; already useful on a limited scale. More active compounds being developed for commercial use.	Triadimenol useful on some irrigated spring wheat in northern USA (D.E. Mathre, pers. comm.). New products expected for use in Europe.	Application technology already in use; small amounts needed; targeting reasonably accurate.
Other chemicals	Poor prospects unless chemicals with ability to alter host resistance found.		Little research directly related to this.

4.2.1 Fumigants

Soil fumigants will decrease Ggt to insignificant levels, but amongst other drawbacks they are costly and therefore not commercially practicable (Heim *et al.*, 1986). They are non-selective and following the year of treatment there may be a rapid build-up of Ggt (Ebbels, 1969), probably because of a decreased antagonistic microflora in the soil. Consequently, severe infection often results if another cereal crop is grown without re-application of the chemical. The large volumes and high concentrations needed, and hence the risk of contaminating ground water, also preclude fumigation as a practical treatment for controlling take-all.

4.2.2 Fungicides

A few reports of successes with fungicides, especially through the 1980s, kept this avenue open. A disease such as take-all, caused by a soil-borne, root-infecting fungus, is a difficult target for control by fungicides and is out of reach of most fungicide applied by conventional means. At the time of writing, few products, and none in the UK, are available with specific recommendations for controlling take-all, but new developments in the agricultural chemical industry indicate that this situation is set to change. One new product is likely to be available first in the UK and to be formulated as a seed treatment (Anon., 1997), but much of the research evaluating the efficacy and potential of modern fungicides involved other methods of application and this will be described first.

Soil fungicides
A research programme at Rothamsted investigated the effects on take-all of soil treatment with conventional fungicides that are more selective than the fumigants tested in most earlier research. Following preliminary screening to establish active compounds (Bateman, 1980), the first report of a significant amount of control of naturally occurring take-all was made, although using an unacceptably high rate (35 kg ha^{-1}) of benomyl (Bateman, 1981). Inadequate persistence was one likely problem with this fungicide (Bateman, 1984b) and experiments showed how it might be improved by slow-release formulations (Bateman, 1981, 1982). A principal objective was to determine how the fungicide should be distributed in the soil, and how to achieve redistribution from the site of application to the location of the inoculum or the site of infection. This would minimize the amounts needed and maximize performance. The need for fungicides with little mobility in soil, located close to the seed to protect the seminal roots from early infection or in the region of the crown to protect developing nodal roots, was demonstrated (Bateman, 1985). The distribution of fungicides applied to the soil as drenches can be improved by the use of surfactants (Bateman and Nicholls, 1982), but this did not improve the activity of benomyl in the field (Bateman, 1984a).

During this work the sterol biosynthesis-inhibiting (SBI) fungicide nuarimol came to be used as a standard because of its consistent, though partial, effects against take-all. It decreased take-all in field experiments, on a range of soil types, each year from 1980 to 1987. In one experiment a yield increase in winter wheat from 7.5 t ha^{-1} in untreated plots to 9.0 t ha^{-1} was achieved through decreasing take-all with nuarimol incorporated at 1.1 kg ha^{-1} into the seedbed using a rotary harrow (Bateman, 1984b). Triadimenol showed even more consistent activity than nuarimol in later field experiments (Bateman *et al.*, 1994). Such fungicides applied in this way to the clay loam soil at Rothamsted are unlikely to perform well at application rates smaller than those tested, but even 1 kg ha^{-1} compares favourably with those of some fungicides applied as sprays, often several times in a season, to control foliar diseases. Also, such a soil fungicide may not need to be applied annually in consecutive cereals, but only in years when the disease risk is high because of the previous cropping (this might be predicted from the expected disease build-up and decline in successive crops illustrated in Fig. 2.16).

Most recently, the emphasis in this work has been on the importance of the physicochemical properties of SBI fungicides when applied to soil (Bateman *et al.*, 1990b). Experiments using pot-grown wheat plants suggested that soil-applied fungicides, as well as needing to be highly fungitoxic with minimal phytotoxicity, should have moderately low lipophilicity. This would avoid excessive sorption to soil particles and allow some mobility and redistribution via the soil water. Redistribution of more volatile compounds as vapour was found not to be useful. Moderate persistence is also required. The importance of redistribution in the water phase rather than as vapour was endorsed by synthesizing and testing a series of volatile analogues of SBI fungicides (Chamberlain *et al.*, 1991); the compound most active as vapour from soil using *in vitro* tests was ineffective against unevenly distributed inoculum in pot tests.

The requirement for the physicochemical properties described above was largely confirmed by field experiments using naturally infested plots (Bateman *et al.*, 1994). Whilst no fungicide with all the appropriate properties was identified, it should be possible to synthesize new molecules which have a balance between mobility and persistence. Where persistence limits are dictated by registration requirements, the need for persistence may be met better by the use of slow-release formulations.

The use of small plots and compact blocks replicated many times throughout relatively large experiment sites (see 6.3.1) allowed fungicide performance to be tested at different levels of disease intensity (Hornby *et al.*, 1993; Bateman *et al.*, 1994). The most effective, immobile and persistent compounds, such as nuarimol and triadimenol, were found to be sometimes relatively more effective where disease severity was greater. This probably resulted from localized protection of the nodal root system, which is often subject to a damaging phase of disease development from spring onwards.

Triadimenol, the most effective fungicide in the soil treatment experiments at Rothamsted, was used as a granular formulation in experiments in Chile, in which interactions with sowing date were investigated in winter wheat crops (Andrade, 1995). With artificial infestation, triadimenol at 225–400 g ha^{-1} increased yields (in some cases by as much as 300%) as a result of controlling take-all. The effect was greater in crops sown late (August) than in those sown early (June or July) and yields in the later-sown crops sometimes approached those in non-infested crops. A similar effect of sowing date (July vs. June) was seen in naturally infested crops using 300–600 kg triadimenol ha^{-1}, where maximum yield increases approached 50%.

Fungicide screening and field trials in Victoria, Australia, also revealed the potential of SBI fungicides such as triadimefon (Ballinger and Kollmorgen, 1986a,b). The related fungicide flutriafol was identified as a potential in-furrow treatment, especially when formulated as a coating on superphosphate granules (Ballinger and Kollmorgen, 1988; Brennan, 1989c; Coventry et al., 1989). Such treatments can increase yield by up to 60% (Coventry et al., 1989). Although similar amounts of yield can be saved in Britain (see above), these normally represent much smaller proportions of the total yield. The proportional yield benefits, the consistency of the treatment effects and the convenient application method would seem to make existing fungicides a more attractive proposition for controlling take-all in Australia than in Britain.

In very different growing conditions in Britain flutriafol has not shown the same promise. After soil incorporation it was often less effective than other SBI fungicides in experiments at Rothamsted (Bateman et al., 1994). As an in-furrow treatment it decreased take-all in some ADAS trials in which grain yields were increased at three out of seven sites, but establishment was severely delayed and yield was decreased at two of the sites (R.W. Clare, personal communication). The variability of the results obtained and the difficulty of predicting the locations at which good responses are likely is illustrated in Table 4.2. Second wheat crops sown in late September were grown on Old Red Sandstone soil in Herefordshire (ADAS Rosemaund) and on very light Bunter Sandstone soil in Nottinghamshire (ADAS Gleadthorpe). Crops on these soils can suffer very severely from take-all and therefore a protective treatment against the disease would seem to be fully justified. However, application of flutriafol at the Rosemaund site would not have been cost-effective and at Gleadthorpe there was an interaction between fungicide and nitrogen treatments, suggesting that the soil-applied fungicide was better able to control take-all where its effects were not being aggravated by a low level of soil nitrogen (Table 4.3). (Interactions of fungicide effects with those of nitrogen have been found in barley, but rarely established statistically in wheat: see 'Seed treatment' in this section and 4.4.1.) Severe take-all such as that at Gleadthorpe is most likely to occur where disease becomes well established in the autumn, and autumn

Table 4.2. Effects of soil-applied flutriafol in ADAS trials on winter wheat at Rosemaund (RM) and Gleadthorpe (GT).

Soil fungicide	Take-all index (0–100)				% Patches[a] in July		Grain yield (t ha^{-1})	
	Spring		Summer					
	RM	GT	RM	GT	RM	GT	RM	GT
–	14.0	24.0	24.2	55.2	2.8	46.8	8.40	2.72
+	13.8	19.9	22.2	50.9	2.4	30.1	8.26	3.23
SED	1.38	1.79	2.16	2.55	0.35	3.38	0.087	0.118
CV (%)	39.6	32.5	37.3	19.2	52.9	35.1	4.2	15.8

[a]Areas of premature ripening (whiteheads).

Table 4.3. Effects of soil-applied flutriafol and nitrogen fertilizers on take-all patches (% area prematurely ripened) in an ADAS trial on winter wheat at Gleadthorpe.

Soil fungicide	No nitrogen	Calcium nitrate	Ammonium nitrate + urea	Ammonium nitrate + urea + Didin[a]	SED	CV (%)
−	66.7	35.3	34.5	50.8	6.76	35.1
+	58.3	21.9	18.2	22.0		

[a]Nitrification inhibitor.

infection may be most affected by the in-furrow fungicide. Late development of disease, which would be aggravated by insufficient spring-applied nitrogen, may be less susceptible to control.

Foliar treatment

To avoid unnecessary treatment, application of fungicides to the seed or seedbed would require more accurate predictions of severe take-all than is possible at present. A possible advantage of a conventionally applied spray relatively late in the season would be the opportunity to make a better assessment of the take-all risk before application. Unfortunately sprays such as those applied at that time to control eyespot are not normally effective against take-all. The availability of a foliar-applied, phloem-translocated fungicide would decrease the need for early prediction, possibly decrease the problem of persistence and avoid much of the wastage that is inevitably associated with applying pesticides to soil. Phloem-translocated fungicides currently available are active against oomycete pathogens; none are active against Ggt, and none are currently approved for controlling root diseases in the UK. However, there is an increasing understanding of the physicochemical properties required by such fungicides, raising the possibility that fungicides with a broader spectrum of activity can be synthesized.

Although fungicides that are not phloem-translocated are not expected to affect root diseases when applied to leaves, autumn spraying has sometimes been associated with decreased take-all (R.W. Clare, personal communication). In such cases sufficient fungicide may have penetrated the soil, perhaps by running off the plant surface, to afford some control. Moreover, even when an autumn-applied fungicide has no obvious effect on take-all itself, it can sometimes increase yield substantially where take-all is severe – more than would be expected solely from the control of other pathogens against which the fungicide has activity. An example is provided by a trial at ADAS Rosemaund (Herefordshire) in 1992/93. The trial involved two sowing dates, 15 September and 15 October, the earlier of which resulted in severe take-all. The effect on yield of triadimefon plus carbendazim (Bayleton BM) applied in autumn is shown in Table 4.4. By controlling early-season foliar diseases, the autumn-applied fungicide may have allowed the plants to produce more roots so that they were able to cope better with the autumn take-all. This parallels results from Rothamsted using seed treatment with triadimenol (see below), but effects on other diseases could not be implicated in yield benefits.

Table 4.4. Effect of autumn-applied fungicide (triadimefon + carbendazim) on take-all and yield in an ADAS trial on wheat at Rosemaund.

Sowing date	Take-all index (0–100) at harvest	Grain yield (t ha^{-1})	
		Autumn fungicide	No autumn fungicide
15 September	90	3.21	2.07
15 October	75	5.06	5.05
SED		0.516	

Seed treatment

The first fungicide to show promise for controlling take-all when applied as a seed treatment was the SBI compound triadimenol. It was found to be useful in experiments in the USA by providing early protection of the seminal root system in early-sown winter wheat (Bockus, 1983). It was effective in delaying take-all development on irrigated spring wheat in Montana (Mathre *et al.*, 1986) and a triadimenol-containing seed treatment was marketed in Montana partly for this purpose (D.E. Mathre, personal communication). Its usefulness there seems to depend on a short growing season in which the disease is able to develop quickly. The beneficial effects may be partly a consequence of modified plant growth or physiology as well as direct fungitoxic activity. Variable performances in Canada (e.g. Conner and Kuzyk, 1990) suggest that triadimenol is not universally effective in spring wheat, but results are often difficult to interpret because of the use of artificially prepared inoculum mixed with the seed in trials and, sometimes, varying levels of drought affecting disease development. The good performance of triadimenol in irrigated spring wheat contrasts with poor control achieved in autumn-sown wheat in other parts of the USA, notably Georgia where little disease development occurs early on but is most rapid after early spring (Rothrock, 1988a). This is comparable to the situation in Britain where winter wheat predominates and prolonged protection from later infection of the nodal root system in spring and early summer is often more important. Situations in which Baytan (triadimenol plus fuberidazole) seed treatment has been effective in controlling autumn infection in early-sown winter wheat in Britain are discussed later in this section. Perhaps prolonged protection was achieved in France, where improved control resulted from the application of a high dose of triadimenol to seed using an experimental coating process (Cavelier and Lucas, 1985).

The seed coating tested in experiments in France was applied by an experimental 'rolling' process (Fraselle and Schiffers, 1982). Seeds were pelleted with a mixture of inert powders, stickers and systemic fungicides, to achieve greater doses of fungicides than allowed by conventional seed treatment. Coating the seed at the normal, recommended dose of fungicide (0.3 g a.i. kg^{-1} seed) resulted in TARs at tillering, stem elongation and anthesis being, respectively, 70, 79 and 84% of the TAR in the control treatment, whereas the same dose applied by conventional seed treatment was ineffective beyond the stem elongation stage. At 1 g a.i. kg^{-1} applied as a seed coating, TARs were 36, 60 and 60% of the control, respectively. Grain yield was increased significantly by the treatments, from 3.8 t ha^{-1} in the control to 4.4 and 4.7 t ha^{-1} for conventional seed treatment at 0.3 g a.i. kg^{-1} and seed coating at 1 g a.i. kg^{-1}, respectively. Some phytotoxicity was seen at early plant growth stages: only 18% of the plants produced tillers in the axils of the first leaves (and 48% in the axils of the second) when treated with triadimenol at the highest dose, compared with 64% (and 90%) in the control. Compensation occurred later and the number of plants with a tiller in the axil of the fourth leaf was greater in treated plots than in the control, resulting in similar numbers of stems per length of row in both treated and untreated plots at the beginning of stem elongation (Lucas *et al.*, 1988). Seed coating that would allow slow release of fungicide throughout plant growth should be a way of avoiding absorption by the seed or the roots of high concentrations of fungicides in too short a time and of protecting the plant for longer without phytotoxic effects. There would seem to be, therefore, opportunities for research on the use of seed coatings to allow controlled release of fungicides. Even so, more effective chemicals will be required as the results, even with triadimenol, were somewhat erratic from year to year (P. Lucas, unpublished results).

At Rothamsted, Baytan can be beneficial as a conventionally applied seed treatment when a winter wheat crop is sown very early and severe take-all develops subsequently and causes premature ripening (Bateman, 1986). Table 4.5 shows how a yield benefit can result in such conditions. The poor control of take-all symptoms on the roots (shown as TAR) suggests that the fungicide may have induced a physiological change which allowed the plants to tolerate the root infection so that patches of early-ripened plants did not develop. Little or no benefit occurred after this treatment in 1995, when severe take-all developed very rapidly in drought-stressed plants in the warm, dry summer, following relatively late sowing (mid-October) and little development of take-all on roots early on. Factorial experiments on winter wheat at Rothamsted (see 4.4.1) confirmed the unreliability of seed treatment with triadimenol plus fuberidazole as a control measure against take-all.

A number of ADAS trials have shown a positive association between the yield response to triadimenol seed treatment and the take-all index of untreated plots. The effects on root infection itself may be greater in more densely sown crops, but generally the treatment has been unreliable. In trials from 1984 to 1986 (Jones, 1987), yield increases sometimes occurred where take-all was negligible in the crop or was not controlled. In an ADAS trial in 1985 in which there was a large yield response (from 4.21 t ha^{-1} with organomercury to 6.78 t ha^{-1}), take-all was

Table 4.5. Effects of seed treatment with triadimenol (as Baytan) on take-all in winter wheat.

Sowing date	Seed treatment	Take-all rating (0–300)	% Area prematurely ripened in mid-June	Yields (t ha^{-1})
8 Sept.	None	268	70	5.72
	Triadimenol	248	43	8.44
7 Oct.	None	218	44	7.43
	Triadimenol	213	44	7.75
SED (15 df)		16.1	9.5	0.655

decreased, but analysis of yield using incidence of take-all as a covariate indicated an effect on yield additional to that which resulted from controlling take-all.

In factorial experiments on winter barley at Rothamsted and Woburn (Jenkyn et al., 1991b), the effects of triadimenol plus fuberidazole on take-all symptoms on the root were not great, but they were often significant and usually persisted until June and appeared to be more consistent than on wheat. When results of seven experiments over 3 years (1984–1986) were combined and interpreted cautiously, there was no overall effect of sowing date on disease, but the effect of seed treatment in decreasing take-all was greater with earlier sowing (Fig. 4.1). Increased take-all in untreated plots and increased effectiveness of treatment were each associated with an increased effect on yield. In a separate, simultaneous series of multifactorial experiments on winter barley at Rothamsted (also referred to in

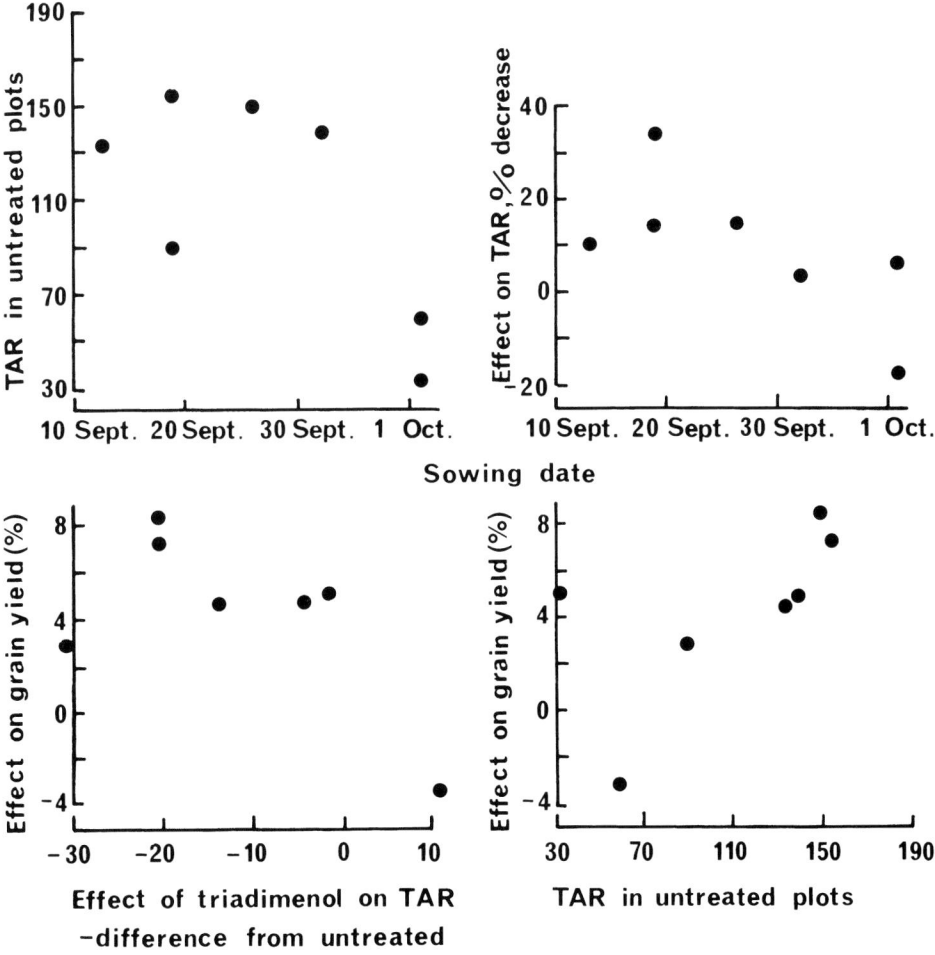

Fig. 4.1. Relationships among take-all rating (TAR) in June samples, effect of triadimenol seed treatment and yield in field experiments on winter barley at Rothamsted and Woburn.

4.4.1), seed treatment with triadimenol plus fuberidazole again had small but consistent effects on take-all (Jenkyn et al., 1992a). Of the factors investigated, seed treatment was most commonly involved in significant interactions between factors in their effects on grain yield. Interactions with seed rate were inconsistent: seed treatment sometimes increased yield more with a higher seed rate and sometimes with a smaller rate. Seed treatment was also more effective when smaller amounts of N were available during the winter (as soil N or applied as fertilizer). There was a strong relationship from year to year between the responses of grain yield and of take-all severity rating in June to seed treatment.

In 1996, Baytan seed treatment of wheat and barley was reported to have been used widely on a Somerset farm, 'mainly to counter take-all' (Blake, 1996a).

In the last few years, tests of new seed treatments for take-all have been proceeding in various parts of the world under confidentiality agreements and an announcement appeared in the British press concerning a new benzamide fungicide from Monsanto, code name MON 41100 (Headrick and Bockus, 1996, who reported that it could be applied as a soil as well as a seed treatment), and its expected launch in the UK in the year 2000 (Abel, 1995). It is claimed to have increased yields by 0.5–2.25 t ha^{-1} in wheat and by over 0.3 t ha^{-1} in barley and these increased outputs may be possible with a '30% cut in nitrogen needs'. A yield increase of 2 t ha^{-1} resulted from seed treatment with one formulation in two trials in France in which take-all infection was at a high level and other diseases had been controlled by foliar fungicides (Lucas et al., 1994). Such performance could influence thinking in regard to rotations, in that more cereals, previously considered at risk, could supplant less profitable break crops. More detail of the chemistry of Monsanto's fungicides for controlling take-all can be found in the patent documents listed in the section 'Chemical (fungicides)' in Table 4.6.

Testing fungicides
Establishing the activity of fungicides against the take-all pathogen presents no problems: it grows readily on agar-based media on which standard 'poisoned-food' assays can be conducted. In Australia, control by the fungicide flutriafol applied to soil was thought to result largely from decreased saprophytic growth of the take-all fungus (Cotterill, 1991); saprophytic growth seems to have greater significance in Australian conditions than in UK conditions (see 2.4.3, 2.7 and 4.7). These findings prompted tests of a simple screening procedure (Cotterill et al., 1992) based on the 'soil-sandwich' technique mentioned in Section 2.4.3. However, the results of tests measuring saprophytic growth in soil, as in agar plates, did not always relate to results for plant infection.

Conventional glasshouse screening procedures may not provide useful information, because tests in soil in pots often do not give an accurate prediction of what will happen in the field (Bateman, 1989). Field performance may be predicted to some extent by knowledge of a compound's intrinsic toxicity and physical properties (Bateman et al., 1994; and 'Soil fungicides' in this section). In tests of fungicides there is no agreed standardization of methods of assessing disease, of which there are numerous variations (see 6.3.3), nor is there for tests to determine pathogenicity or confirm the identity of isolates (see 5.4.1).

Table 4.6. Some patents or patent applications concerned with the control of take-all.

Type of control	Involves	Inventors working in	Patent reference
Biological (bacteria)	Method for screening, application in commercial setting, fluorescent pseudomonad strains.	USA	Weller et al. (1984)
	Pseudomonas putida strain IMET 11286, seed treatment.	East Germany	Zaspel et al. (1989)
	Bacillus subtilis strain IMET 11327, seed treatment.	East Germany	Zaspel et al. (1990)
	Strain of Arthrobacter sp., seed treatment.	East Germany	Zaspel et al. (1991a)
	Antibiotic-producing strain of Pseudomonas, IMET 11422.	East Germany	Zaspel et al. (1991b)
Biological (fungi)	Hypovirulent Ophiobolus graminis isolated from slightly infected plants following a peak of disease in a wheat monoculture; on particulate medium.	France	Jouan and Lemaire (1976)
	Microdochium bolleyi applied to seeds or soil, polymeric seed coating.	UK	Foxroberts and Deacon (1988)
	'Myzelgranulats'.	East Germany	Augustin (1992)
	Phialophora graminicola IMET 43895.	East Germany	Augustin et al. (1990)
	Phialophora sp. (lobed hyphopodia) IMET 43910.	East Germany	Augustin et al. (1991a)
	Avirulent Gaeumannomyces graminis.	East Germany	Augustin et al. (1991a)
	Ggg strain 90/3B and carrier, Ggg tolerant to fungicides.	Australia	Wong (1993)
	Phialophora sp. (lobed hyphopodia).	Australia	Wong (1995)
Chemical (antibiotics)	P2–79, a dimer of phenazinecarboxylic acid from Pseudomonas fluorescens.	USA	Gurusiddaiah et al. (1986a)
Chemical (fertilizer)	Special fertilizer containing N, P, K, Mn, Zn, B, Cu, Na and Cl.	China	Zi et al. (1991)
Chemical (fungicides)	Aryl fungicides, preferably as seed coating.	Europe[a]	Phillion et al. (1993)
	New fused aryl and heteroaryl derivatives applied to seed or soil.	Canada[a]	Graneto et al. (1994)
	A compound applied to seed, agronomic salt thereof, methods of use, processes of preparation.	USA[a]	Phillion et al. (1996a)
	A thiophenecarboxamide, compositions and method.	USA[a]	Phillion et al. (1996b)

[a]Where basic patent granted.

In field trials, artificially produced inoculum (see 6.4) may give results that differ from those produced with natural inoculum, especially if seed treatments are tested against inoculum drilled with the seed. However, where artificially produced inoculum applied to the seedbed simulates rapidly developing take-all in spring wheat, as for example in Montana, it is likely to provide a fair test of seed treatments (D.E. Mathre, personal communication). Inoculating a trials area and growing a susceptible crop the year before the area's intended use may achieve uniform infection that arises more naturally when a second crop is grown. However, even after drilling inoculum (colonized, killed oat grains) with seed of winter wheat, some patchiness has still occurred (Bateman and Hornby, 1995). Whereas approaches relying on artificially produced inoculum may not reproduce realistic farm conditions adequately, natural infection is notoriously unreliable and even preliminary trials normally require replication on several sites over several seasons. Also, patchy distribution of naturally occurring disease detracts from the reliability of traditional large-plot trials. The testing procedures involving small plots, described above and in section 6.2.1, were introduced at Rothamsted for these reasons.

Recent patent documents ('Chemical (fungicides)' section of Table 4.6) describe screening procedures used in the development of a fungicide by an agricultural chemical company (and emphasize the reliance on data from tests using plants inoculated artificially in obtaining supporting data). Typically, 'poisoned-food' agar plate tests were followed by seedling tests in growth chambers in which the chemicals were tested in artificially infested soil in small pots. Slightly longer but similar tests were then made using treated seed in a glasshouse. Next came field tests of a range of concentrations of the fungicide and various adjuvants and carriers. From this point the most promising formulations would be tested on many sites, most with natural infestation, and often by independent contractors. In the patents referred to it is not clear whether the first field tests used fields that were infested artificially, or ones that were infested naturally. Also, no comparison with other chemicals is reported, presumably because no standard treatment exists, even though comparisons with triadimenol, on which there is much published information, have been made and therefore would have been of interest.

Conclusions and prospects
Development of a soil treatment, for use in Britain at least, would need further research on appropriate formulations or carriers. In the absence of a convenient carrier, such as the superphosphate granules used in phosphate-deficient Australian soils, additional operations and materials would be needed. From a practical standpoint, therefore, current interest in fungicides centres on seed treatment, which uses existing methodology and equipment. Seed treatment also requires much smaller amounts of chemical and ensures accurate placement, although neither of these, as achieved by conventional seed treating methods, are necessarily ideal for the control of take-all. Significant increases in yields of winter wheat have been achieved by using triadimenol in some UK situations where crops were sown early and the season favoured the extensive development of whiteheads. Unfortunately, predictions of the epidemic of take-all and contributory weather conditions are not reliable enough to identify beforehand situations where treatment would be cost effective. The development of a treatment that performs more consistently than triadimenol, either through greater fungitoxicity, greater mobility in the soil, the root, the rhizosphere or the sub-crown internode, or by altering the host so as to increase its resistance, are reasonable targets that should be achievable. (It will be interesting to see whether any have already been achieved

by new products under development.) Additional effects against other diseases such as eyespot would improve the likelihood of such a product being commercially successful. Some of the properties required by a soil fungicide discussed above may also be appropriate for a seed treatment. The need for fungicide treatments should, ideally, be evaluated by: (i) surveys to assess the importance of the disease; (ii) determining the reliability of risk assessments; and (iii) improving estimates of yield loss from take-all, thus allowing an economic evaluation of treatments to be made.

Despite the fact that public funding has not been made available to meet these needs adequately (see Preface, Fig. 1.1 and Chapter 7), the obvious importance of take-all has encouraged several agricultural chemical companies to proceed with product development. To make best use of prospective treatments, their effects on disease at different intensities (see 6.2.1) and with different sowing dates, host crop species and cultivars, and levels of available nutrients will need to be determined.

4.2.3 Herbicides and plant growth regulators

Herbicides may have indirect effects on take-all, more often increasing rather than suppressing the disease (Yarham, 1981) by, for example, predisposing the plants to attack through phytotoxicity (Huber, 1981b). They may also suppress take-all by reducing competition for plant nutrients (Huber, 1981b), or controlling perennial grass weeds that act as carriers of Ggt (Yarham, 1981). The importance of, and reasons for, controlling perennial grasses well in advance of the sowing of a susceptible cereal, particularly where the phloem-translocated herbicide glyphosate is to be used on couch grass (*Elytrigia repens*), are discussed in Section 3.5.5. Some herbicides have been reported as being directly toxic to Ggt, but there is no evidence that this is important in practice.

Using herbicides to decrease the grass component, which provides hosts for Ggt, in legume pastures in pasture/wheat rotations in Western Australia was shown to decrease take-all in the wheat (MacLeod *et al.*, 1993). An increase in yield of the wheat may have arisen through increased nitrogen from the increased amount of legume in the pasture as well as from decreased carry-over of inoculum of Ggt. Because of the rotations used and the grass species involved, such observations have no direct bearing on the management of take-all in Europe.

In experiments in Oregon, USA, the herbicides diclofop, difenzoquat, dinoseb or mecoprop were incorporated into soil in field plots. Inoculum in the form of oats colonized by Ggt was then also incorporated before winter wheat seed was drilled (Geddens *et al.*, 1990). Take-all severity, determined by assessing whiteheads, was decreased by each herbicide. In the absence of any satisfactory explanation for these effects, the main conclusion from this work was that, contrary to the suspicions of farmers, the herbicides were unlikely to exacerbate the effects of take-all in Oregon.

Some of the effect of the fungicide triadimenol against take-all when applied as seed treatment may be due to its plant growth regulatory activity (Bateman, 1986). It may be possible, therefore, to target such activity, in other compounds, against take-all. The herbicide diallate was reported to increase the tolerance of wheat to infection by *Fusarium* spp., probably by strengthening the cell walls (Paul and Schönbeck, 1976; Schönbeck *et al.*, 1977). Thickened cell walls in roots, especially in the endodermis, may obstruct the entry of Ggt into the stele; the weak pathogen, Ggg, is less able to cross this endodermal barrier (Hornby and Fitt, 1981).

4.3 Biological Control

Biological control by antagonistic microorganisms is an alternative strategy, which in certain circumstances might be integrated with other strategies (see 4.4). Diseases that are caused by soil-borne, root-infecting fungi are a popular target for biological control by introduced organisms because control is often difficult by other means and because there is a precedent in that some of these diseases may already be subject in some situations to naturally occurring biological control by resident soil organisms (e.g. TAD). By being intractable in the traditional areas of disease control (fungicides, breeding for resistant host cultivars), take-all became one of the main prospects and challenges for biological control (defined as a self-perpetuating, safe, inexpensive, environmentally sound disease control practice) of a plant disease at a time when biotechnology was 'poised to make a major impact on society' (Hemming and Houghton, 1993). However, soil is a complex medium in which manipulating or augmenting populations of microorganisms has proved to be difficult. Less demanding targets than take-all for introduced biological control agents (BCAs) are likely to be damping-off diseases and post-harvest diseases (Wilson and Wisniewski, 1992). Seed coatings offer a good prospect for controlling damping-off, because the BCAs need to colonize only a small area of mainly new and microorganism-free tissue and occupy it for only a short time (Paulitz, 1992).

An upsurge in biological control interest in take-all research starting in the 1960s has been marked by the demonstration and naming of TAD, the discovery of *Phialophora* spp. which seemed to suppress the onset of take-all, and an infatuation with introduced BCAs that have been the subject of the bulk of the research since. Subsequently, a tactical withdrawal from the promotion of single BCAs in the face of inconsistent and small effects to a more realistic position was forced on the enthusiasts. In the past 10 years or so a favourable research milieu for introduced BCAs has arisen because of trends towards greater sustainability in agriculture, increased public concern about the use of synthetic pesticides (which are perceived as less environmentally acceptable for controlling diseases than is biological control) and rapidly evolving technologies in molecular biology and genetics (Thomashow and Weller, 1996). In the past, the BCAs for take-all have shown unwelcome traits such as not working in soils with high clay content. Research on genetic engineering, new application strategies and the use of biological control as part of integrated control are giving the subject a new lease of life. Already, the investigation of mechanisms, such as evidence for antibiotic production by BCAs *in situ* (Thomashow and Weller, 1988; Thomashow *et al.*, 1990), has taken the place of straightforward screening and testing of organisms as the main area for research. The stage may now be set for overcoming the long-standing obstacles of inadequate colonization of the target site by BCAs and the variability of their suppressive mechanisms (Thomashow and Weller, 1996).

4.3.1 *Natural biological control*

TAD soils are the best known of several types of take-all-suppressive soil (see 2.4.1). TAD develops after severe infection and, once induced, can provide an effective form of natural biological control that may have been exploited in long runs of cereals for decades by some farmers in the UK (see 3.3.3). The most popular explanation for TAD is microbial antagonism. Many have searched for specific antagonists and, with much publicity, several microorganisms have emerged that have decreased take-all build-up when introduced as BCAs (see 4.3.2). However, these BCAs have rarely reached the level and consistency of control achieved by TAD. This may be because TAD is the expression of a collection of BCAs and control mechanisms which ensures operation over a wider range of conditions than that within which single BCAs operate. Analyses of disease progress curves at Rothamsted, based on large datasets (see 2.5.4), have recently revealed a manifestation of TAD in crops near anthesis that is affected by changes both in inoculum of Ggt and in the environment. Furthermore, there is the possibility of changes in Ggt itself during consecutive crops of wheat. The findings from a recent examination of this are summarized in Table 5.1. Additional evidence is the observation that a diffusible fungal growth inhibitor on buffered media at pH 3.5–5.0 was associated more with isolates from TAD soils than with those from non-decline soils (Romanos *et al.*, 1980). Also, 27% of 'fresh' Ggt isolates from a continuous sequence of wheat produced the inhibitor on PDA buffered at pH 4.0, whereas no inhibitor was produced by isolates from cereals in rotation (Slope and Gutteridge, 1982).

Once TAD is established it is usually persistent and effective, though there have been reports of take-all patches occurring in decline situations (an example is given in Section 1.2.3). These patches usually occur in years that are exceptionally favourable to take-all. Suggestions that this indicates the breakdown of decline may or may not be justified. However, some of the implications of losing TAD in the UK are given in Section 2.4.1 and possible reasons for an apparent or real breakdown are the following.

- **Weather.** In years with weather unfavourable to take-all, the amount of disease may not be sufficient to stimulate TAD, or it may provide a mild stimulus only, resulting in a weak expression of the phenomenon in subsequent crops. There may also be seasons that are more directly unfavourable to antagonistic organisms that cause, or contribute to, the TAD effect.
- **Soil conditions.** Application of lime, for example, may favour the pathogen. Take-all of bent turf (*Agrostis* spp.) occurs after heavy liming to correct excessive acidity (usually following excessive use of ammonium fertilizers), or in new turf established after fumigation. Normally, resident BCAs hold it in check (Deacon and Berry, 1993).
- **Different cereals.** TAD built up under one cereal species may not be of sufficient strength to suppress disease on a more susceptible cereal (see 2.4.1).

- **Soil type.** Experiments at Rothamsted (on flinty clay loam) and Woburn (sandy loam) showed that the patterns of take-all development were similar until TAD was apparently established. At Rothamsted the disease then remained at low levels, but at Woburn it increased again (Prew *et al.*, 1995). No contemporary comparisons with crops at peak disease were available and so it was not possible to confirm breakdown of TAD. However, the sudden surge in disease suggests that TAD may be more unstable on the less fertile soil at Woburn than at Rothamsted.

Alternative hosts can influence the development of TAD and its usefulness. In an experiment at Rothamsted (already referred to in 3.3.3), the grass *Holcus lanatus* (a known carrier of Ggt) was grown for 2 years, followed by three consecutive winter wheat crops. Severe take-all was observed in all the wheat crops. In comparison, where three consecutive winter wheat crops followed two crops of spring cereals, severe take-all was observed in the first two winter crops, but not in the third. Also, natural antagonists can influence the rate of development of take-all and the onset of TAD. For example, large populations of *Phialophora graminicola* were present under grass leys and were associated with a delay in the build-up of severe take-all (Slope *et al.*, 1979). In pots, addition of *P. graminicola* to soils from different cropping sequences was shown to decrease the severity of take-all (Gutteridge and Slope, 1978).

Johnston (1994) reported that maximum grain yields of first wheats after a 2-year break on Broadbalk field at Rothamsted were 1.4–1.9 t ha^{-1} more than in continuous wheat. Third wheats after the 2-year break yielded less than continuous wheat and this was attributed to the build-up of take-all following the loss of TAD because of the break, but no disease data were presented. The disease data associated with the yields in Table 4.7 are what could be gleaned from the Rothamsted archives and suggest that take-all was not severe during the period 1985–1990. TAD has often been used as a convenient explanation for yield recoveries in continuous cereals, despite unconvincing levels of disease, or even the absence of any disease data. Frequently, such information as is available may

Table 4.7. Maximum grain yields of, and take-all in, first, second, third and continuous crops of winter wheat, cv. Brimstone, on Broadbalk, Rothamsted, 1985–90.

| | Wheat after a 2-year break | | | | | | | |
| | First | | Second | | Third | | Continuous wheat | |
Treatment	(t ha^{-1})	TAR	(t ha^{-1})	TAR	(t ha^{-1})	TAR	(t ha^{-1})	TAR
NPK[a]	8.61	0	7.85	62	6.47	110	6.69	63[d]
FYM[b]	7.89	–	5.86	–	5.37	–	6.17	–
FYM[b] N[c]	9.36	0	8.64	38	7.59	86	7.93	52

Means for first and second wheats based on 6 years; means for third wheats based on 5 years. NPK, fertilizer; FYM, farmyard manure; TAR, take-all rating (0–300).
[a]Best yield with 288 kg fertilizer N ha^{-1}; [b]35 t ha^{-1} FYM; [c]96 kg fertilizer N ha^{-1}; [d]240 kg fertilizer N ha^{-1} (nearest treatment to that used for yields).
Based on a table of yields supplied by A.E. Johnston.

be interpreted in a number of ways. In Broadbalk, the cropping of the plots that ultimately had 2-year breaks could also have contributed to the loss of TAD: five cycles of fallow/wheat/wheat or beans/wheat/wheat, started in 1968, preceded the 2-year break. The wheat referred to as continuous wheat in Table 4.7 was grown on plots that had grown wheat annually since at least 1959. Before this date there had been a regular cycle of 4 years in wheat, 1 year fallow (started in 1931 to control weeds after mostly wheat with only very occasional fallows since 1843). The yield recovery after the third wheat, shown by comparing yields with those of continuous wheat, ranged from 0.22 to 0.80 t ha^{-1}, according to the fertilizer regime. Although the yields from continuous wheat were in the range 78–85% of the first crop after the break, which is what might be expected for crops protected by TAD at Rothamsted, the improvement in yields in relation to those of third crops after the break was not particularly impressive. One way of reconciling the points raised by these observations is to propose that the effectiveness of TAD lessens during long runs of wheat. In support of this, there is limited evidence from Broadbalk to suggest that populations of Ggt with 'decline' characteristics change, or revert, in the long term (Hornby and Gutteridge, 1995).

Factors that influence the rate of development of the take-all epidemic and its severity affect, in consequence, the extent and timing of TAD. Manipulating these factors, which include crop sequences, sowing date and fertilizer regimes, can affect the disease and its natural biological control (Chapter 3 and Section 4.1), but not always in the desired direction for both, because, depending on the stage of the epidemic, decreasing the disease may depress or delay the onset of the biological control.

The most direct form of natural biological control is to disrupt the epidemic by introducing a non-susceptible crop into a rotation. This depletes the inoculum of Ggt and perhaps changes its nature. Disrupting epidemics in crop sequence experiments is usually achieved by growing an oat break crop, which depletes inoculum almost completely, whilst leaving nutrient residues in the soil similar to those left by wheat. Some grass species provide a benefit additional to the above through being hosts to antagonistic *Phialophora* spp. that continue to suppress Ggt when cereal cropping is resumed (Deacon, 1981; Wong, 1981). Although the phenomenon of cross-protection by antagonistic *Phialophora* spp. is well known, the manipulation of these fungi in natural populations to control take-all is not part of agricultural practice in the UK, presumably because of the hitherto economic unacceptability of growing grass in cereal-growing regions. Where leys have been introduced for 1 or 2 years, it has seldom been specifically for the purpose of increasing biomass of *Phialophora* spp., but it is often assumed by researchers and advisers that such a build-up contributed to those ley grasses acting as effective break crops. The implications for this form of biological control of grasses being grown in some fields that are set-aside in accordance with EU regulations are not understood (see 3.3.1 and 3.3.4). *G. cylindrosporus* (the teleomorph of *P. graminicola*) has been recorded on *Poa annua* and *Lolium perenne* in the USA (Jackson and Landschoot, 1984), but the natural biological control referred to as '*Phialophora*' in the UK does not seem to have been recorded there, or in Australia (hence the entry in Fig. I.1).

The idea of combining the effects of different natural biological control phenomena seems not to have been pursued. Some experimental work at Rothamsted in the late 1970s suggested that allowing *P. graminicola* to build up before a sequence of wheat crops not only delayed the peak in take-all, but also decreased it compared with a peak that developed in the absence of that fungus. This is discussed further in Section 7.3.4.

4.3.2 Biological control by introduced organisms

The biological control of take-all by bacteria and fungi was demonstrated in pots of sterilized soil in Canada as long ago as the 1930s (Sandford and Broadfoot, 1931). Much effort since the early 1970s has been devoted to screening and testing microorganisms, particularly from TAD situations and so-called 'suppressive' soils (see 2.4), for activity against Ggt. 'Specific' suppression is transferable and destructible (Shipton, 1972; Shipton *et al.*, 1973; Cook and Rovira, 1976). It usually occurs in response to the presence of a severely diseased host and virulent Ggt and is thought to be due to a build-up of specific antagonistic microorganisms. It contrasts with non-transferable 'general' suppression caused by increased microbial activity associated with organic amendments, improved fertility, high soil temperature and ammonium-nitrogen uptake by roots.

In the last 25 years, biological control has emerged as a possible solution to the control impasse and much has been promised. However, to a large extent the promises have not materialized, many questions remain unanswered and many researchers remain sceptical. It needs to be asked whether the expectations were realistic for the UK in the first place, and whether there has been an uncritical expectation that findings elsewhere translate readily to northern Europe. Raised expectation of biological control of take-all by introduced BCAs has been a worldwide phenomenon. Many of the claims have come from the USA and Australia, although there are a few European ones. There are now several patents, or patent applications, on some organisms and processes (Table 4.6). One commercial biocontrol product against soil-borne diseases is GUS 4000 (USA) based on *Bacillus subtilis*, recorded as effective against *Fusarium*, *Rhizoctonia* and *Gaeumannomyces* on cereals (Becker and Schwinn, 1993). However, as yet nothing has emerged that can be used to decrease take-all reliably and effectively in UK fields (nor, at the time of writing, that is widely used commercially elsewhere). Wong (1994) gives the reason that no commercial product specifically for take-all is on the market is the low to moderate control achieved by BCAs and their considerable variability.

Bacterization of wheat seed by species of *Pseudomonas* and *Bacillus* have given yield increases of 10–20% in fields throughout the world (Wong, 1994). In the USA fluorescent pseudomonads have been applied as seed treatments to wheat to control take-all and pythium root rot (Thomashow and Weller, 1990). In wheat suffering from take-all, yield increases averaged 17% in experimental plots and 11% in commercial-scale tests. Performance in the field has, however, been inconsistent and significant improvements in yield have occurred in only 60% of treatments. This major impediment to wide-scale commercial use was explained as follows:

- Inoculum density of Ggt was often too low.
- Environmental conditions were unsuitable for disease development.
- BCAs have no ability to stimulate growth directly.
- Non-target pathogens offset beneficial effects.
- Variable colonization of wheat roots may be a critical factor.

In the Pacific Northwest of America it was estimated (Heim *et al.*, 1986) that some 624,850 acres (252,877 ha) were affected by take-all, 43.4% of which had moderate or severe disease. If seed coating technology, using *P. fluorescens*, were introduced, it was predicted that yield increases would be between 3 and 25% (mean 5–10%). It was assumed that overall this would generate US$13.6 million (maximum US$37.9 million), but it was also noted that if yield responses and economic values of wheat were low, then farmers could actually lose money by adopting this technology.

Considerable public and commercial sector funding has been allocated to biological control of take-all in Australia and the USA, although it is now generally accepted that no single organism or combination of organisms is likely to be effective under all conditions of wheat production. In the UK, *P. fluorescens* strains 2-79 (NRRL B-15132) and 13-79 (NRRL B-15134) (see 4.3.3) showed good persistence in soil of low microbial, chemical and physical complexity (Paveley *et al.*, 1992), but much wheat in the UK is grown on complex, fertile soils.

The main research approaches to controlling take-all by BCAs worldwide are:

- To elucidate the mechanisms of action of BCAs with regard to root colonization, antibiotic production and genetic control.
- To assess the performance of combinations of BCAs, using up to four bacterial strains and possibly fungi.
- To study interactions among BCAs in combinations (and the role of 'promotor' strains, which on their own have no apparent biological control activity).
- To relate performance of BCAs to a range of soil variables with the aim eventually of identifying optimal combinations for specific situations.

Before a BCA can be a commercial proposition it must meet several requirements. There must be a market opportunity. Until recently, the absence of an effective fungicide and of a wheat cultivar with resistance contributed strongly to the impression that there was indeed an opportunity for a BCA that could control take-all. However, the case for such a BCA must now accommodate the possibility of new chemical seed treatments (see 'Seed treatment', Section 4.2.2). In its favour, the use of a 'natural agent' tends to be perceived as more environmentally desirable than the application of a chemical treatment. Ultimately, the BCA must carry with it a guarantee of reliability and must be patentable, registerable and cost-effective.

It has been argued (Ryder and Rovira, 1993) that, in practice, mixtures of BCAs or integrated control involving BCAs would be necessary to control root disease problems comprising several diseases in Australia (e.g rhizoctonia bare patch, common root rot and fusarium crown rot, as well as take-all). Some combinations might be quite unexpected, such as the bacterium *Pseudomonas corrugata* 2140R and the earthworm *Aporrectodea trapezoides*, which gave a significant increase in shoot weight in a greenhouse study, possibly because the worm increased root colonization by the seed-inoculated bacterium and so increased the bacterium's ability to inhibit take-all (Stephens *et al.*, 1994b). In experiments using a *lux* marked strain of *P. corrugata* as a BCA, suppression of primary infection (see 2.5.4) was predicted and demonstrated, but durable control of take-all will also require suppression of secondary infection, which may be achieved only after extensive colonization of the nodal root system by the antagonist (D. Bailey, personal communication). In practice, the most fundamental requirement for a BCA is that it attains a population of sufficient size and activity to control its target (Thomashow and Weller, 1996). Knowledge of the mechanisms and factors involved in this is growing, but questions remain about signals and control mechanisms, and the involvement of the host plant needs to be clarified.

4.3.3 *Bacteria as biological control agents*

Research worldwide
The work on biological control of take-all, especially using *Pseudomonas* spp. as applied BCAs, in the Pacific Northwest of the USA under the leadership of R.J. Cook of the USDA Agricultural Research Service, Pullman, Washington, has been very influential (recent reviews: Cook, 1992a, 1994). The wider objectives were to control the root disease complex of take-all and rhizoctonia and pythium root rots, and to maximize yields in irrigated wheat in rotations that favour take-all in arid and semi-arid regions (Cook, 1994). These objectives emphasize circumstances different from those in northwest Europe, where cereals are rarely irrigated and take-all is less obviously part of a root disease complex. In Washington, earlier field results with take-all were variable, although there were some promising reports of suppression, with a concomitant 20% increase in yield (Weller, 1983; Weller and Cook, 1983). Partial and inconsistent effects of bacterial BCAs identified the need to evaluate the 'total management system' for take-all control (Cook, 1994). A 'customized' approach to the use of BCAs in integrated control was developed for biological control of the root disease complex on a local basis (Cook, 1992a). This, it was claimed, had nearly achieved yields in continuous wheat that were formerly possible only for wheat in a 2-year rotation, and yields in a 2-year rotation that were formerly possible only in a 3-year rotation. The identification of agronomic factors, including herbicide use, fertilizer placement and sowing date, to '(a) favor resident antagonists, (b) favor introduced antagonists, (c) reduce the inoculum potential of the fungus, and/or (d) minimize predisposing stresses on the host' was vital in this. Research begun with genetically engineered bacteria in the field in Washington in 1988 has now terminated (e.g. Cook *et al.*, 1990). The fluorescent pseudomonad strain 2-79RN$_{10}$ had been altered with the *lac*ZY tracking system and named 2-79RNL3. The altered strain did not differ in behaviour from the unaltered strain and its spread was not more than about 30 cm during 326 days

of the experiment, although volunteer lentil plants were colonized and contamination of the planting machinery occurred. None of the bacterial treatments controlled take-all in this experiment.

Some of the materials and approaches developed in Washington were taken up in South Australia (Rovira *et al.*, 1990) and elsewhere, and there have been numerous searches for local pseudomonad strains with activity against take-all (e.g. Sivasithamparam and Parker, 1979; Mróz *et al.*, 1994).

D93, a Tn5 mutant of *P. fluorescens* (strain CN12 from a take-all-suppressive soil in the Ningxia autonomous region of China), acquired enhanced biological control of take-all on wheat from a single transposon insertion (Peng *et al.*, 1994). In field trials conducted with winter wheat and spring wheat during 1988–1992, seed treatment and foliar spray decreased whiteheads by 65–90% and increased yield by 15–44%. In trials at three locations, D93 increased yield by 27.2% and CN12 by 12.3%. Preliminary commercialization started in China in 1991 when the BCA was used on over 3800 ha, giving an average yield increase of 20%. However, levels of disease control and yield increase varied with season, location and disease severity.

P. fluorescens strains 2-79 and 13-79 from Washington state, USA, have been tested in the UK and the results (see section on 'Experimental results' below) led to the conclusion that these isolates have potential to decrease take-all and increase the yield of wheat in the UK, but the beneficial effects were inconsistent (Capper and Higgins, 1993). Perhaps because of such results, this kind of biological control research is not now pursued to any extent in the UK, whereas it is actively continuing in other countries, particularly the USA, where in recent years expectations have been high that farmers in the Pacific Northwest would soon have a commercial seed treatment using *P. fluorescens* (Cook, 1995). Bacteria other than pseudomonads have also been reported to show antagonistic activity to take-all in the field (see section on 'Experimental results' below).

A number of hypotheses proposed for the mechanisms responsible for the specific suppression of take-all were summarized by Rovira and Wildermuth (1981), who provided evidence for a massive proliferation of 'asporogenous' bacteria on infected roots as a forerunner to TAD. These bacteria were most probably Gram-negative fluorescent pseudomonads (Weller, 1983). Such organisms are seen as strong candidates for biological control because of their ecological and physiological characteristics. They grow quickly in the rhizosphere which is their natural habitat (Rouatt and Katznelson, 1961), are nutritionally versatile (Stolp and Gadkari, 1981), produce antibiotics (Leisinger and Margraff, 1979) and siderophores (Misaghi *et al.*, 1982) that inhibit plant pathogens *in vitro* and produce other toxic secondary metabolites such as cyanides (Ahl *et al.*, 1986). Fluorescent pseudomonads quickly colonize roots and are present in lesions caused by Ggt (Rovira and Wildermuth, 1981). They can be introduced and become successfully established in the rhizosphere and on the rhizoplane of wheat (Weller, 1983). More recently, antibiotic production by fluorescent pseudomonads has been demonstrated in the rhizosphere of wheat (Thomashow *et al.*, 1990). In tests in pots of virgin bushland soil from the wheat belt of Western Australia, the ability of successive generations of root-piece inoculum of Ggt to cause severe take-all decreased (Charigkapakorn and Sivasithamparam, 1987). Diseased roots had more bacteria in total and more fluorescent pseudomonads (especially *P. fluorescens* biovar V), although more *P. putida* occurred on healthy roots. The fluorescent pseudomonads, of which a third were antagonistic to Ggt *in vitro*, represented up to 9.3% of the total of aerobic bacteria.

The production of secondary metabolites that are antagonistic to Ggt is an important trait (Weller and Cook, 1983; Gurusiddaiah *et al.*, 1986b). Antibiotic production plays a significant role in defence of plants by rhizobacteria and in the ecological competence of these bacteria (Cook *et al.*, 1995). Production of phenazine (Thomashow and Weller, 1988) and phloroglucinol antibiotics (Keel *et al.*, 1990) account for most of the natural defence provided by fluorescent *Pseudomonas* strains (including *P. aureofaciens*, strain 30-84; Thomashow *et al.*, 1990) associated with TAD. Plant defence of this type may be seen as analogous to defence by phytoalexins, even to the extent that similar enzymes are involved in the production of phytoalexins and 2,4-diacetylphloroglucinol. The selection pressure imposed on plants by soil-borne pathogens may be towards plants that support and respond to rhizosphere microorganisms antagonistic to these pathogens (Cook *et al.*, 1995).

Bacterial isolates from the rhizosphere of wheat – some inhibitory to Ggt *in vitro*, some not – were tested in Australia for their ability to control take-all in pots in a natural, non-sterilized soil–sand mixture with added Ggt inoculum (Ryder and Rovira, 1993). A group of non-fluorescent pseudomonads (identified as most closely related to *P. corrugata*) gave significant and reproducible suppression. One strain, 2140, controlled take-all *in vivo* at 15°C and inhibited Ggt *in vitro* in the range 10–25°C. Another non-fluorescent Australian *Pseudomonas* sp. (AN5) was effective on plates, in pot experiments and in field trials at dryland sites. Antibiosis was the most significant factor in biological control and bacterial exopolysaccharides (EPS) may be important (Nayudu *et al.*, 1994). *P. fluorescens*, strain 2-79, decreased disease severity but did not improve crop growth. In these tests, the fungicide triadimenol was not as good as the best bacterial strains. Strain 2140 effectively colonized primary and secondary roots of field-grown wheat following seed inoculation. In other work (Zaspel, 1992), rhizosphere bacteria inhibited pathogens *in vitro* and stimulated the growth of wheat. Outdoors, *P. putida* strain 27 increased yields by 8.1–19.4% (where take-all infection was 2.7–15% and the control of infection 6.3–25%), whereas a *Bacillus* sp. achieved a 9.5–10.8% increase in yield. Seed treatment was more efficient than peat preparations, although successful colonization of wheat roots in autumn by BCAs did not guarantee increased yields.

P. fluorescens strain CHA0 has received particular attention in Europe (e.g. Wüthrich and Défago, 1991), where it was reported to be an effective BCA against take-all and several other soil-borne diseases in the glasshouse (Maurhofer *et al.*, 1994). Research reports, some apparently contradictory, present evidence of more than one mode of action. Strain CHA625, obtained from CHA0 after Tn5 mutagenesis, did not produce the antibiotic 2,4-diacetylphloroglucinol and was less able to suppress take-all under gnotobiotic conditions. A cosmid from the genomic library of CHA0 restored the ability of CHA625 to produce the antibiotic and

Table 4.8. Mechanisms of bacterial BCAs for controlling take-all that have been favoured by research in different regions.

Interests	Countries	Interest in natural biological control
HCN, secondary metabolites[a]	Switzerland, France	As a source of BCAs
Siderophores	USA, The Netherlands	As a source of BCAs
Antibiotics	USA	As a source of BCAs
Combinations	Australia	As a source of BCAs No TAD?
Little interest (BCAs disappointing)	UK	TAD, Phialophora

[a]Interactions with clay minerals.

partially restored the ability to suppress take-all (Keel *et al.*, 1990). A functional global regulator gene *gacA* is necessary for CHA0 to protect dicotyledonous plants against root diseases, but not for the protection of *Gramineae* (Schmidli-Sacherer *et al.*, 1997). A GacA⁻ mutant failed to produce 2,4-diacetylphloroglucinol, pyoluteorin and hydrogen cyanide, but overproduced the iron chelators, pyochelin and pyoverdine (Pvd), and was still able to protect wheat against Ggt. Pyochelin and/or its precursor, salicylic acid, seemed to be involved in the ability of a GacA⁻ Pvd⁻ mutant to protect against Ggt. (Besides production by microorganisms, salicylic acid has several roles in higher plants (Raskin, 1992); it is considered in the context of systemic acquired resistance in Section 7.1.3.)

None of 48 isolates of fluorescent pseudomonads isolated from a soil in France fertilized with ammonium sulphate ('low receptivity' level, i.e. take-all developed poorly in the soil) or with calcium nitrate ('high receptivity' level) produced phenazine, indicating that this antibiotic was not involved in suppressing take-all, unless there was another form produced *in situ* which was not detected by the method used (Sarniguet *et al.*, 1992b). However, a large proportion of the strains produced HCN and all the highly cyanogenic strains from the soil receiving ammonium sulphate were highly antagonistic to Ggt *in vivo*. Highly cyanogenic strains also increased root growth in the presence of Ggt, compared with the control plants grown in non-bacterized soil in the absence of Ggt. Cyanides act as toxic metabolites directly against Ggt, but can also modify the plant's physiology and enhance plant defence reactions (Défago *et al.*, 1990).

Antagonists may cause lysis of hyphae of Ggt on wheat roots (Campbell and Faull, 1979) or even inhibit the trophic response of the fungus to the wheat root (Pope and Jackson, 1973). Some mechanisms may be non-specific, e.g. factors related to colonization such as competition for sites and nutrients (Weller, 1983) and root stimulation (Weller, 1985), or production by bacteria of siderophores which have a higher iron-binding ability than fungal siderophores (Wong, 1985). Table 4.8 is a simple summary of mechanisms of major interest and the location of research groups working on them.

Factors affecting BCAs
It is thought unlikely that the same mechanisms of suppression and cross-protection (reviewed by Wong, 1981) operate in all environments. The build-up and maintenance of suppression in natural conditions will probably depend on soil fertility, the quantity of plant residues and the levels of pathogens and antagonists that they carry, soil temperature and moisture and the length of time between crops. Much of the experimental work which attempts to determine the mode(s) of action of particular BCAs has been carried out under controlled conditions in the laboratory or glasshouse. Whilst it is important to understand how a BCA accomplishes its control, it is also important to understand how that control is affected by environmental factors such as temperature, soil texture, pH, soil matric potential and the activity of other microorganisms. For example, soil matric potential can affect root colonization by fluorescent pseudomonads, soils with very high available water probably providing inadequate oxygen for bacterial multiplication, whilst in very dry soils there is inadequate water available to maintain cell turgor (Howie *et al.*, 1987). Soil pH in particular affects colonization (Weller, 1988) and the degree of toxic activity of antifungal compounds produced by microorganisms is also affected by pH (Brisbane and Rovira, 1988).

It has been known for some time that the form of nitrogen fertilizer affects disease expression (Huber *et al.*, 1968; Section 3.4.1) and that the percentage of fluorescent pseudomonads which are antagonistic to Ggt *in vitro* is greater in the rhizosphere of wheat fertilized with NH_4^+-N than in the rhizosphere of wheat fertilized with NO_3^--N (Smiley, 1978). The uptake of NH_4^+-N decreases pH at the root surface and enhances the environment for take-all-suppressive organisms (Christensen and Brett, 1985). Bioassays of soil in France, from fields fertilized with either ammonium sulphate or calcium nitrate, revealed different levels of receptivity to take-all 40 days after fertilization (Sarniguet *et al.*, 1992a). Isolates of pseudomonads from these two soils were tested *in vivo* in a soil artificially infested with Ggt and were shown either to decrease take-all (beneficial isolates), aggravate the disease (deleterious isolates) or have no effect on the disease (neutral isolates). Principal component analysis and factorial discriminant analysis showed that the fluorescent pseudomonads which were antagonistic *in vivo* were also more prevalent in the soil fertilized with NH_4^+-N and that the deleterious bacteria dominated the soil receiving NO_3^--N. These results indicate a role of nitrogen in antagonistic microbial activity in soils and draw attention to the complex interactions between soil microorganisms.

The impact of microbes on take-all is, therefore, the result of divergent activities, rather than just the consequences of the presence and activity of antagonistic bacteria. Moreover, after soil receptivity decreased there was a delay before there was a decrease in take-all in the field with NH_4^+-N fertilization. For this reason, the establishment of an antagonistic microbial population may be regarded as a dynamic process, induced early on seminal roots, stimulated by external factors such as form of nitrogen, and with root necrosis playing a central role by providing a niche for the multiplication of pseudomonads. Later on, the nodal roots growing out from the crown over the seminal roots benefit from these changes (Sarniguet *et al.*, 1992a).

Strain 2–79 of *P. fluorescens* can give excellent early control of take-all, but control breaks down by mid-season when the inoculant population has declined. Its numbers rise again in infected regions of the root, presumably because the pseudomonad can exploit the locally abundant nutrients from disease lesions. *P. fluorescens* can antagonize pathogens in montmorillonite, but not illite clay soils, because only the former make iron readily available. Soil type, therefore, affects the success of inoculant pseudomonads for take-all control (Deacon and Berry, 1993).

Experimental results
Some successes in field trials (Weller and Cook, 1983; Capper and Campbell, 1986) have been mentioned already. In ADAS trials (mainly unpublished), suppressive bacteria applied either as a seed treatment or in alginate beads have produced variable results in terms of significant yield benefits. Positive effects (i.e. decrease in root infection or increase in yield) were observed in approximately 80% of treatments over a 7-year period, but the differences were not always statistically significant. Soil pHs have been generally higher in UK trials than those in the USA, e.g. Weller and Cook (1983).

Seed coating proved to be an appropriate method of application in ADAS trials and this allowed intimate contact between BCAs and the radicle as it emerged. Although such applications of BCAs readily achieved about 10^8 cfu per seed, they afforded little protection to the organisms themselves. Also, interactions with seed treatment fungicides – e.g. triadimenol and fuberidazole (Baytan) – resulted in increased phytotoxicity, thus placing constraints on chemical applications. Biological and chemical seed treatments need not necessarily be mutually exclusive, but the order of treatment is important; for example, the seed must be coated with bacteria before being treated with Baytan, otherwise phytotoxic effects in the form of reduced emergence may occur if fungicide penetrates the seed coat during bacterization. Additionally, asporogenous bacteria such as pseudomonads, if used for seed coating, need protection from desiccation if they are to survive in adequate numbers for any period of time, e.g. for up to 1 month.

Perlite granules were found to be unsuitable as carriers for bacterial BCAs in ADAS trials because they could not be coated with sufficient numbers of bacteria. Alginate beads showed more promise, with initial levels of *c.* 10^7 cfu of bacteria per bead, decreasing to 10^4–10^5 cfu after storage at 22°C for 6 months. The beads also acted as foci of inoculum which gave a sustained slow release of the organisms. In pot experiments the levels of take-all infection and the reduction in infection achieved with BCAs were greater in cvs Avalon and Wembley than in cvs Axona, Brimstone, Brock, Fenman, Mercia, Minaret or Norman. *P. fluorescens* strain 13-79 applied either in alginate or as a seed treatment significantly reduced infection in both Avalon and Wembley. Significant ($P < 0.05$) decreases in infection were also achieved in cv. Avalon using *Bacillus* spp. in alginate beads. Preliminary studies indicated that there may be a differential response between wheat cultivars to colonization by BCAs.

Results of ADAS field experiments are summarized as follows.

- 1983: Promising results were obtained by the addition of *Bacillus pumilus* to spring wheat at sowing on a very organic soil where take-all occurred naturally (loamy fen peat, pH 7.5) in Cambridgeshire. A significant reduction in infection and an increase of 50% in yield were observed in this very dry year (Capper and Campbell, 1986).
- 1985/86: *Bacillus pumilus* and *B. cereus* var. *mycoides* gave slight decreases in root infection and increases in yield. Variability was such that yield increases up to 16% were not statistically significant.
- 1987: *P. fluorescens*, strain 13-79, showed a slight but not significant yield increase when applied as a seed coating to wheat sown in London clay soil, pH 6.9–7.5. The use of alginate beads for inoculum delivery was promising in that 50% of alginate treatments out-performed the control.
- 1988: At Arthur Rickwood EHF, on fen peat soil in Cambridgeshire, TAR was significantly reduced ($P < 0.05$) by *P. fluorescens*, strain 13-79, applied as a seed coating in carboxymethyl cellulose and by *P. fluorescens*, strain 2-79, applied in alginate beads. As peat-based microgranule formulations, strains 2-79 and 13-79 decreased take-all in spring wheat on fen peat soil, but the decrease was significant only with strain 13-79. In a trial in Northamptonshire on loamy drift over Oxford clay, pH 7.0, seminal root infection was significantly reduced by strain 2-79 in alginate beads (with or without strain 13-79), or as a seed coating.

At Rothamsted farm (clay loam with flints soil) and Woburn Experimental Farm (sandy loam soil), seed coatings with strains 2-79 and 13-79 of *P. fluorescens*, prepared by the method of Weller and Cook (1983), and in-furrow treatments with *Bacillus* spp. from Bristol University were tested against naturally occurring take-all in much-replicated small plots (see 6.3.1) in 1983–1985. These bacteria had variable, short-term effects and overall were ineffective (Hornby, 1987; Hornby *et al.*, 1993). Of the control measures that have been tested on winter wheat in the field at Rothamsted, BCAs have usually been ranked below other factors (sowing date > fungicides > fertilizers; Hornby, 1989) that have decreased take-all. TAD, on the other hand, has been one of the more effective and reliable ways of decreasing disease. This practical experience conflicts with the more positive impression of BCAs created by many research reports from overseas and popular articles that have promulgated the views expressed in such reports.

4.3.4 Fungi

Hypovirulent isolates of Ggt may invade host roots in the same way as virulent isolates, but more slowly, and they could be common in some field conditions (Deacon, 1981; Wong 1981), although Hornby *et al.* (1990b) found only 11 hypovirulent isolates in a collection of 527 isolates from Rothamsted and Woburn fields. There is a possibility that hypovirulent isolates have been underestimated in the past because of a tendency to isolate Ggt from lesioned roots. Hypovirulence appears to be under genetic control and such isolates are significant in their abilities to compete with, and to induce resistance to, virulent isolates (Asher, 1981; Wong, 1981). Tivoli *et al.* (1974) hypothesized that changes in pathogenicity within populations of Ggt may explain TAD, after some hypovirulent strains of Ggt had been found associated with a long-term monoculture of wheat. This suppressive property was utilized in further biological control studies in France (Lemaire *et al.*, 1979a,b). Non-aggressive, auxotrophic mutants obtained from nitrosoguanidine-treated protoplasts of Ggt retained their ability to immunize wheat against take-all, as did mutant strains resistant to the fungicides benomyl and carboxin derived from a hypoaggressive parent (Rochefrette *et al.*, 1979).

Non-aggressive *Phialophora* spp., commonly found on grass roots and on cereal roots following grass and other break crops, have the ability to suppress take-all. The relationships between Ggt and weakly pathogenic *Phialophora* spp. and Ggg were discussed in detail by Deacon (1981). The first experiments towards developing them as applied biological control agents were in the early 1970s (Wong and Southwell, 1979; Wong, 1981) and recently a strain of Ggg (see 5.2 for the relationship between these fungi) has been patented in Australia for use in take-all control (Table 4.6). It was claimed that applications of inocula of Ggg and *Phialophora* sp. (lobed hyphopodia) mostly increased yields by over 20%. Fungi can remain active in soil or in wheat roots at moderate water potentials that are sufficiently low to inhibit bacterial activity. One cold-tolerant isolate of Ggg (90/3B) increased wheat yields by 30–45% over three seasons in New South Wales. It was this isolate and another cold-tolerant isolate, *Phialophora* sp. (lobed hyphopodia, isolate KY), and their use for take-all control that were patented and reported as being developed for commercial use (Wong, 1994). Cold-tolerant isolates had at least comparable growth rates to Ggt at 5°C and effectively controlled take-all arising from artificial inoculum in field experiments when applied on colonized oat grains (Wong *et al.*, 1996). Demonstration of their efficacy against naturally occurring take-all is still required and cheaper and more effective formulations are needed to make this biological control more cost-effective. There have been trials using a strain of Ggg in the Pacific Northwest, USA, both alone and in combination with fluorescent pseudomonads (Duffy and Weller, 1995). Local *Phialophora* strains have been tested on a small scale in the UK, but with no great success. Earlier experimental work, much of it done in Australia, using *Phialophora* sp. (lobed hyphopodia)/Ggg, often with artificially applied Ggt, gave the best control where take-all was less than severe. Work in Europe has also found that control of take-all in the field was only partial (Martyniuk and Myśków, 1984; Speakman, 1984a).

There is a history of investigations of the effects on take-all of fungi, mycoparasites and inhibitors, other than the closely related *Phialophora* spp. Since 1981 *Idriella (Microdochium) bolleyi* (Kirk and Deacon, 1987; Paveley *et al.*, 1992), a sterile black fungus (Speakman and Krüger, 1984) and *Trichoderma harzianum* and *Gliomastix murorum* (Mróz *et al.*, 1994) have been researched in Europe. Outside Europe, other sterile fungi (e.g. by Dewan and Sivasithamparam, 1988; Narita and Suzui, 1991) and *Trichoderma* spp. (e.g. Maas and Kotzé, 1987; Simon, 1989; Ownley *et al.*, 1992) have been studied. Much of this work, especially on *Trichoderma* spp. and their mechanisms of antibiosis, has been done in Western Australia. A sterile red fungus (SRF), applied on colonized ryegrass seeds, was able to colonize wheat roots in soil when placed at up to 3 cm away from the roots, but the best control of Ggt was achieved when inoculum of the SRF was placed close to the seed at sowing (Dejong *et al.*, 1993).

Frequently BCAs have achieved control of take-all in the early part of the growing season only, as was illustrated by a seed treatment with the fungus *I. bolleyi*, which significantly decreased early season take-all in an experiment on a silty clay loam (pH 7.1) in the UK (Paveley *et al.*, 1992). [In pot tests, *I. bolleyi* decreased the concentration of K in the shoot, the shoot water content and the plant dry weight 1 week after planting wheat and disrupted the stele in most seminal roots; however, the plants recovered rapidly because *I. bolleyi* did not penetrate beyond the epidermis of new roots (Fitt and Hornby, 1978). There is also a report that *I. bolleyi* may increase the likelihood of take-all after freezing temperatures in Canada (Sturz and Bernier, 1991).] The effects of host admixtures with oats and organomercury seed treatment also did not persist to mid-season. BCAs and organomercury improved establishment, possibly due to effects on fusarium seedling blight, and oat admixture reduced yield, presumably due to plant competition. High soil pH, use of peat and BCAs that were not adapted might explain the poor persistence or absence of control in these experiments.

Phialophora graminicola and *I. bolleyi* control take-all only when their populations greatly exceed that of Ggt, particularly when Ggt must infect from a critically low food base. Typically this occurs after a break, when residues containing Ggt are progressively fragmented and decomposed. Ggt may then need to exploit the naturally senescing root cortex of the succeeding cereal crop to rebuild its food base for infection and the weak parasites competitively exclude it (Deacon and Berry, 1993).

In vitro antagonism to Ggt was shown by *Trichoderma lignorum*, *Penicillium waksmanii*, *P. restrictum*, *P. oxalicum* and *P. multicolor* isolated from soil from non-irrigated plots of wheat that had been used to grow wheat for another 2 months after harvest (Amein, 1988). However, notable fungal antagonists such as *Trichoderma* and *Penicillium* did not form a large percentage of any of the wheat root mycofloras investigated in a Canadian study (Sturz and Bernier, 1991). In the UK, various root-colonizing fungi that were strongly antagonistic to Ggt in culture were less effective in pots in protecting wheat seedling roots from infection by Ggt inoculated subsequently, or already present naturally in soil, than were isolates of *Phialophora* sp. (lobed hyphopodia) or the slower growing (and less effective) *P. graminicola*, which are not antagonistic to Ggt in agar culture (G.L. Bateman, unpublished data). The often poor relationship between inhibitory effects in culture and an ability to control take-all, even in pots, has been well known since research in the early 1930s (Asher, 1981). A recent example of suppression of take-all and common root rot on wheat due to competitive root colonization by PGPF (plant growth promoting fungi) involved a *Phoma* sp. and a non-sporulating fungus from the rhizosphere of zoysiagrass, *Zoysia tenuifolia* (Shivanna *et al.*, 1996). In this work, barley kernels colonized by Ggt were used as inoculum in glasshouse tests in which the two PGPF suppressed take-all and increased plant growth. *Zoysia* spp. are used at lower latitudes in lawns and public parks (e.g. *Z. japonica*, *Z. matrella* and *Z. tenuifolia*) and in Florida and southern France as rangeland grasses (Chapman, 1996). It has been usual in the past to seek antagonists of Ggt where take-all is known to be suppressed (e.g. TAD), but it is not clear from the report of Shivanna *et al.* (1996) that the isolates from zoysiagrass were in any way associated with a Ggt-suppressive or take-all-suppressive soil. Relatives of Ggt are known to occur on zoysiagrass; for example, a fungus presumed to be Ggg was associated with diseased zoysiagrass turf in Illinois (Wilkinson and Kane, 1993) and *Gaeumannomyces incrustans* was associated with large patch disease of zoysiagrass, caused by *Rhizoctonia solani* (AG-2-2) in Kansas (Green *at al.*, 1992).

Mechanisms
Cross-protection, or immunization, of roots by *Phialophora* spp. occurs when the antagonist colonizes the root cortex before the pathogen and the process benefits from the phenomenon of natural root cortex death (Deacon, 1981; and see 2.1.1). Greater lignification and suberization of the endodermis and xylem vessels may also be involved (Cowan, 1978; Wong *et al.*, 1996). Competition for resources in naturally senescing cereal and grass root cortices seems to be a major mechanism of biocontrol of take-all by weakly parasitic fungi that naturally exploit these tissues (e.g. *P. graminicola* and *I. bolleyi*). These fungi are closely related ecologically and, in one case, taxonomically to Ggt. Control does not operate if Ggt has adequate nutrients in its inoculum food base. Nevertheless, it is impossible to exclude a contributory role of induced host resistance, especially when this is strictly localized to the vicinity of the control agent (Deacon and Berry, 1993).

Antibiotics that affect Ggt have been found in strains of *Trichoderma harzianum* (Almassi *et al.*, 1991) and *T. koningii* (Simon *et al.*, 1988a; Dunlop *et al.*, 1989). The effects varied amongst isolates; the antibiotic effects of some strains were associated with the presence of pyrones, but other mechanisms may be involved in some antagonistic strains of *T. harzianum* that did not produce pyrones (Ghisalberti *et al.*, 1990). Soaking seeds in trichodermin-3 decreased disease, apparently caused by Ggt, more than did seed treatment with a spore suspension of *T. lignorum* (*T. viride*) (Ponomareva, 1965).

Field results in northern Europe
When wheat was inoculated with hypovirulent strains of Ggt in France, plants became more resistant to further attacks by a virulent isolate of Ggt. There was a reduction in the size of take-all patches where a hypovirulent fungus was either incorporated in the soil or applied on the seed (Tivoli *et al.*, 1974). Research in Belgium involving inoculation experiments in the field showed that a strain with low virulence could benefit successive wheat crops for up to 2–3 years (Jadot *et al.*, 1982; Raepsaet and Defosse, 1982).

Hypovirulent strains were used in field trials conducted between 1980 and 1983 by French farmers. The treatment consisted of seeds coated at the farms with ground inoculum produced on autoclaved grains of barley. Yield increases were observed in about 60% of the trials, sometimes exceeding the control by over 20%, but more often ranging between 1 and 10% (Lucas *et al.*, 1986). Depending on the year, yield decreases occurred in 20–30% of the experiments, mainly where the treatment was applied to seed for a first wheat crop. Seeds needed to be moistened before coating with the ground inoculum and then air-dried, so the decreases could have been due to faulty procedures, rather than direct effects of the hypovirulent fungus. Also, inhibition of the protective effect of the fungus may have occurred where seeds had been treated previously with a fungicide, or when seeds were treated in damp weather and kept wet for too long.

The importance of soil type was explored using samples from 58 experimental sites in the mid-north part of France in 1981/82 (Lucas and Nignon, 1986). In bioassays, pots of these soils were infested with a virulent strain of Ggt, or with a hypovirulent strain, or not infested at all. After growing wheat in the soils for 3 months, disease index (Table 6.4) and root volume were assessed. The disease index was negatively correlated with sand and potassium contents and positively correlated with silt content. In some soils, however, the virulent or hypovirulent strain induced more root development than occurred in the uninfested control. This capacity for the plant to recover from take-all, or to react to infection by a hypovirulent strain, was correlated positively with the manganese content of a soil. Earlier studies by Lemaire *et al.* (1979b) had shown that different cultivars of wheat responded differently to inoculation by a hypovirulent strain of Ggt. Attempts have even been made to breed wheat with an increased ability to cross-protect against Ggt (Lemaire *et al.*, 1982). Cultivation of different wheat cultivars is another possible explanation for the variability observed in the protection afforded by hypovirulent Ggt by Lucas and Nignon (1986).

4.3.5 *Other organisms*

Other microorganisms have been associated with biological control of take-all. Much work on soil amoebae (Old and Patrick, 1979; Chakraborty and Old, 1982; Chakraborty, 1983; Chakraborty *et al.*, 1983; Chakraborty and Warcup, 1983, 1984, 1985) originated in Australia. In particular, several species of mycophagous amoeba were found in the take-all-suppressive soil of the permanent pasture plot at the Waite Agricultural Research Institute in South Australia in numbers greater than in non-suppressive soil. The organisms were thought to be contributors to a suppression founded on several different mechanisms functioning simultaneously. Also in Australia, earthworms have been implicated in the suppression of take-all (Stephens *et al.*, 1994a,b; Doube *et al.*, 1995; Stephens and Davoren, 1995). The mechanisms involved are not known, but increasing the amount of nutrients in soil available to plants, disturbing soil and ingesting hyphae have been suggested. The effectiveness of earthworms depends on rainfall, because they decrease the disease only at low soil matric potentials.

Other organisms have been positively associated with take-all, e.g. myxobacteria (which have bacteriolytic and cellulolytic capabilities, but little is known about their natural populations) and nematodes (Feest and Campbell, 1986). Indeed, Smiley *et al.* (1994) found that the combined damage of cereal cyst nematode and Ggt resulted in the highest overall yield loss in winter wheat, which caused them to draw attention to lack of information on the efficacy of biological control for plants affected simultaneously by two or more pathogens.

4.3.6 Prospects

In the now considerable literature on the biological control of plant diseases, it is difficult to avoid the observation that numerous laboratory studies and subsequent development have led to only a handful of commercial products for specific use and that these constitute a minute fraction of the sales value of plant protection products (Fig. 7.2). No one doubts the enormous antagonistic potential amongst microorganisms in or on dead substrates and on plant surfaces, but with one or two notable exceptions success in realizing it to control disease seems to be inversely proportional to the complexity of a crop's environment.

Reliance on one of the current putative BCAs to control disease in the field is at worst naïve and at best too optimistic. Future research needs to pay more attention to how natural biological control phenomena such as TAD, *Phialophora* and suppressive soils decrease take-all more reliably and effectively than the vast majority of introduced BCAs. This is likely to reveal that most natural phenomena have several components that together ensure relative robustness and should give pointers on how best to use BCAs. Attention should also be given to the possible exploitation of growth-stimulating rhizobacteria to enable plants to cope better with take-all.

Inconsistent field results are perhaps to be expected until there is better understanding of the ecology of BCAs, the interactions of BCAs with the root and the pathogen in the rhizosphere, and the interactions of BCAs with environmental factors, including those that are controllable to some extent (e.g. soil pH) and those that are uncontrollable (e.g. weather patterns). Much fundamental work still has to be done in such areas, but at the time of writing there appears to be little funding available for this in the UK.

Some results shown in Fig. 6.10 emphasize the problem of applying microorganisms to already complex soils. In this example, a large amount of inoculum of Ggt grown on sterilized oat grains was added to field plots, usually at 200 kg ha^{-1}. Although this usually resulted in moderate to severe take-all in a wheat crop then grown in the plots, it appeared that the naturally occurring microflora began to predominate in subsequent wheat crops. Therefore, rather than a robust TAD setting in after the artificially created peak of take-all, a further peak developed at the same time as occurred in adjacent plots with only natural infestations. This occurred in the fourth wheat after inoculation in one experiment (Fig. 6.10) and in the second year after inoculation in another experiment on a different site (not shown). In each experiment, TAD followed this natural peak. The ability of the natural microflora to exert its effects so soon after applying large amounts of a single fungus makes it unsurprising that little long-lasting disease control has been achieved by applying biological control agents to field soil.

The commercial requirement of 'reliability' will not be achieved until the mysteries of root colonization, host specificity of effective agents, sites of action and the biological characteristics that contribute to the ability of a BCA to suppress the pathogen are resolved for the practical situation. If these characters can be determined it might then be possible to enhance desirable properties in an agent by genetic manipulation techniques and also to tailor organisms to environments, because it is unlikely that isolates will transfer readily from one environment to another.

The alternative to enhancing the BCAs that are currently available is to continue to screen potential BCAs in the laboratory and field for antagonistic activity. This is time-consuming and laborious, but it is the method most likely to produce new antagonists in the short term and ideally should be done in parallel with the fundamental work.

Effective delivery systems for the BCAs must also be found. Seed coating (although effective) can be restrictive, and alternative means of delivery may be needed to provide easy handling and an acceptable shelf-life.

Finally, there is a need to investigate the use of root-stimulating organisms to decrease the effects of disease, rather than to control it, and to study manipulation of the soil environment to stimulate the development of antagonistic organisms (e.g. Lennartson, 1990).

4.4 Integrated Control

Despite optimism in the literature for such measures as seed treatment, particular types of fertilizers or biological controls, no single treatment has emerged for the effective, universal control of take-all in second or subsequent cereal crops. Integrated control may utilize different methods of control that affect different stages of pathogen development in a cumulative and beneficial manner that may also decrease other diseases. Precedents for the success of such systems can be found in the literature, e.g. in the control of onion white rot (Abd-El Moity et al., 1982). The various components in an integrated system must not affect each other deleteriously and should preferably enhance each other. Thus, research to develop practical integrated approaches involving BCAs is unlikely to proceed far until effective BCAs with established modes of action are developed for the field, and more is known about the effects of agronomic practices and agrochemicals (fertilizers and herbicides as well as fungicides) on the biological balance in the soil. It is important that BCAs remain effective under the normal agricultural practices to which they will be exposed so that further constraints on crop production are not introduced. From an environmental viewpoint the effects of chemicals and any microbial introductions (particularly of genetically manipulated organisms) on the soil microflora must be ascertained in order to determine appropriate usage and allow for informed legislative control.

4.4.1 Recent research in the UK and France

A series of experiments was begun at Rothamsted in the 1985/86 season to compare some of the most promising treatments and farming practices for decreasing take-all (Table 4.9), separately and in combination, on sites considered to be at risk from take-all (Hornby et al., 1990a). Although the practicalities of applying the treatments on a farm scale were not a consideration at that stage, it was envisaged that any treatments which consistently achieved significant decreases in combination would form the basis of a package of treatments for recommending

Table 4.9. Treatments tested in multifactorial experiments at Rothamsted, 1986–1988.

Factor	Standard treatment	Test treatment
Sowing date	Early (September)	Late (mid-late October)
Soil fungicide	None	Nuarimol (1.1 kg ha^{-1})
Seed treatment	Organomercury	Triadimenol plus fuberidazole
Autumn N	None	60 kg N ha^{-1} as ammonium nitrate (Nitro-Chalk) at sowing
N time	Single application of 200 kg ha^{-1} in April	Divided application: 40 kg ha^{-1} in February/March followed by 160 kg ha^{-1} in April
N form	Ammonium nitrate (Nitro-Chalk) in spring	Ammonium sulphate in spring

Table 4.10. Single factors ranked according to their effectiveness against take-all in experiments at Rothamsted.

	Mean ranking: all assessments				Number of significant, positive main effects out of a total of eight assessments, including % and logits of %		
Test treatment	1986	1987	1988	All years	1986	1987	1988
Late sowing	1	1	1	1	7	8	8
Soil fungicide	2	2	4	2	0	6	0
Triadimenol seed treatment	6	3	3	3	0	6	1
Autumn N	5	4	2	4	0	3	6
Divided N[a]	3	6	6	5	1	0	0
Ammonium sulphate	4	5	5	6	0	0	0

[a]Rankings include Feb/Mar samples, taken before divided N applications were made.

to farmers. Table 4.10 ranks six factors according to their effectiveness against take-all over three seasons in the Rothamsted experiments. The value of delayed sowing is emphasized. The soil fungicide (see also 4.2.2) was consistently the next most effective treatment, but the seed treatment was more erratic in its effects. Autumn nitrogen was sometimes effective, and other effects of fertilizers were slight and infrequent. There were relatively few significant positive interactions among treatments, and these mostly involved sowing date. In fact, interactions which increased disease were more frequent. Table 4.11 lists the number of occasions in two seasons in which the test treatment contributed to least disease (measured as TAR) and greatest yield in any three-factor combinations, regardless of whether or not the interactions were statistically significant. (Such interactions were not determined in 1986 because a full replicate of the factorial experiment was not used.) In 1987, late sowing resulted in poor yields because of poor plant emergence; in other years late sowing resulted in greater yields because of less take-all. All test treatments except divided N application in spring contributed to least disease in the majority of combinations, but it is evident from these experiments

Table 4.11. Number of occasions (out of a maximum of 10) that each test treatment occurred in three-factor combinations that resulted in the smallest take-all rating (TAR) or greatest yield in experiments at Rothamsted.

Test treatment	1987		1988	
	TAR	Yield	TAR	Yield
Late sowing	10	0	10	10
Soil fungicide	10	10	10	1
Triadimenol seed treatment	9	3	10	6
Autumn N	7	10	10	10
Divided N	5	4	0	10
Ammonium sulphate	7	1	10	8

that recommendations for a package of treatments for minimizing take-all can not be made with confidence. The priority recommendation is the avoidance of early sowing of crops at risk from take-all.

A parallel series of experiments was conducted in different conditions at Le Rheu, France, in 1987–1989 (Hornby et al., 1990a), following somewhat similar experiments with broader objectives (Lucas et al., 1988). These experiments also showed the value of delayed sowing (tested in 1987/88) and fungicides. A late effect of ammonium sulphate, tested as an autumn treatment at Le Rheu, in early-sown plots may have resulted from a gradual build-up of antagonistic microorganisms. In France, promising decreases in take-all were achieved using ammonium fertilizers to decrease soil receptivity (conduciveness), applying fungicide to seed (triadimenol) or the seedbed (nuarimol) to delay infection, and selecting a wheat variety (cv. Arminda) to delay infection and retard lesion development (Cavelier, 1989).

Take-all assessments were also made in a similar series of factorial experiments at Boxworth EHF in the UK in 1984–1986 (Werker and Gilligan, 1990). The results were generally similar, although take-all was less severe and differences between early and late-sown plots seen in early samples were not maintained through to the summer. When a field experiment was simulated as nearly as possible in pots (Brassett and Gilligan, 1990; also discussed in 2.6.2), sowing date and the density of inoculum applied artificially (millet seed colonized by Ggt) had the greatest effects on disease severity and also influenced disease resulting from the carry-over of inoculum on 'volunteer' seedlings. Controlling this carry-over, on volunteer plants and by other means, should be a consideration in any integrated control package (see also Chapter 3). There was no support from this experiment for increased carry-over on volunteers killed by the herbicide glyphosate, which has been implicated in occasional outbreaks of severe take-all (see 3.5.5).

Two other factorial experiments at Rothamsted are relevant. In one, during 1988/89, take-all in June was on average only slight, and the seed treatment, N timing and N form factors discussed above were replaced by seed rate (200 and 100 kg ha^{-1}) and spring N rates (100, 150, 200 and 250 kg ha^{-1}). The disease

Table 4.12. Effects of seeding rate, nitrogen applied in autumn or winter, and seed treatment fungicide (triadimenol plus fuberidazole) on take-all and grain yield in winter barley at Rothamsted.

Treatment	Take-all rating in June (0–300)		Grain yield (t ha^{-1})	
	Untreated control	% Decrease with treatment	Untreated control	% Increase with treatment
1984				
Low seed rate[a]	79	10.1	9.07	0.2
'Winter' N	75	−1.3	8.97	2.6
Seed treatment	90	33.3	8.96	2.8
1985				
Low seed rate[a]	131	6.1	6.48	2.9
'Winter' N	149	29.5	6.34	7.9
Seed treatment	134	10.4	6.43	4.5
1986				
Low seed rate[a]	131	−8.4	5.98	−1.3
'Winter' N	136	−0.7	6.02	−2.7
Seed treatment	139	3.6	5.80	4.8

[a]300 seeds m^{-2}; 450 seeds m^{-2} being the comparison under 'untreated control'.
Data from Jenkyn *et al.* (1992a).

was affected less by late sowing than by soil fungicide and reduced seed rate. Late sowing decreased the incidence of infection overall, but moderate and severe infections were decreased only at the higher seed rate. Other interactions were rare. In factorial experiments over 3 years in winter barley crops at risk from take-all, three treatments had most effect on decreasing take-all and increasing yield (Table 4.12 and Jenkyn *et al.*, 1992a). Ranked in order of effectiveness these were: seed treatment with triadimenol plus fuberidazole (Baytan) > winter nitrogen > decreased seeding rate. Other factors in the experiments were plant growth regulator, amounts of spring-applied nitrogen, insecticide in autumn and fungicides in summer. Sowing date, the most important factor in the winter wheat experiments, was not tested.

4.5 Decision Making and Forecasting

4.5.1 Short-term: before sowing

Assessment of take-all in a crop may indicate the likely damage to that crop, but it does not reliably indicate the risk to a following cereal crop. At Rothamsted, the disease in a first wheat crop grown after a 'clean' break-crop (i.e. a break without grass weeds or self-sown hosts of Ggt) is usually less than 10% of plants infected with slight take-all. This seems insufficient to account for the severe infection that often occurs in a following (second) wheat crop. It is possible that in the late stages of the mature plant Ggt colonizes roots without causing lesions. Standard assessments, based on lesions on main root axes, exclude runner hyphae and infected branch roots, which are likely to be lost on sampling or when the roots are washed. Any fungus active in the rhizosphere soil would also be overlooked. The absence of direct methods for quantifying inoculum of Ggt in the soil means that baiting with a suitable host has been a standard procedure for decades (Hornby, 1978b). A bioassay of soil (Slope *et al.*, 1979) from under first and second wheat crops grown after oats in dry years revealed relatively small increases in root infection until June/July and thereafter a tendency for the soil to become less infective (Fig. 4.2). By September, soil infectivity was little more than at the start of the wheat crop and severe disease in the next crop was unlikely (Slope *et al.*, 1977). In wetter seasons, however, soil infectivity increased from May to September (Fig. 4.3) and

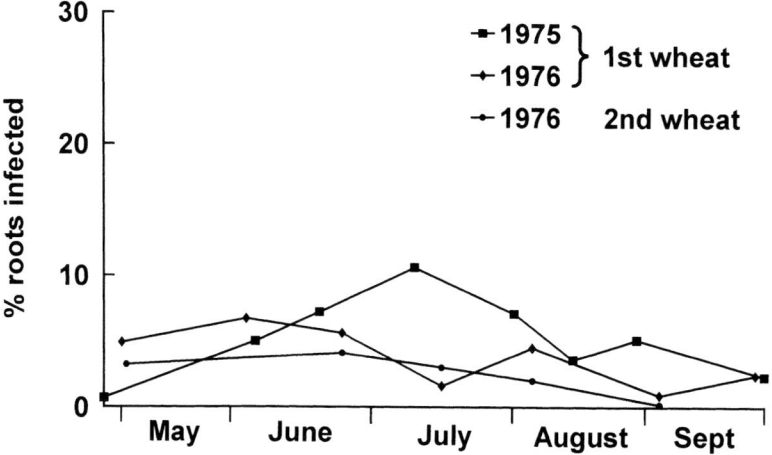

Fig. 4.2. Percentage of roots with take-all lesions in bioassays of soil cores from first and second winter wheat crops in dry years.
Each percentage is based on counts of infected roots on 500 plants (10 plants for each of 50 soil cores sampled in an area of 0.02 ha).

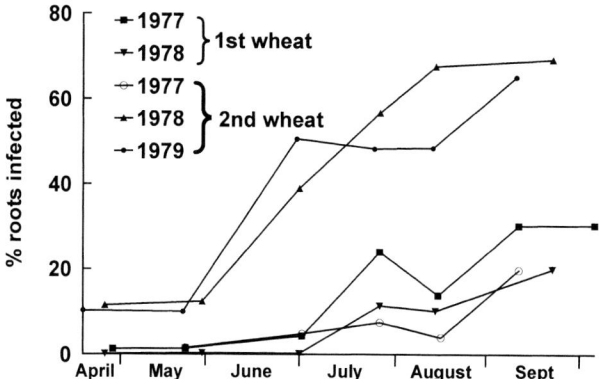

Fig. 4.3. Percentage of roots with take-all lesions in bioassays of soil cores from first and second winter wheat crops in wetter seasons.
Each percentage is based on counts of infected roots on 500 plants (10 plants for each of 50 soil cores sampled in an area of 0.02 ha).

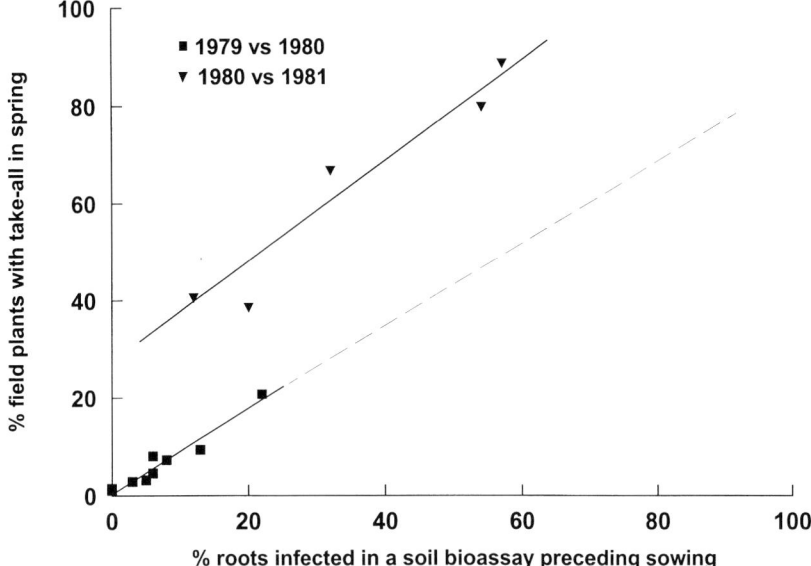

Fig. 4.4. Relationship between percentage of roots infected by take-all in a soil bioassay after harvest and percentage of plants infected by take-all in April in a following winter wheat crop.
1979 vs. 1980, $y = 23.01 + 1.112x$, 90.8% of the variance accounted for; the linear relationship is extrapolated beyond the data points as a dashed line.
1980 vs. 1981, $y = 0.501 + 0.8191x$, 93.9% of the variance accounted for.

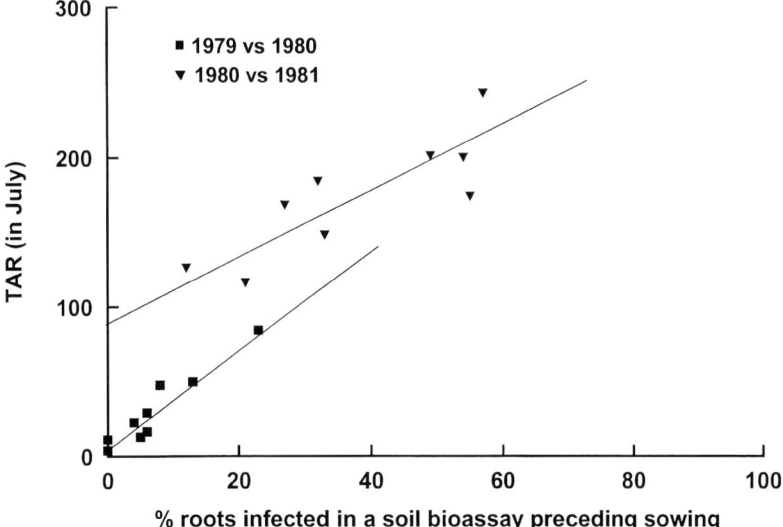

Fig. 4.5. Relationship between percentage of roots infected by take-all in a soil bioassay after harvest and the take-all rating of a following winter wheat crop: disease build-up phase.
1979 vs. 1980, $y = 6.04 + 3.372x$, 90.4% of variance accounted for.
1980 vs. 1981, $y = 94.6 + 2.04x$, 66.3% of variance accounted for.

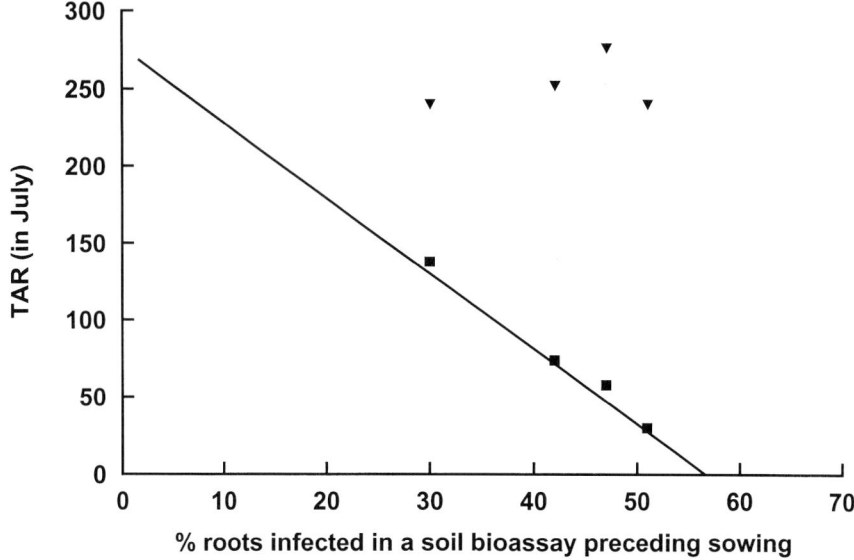

Fig. 4.6. Relationship between take-all ratings in July in a winter wheat and the percentage of roots infected by take-all in a bioassay of the soil in September before sowing (■).
TARs in July in the previous crop (▼) had been high and plots which subsequently showed a decrease of 100 TAR units or more were assumed to be exhibiting TAD.

the likelihood of severe disease in a following crop was greater (Slope and Gutteridge, 1979). Subsequently, in an experiment at Rothamsted where take-all was expected to develop differentially because of treatment differences, the percentage of roots infected in bioassays in September was found to be positively correlated with the incidence of take-all (expressed as the percentage of plants infected) in spring (Fig. 4.4). It was also positively correlated with the TAR in July (Fig. 4.5). Such findings would only be applicable to assessing risk in crops in which take-all was building up: when the disease is at a peak, the correlation between the bioassay after harvest and disease the following year could be negative rather than positive (Fig. 4.6). There are many problems in trying to forecast take-all (Hornby, 1978b) and the above examples may be the exception rather than the rule. However, experimental data collected over many years from Rothamsted suggest that if plants in a post-harvest soil bioassay have 20% of their roots infected, then the risk of take-all patches in the subsequent crop is high. The more this 20% is exceeded, the higher the risk and the more widespread the disease. Conversely, if little or no inoculum can be detected, then the risk to a following susceptible crop would be very slight (R.J. Gutteridge, unpublished data).

4.5.2 *Short-term: during the crop*

If conditions remain favourable for disease development, then the more roots that are infected in the autumn, the greater will be the likelihood of severe disease developing in July. Root infection develops most rapidly between May and July; for instance, TAR, a measure of the intensity of the disease that takes into account disease incidence and severity, can double between June and July (Hornby and Gutteridge, 1995).

4.5.3 Longer term

In the last 15 years take-all has waxed and waned in intensity and has developed a disturbing tendency to reach damaging levels more frequently in second wheats than in earlier times. Weather, early sowing and oilseed rape in rotations may all have had a bearing on this (see Sections 2.3.1–2, 2.6.2, 2.6.8, 3.3.1–2, 3.3.4, 3.4.3, 3.5.1 and 3.6), but the changes have never been properly explained. Until recently, epidemics of the disease were monitored quite intensively at Rothamsted and Woburn. One Woburn experiment (Hornby and Henden, 1986) is of particular interest because the observed fluctuations were thought not to be complicated by previous cropping, which means they were likely to be more directly related to changes in the weather. Although identifying which factors have been important is difficult, potential soil moisture deficit (PSMD) is often considered to be important. In the 20-year period 1969–1989, annual averages for PSMDs were dominated by a very dry year (1976), otherwise they show a weak trend towards wetter soils. Increased disease in the 1980s may have been affected by changes in PSMD (Fig. 4.7), but there is clearly a need for much more detailed analyses of weather and take-all data.

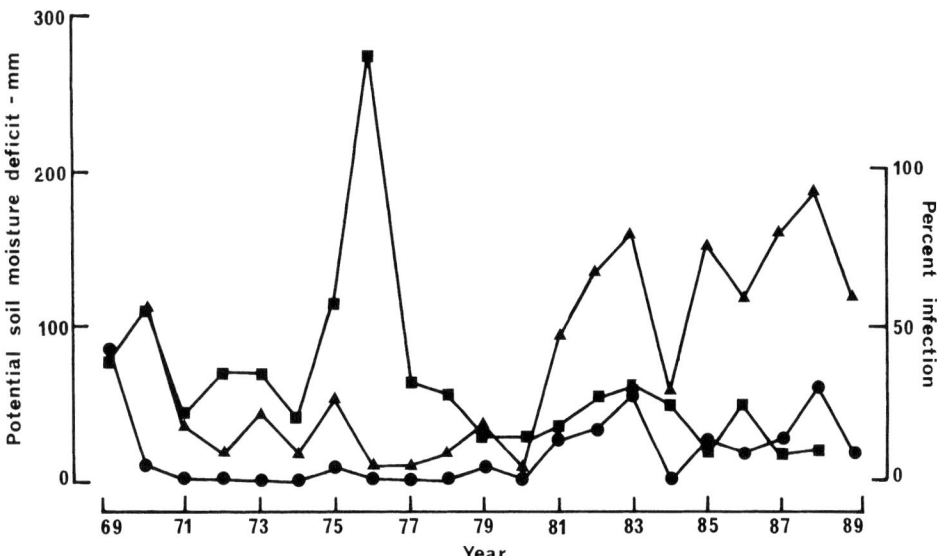

Fig. 4.7. Changes in potential soil moisture deficit (annual averages) and the incidence and severity (in July) of take-all in continuous spring barley in a field at Woburn, Bedfordshire, 1969–1989.
■, Potential soil moisture deficit, ▲, % plants infected, ●, % roots infected.

4.6 Recommendations for Take-all Control

Advice on take-all is available to farmers in the UK from a variety of sources such as ADAS, growers' associations, technical representatives and professional advisers, although this is less freely available than hitherto because of commercial pressures. In the 19th edition of *The Agricultural Handbook*, six points are listed under the heading of 'cultural and organic control' and two under 'chemical control' (Moule, 1995). Most of these points are valid and are dealt with below, but bald statements that disease build-up is avoided by growing not more than 3 years of continuous wheat or barley, and that direct-drilled crops are affected less than traditionally sown crops, are contentious (an observation conflicting with the latter assertion is mentioned in Section 3.5.2). Current ADAS advice on minimizing losses due to take-all can be summarized as follows.

1. *Local knowledge.* It is important that a farmer should know his soils and be able to assess the risks of take-all on them. Whilst data are difficult to come by, experience suggests that soils can be ranked tentatively with regard to their disease risk, in descending order from highest to lowest risk, as follows:

- fen peats;
- 'black sands' low in manganese;
- mineral soils where long-term cropping with grass has resulted in a high organic matter content;
- light, alkaline soils;
- sandy silt loam;
- non-calcareous or decalcified clays (serious problems can occur when a wet season aggravates structural problems);
- silt loams/silty clay loams;
- well-structured calcareous clay loams.

2. *Avoid highest risk.* On the highest risk soils wheat should never be grown as a second or third cereal. On the lowest risk soils careful attention to husbandry practices will enable such crops to be grown with relative impunity. The higher the risk, the greater is the importance of adopting the control measures outlined below.

3. *Rotation.* Short rotations which prevent the field from ever developing severe take-all, or continuous cropping with a single cereal species which exploits the phenomenon of TAD, are preferable to the intermediate rotations that are so often practised. If, for example, a break crop is taken every fifth year, then at least two of the four cereal crops in the rotation are likely to suffer from moderate to severe attacks of the disease.

Because barley suffers less severely from take-all than does wheat, it is safer than wheat where cereals have to be grown in high-risk situations. For the highest risk soils, triticale or rye offer even safer options. Oats, being resistant to Ggt, make an acceptable break crop except in areas where Gga is known to be a problem.

Leys, free from perennial grass weeds, also make acceptable breaks. Although ley grasses may carry Ggt, they also carry a fungal antagonist, *Phialophora graminicola*, which is able to suppress the pathogen's development in a subsequent cereal crop. It is the latter property that seems most often to influence matters. It offers an explanation of why second wheats are safer after clean, 2-year leys than after non-grass breaks which are likely to decrease inoculum of Ggt, but do not build up the antagonist.

4. *Weed control.* Perennial grass weeds should be controlled before a break crop is taken, because control before a cereal may actually increase the risk of infection. Ggt can survive on dying rhizomes and it is also reported that it can build up on them. Severe infection frequently occurs in first wheats on couch-infested land, irrespective of spraying-off couch with glyphosate in the previous autumn. Spraying before a break does not always have the desired effect if a dry season follows. This is attributed to dead rhizomes, colonized by the pathogen, surviving unrotted through the break year. It is sensible never to allow a field to become seriously infested with couch.

5. *Soil management.* Drainage or structural problems that may restrict rooting should be corrected. Firm seedbeds will hinder ectotrophic spread of the pathogen's mycelium over the roots. Ploughing is preferable to tine cultivation for the incorporation of straw.

6. *Sowing date.* First wheats should be sown first, long-term wheats next and the high-risk wheats (second, third and fourth wheats) left until last and never sown before October.

7. *Sowing rate.* High seed rates which reduce root development by interplant competition and favour the disease should be avoided.

8. *Crop nutrition.* Fields should not be allowed to become deficient in major nutrients, particularly phosphate. If it is intended to take a second wheat on high-risk soils, a phosphate index (Ministry of Agriculture, Fisheries and Food, 1980) of at least 2 should be achieved before the first wheat is grown, and then maintained. Adequate nitrogen is important and an early application of part of the nitrogen in late February/early March, providing at least 40 kg ha^{-1}, is often beneficial. In high risk situations, ammonium sulphate is very likely to decrease take-all more than other sources of nitrogen.

9. *Control of pH.* Liming to prevent the development of acid patches is important, but overliming should be avoided. Lime should be applied before a break crop or first cereal, rather than before a second or third wheat or barley crop.

10. *Chemical control.* Although no currently available fungicide is reliable, triadimenol seed treatment will sometimes decrease infection and increase yields in the presence of high levels of the disease.

Measures to prevent or decrease take-all in Germany (Mielke, 1988) are:

- Rotation with 'leaf' crops.
- Not exceeding 60% of susceptible cereals in the arable acreage.
- One-year breaks to minimize risk.
- Utilizing microbial antagonism through using green manure.
- Sowing cereals after grass–clover mixtures to utilize biological control by *Phialophora*.
- Controlling couch.
- Using cultivars with better rooting.
- Using rye in preference to wheat or barley.

- Careful working of stubble from diseased crops.
- Deep ploughing.
- Late sowing.
- Shallow sowing and lower seed rates.
- An additional 60 kg N ha^{-1} early in the year.

Because of saprophytic growth of Ggt on couch rhizomes following glyphosate application, the herbicide should be used in the non-cereal rotation crops. Oilseed rape is claimed to be a carrier of Ggt and so cereal volunteers should be controlled after sowing rape. Plates 8.19 and 8.20 show examples of Ggt growing on both couch and rape, respectively, and causing lesions. The most effective measures in controlling take-all in Germany are choice of site, rotation and previous cropping.

In Australia, BCAs are being investigated but are not yet available, and so recommendations for the control of take-all (Wallwork, 1996) are as follows.

- *Plan rotations.* These should include disease breaks such as pulses, oilseed crops or pasture legumes to avoid inoculum building up. Oats may be used where oat-attacking strains of *G. graminis* are not a problem.
- *Remove grass hosts.* Herbicides should be applied in the year prior to cereal crops, by late June in low rainfall areas and by late July elsewhere. 'Spraytopping' (late spraying) to prevent seed set is useful if applied 2 years before a cereal crop.
- *Control volunteer grasses.* If done after harvest this will decrease early build-up of inoculum and take-all damage.
- *Sow paddocks at risk from take-all last.* This maximizes the time for stubble to break down and for inoculum to decrease.
- *Adequate nutrients.* These will promote plant vigour. Nitrogen, phosphorus and manganese will decrease susceptibility to take-all.
- *Ammonium nitrogen.* This form of nitrogen reduces the risk of take-all in some areas.
- *Liming.* Favours take-all.
- *Fungicides.* Decreases in take-all can be achieved by in-furrow applications but are often uneconomic.

In the Pacific Northwest of the USA recommendations for agronomic practices to decrease the root disease complex there and complement seed treatment by bacteria (Cook and Veseth, 1991; Stelljes and Hardin, 1995) are as follows.

- Delay seeding for as long as possible.
- Kill grass weeds and volunteer wheat well in advance of planting.
- Plant, fertilize and loosen soil in the seed row in one operation.
- Plant wheat in paired rows, 10–16 inches (25–40 cm) apart.

4.7 Take-all Control in a Worldwide Context

In Australia there are more reports of successful use of fungicides such as benomyl and triadimefon against take-all than in Britain. Bateman (1989) suggested that

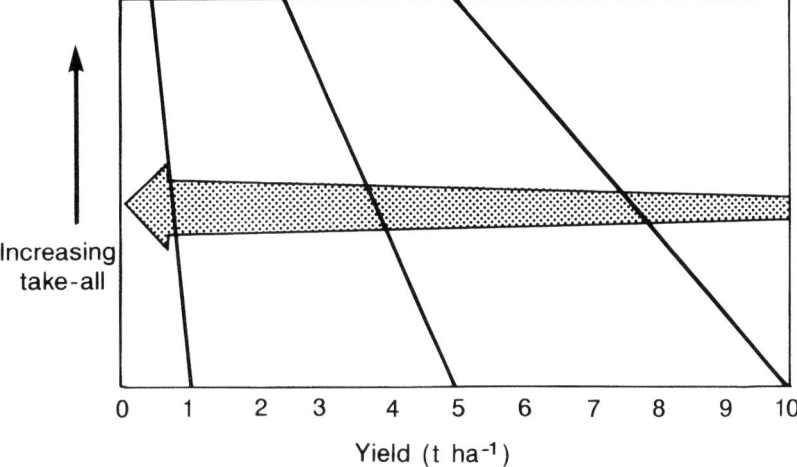

Fig. 4.8. Regional yields and the impact of control measures.
In regions where take-all is a problem its effects on yield tend to be proportionally similar, and control measures (stippled arrow) are generally more effective in lower-yielding regions (details in text).

similar fungicide effects in Britain's fertile soils, with their greater yield and proportionally smaller losses from take-all, were less noticeable. However, the question of whether losses in infected crops in Britain are proportionally smaller is debatable. Figure 4.8 incorporates some ideas arising from work with artificial inoculum on spring wheat (Hornby and Bateman, 1990). At that time the yields of uninfected crops in Australian experiments were about 1 t ha^{-1} and the most infected crops yielded about half that. Work at Rothamsted also showed almost a halving of yield over the range of disease created by using artificial inoculum (see 6.5.2), when uninfected crops were yielding 7 t ha^{-1}. Figure 4.8 also shows what 50% reductions would look like for a good winter wheat yield and a cereal giving 5 t ha^{-1}. If similar proportional disease effects occur in different countries, the absolute losses per hectare of infected crop will be much greater in high-yielding regions. In the UK, generally larger yields, unawareness of a take-all problem in fields not showing patches and recent grain surpluses may have conspired to mask this point. Maas *et al.* (1989) reported even bigger yield reductions of up to 72% in South Africa using relatively huge amounts of artificial inoculum, but their unusual method of assessing disease makes comparison with other results difficult.

There are various ways of looking at regional differences (e.g. Fig. I.1; Table 4.13). The accumulating experience that control measures that worked in Australia and elsewhere were ineffective or unimpressive when applied in the UK is one aspect worth further consideration (Table 4.14). Cereal soils in the UK (and in much of northern Europe) are generally more fertile than those in many other cereal-growing regions. Further, the UK cereal production system uses much greater inputs. In 1989, an average of 189 kg N ha^{-1} was used on winter wheat, compared with relatively little or none in Australia, and there was much more use of pesticides compared with the USA, where 52% of wheat was produced without

Table 4.13. A subjective assessment of research input by non-industrial organizations and of commercial achievement.

Approach to control	*	Europe[a]	North America	Australia
BCA: bacteria	A	Moderate	Large	Moderate
	B	None	None (some patents)	None
BCA: fungi	A	Slight–moderate	Slight	Moderate
	B	None (some patents)	None	None (a patent)
Chemical fungicides	A	Moderate	Slight–moderate	Moderate
	B	None	Triadimenol seed treatment of irrigated spring wheat locally (see 4.2.2 'Seed treatment')	Flutriafol-coated superphosphate granules
Husbandry (rotations, cultivations, etc.)	A	Large	Moderate	Moderate
	B	Only option until recently; now partly imposed for reasons of over-production not concerning take-all	'Triple play' system in Pacific Northwest[b]	Limited cropping options
Mixed	A	Moderate	Moderate	Small
	B	None; factors mainly independent	Nothing specifically for take-all	

*A, research input; B, product marketed or practical application achieved.
[a]Based mainly on information from the UK and France.
[b]Cook and Veseth (1991).

pesticides (Scott, P.R., 1990). The 1,802,191 ha of wheat grown in 1994 accounted for 38% of the total area of arable farm crops in Great Britain, but 54% of the total pesticide treated area. Hardly any wheat crops were untreated and, by chemical group, the treated wheat areas were: insecticides, 1,552,960 ha; fungicides other than seed treatments, 6,525,831 ha; herbicides, 5,181,935 ha; growth regulators, 1,765,566 ha; molluscicides, 485,954 ha; and all seed treatments, 1,859,589 ha (Garthwaite et al., 1995). These figures clearly indicate that British wheat crops received more than one spray round of both fungicide and herbicide.

Because spring wheats may suffer from take-all more than winter wheats and may rely more on their seminal roots, the relative importance of the two kinds of wheat in different countries also needs to be taken into consideration. The national average yields of wheat given in Table 4.14 are a crude categorization, pooling all sorts of data: the figure for the USA, for example, contains winter, spring and durum wheat, and average winter wheat yields for individual states range from 1.7 t ha^{-1} in New Mexico to 6 t ha^{-1} in Arizona.

There is some indication that reports of successful control of take-all may be related to regional differences in fertility and inputs, as suggested in Table 4.14. At the high fertility end, TAD is well documented and of the classical kind that follows

Table 4.14. A simple scheme showing some characteristics of wheat production in different parts of the world in relation to some factors which affect take-all (full explanation in the text).

System	Trends			
Inputs[a]	High ←———————————————————→ Low			
Fertility	High ←———————————————————→ Low			
Yields[b] (t ha^{-1})	6.0[c] (UK)	2.3 (USA)	1.9 (S. Am.)	1.5 (AUS)

Factors	Effects in relation to trend above
TAD	Classical — Different or none
BCAs	Weak — Some large
Other[d]	Small — Some large

AUS, Australia.; S. Am., South America.
[a] For example, in 1988 nitrogenous fertilizers applied on arable land in the UK, USA and Australia were 20.9, 5.1 and 0.8 t km^{-2}, respectively (OCDE/OECD, 1991).
[b] Average yields about 1989.
[c] Lowest value since beginning of 1980s; actual value in 1995 was 7.66 t ha^{-1} (National Institute of Agricultural Botany, 1996).
[d] E.g. fertilizer treatments (Mn^{2+}, NH_4^+, Cl^-).

a peak of disease in consecutive cereal crops. Moving towards lower fertility, TAD becomes weaker, less frequent, ephemeral, or is absent. Many of the other factors that affect take-all do so most convincingly in regions that would be located some way along the scale in Table 4.14. Forms of N, chloride fertilizers and Mn deficiencies are all factors that have had greater effect in the USA than in the UK. Systems for growing wheat that are mostly alien to British farmers (irrigation, double-cropping with soya beans and alternating cereals with self-regenerating legume pastures) are also encountered along this trend. Although long regarded as an ecologically obligate parasite with a low competitive saprophytic ability, Ggt is now reported to grow through soil in Australia (Grose, 1989; Grose *et al.*, 1984). Glenn *et al.* (1988) found that Ggt was capable of combative migration through soil and of growing saprophytically on organic residues present in natural soil, although according to Grose (1989) combative migration is not yet accepted as part of the life of Ggt (cf. Fig. 2.3). Australian soils tend to be low in organic matter and total nitrogen and have low cation exchange capacities. Saprophytic growth of an isolate of Ggt from a very nutrient-leached acid sand and in the absence of a host was favoured by low nitrogen and was affected slightly by the form of nitrogen (Grose, 1989). (Some opposite effects of nitrogen in influencing take-all are discussed in Section 3.4.1.) Western Australian soils seem to support more Ggt and the fungus seems to be more infectious than in UK soils (see 2.7). Claims for simple systems that predict take-all and conviction that resistance exists amongst cultivars (see Fig. 3.3) tend not to be associated with high fertility. It may be that the high fertility–high input system presents a biological buffer that is not easily overcome by single BCAs, which appear to achieve effects in less fertile regions. TAD may be a phenomenon more characteristic of high fertility regions, where only robust, multi-component systems can influence take-all significantly. In a discussion of cereal diseases in Australia it was suggested that TAD is unlikely to occur in dry environments because the causal suppressive organisms apparently fail to survive dry conditions between cropping seasons (Wallwork, 1996).

Usually liming increases take-all and ammonium sulphate fertilizer decreases it (Table 4.15). The mechanisms for this are not known, but explanations for the effects of liming are changes in soil pH and predisposition of the host as nutrients such as Cu, Mn, Fe, Zn and Mg become less available at higher pHs. The effect of $(NH_4)_2SO_4$ has usually been explained as an interaction between some or all of these. The following examples, dealt with in more detail elsewhere in this book, will suffice to make the point that many factors interact:

Table 4.15. The effects on take-all of applying lime or ammonium sulphate.

	Lime	Ammonium sulphate
Take-all	Increased	Decreased
Proposed mechanisms	pH Host disposition	Interactions among microbes, nutrients and pH

Table 4.16. Conditions affecting take-all and their effects on nitrification and manganese availability.

	Effect on:		
Condition	Take-all	Nitrification	Mn availability
Nitrate nitrogen	I	I	D
Liming	I	I	D
Animal manures	I	I	D
Previous crop – soya bean	I	I	D
– lucerne	I	I	D
Alkaline soils	I	I	D
Loose seedbed	I	I	D
High soil moisture	I	I[a]	D[a]
High plant populations	I	–	D
Short runs of cereals	I	–	D
Plant stress	I	–	D
Ammonium nitrogen	D	D	I
Fertilization – manganese	D	–	I
– chloride	D	D	I
Nitrification inhibitors	D	D	I
Acid soils	D	D	I
Previous crop – oats	D	–	I
– lupins	D	D	I
Tolerant cultivars	D	–	I
Late sowing	D	–	I

I, increased; D, decreased.
[a]Variable with some other conditions.
–, unknown effect.
From Huber (1990).

- In Australia, applications of ammonium sulphate favour the antagonist *Trichoderma*, which suppresses Ggt.
- In Switzerland, certain fluorescent pseudomonads produce HCN, which suppresses take-all. The production of HCN is induced by iron and the availability of iron depends on the type of clay mineral in the soil (Fe is available in vermiculitic but not illitic soils). Hence there is a soil type effect.
- In the USA it has been suggested that the availability of Mn is a unifying concept that accommodates the effects of pH and forms of nitrogen in plant disease (Huber, 1990). The solubility of Mn and its equilibrium concentration in soil are determined largely by pH and redox potential. Reduction of Mn is favoured by a surplus of protons (H^+) and electrons ($\bar{e}$). Consequently, the dissolution of Mn in soil as Mn^{2+} (manganous ion) is promoted by low pH and low O_2 tension. The transformation is also affected by NH_4^+-N, superphosphate fertilizer and organic matter, all of which have been reported to affect take-all. Mn may be unavailable to the plant because of microbial oxidation. Mn-oxidizing populations increase in response to NO_3^--N. Table 4.16 shows many associations that have been observed for take-all, nitrification and manganese availability.

The complexities of such interactions have added greatly to both the fascination and the frustration of take-all research and help to explain why seemingly contradictory results have often been achieved by workers in different parts of the world.

The Pathogens and Related Fungi   5

5.1 Introduction

This chapter describes recent developments in understanding the biology of the fungi that cause take-all – Ggt, Gga and now *Gaeumannomyces graminis* var. *maydis*, the cause of 'maize take-all' (Yao *et al.*, 1992) – and new techniques for their study and identification. Advances in molecular biological techniques have paved the way to improved identification of closely related fungi as well as to a better understanding of their evolutionary relationships. New methods are also being applied to studies of fungal behaviour and ecology, including the processes of pathogenicity and host defence.

Take-all of cereals, Ggt and the perceptions of researchers are all influenced by a group of closely related fungi, the *Gaeumannomyces–Phialophora* complex, that occurs on the roots of graminaceous plants and often on roots of plants infected by take-all. The appearance of the then-known members of this complex of fungi on host plants and in culture was described respectively by Deacon (1981) and Cunningham (1981). Therefore, before discussing new developments as they affect the fungi involved in take-all, some consideration needs to be given to this complex and especially to taxonomic and ecological research published since Asher and Shipton (1981). This provides essential background to the review of experimental approaches that follows and reveals some of the reasons behind the effort that has gone into methods for identifying the pathogens and other members of the complex. Experimental techniques relating to the disease, rather than to the pathogen, are the subject of Chapter 6. Whilst much discussion in this chapter concerns molecular biological methods, more conventional methods, some of which are included in a short review of methods for studying *Gaeumannomyces* (Mathre, 1992), are introduced where relevant. The later sections of this chapter include accounts of recent genetic and biochemical studies.

5.2 The *Gaeumannomyces–Phialophora* Complex

Gaeumannomyces v. Arx & Olivier {emend. Walker} is a genus of ascomycete fungi with five known species: *G. caricis, G. cylindrosporus, G. graminis, G. incrustans* and *G. medullaris*. Those species that occur on *Gramineae* (= *Poaceae*) tend to have *Phialophora*-like anamorphs (asexual states) and produce characteristic ectotrophic mycelial growth and growth-cessation structures on roots and hyphopodia (appressorium-like hyphal swellings) on coleoptiles (examples of these structures are illustrated in Figs 2.7, 5.1–5.3 and 5.8–5.11). The *Gaeumannomyces–Phialophora* complex on *Gramineae* comprises these holomorphs and some *Phialophora*-like anamorphs apparently without perfect states. This complex contains the fungi of interest in this book and provides a convenient way of referring to them. The review by Walker (1981) is a thorough taxonomic coverage of the *Gaeumannomyces–Phialophora* complex for the state of knowledge at that time and it continues to serve as a valuable cornerstone for subsequent developments. However, the concept may be beginning to unravel, because these two genera no longer accommodate all the strands of relationship. A more recent review of the taxonomy and biology of ectotrophic, root-infecting fungi associated with patch diseases of turf grass covers the complex in part and includes members discovered since 1981 (Landschoot, 1993). Figure 5.13 shows how the *Gaeumannomyces–Phialophora* complex might be conceived in a modern context.

Often plant pathologists are exasperated by what they regard as the machinations of taxonomists, which seem merely to make life more complicated. Those studying take-all have been no exception and it took well over 20 years for most plant pathologists to give up the misapplied name *Ophiobolus graminis* for the take-all fungus in favour of *G. graminis* var. *tritici* (Ggt), thereby losing in brevity and euphony what may have been gained in piety, according to Garrett (1981), mourning the passing of a familiar name. However, accurate nomenclature is important, particularly if confusion is to be avoided. For example, during the transition period from widespread use of *Ophiobolus* to general use of *Gaeumannomyces*, a colleague from a region with little take-all research assumed that toxins were involved in take-all because *O. miyabeanus* (syn. *Cochliobolus miyabeanus*, the cause of brown spot of rice) was known to produce a toxin. Had Ggt been in universal use, that flawed connection perhaps would not have arisen.

The classification and nomenclature of the take-all fungus have changed on several occasions (Walker, 1981) and new changes are still being proposed. *Gaeumannomyces* (the genus was established in 1952) has been classified historically in the order *Diaporthales*. This was the classificatory hierachy adopted by Landschoot (1993): division *Amastigomycota*, subdivision *Ascomycotina*, class *Ascomycetes*, subclass *Hymenoascomycetidae I* (*Pyrenomycetes*), order *Diaporthales* (*Sphaeriales*), family *Diaporthaceae*. However, links with the *Magnaporthe* assemblage, for which the family name *Magnaporthaceae* was introduced, have been emphasized recently (Cannon, 1994). This new family, which is yet to be assigned to an order but is probably not close to the *Diaporthales*, includes the genera *Magnaporthe* (which includes the rice blast pathogen), *Buergenerula, Clasterosphaeria, Gaeumannomyces, Herbampulla, Omnidemptus* and *Pseudotracylla*. These fungi

continued on p. 220

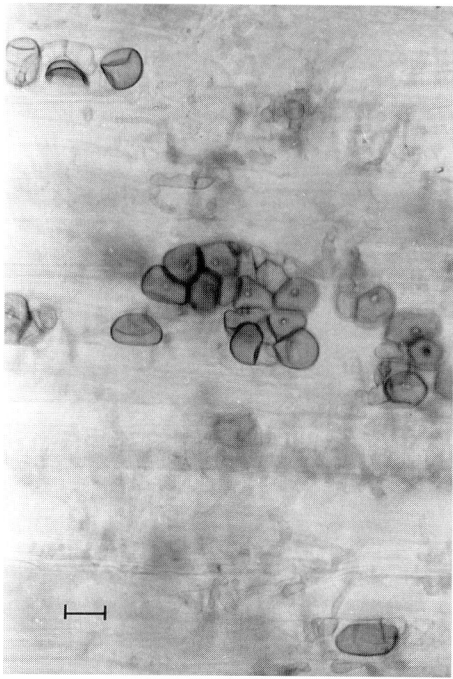

Fig. 5.1. *Phialophora graminicola*. Swollen cells with pores (known as growth cessation structures or vesicles) in a wheat root. Bar = 10 µm.

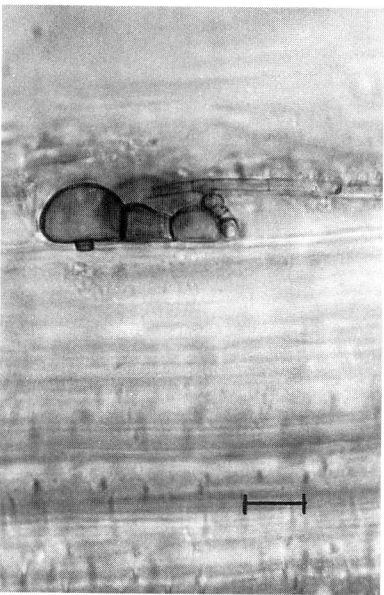

Fig. 5.2. *Phialophora graminicola*. Swollen cells, one showing a short side branch (infection hypha) in a wheat root. Bar = 10 µm.

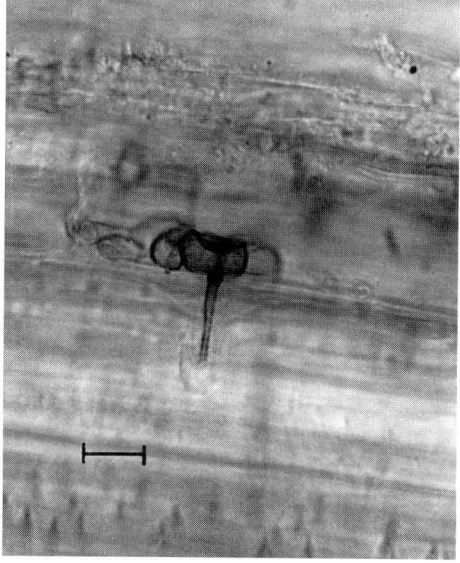

Fig. 5.3. *Phialophora graminicola*. Swollen cells, one showing a long infection hypha contained by a lignituber (or papilla or callosity) formed by the challenged cell of a wheat root. Bar = 10 µm.

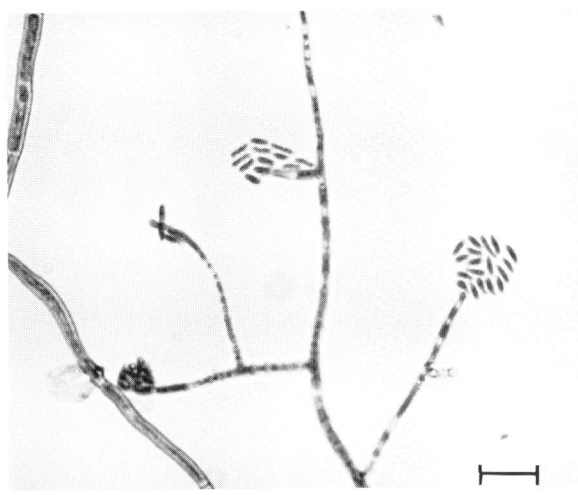

Fig. 5.4. *Phialophora graminicola*. Phialides and phialospores on PDA. Bar = 10 μm.

Fig. 5.5. *Gaeumannomyces cylindrosporus*. Perithecium produced between 3 and 6 months on excised wheat roots in a rotting test[a]. Bar = 50 μm.
[a]Holden and Hornby (1981).

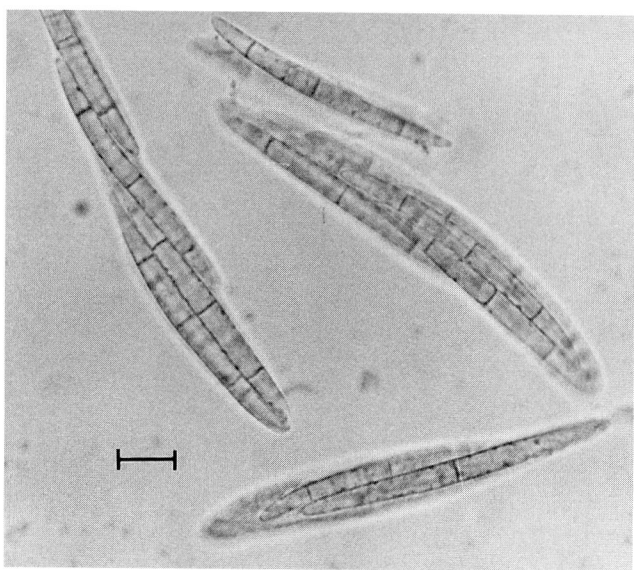

Fig. 5.6. *Gaeumannomyces cylindrosporus*. Asci and ascospores from the perithecium in Fig. 5.5. Bar = 10 µm.

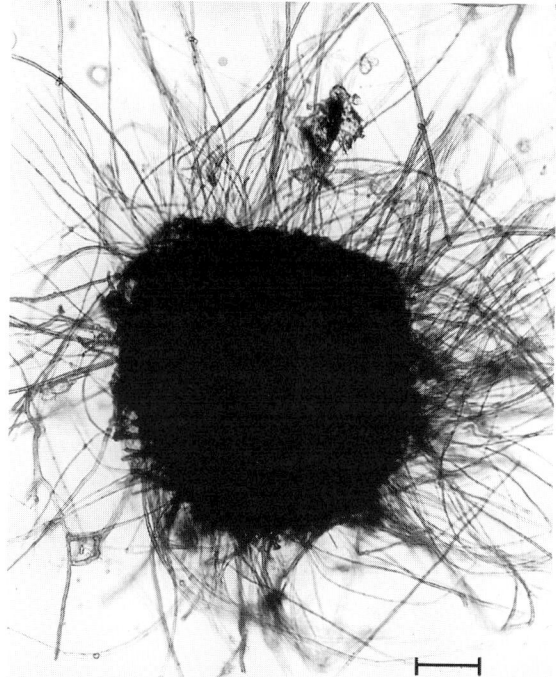

Fig. 5.7. *Phialophora* sp. (lobed hyphopodia). Hirsute structure (sclerotium) produced between 6 and 8 weeks on excised wheat roots in a rotting test[a].
Bar = 50 µm.
[a]Holden and Hornby (1981).

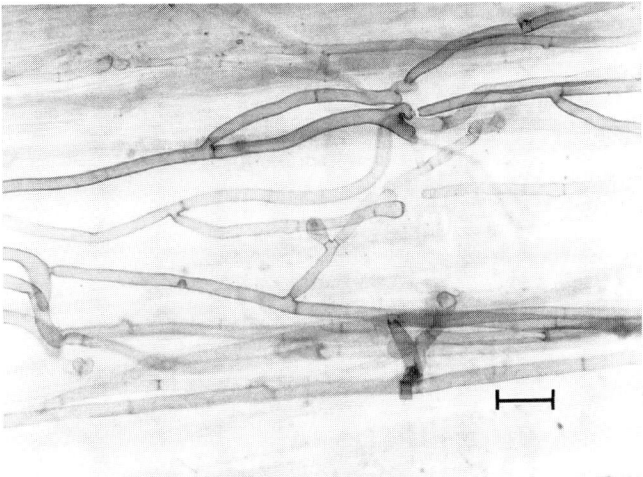

Fig. 5.8. *Phialophora* sp. (lobed hyphopodia). Runner hyphae on a wheat root. Bar = 10 µm.

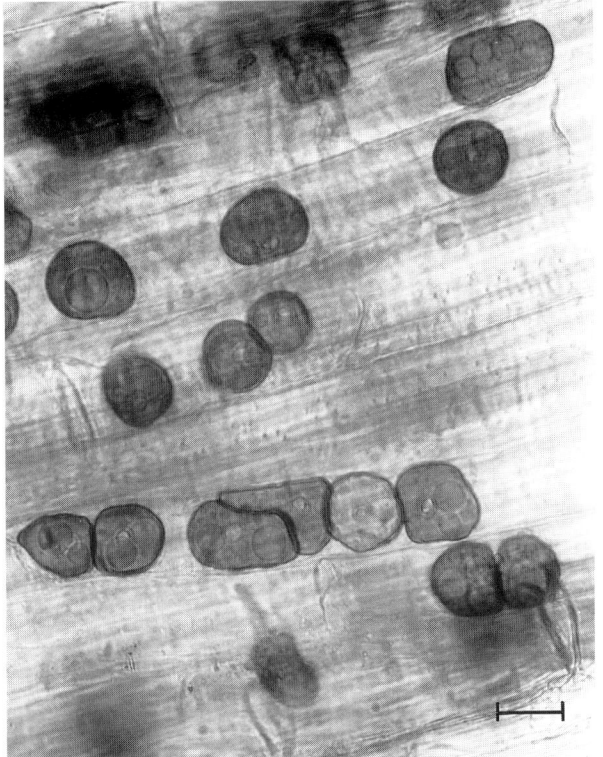

Fig. 5.9. *Phialophora* sp. (lobed hyphopodia). Swollen cells[a] in wheat roots. Bar = 20 µm.
[a]See Fig. 5.1 for alternative names.
In British isolates these cells tend not to be lobed, as here, whereas in some Australian Ggg isolates examined at Rothamsted they were slightly lobed.

218 Chapter 5

Fig 5.11. *Phialophora* sp. (lobed hyphopodia). Scanning electron micrograph of a lobed hyphopodium on the surface of a wheat coleoptile. Bar = 10 μm.

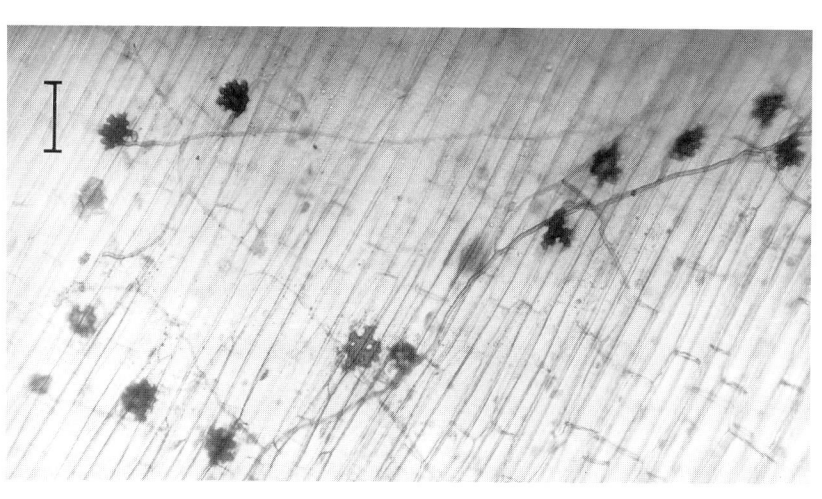

Fig 5.10. *Phialophora* sp. (lobed hyphopodia). Lobed hyphopodia on the coleoptile of a wheat seedling. Bar = 50 μm.

Fig. 5.12. *Phialophora* sp. (lobed hyphopodia). Phialides and phialospores on PDA. Bar = 10 µm.

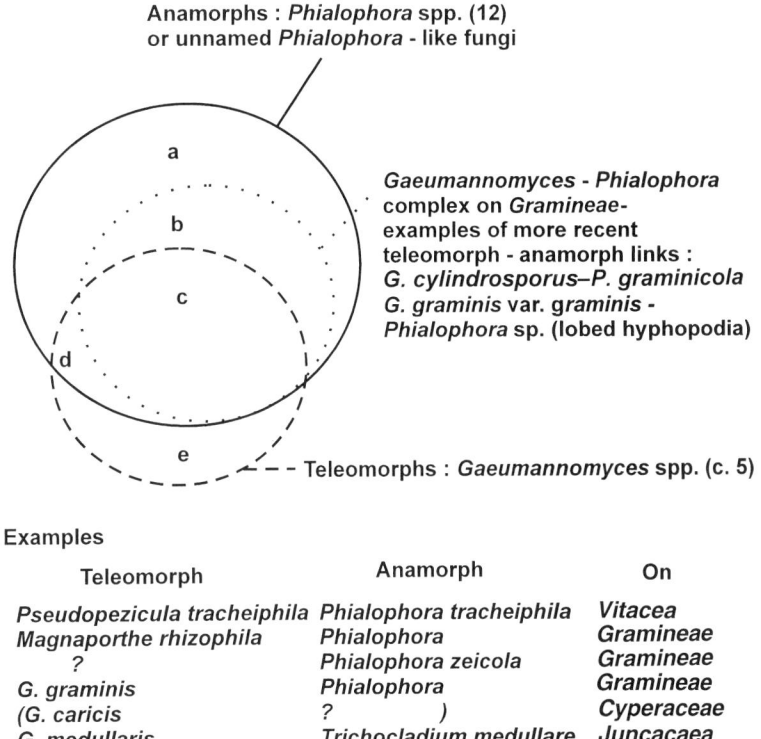

Anamorphs : *Phialophora* spp. (12) or unnamed *Phialophora* - like fungi

Gaeumannomyces - Phialophora complex on *Gramineae* - examples of more recent teleomorph - anamorph links :
G. cylindrosporus–P. graminicola
G. graminis var. *graminis* - *Phialophora* sp. (lobed hyphopodia)

Teleomorphs : *Gaeumannomyces* spp. (c. 5)

Examples

	Teleomorph	Anamorph	On
a	*Pseudopezicula tracheiphila*	*Phialophora tracheiphila*	*Vitacea*
b	*Magnaporthe rhizophila*	*Phialophora*	*Gramineae*
	?	*Phialophora zeicola*	*Gramineae*
c	*G. graminis*	*Phialophora*	*Gramineae*
d	*(G. caricis*	? *)*	*Cyperaceae*
e	*G. medullaris*	*Trichocladium medullare*	*Juncacaea*

Fig. 5.13. How the original concept of the *Gaeumannomyces–Phialophora* complex of fungi on *Gramineae* (a set indicated by the dotted circle) relates to the genera *Gaeumannomyces* (a set indicated by the dashed circle) and *Phialophora* (a set indicated by the solid circle). Named examples are given for each of the enclosed zones; the example for zone d is tentative.

are mostly necrotrophic parasites that attack root systems, particularly of *Gramineae* and *Cyperaceae*. The take-all fungus, under the name *Ophiobolus graminis*, was associated with the family *Gnomoniaceae* as early as 1932 and recently *Gaeumannomyces* was accepted into the enlarged family *Valsaceae* (Eriksson and Hawksworth, 1993). Both these are families of the *Diaporthales*, but *Gaeumannomyces* species do not have the typical nutritional status of members of the *Diaporthales*, which are either saprophytes, endophytes fruiting on moribund tissue, or systemic pathogens causing cankers or dieback. Also, true members of the *Diaporthales* do not have appressoria (term preferred to hyphopodia by Cannon, 1994), or *Phialophora* anamorphs. Although some anamorphs of *Magnaporthe* spp. are *Pyricularia*, others are *Phialophora*, thus blurring further the distinction between these two genera with similar teleomorph (sexual state) morphology. Recent ribosomal DNA sequence analysis (see 5.5.5; Bryan *et al.*, 1995; Bunting *et al.*, 1996) suggests a close relationship between *G. graminis* and *Magnaporthe*.

G. graminis (Sacc.) v. Arx & Olivier var. *avenae* (Turner) Dennis (referred to throughout this book by the abbreviation Gga) has a wide host range within the *Gramineae*. It has, for instance, been recorded on *Deschampsia caespitosa* at Berriedale, Hoy, Orkney, the sole surviving relic of natural woodland in all the northern isles of Scotland (Dennis and Spooner, 1992). Gga has traditionally been regarded as the main cause of take-all of oats in Great Britain and it is also able to attack wheat and barley. In culture, Gga and Ggt resemble each other closely and the perithecia (Plate 1.3) are similar, although Gga produces ascospores that are longer on average than those of Ggt. Since the ranges of ascospore lengths overlap, this is not a reliable diagnostic. The current status of Gga in the British Isles, particularly with regard to cereals, is unclear and raises the following questions:

- *What is its distribution and host range?* It was thought to be restricted to the north and northwest, but it may be spreading as *Agrostis* spp. are planted for new golf course greens (York, 1996). It is now rarely recorded from cereals, except oats in Ireland.
- *What factors affect its distribution?* Limited survey work suggested that its absence from eastern Britain, where soils are often less acid, was associated with the presence of the competitive, avirulent parasite *Phialophora graminicola* (Deacon, 1981). This parasite is also a competitor of Ggt, which does occur in these regions.
- *Could Gga become a problem on cereals other than oats?* This might become more of a concern should oats become more commonly grown, perhaps because of their well publicized health value, or should it be possible to transform wheat to produce avenacin (see 5.8.3).
- *What are the characteristics, other than being able to attack oats, that reliably distinguish Gga and Ggt?* Exceptions that do not fit the traditional criteria are occurring increasingly; refer to the discussion on varietal variants in Section 5.5.5.

Since *G. graminis* and *G. cylindrosporus* (which infect cereals and grasses) and *G. caricis* (which occurs on sedges) were described, *G. incrustans* has been isolated from turf grass roots in the USA (Landschoot and Jackson, 1989). Even more recently a fifth species, *G. medullaris*, has been described from the lower parts of the culms of the rush, *Juncus roemerianus*, in salt marshes in North Carolina on the Atlantic Coast of the USA and considered to be an obligate marine species (Kohlmeyer *et al.*, 1995). *G. medullaris* is unusual in that its anamorph is *Trichocladium medullare* and not a *Phialophora* species like the known anamorphs of other *Gaeumannomyces* species. There are also reports of some incompletely characterized fungi, which may represent other species yet to be described. These include two poorly defined taxa from sedges (Walker, 1980, 1981), one of which, *G.* tax. spec. 3 from *Carex*, is similar to a *Gaeumannomyces* species found on the mud rush, *Juncus geradii*, in Orkney, Scotland (Dennis and Spooner, 1992), except that it has longer spores.

The size and shape of ascospores are important in differentiating amongst these species. The three best-known varieties of *G. graminis*, vars *graminis* (Ggg), *tritici* (Ggt, the take-all fungus of wheat and barley) and *avenae* (Gga), are differentiated by ascospore size and hyphopodial structure (the hyphopodia of Ggg are characteristically lobed) as well as by their different host preferences. Ggg is widespread on grasses, is pathogenic on bermudagrass, causing bermudagrass decline (McCarty and Lucas, 1989; Elliott, 1991a), and causes a sheath rot in rice (Ou, 1972; Datnoff *et al.*, 1993); Ggt is most commonly associated with wheat and barley and Gga with oats and turf grasses. Some oat-attacking isolates of *G. graminis*, but with shorter ascospores than typical Gga and more characteristic of Ggt, have been described from Australia (Yeates, 1986). (The take-all that occurred on oats at ADAS Gleadthorpe in 1990 – see 1.2.3 – appeared to be caused by typical Ggt as a result of great inoculum pressure and very conducive conditions.)

A fourth variety of *G. graminis*, var. *maydis*, has been proposed recently for a maize pathogen in China (Yao *et al.*, 1992; Yao, 1993). It was found in Liaoning Province in 1986 and causes an early blight of maize and browning or necrosis of roots. Its asci and ascospores are shorter than those of other varieties of *G. graminis* and it grows faster than Ggt and Ggg between 20°C and 30°C. The perithecia are described as black, pear-shaped, 200–450 µm; ascospores are threadlike and curved, 55.5–85.0 × 2.4–4.0 µm, with one end pointed. In culture it forms oval, germinable phialospores and crescent-shaped, non-germinable phialospores. The hyphopodia are simple. In tests it caused slight infections of wheat, barley, sorghum, rice and some grasses, but it did not infect oats or soya bean.

Whilst rye is generally less susceptible to Ggt than are wheat or barley (see 3.2.1), isolates with increased pathogenicity to rye seedlings (R-isolates) have been identified in cereal crops in Britain and other parts of Europe (Hollins and Scott, 1990). They seem not to show particular preference for rye crops, however, and occur in mixed populations with normal isolates (N-isolates) and isolates with intermediate pathogenicity. The taxonomic and ecological significance of the distinctness of the R-isolates from the N-isolates is unknown. Some of these isolates have been shown to be distinct in their effects on disease and yield in the field (T.W. Hollins, personal communication).

Landschoot (1993) gave the following classification of the form-genus *Phialophora*: division *Amastigomycota*, subdivision *Deuteromycotina*, form-class *Deutero-*

mycetes, form-subclass *Hyphomycetidae* (*Hyphomycetes*), form-order *Hyphomycetales* (*Moniliales*), form-family *Moniliaceae*. The genus is heterogeneous and contains anamorphs of widely divergent ascomycetes (e.g. *Capronia*, *Mollisia* and *Gaeumannomyces*); some of the species are pathogenic on humans, causing mycoses, some are plant pathogens and *P. parasitica* is both (Hoog and Guarro, 1995). Consequently, not all species of *Phialophora* are of interest within the context of the *Gaeumannomyces–Phialophora* complex. *Phialophora graminicola* (Deacon) Walker (Figs 5.1–5.4), frequently found on grass and cereal roots and associated with a natural biological control of take-all (see 4.3.1), is the anamorph of *G. cylindrosporus* (Figs 5.5–5.6). It took more than a decade to achieve a full demonstration of this relationship (Hornby *et al.*, 1990b). *P. graminicola* occurred under all 15 species of field-grown grass investigated in Poland, whereas *Phialophora* sp. (lobed hyphopodia) occurred under only seven of them (Martyniuk, 1987). The different grasses supported different populations of *P. graminicola*, according to estimates based on percentage of roots colonized in wheat seedling assays. *P. graminicola* grows well as runner hyphae on the roots of such grasses, but does not grow on their stem bases and grows only poorly on grass rhizomes (Deacon, 1974). Invading hyphae are restricted to the epidermis and cortical cells of roots and produce characteristic swollen cells, but the fungus causes no damage to cereals in the field. In Australia and South Africa, *Phialophora* isolates with lobed hyphopodia are regarded as anamorphs of Ggg, and some produce perithecia. Perithecia of Ggg have been induced less frequently on plants inoculated with English isolates of *Phialophora* sp. (lobed hyphodia) (Figs 5.7 to 5.12) (R.J. Gutteridge, unpublished data). Scott (1989) suggested that South African isolates of *Phialophora* sp. (lobed hyphopodia) from Italian millet were states of Ggg in which the ability to produce perithecia had been lost as a result of adaptation to dry conditions; this is unlikely to be the case for non-perithecial isolates in Britain, where the conditions are wetter. *Magnaporthe rhizophila*, isolated from wheat grown in soil from a millet field in the Transvaal, South Africa, was found to have a *Phialophora* conidial state indistinguishable from that of Ggt (Scott and Deacon, 1983). Another *Phialophora* sp., *P. zeicola* (Deacon and Scott, 1983), from maize roots in South Africa was described in the absence of a perithecial (teleomorphic) form, although the possibility of a link with the more recently described *G. graminis* var. *maydis* (Yao *et al.*, 1992) needs investigating.

A variety of *in vitro* and *in vivo* methods are available for inducing perithecial production in fungi of the *Gaeumannomyces–Phialophora* complex and occasionally unusual forms of perithecium are encountered (Plate 1.4). The different methods suit different fungi and some isolates which fail to produce perithecia will remain of uncertain affinity (Holden and Hornby, 1981) until subjected to newer diagnostic methods. *Phialophora* sp. (lobed hyphopodia) and *P. graminicola* can often be differentiated *in vivo* by their vesicles or terminal hyphal swellings in the root cortex; those of *P. graminicola* are the smaller (Deacon, 1981). However, vesicles of intermediate size have been observed occasionally in cereal roots from some fields at Rothamsted and Woburn (Hornby *et al.* 1979; R.J. Gutteridge, unpublished data), suggesting that vesicle size may not always be diagnostic, or that intermediate or otherwise different forms exist. In Landschoot's (1993) review of the taxonomy and biology of ectotrophic fungi associated with patch diseases of turf grasses there is useful descriptive and illustrative material on many members of the complex. There is also an interactive CD-ROM (Schumann and MacDonald, 1996), which with over 350 images is another useful source of information.

5.3 Isolation and Culture Maintenance

5.3.1 Isolation

No method has been developed that consistently and reliably isolates take-all fungi directly from soil. Isolating from root debris collected by washing infested soil through sieves has met with some success. However, the most widely used methods for isolating take-all fungi from the soil are based on growing susceptible plants as baits. Traditionally, this has involved removing soil from the field, sowing it with seed of a susceptible host (e.g. wheat for Ggt) and allowing the seed to germinate and grow, preferably under conditions that favour infection (Hornby, 1981). Susceptible crop plants growing in the field also serve as bait, at least until they are too old and senescent. Whichever way is chosen, the fungi are subsequently isolated from the roots. It often helps to select lesioned roots to increase the success rate.

Isolation of take-all fungi from infected roots is best achieved after the roots have been surface sterilized. There is a range of surface sterilants available. A traditional one that is effective where Ggt is concerned is 1% silver nitrate, followed by washes in 5% sodium chloride and sterile water (Slope et al., 1978). Another, commonly used at Rothamsted, is sodium hypochlorite. Segments of surface-sterilized root are then placed on agar. For general work, most common mycological media with antibiotics (e.g. aureomycin, or combinations such as penicillin, streptomycin sulphate and terramycin (oxytetracycline)) to suppress bacteria suffice for this. PDA is used most frequently at Rothamsted. Characteristic pale growth of Ggt, with its curling back of the hyphal tips, usually appears after 2–4 days at 20°C. To get pure cultures, subcultures may be made from the hyphal tips. As cultures age on PDA, some become darker, varying from pale grey to black. Some information about variation amongst isolates may be found in Table 5.1 and Section 4.3.1. Where sectoring occurs (Table 5.1), the morphology of the sectors may differ markedly from that of the parent. In general, those isolates without a tendency to sector remain relatively true to type morphologically in culture for months, and in some cases years. More detail on isolation from surface-sterilized tissue, including its use in conjunction with selective medium (see 5.4.2), is available in Mathre (1992).

The less virulent members of the *Gaeumannomyces–Phialophora* complex can usually be isolated readily from infected roots of both grasses and cereals, using the same methods as for Ggt. With P. graminicola, for example, growth from root pieces becomes visible to the eye on PDA after about 5 days and after a further 4 days the culture may become pigmented in colours ranging from orange to grey, but this is very variable. Wild-type isolates from roots tend to be less stable than those of Ggt or Gga and subcultures of one isolate may give rise to almost the full range of

Table 5.1. Characters used in comparisons of isolates of *Gaeumannomyces graminis* var. *tritici* and assessments of the strengths of associations with TAD or other stages in take-all epidemics.

Observed character	Strength of association using isolates from a crop sequence experiment[a]	Evidence from the literature
RFLP type	Weak, poor statistical support	No data.
dsRNA/viruses	Weak; also weak association with RFLP type	Very weak (Rawlinson and Buck, 1981; Buck, 1986); the analytical methods used were unsuited to populations.
Perithecial production	Weak; moderate association with RFLP type	No data, but association with pathogenicity (Chambers and Flentje, 1967; Rawlinson *et al.*, 1973).
Pathogenicity	None	Moderate in spring cereals (Cunningham, 1975); none (Asher, 1980; Cook and Naiki, 1982).
Colony growth and morphology	None	Darkly pigmented isolates most frequent in collections from continuous cereals (Chu Chou and Hornby, 1972); in intermediate runs where take-all was at a peak, pale cultures dominant (Cunningham, 1975). Strong on defined (Lilly and Barnett's) medium (Slope and Gutteridge, 1982): more isolates with slow (see 4.3.1) and floccose growth from continuous cereals.
Colony-producing sectors	Weak	Strong; 41% of isolates from a continuous cereal sequence (10 years of cropping) compared with 17% from cereals in rotation (Slope and Gutteridge, 1982).

[a]Further information in the text.

variation in colour and growth characteristics. Growth of *Phialophora* sp. (lobed hyphopodia) from root segments on agar becomes visible after 2–4 days and is similar to that of Ggt, except that behind the margin of the colony aerial hyphae are much more conspicuous. Colony coloration is similar to Ggt, but sectoring is uncommon, and older cultures of this *Phialophora* develop small, or sometimes large, sclerotium-like bodies in the agar. Plates 1.1 and 1.2 show colonies of some members of the *Gaeumannomyces–Phialophora* complex on agar.

5.3.2 Maintenance

Many researchers store cultures of the *Gaeumannomyces–Phialophora* complex fungi on agar (e.g. refrigerated at 3–5°C on agar slopes, on agar under mineral oil at similar temperatures, or on agar blocks in sterile water at room temperature). Other methods, some of which have been used for these fungi, are storage on anhydrous silica gel or in soil, freeze-drying or storage in liquid nitrogen (Smith, 1984). Not all methods are applicable to all fungi and most methods have some disadvantages in relation to survival or stability (in particular the loss of pathogenicity) of the stored isolates. For example, isolates of Ggt varied in loss of pathogenicity whilst maintained on agar either for storage purposes or during repeated subculturing (Naiki and Cook, 1983a; more detail in 5.8.2). The capacity of *G. graminis* to produce perithecia may also be affected during long periods of maintenance and there is a report of loss of virulence in freeze-dried cultures of one isolate of Ggt (Holden and Hornby, 1981). Storage in liquid nitrogen is expected to induce least adverse change in fungi and this is supported by the experience that all isolates of Ggt and Gga in a collection of T.W. Hollins (personal communication) returned to culture after varying periods of up to 12 years in liquid nitrogen had retained their pathogenicity.

Plant pathogenic fungi of many genera stored on agar blocks in tubes of sterile water at room temperature survived morphologically unchanged for 6 or 7 years (Boesewinkel, 1976). Using this method, a collection of fungi isolated from cereal and grass roots has been maintained in water for more than 20 years at Rothamsted. Most isolates have survived in their original tubes, or in replacement tubes after subculturing. Occasionally cultures have died; some of these have duplicates that survived. Maintenance of pathogenicity is not expected after storage in this way for more than a year or two, but there have been notable exceptions. An isolate of typical Ggt (M28, from a bioassay of a soil growing a first maize crop at Rothamsted) was subcultured after storage for almost 19 years in water and was found to be highly pathogenic on wheat.

5.3.3 Culture collections

The provenance of all fungal isolates taken from culture and used in experimental work needs to be known and documented. This has become increasingly apparent as molecular biological work has concentrated on individual isolates and exchanges of isolates between different research groups have taken place. In such circumstances it is important that the identity of each isolate remains unchanged. Most laboratories have their own coding systems for naming isolates. Whilst these may be useful within the laboratory, only the original code name or number, or that first used in a publication, should be used in new publications.

Many of the large mycological herbaria (e.g. DAR, CBS, IMI) hold living cultures of fungi belonging to the *Gaeumannomyces–Phialophora* complex. The Crop and Disease Management Department of the IACR holds a specialist collection of over 600 living cultures of soil-borne fungi from around the world. This collection was created by the cereal root pathology group of the former Plant Pathology Department and consequently a large majority of the isolates is cereal-root-infecting fungi, with the *Gaeumannomyces–Phialophora* complex being well represented. New isolates added to the collection are now identified by a two-letter prefix (GP, the initial letters from *Gaeumannomyces–Phialophora* complex) followed by an accession number. This simple system was proposed by Yoder *et al.* (1986). Isolates already in the collection have not been allocated a GP number, their original identification codes being retained. The collection's computer database, named GPDATA, was originally created in Datatrieve, but it has now been transferred to Oracle. The collection catalogue can be browsed via World Wide Web forms at the location http://www.res.bbsrc.ac.uk/plantpath/cultures/

5.4 Identification by Conventional Methods

5.4.1 *Pathogenicity tests, infection structures and perithecia*

The traditional way of confirming that an isolate is Ggt is to show that it causes characteristic symptoms of take-all on wheat. This is normally done in seedling infection tests, also referred to as pathogenicity tests. These tests can provide useful information for identifying other members of the *Gaeumannomyces–Phialophora* complex. The following procedure is still used routinely at Rothamsted.

A test fungus is grown on PDA at 20°C for 7–10 days (the shorter period is usually sufficient for Ggt, the longer period for the slower-growing *P. graminicola*). To coincide with the end of this period, plump, healthy wheat grains (loosely referred to as seeds hereafter) are immersed in tap water for 2–3 hours and then spread on wet absorbent paper in a propagating tray to incubate for 2–3 days at an ambient temperature of 20°C. A layer of colonized agar from a plate culture is placed on coarse sand in each of a series of partially filled, square plastic pots with a depth of 5.5 cm and a capacity of about 200 cm^3. This inoculum is then covered by a 1 cm depth of coarse sand, on which two to five germinated seeds are placed separately, covered with coarse grit and watered. Alternatively, the pots may be filled with coarse sand in which depressions are made to take 12 mm diameter pieces of agar culture. A germinated seed is placed on each agar disc and the depressions are filled with more sand, before a layer of grit is added and the whole is watered. After this, the pots are kept in controlled environment rooms with a 16-hour day length and day/night temperatures of 15/10°C for 5 weeks, or at a continuous 19°C for 4 weeks (Hornby, 1981). Water is applied twice weekly. After incubation, the plants are knocked out of the pots and the root systems washed free of sand in a jet of water. The roots are then examined whilst being held under water against a white background; the same procedure is used for washed field plants. (MacNish *et al.*, 1986, assessed length of vascular discoloration against an orange background to eliminate the effects of dark runner hyphae on visual symptoms.) The numbers of healthy and infected plants and of healthy and infected main root axes are recorded. A root axis is scored as infected with Ggt if it has dark lesions that penetrate to the stele (Plates 2.7 and 3.8). On occasions *P. graminicola* may cause host roots to look grey because of much mycelial growth (Plate 3.10, as teleomorph, *G. cylindrosporus*), but sometimes there is little colonization by this fungus, except close to the inoculum. *Phialophora* sp. (lobed hyphopodia) renders the roots of test plants light to medium brown in colour depending on the intensity of infection (Plate 3.9 illustrates Ggg, the probable teleomorph). Stem bases also are often discoloured, but not as intensely as with Ggt, and lobed hyphopodia appear on the coleoptile and first leaf sheath. Other records routinely made in host-infection assays are: a disease severity score for the root system (0–3, indicating no symptoms and slight, moderate or severe disease); length of shoot; number of leaves; number of leaves yellowed or necrotic; assessment of stem-base blackening (0–3: none, slight, moderate or severe); weights (may be wet or dry weights of whole plants and/or roots or shoots, depending on requirements).

An example of a 0–5 severity rating on seedlings, based on runner hyphae and lesions, is: 0 = no runner hyphae (RH) or lesions; 1 = RH and a few small lesions on < 10% of the root system; 2 = RH and small lesions on 10–25% of root system; 3 = RH and lesions on 25–50% of root system; 4 = RH and many large, coalescing lesions on > 50% of root system; 5 = root system and culm completely inundated with lesions and RH (Phillion *et al.*, 1994).

Pathogenicity aside, there are signs that help to identify putative *Gaeumannomyces*–*Phialophora* complex fungi in seedling infection tests. These include dark runner hyphae growing longitudinally on root surfaces, subepidermal vesicles within roots (*Phialophora* spp.) and lobed or unlobed hyphopodia developing on coleoptiles. Apart from establishing the appearance and sizes of such structures, each type may be assessed on a scale of 0–3, where 1 indicates that the structures are present but difficult to find using a dissecting microscope, and 3 indicates that they are abundant. Pictorial keys used for assessing the abundance of vesicles of *Phialophora* spp. in roots are shown in Fig. 5.14. Infected assay plants have a further important use in that perithecia can often be induced to develop after appropriate treatment. In such a treatment that has worked well at Rothamsted, plants from each pot are kept together, their shoots cut off to within 2 cm of the seed and the moist roots and shoot bases pushed into a 1 cm diameter test tube, which is then lightly plugged with cotton wool. After about 6 weeks in the 15/10°C growth room described earlier in this section, the rotting roots are examined at weekly intervals for perithecia. The great majority of Ggt isolates will produce perithecia in these conditions, usually within 6 weeks; those of *G. cylindrosporus* (Hornby *et al.*, 1977), or their initials, may be observed after 10 weeks, but usually take up to 20 weeks. Description of shapes and measurements of mature ascospores, preferably from at least ten perithecia, provide additional useful information.

Pathogenicity tests of the kind described above are adequate for confirming pathogenicity, and hypovirulence in some cases, and provide strong evidence as to the identity of the test fungus. Demonstrating differences in pathogenicity amongst Ggt isolates may require some modification and has been achieved successfully by using cultures grown on a sand–maizemeal mixture (500 g sand, 2 g maizemeal, 65 ml water) as inoculum (Hollins *et al.*, 1986). This inoculum is placed in pots of sand as a 0.5 cm layer, 3 cm below the layer of seed. The method showed that some isolates (R-isolates) had greater pathogenicity on rye seedlings than did others (N-isolates) (Hollins and Scott, 1990), a classification of isolates by pathogenicity that was consistent with subsequent results from DNA testing (O'Dell *et al.*, 1992).

Some laboratories use simpler pathogenicity tests than those above to check the identity of isolates. These may involve gnotobiotic systems with both wheat and test fungus growing on agar, or inoculation of seedlings growing in test tubes containing soil or vermiculite. Similarly, there is a range of techniques for producing perithecia of *Gaeumannomyces* spp. In comparisons of perithecial production on infected plants and *in vitro* using colonized agar, drained liquid cultures or colonized filter papers, no one method was better than the others for all fungi (Holden and Hornby, 1981). A method of inoculating Ggt on to sterile wheat seedlings growing on water agar (Speakman, 1982) has been simplified further in some laboratories by growing the fungus on agar on which a sterilized piece of leaf

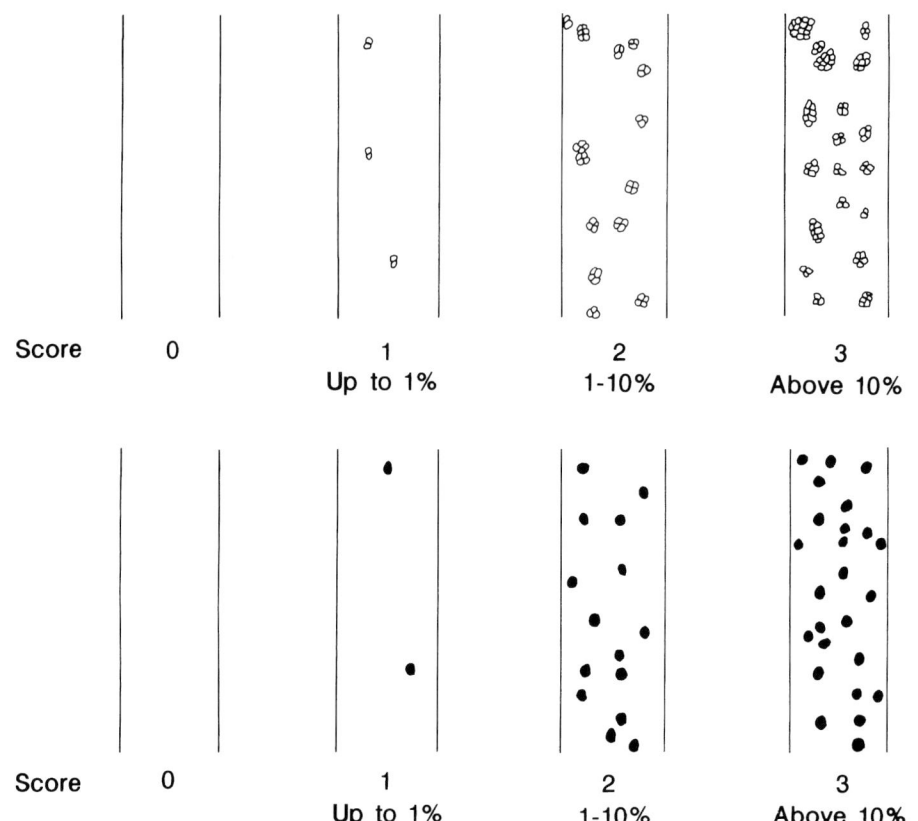

Fig. 5.14. Pictorial keys that have been used at Rothamsted to aid the assessment of abundance on a 0–3 scale of small, clustered vesicles of *Phialophora graminicola* (above) and larger vesicles of *Phialophora* sp. (lobed hyphopodia) (below) in seminal roots of cereals. Segments of root, 1.5 cm long, taken at 2 cm below the seed, were first cleared in hot 10% sodium hydroxide, acidified with 50% hydrochloric acid and mounted in lactophenol for low-power examination under a microscope. The area seen to be occupied by the vesicles is estimated as a percentage of the area of the radial, longitudinal plane of the root.

from a wheat seedling is placed (see 5.7.2). Speakman (1984b) has also described a method of developing perithecia of Ggt and Ggg on malt extract peptone agar containing antibiotics after an initial period of growth in the dark, followed by flooding the plates with sterile distilled water and several weeks' incubation in daylight from a north-facing window. *G. incrustans* differs from *G. graminis* in being heterothallic and perithecial production in this fungus requires two compatible mating types, minimal nutrient medium and host tissue (Landschoot and Jackson, 1989).

5.4.2 *Selective and other media*

A semi-selective medium, SM-GGT3, developed for isolation of Ggt from infected plant roots (Juhnke *et al.*, 1984), is PDA-based and contains antibiotic (streptomycin), antifungal agents (dicloran, for inhibition of *Rhizopus* spp., metalaxyl and HOE 00703, which is 1-(3,5-dichlorophenyl)-3-methoxymethyl-pyrrolidin-2,4-dion) to restrict the growth of competing organisms, and L-DOPA (L-β-3,4-dihydroxyphenylalanine) which is turned black (or dark brown as seen in the agar) by Ggt as melanin is produced. It is used routinely for identification of take-all in some laboratories in the USA and there are modifications, such as SM-GGT4 and SM-GGT7 (Elliott, 1989, 1991b). Avirulent *Phialophora* spp. that infect cereal roots also produce the diagnostic dark coloration when grown on the medium (Fig. 5.15); these fungi are not known to occur in the north and northwestern states of the USA where the medium is mostly used, and so are not considered a problem in diagnosis of take-all. In Europe, the medium has proved useful where take-all symptoms were obscured by other pathogenic fungi, especially *Fusarium* spp. (Hossain and Schlosser, 1986). Isolation on a medium similar to SM-GGT3 (Juhnke *et al.*, 1984), but with cymoxanil instead of metalaxyl, combined with DNA analysis (see below) to identify Ggt and *Phialophora*, had limited success because slow-growing *P. graminicola* was often overrun by other fungi (G.L. Bateman, unpublished data).

Another semi-selective medium, R-PDA, which was reported to aid isolation and identification of Ggt in the presence of other pathogenic fungi and secondary fungal colonists, is based on dilute PDA amended with the antibiotic rifampicin (at 100 µg ml^{-1}) and the fungicide tolclofos-methyl (at 10 µg ml^{-1}) (Duffy and Weller, 1994). On this medium, Ggt altered the colour of rifampicin from orange to purple within 24 hours; Gga, Ggg and *G. incrustans* also caused this reaction. The medium was found to be more effective than SM-GGT3 for isolating Ggt from wheat.

Oat-leaf agar has been used to discriminate between Ggt and Gga (Holden and Ashby, 1981; Bateman *et al.*, 1992) and was more reliable than oat-root agar. However, more recent work suggests that the ability of Gga to grow on oat-leaf agar is not an entirely reliable property (G.L. Bateman, unpublished data). The relationship between this phenomenon and the occurrence of avenacin (see 5.8.3) in the plant material is not clear; extracts of oat roots, not extracts of oat leaves, would have been expected to give better differentiation if the effect was solely dependent on avenacins.

Comparisons of some of the fungi in the *Gaeumannomyces–Phialophora* complex were made in experiments to investigate their responses to different osmotic potentials in agar media (Wong, 1983). The purpose was to aid in the selection of the most suitable fungi for biological control of take-all, especially in the drier wheat-growing areas of Australia. Avirulent fungi were, in general, capable of growing at lower osmotic potentials than were the pathogens. Although this was unlikely to give the avirulent fungi advantage at the osmotic potentials encountered in young wheat tissue, it might make them better colonizers of maturing plant bases. Starting with a basal agar medium (Wong, 1983) at pH 4, containing salts, glucose, yeast extract and malt extract, D. Hornby and K. House (unpublished data) made adjustments to osmotic potentials of −0.69, −1.37, −2.70, −3.85 and −5.20 MPa, using sodium chloride. Whilst most fungi showed increases in colony diameter at −3.85 MPa after 5 days at 25°C, only those fungi in the *Gaeumannomyces–Phialophora* complex identified as Ggg/*Phialophora* sp. (lobed

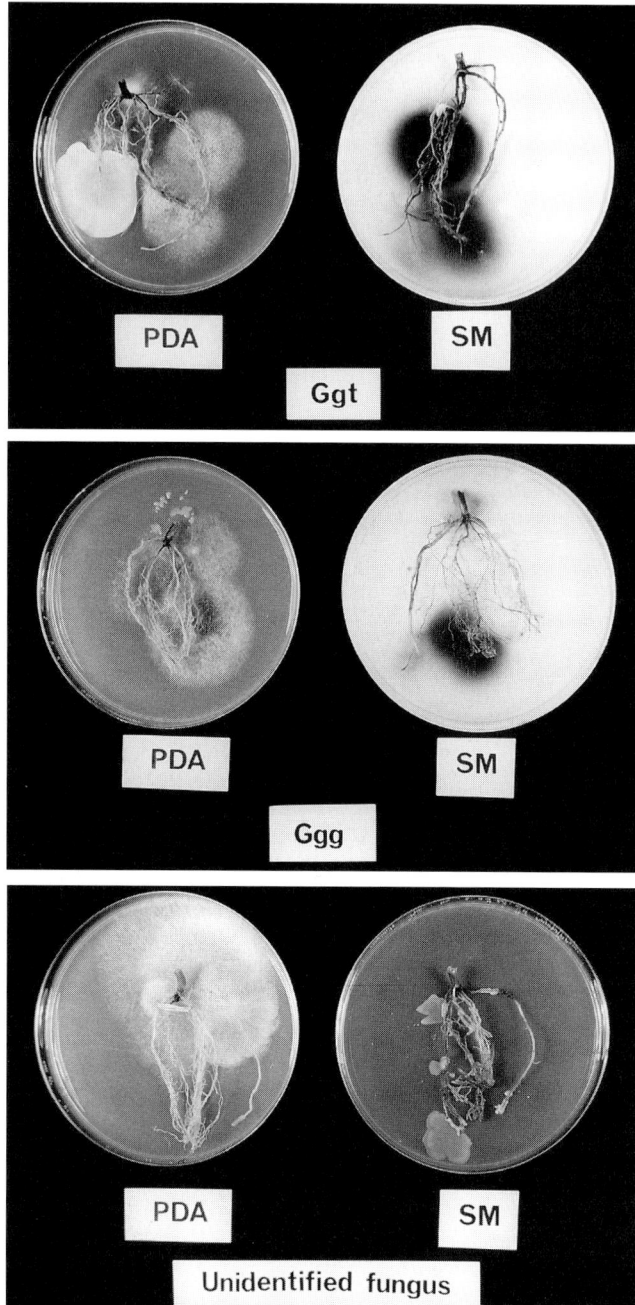

Fig. 5.15. Growth from the roots of wheat seedlings of *Gaeumannomyces graminis* var. *tritici* (Ggt), *G. graminis* var. *graminis* (Ggg) and an unidentified fungus on to potato dextrose agar containing antibiotics (PDA) and SM-GGT3 (SM), a selective medium containing L-DOPA.
A white disk has been placed behind each SM plate on which Ggt or Ggg is growing to show the pigmentation (brown) in the agar.

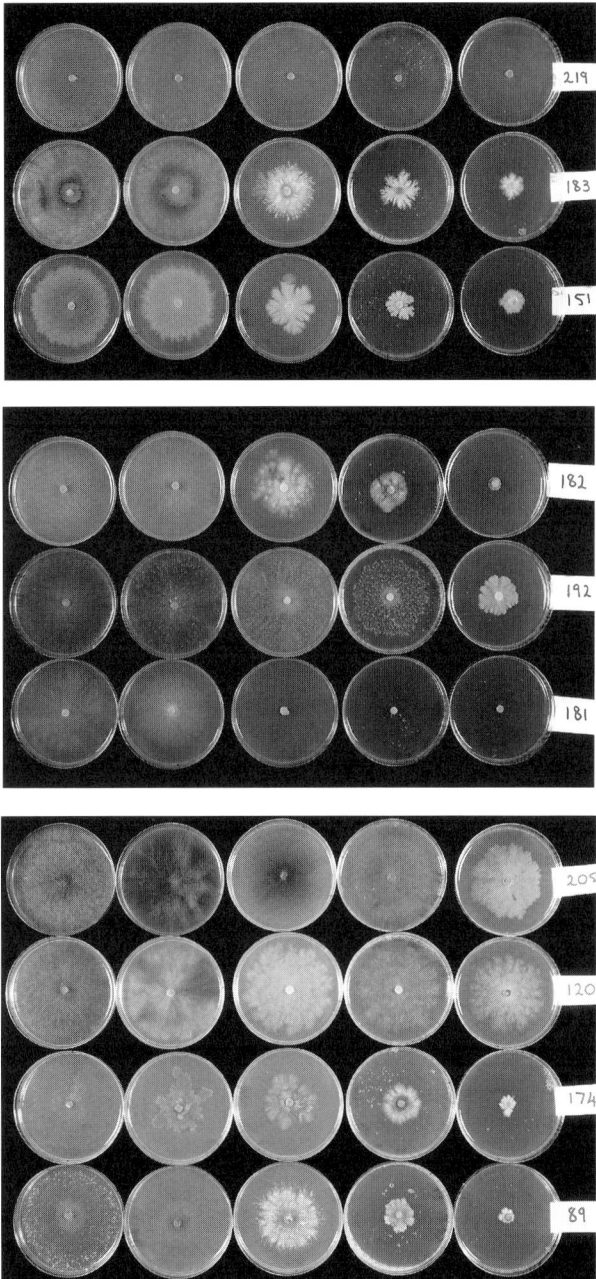

Fig. 5.16. Growth of fungi in the *Gaeumannomyces–Phialophora* complex after 2 weeks on agar at different osmotic potentials.
Osmotic potentials increase from left to right. *G. graminis* var. *tritici*: 89, 182, 183; *G. graminis* var. *avenae*: 174; *G. graminis* var. *graminis*/*Phialophora* sp. (lobed hyphopodia): 120, 205, 219; *P. graminicola*: 151; *Phialophora* species: 181; *Cephalosporium maydis*: 192 (see Table 5.2 for further details).

Table 5.2. Effects of osmotic potential in agar medium[a] on growth after 5 days at 25°C of fungi in the *Gaeumannomyces–Phialophora* complex and *Cephalosporium maydis*.

Fungus	Isolates tested[b]	Colony diameter (mm) at −0.69 MPa	% Inhibition of colony diameter at −3.85 MPa	% inhibition of colony diameter at −5.2 MPa
G. graminis var. *tritici*	89	73	83	100
	182	76	84	100
	183	49	80	100
G. graminis var. *avenae*	174	28	60	100
P. graminicola	139 (CBS 389.81)	80	88	100
	151	41	75	100
G. graminis var. *graminis* – *Phialophora* sp. (lobed hyphopodia)	120	79	71	84
	205	71	73	83
	219	74	65	80
P. radicicola Cain	181 (CBS 296.53)	21	100	100
Cephalosporium maydis	192 (IMI 272212)	80	68	86

[a]Details in 5.4.2.
[b]Isolate codes are those used in GPDATA (see 5.3.3). Alternative codes are shown in parentheses for those isolates lodged in national collections; the isolates used are subcultures from these and have not necessarily retained identical characteristics during transfer and storage.
See also Fig. 5.16.

hyphopodia) showed any growth at the lowest osmotic potential (Table 5.2; Fig. 5.16). A fungus described as *Cephalosporium maydis*, isolated from wilted maize plants, showed similar tolerance of low osmotic potentials. This fungus, a vascular wilt pathogen confined to Egypt and India, is in need of work to establish its correct generic position, which may be in the genus *Phialophora* (Walker, 1981). Preliminary DNA tests indicate a close relationship between it and isolates of *G. graminis* (E. Ward and G.L. Bateman, unpublished data). These limited results, like those of Wong, suggest a clear distinction between the lobed hyphopodial forms (Ggg) and the other varieties within the species *G. graminis*, with the greatest similarity being between Ggt and Gga. The possibility of being able to distinguish between virulent and avirulent varieties using a single discriminating osmotically adjusted medium deserves further investigation. This separation within the species is also made by some of the DNA tests described in Section 5.5.2.

5.5 Identification by Molecular Methods

5.5.1 DNA methods: background

Molecular biological methods for identification of organisms at all taxonomic levels based on differences in their nucleic acids (DNA or RNA) have come very much into prominence in the last decade (Hansen and Wick, 1993; Oliver, 1993; Schots *et al.*, 1994). A brief explanation of the general techniques and terminology used is given before the procedures which relate specifically to the take-all fungi are described.

Differences between organisms can be detected by: (i) determining the sequence of nucleotides (bases) in their DNA; (ii) DNA probing; and (iii) the polymerase chain reaction (PCR). Methods (ii) and (iii) may also involve identifying restriction fragment length polymorphisms (RFLPs).

DNA sequencing
Determining the sequence of the four bases (adenine, thymine, cytosine and guanine) in the DNA and comparing it with others is the most direct way of identifying differences (White *et al.*, 1990; Sambrook *et al.*, 1989b). Sequencing used to be a relatively difficult and laborious technique, but recently it has become simpler with the introduction of automated DNA sequencing equipment. However, these machines and the reagents for the techniques are expensive.

DNA probes
Probes are small fragments of DNA that have been labelled with a reporter molecule and are used to detect complementary sequences of DNA in test samples (Sayler and Layton, 1990; Hansen and Wick, 1993; Oliver, 1993). Usually, the first stage of the method is to separate the two strands of the target DNA by alkali or heat, and to bind them to a nylon or nitrocellulose membrane. This is then bathed in a solution containing the probe which hybridizes (attaches) to complementary sequences on the target DNA, if these are present. Then, after washing to remove unattached probe, the probe is detected by means of its reporter group (label). Originally these were radioactive labels but now non-radioactive labels are being increasingly used. Detection usually involves blackening of X-ray film by radioactivity or light emitted (directly or indirectly) by the label.

DNA for use as a probe can be made by cloning procedures or by PCR (see below). Cloning involves using enzymes (restriction endonucleases) that cut DNA into small fragments. The fragments are then joined enzymically to vector molecules – either plasmids (loops of DNA) or bacteriophages – that are multiplied in bacterial cultures.

PCR
The PCR technique, introduced in the mid-1980s, is used for rapid, sensitive multiplication and detection of specific DNA sequences (Cherfas, 1990; Mullis, 1990; Steffan and Atlas, 1991; Henson and French, 1993; Ward, 1994). PCR involves amplification of specific segments of DNA, the ends of which are defined by two primers, i.e. short pieces of single-stranded DNA designed to match the target sequence precisely. The specificity of the PCR is determined by the primers – amplification can occur only if the primers are complementary to the target DNA and can therefore bind to it. The amplified DNAs are visualized by staining after separation on gels.

RAPDs

One variation of PCR that can be used to discriminate between closely related organisms (species, varieties or isolates) is random amplified polymorphic DNA (RAPD) analysis (Welsh and McClelland, 1990; Williams *et al.*, 1990; McClelland and Welsh, 1995). The PCR generally involves amplification of known sequences using highly specific primers. However, in RAPDs a short arbitrarily chosen oligonucleotide is used to produce DNA 'fingerprints' without the need to know anything about the sequences to be amplified. The technique is also known as arbitrarily primed PCR (AP-PCR), or DNA amplification fingerprinting (DAF).

RFLP analysis

In this method the criterion for identification is not just detecting the presence of complementary DNA (by probing or PCR) but recognition of characteristic patterns of DNA fragments (Bernatzky, 1988). These differently sized fragments are generated by using restriction enzymes, which recognize and then cut DNA at specific base sequences (usually 4–6 base pairs long). Different organisms vary in the frequency and distribution of these sites and this can be used to distinguish between them. If probes are used to detect RFLPs, the technique is known as Southern blotting (Sambrook *et al.*, 1989a). The first stage of this is to cut DNA, extracted from the organisms, with restriction enzymes and electrophorese it on agarose gels to separate the differently sized fragments produced. The DNA is then transferred to a nylon (or nitrocellulose) membrane and incubated with the probe. Those DNA fragments that bind to the probe are then detected as black bands on X-ray film. Differences in the DNAs are detected as different patterns of bands.

RFLPs can also be detected using PCR. In this case a small fragment of DNA is first amplified from the total DNA of the organism using PCR. This is then digested with restriction enzymes to produce the different DNA fragment patterns.

Types of DNA

Some regions of the genome are more useful for identification than are others. The genes coding for ribosomal RNA (rDNA) are especially useful for analysing phylogenetic relationships and also for identification of taxa at various levels. This is because rDNA contains some regions that are well conserved throughout evolution and others that are variable even within a species. Some of the ribosomal RNA genes are present in the DNA of the nucleus and others are present in the DNA of the mitochondrion. Nuclear rDNA in all fungi has a series of genes (subunits) separated by internal transcribed spacer (ITS) regions. Evolution is relatively rapid in the ITS regions and these are useful for studying differences between very closely related organisms; it is slower in the subunits (i.e. the DNA is more conserved). The genes of the mitochondrial DNA are moderately conserved, but can be useful in evolutionary and taxonomic studies.

5.5.2 DNA methods: identification of G. graminis

The first method available for specific detection of *G. graminis* was a DNA probe (pMSU315) derived from mitochondrial DNA (Henson, 1989). This probe recog-

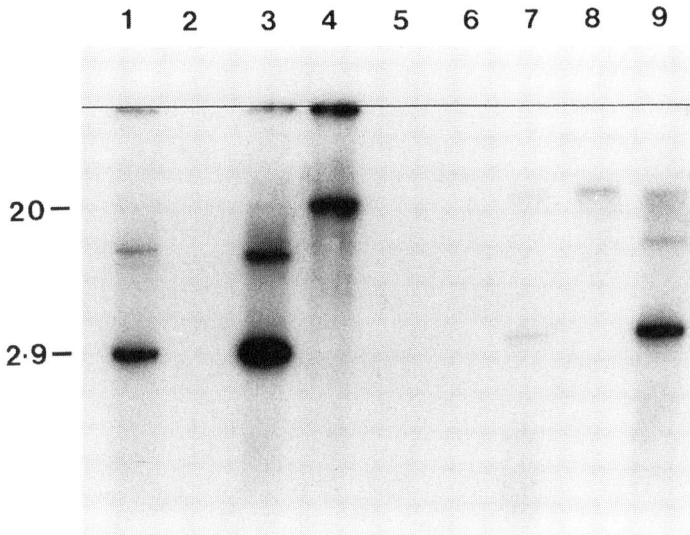

Fig. 5.17. Hybridization of ^{32}P-labelled pMSU315 with *Bam*HI-digested DNA from isolates of Ggt[a], Gga[b], Ggg[c] and *Phialophora* sp. (lobed hyphopodia)[d].
[a]Lanes 1 and 3; [b]lanes 7 and 9; [c]lane 4; [d]lane 8.
Other lanes contain DNA from other, unidentified fungi from cereal roots.

nizes only *G. graminis* (all varieties) and fungi very closely related to it (e.g. *P. graminicola*). It is also possible to differentiate Ggt and Gga from Ggg/*Phialophora* sp. (lobed hyphopodia) and *P. graminicola* (Pg) using probe pMSU315 in RFLP analysis with *Bam*HI as the restriction enzyme. A band of 2.9 kb is seen with Ggt and Gga, but not with Ggg/*Phialophora* sp. (lobed hyphopodia) or Pg (Bateman *et al.*, 1992; Ward and Bateman, 1994; and Fig. 5.17).

Subsequently, using primers designed from DNA sequence data, a PCR assay was developed for the specific detection of *G. graminis* (Schesser *et al.*, 1991). In this study the use of a single primer pair was found not to give a sufficiently specific test for *G. graminis*, because other, non-related fungi (*Cochliobolus sativus* and *Fusarium culmorum*) were also detected. To overcome this, nested primers were used: the first round of the PCR was done using one set of primers (KS1F/ KS2R) and the product of this was used in a second PCR using primers (KS4F/ KS5R) located within the first pair. This modification increased the specificity so that only *G. graminis* was detected (Schesser *et al.*, 1991; Henson, 1992; Henson *et al.*, 1993). A recent study using this method showed that Ggg isolates from different grass hosts gave different bands (Elliott *et al.*, 1993). Other work has shown that the outer primers alone (KS1F/ KS2R) can be used to detect *G. graminis* specifically if the temperature of annealing during the PCR is increased (Ward, 1994, 1995; and Fig. 5.18). Using this modification a band is produced with all three varieties, but that of Ggg is a different size (490 bp) from that of Ggt and Gga (600 bp). No DNA was amplified from *P. graminicola* isolates, unrelated fungi or uninfected wheat roots; presumably no sequence complementary to the primers exists in the DNA of these organisms and so amplification cannot occur.

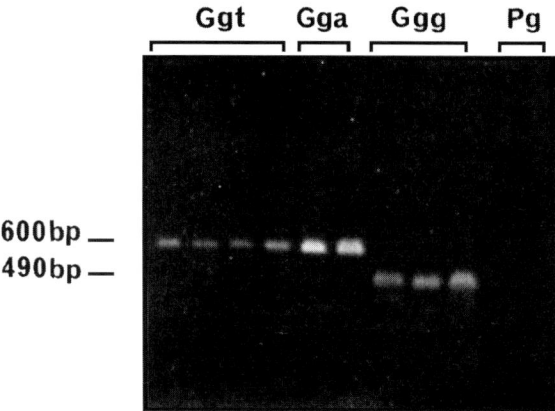

Fig. 5.18. A specific PCR test for *G. graminis*.
Primers KS1F and KS2R (Schesser *et al.*, 1991) were used in PCR assays with DNA from isolates of *G. graminis* var. *tritici* (Ggt), var. *avenae* (Gga), var. *graminis* (Ggg) and *P. graminicola* (Pg). The sizes of the DNA fragments in base pairs (bp) are given on the left-hand side of the figure.

Recently, primers have been developed that are variety specific. Bryan *et al.* (1995) devised a primer pair that will amplify DNA only from Ggt isolates and another pair that will amplify DNA only from Gga isolates. These primer pairs have now been used to develop quantitative competitive PCR techniques (Fig. 5.19) which allow rapid and specific detection of picogram (pg = 10^{-12} g) quantities of Gga or Ggt genomic DNA in root extracts (J. Carter, M.J. Daniels and A.E. Osbourn, unpublished data). Also, Goodwin *et al.* (1995), working on turf grasses, described primers that amplify DNA from Gga and Ggg (the varieties pathogenic on turf grasses) but not from Ggt.

PCR assays are simpler, quicker and more sensitive than tests using probes, because they can be performed on small quantities of relatively poor-quality DNA that can be prepared easily. The tests can be done on DNA released by simply boiling fungal mycelium in Tris buffer and from crowns and roots infected with *G. graminis* (Henson, 1992; Henson *et al.*, 1993; Ward, 1995).

Methods using ribosomal DNA and other repetitive DNA sequences have also been used to discriminate between fungi in the *Gaeumannomyces–Phialophora* complex. O'Dell *et al.* (1992) used an rDNA probe from wheat to distinguish between rye-adapted and non-rye-adapted isolates of Ggt (Hollins and Scott, 1990; see Section 5.4.1), and isolates of Gga. Ward and Gray (1992) used a mitochondrial rDNA probe, pEG34 (GggMR1), to distinguish between each of the *G. graminis* varieties and *P. graminicola* (Fig. 5.20). Using this probe, which was made by PCR from DNA from an isolate of Ggg, at least two distinct RFLP types for both Ggt and Gga were identified, so making it potentially useful in the study of population dynamics of these fungi (see 5.6). Differences in RFLP patterns between the varieties can also be detected more simply by restriction analysis of PCR-amplified rDNAs (Fig. 5.21; and Ward and Akrofi, 1994).

A flow chart summarizing the various stages in a combination of molecular and conventional approaches used at Rothamsted to identify fungi in the

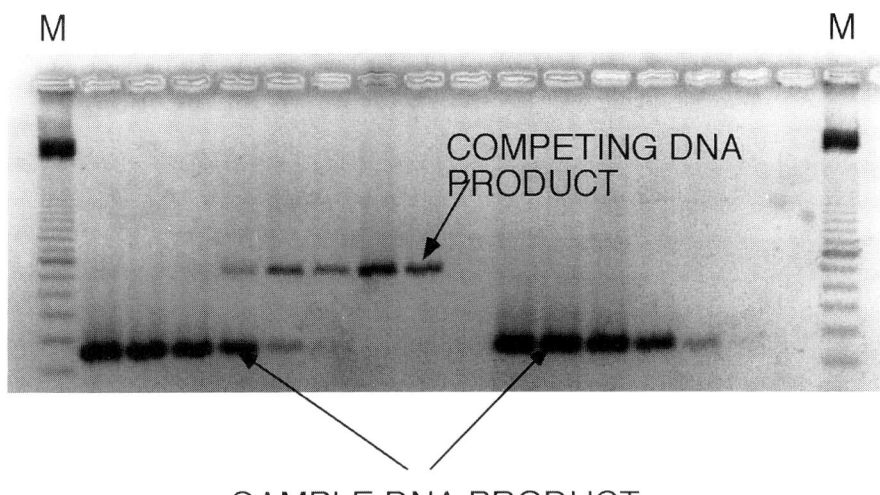

LANE	SAMPLE DNA PRODUCT G. graminis DNA	COMPETING DNA (0.1pg)
1	10ng	+
2	1ng	+
3	100pg	+
4	10pg	+
5	1pg	+
6	100fg	+
7	10fg	+
8	0	+
9	0	−
10	10ng	−
11	1ng	−
12	100pg	−
13	10pg	−
14	1pg	−
15	100fg	−
16	10fg	−

Fig. 5.19. Quantitative competitive PCR assay of *G. graminis* DNA. The competing DNA is a cloned piece of rDNA with a small 'stuffer' fragment inserted into it, so that this gives a larger PCR product than the genomic rDNA. The ratios of the two bands allow quantification of the amount of fungal genomic DNA present in the reaction mix, and the competing DNA provides an internal check to verify that PCR is working. In this example the DNA was from fungal cultures, but the technique works with extracts of infected roots. M = size marker. The prefix symbols n, p and f represent the factors 10^{-9}, 10^{-12} and 10^{-15}, respectively.

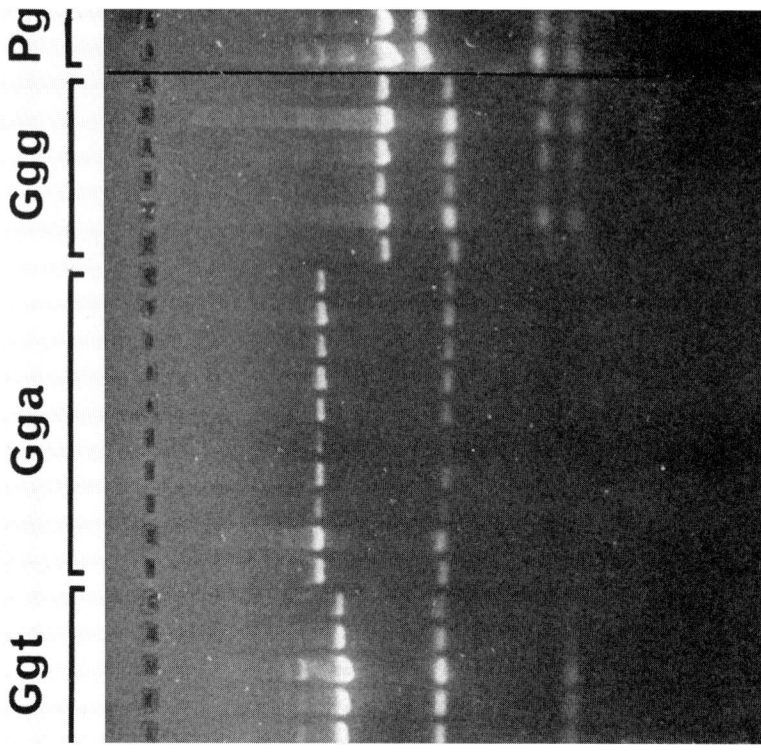

Fig 5.20. RFLP analysis using the mitochondrial ribosomal probe GggMR1. DNA from isolates of the *Gaeumannomyces–Phialophora* complex was cut with *Eco*RI, electrophoresed on a 0.8% agarose gel, blotted on to a nylon membrane and hybridized with probe GggMR1 (Ward and Gray, 1992). Detection was by the non-radioactive digoxigenin system (Boehringer–Mannheim) using chemiluminescence. Lane 1 = Ggt 88/10-1; 2 = Ggt 0g12N; 3 = Ggt 87/7-4; 4 = Ggt 180; 5 = Gga P086/439; 6 = Ggg 89/3-1; 7 = Ggg 148; 8 = Ggg 153; 9 = Pg 74/1736-2; 10 = Pg 85/17-3B; 11 = *G. caricis*. These isolates are described in Ward and Gray (1992).

Fig. 5.21. RFLP analysis of *G. graminis* and *P. graminicola* isolates using ribosomal DNAs amplified by PCR. The PCR primers used were ITS4 + ITS5 (White *et al.*, 1990) and the restriction enzyme for cutting the amplified bands was *Dde*I.

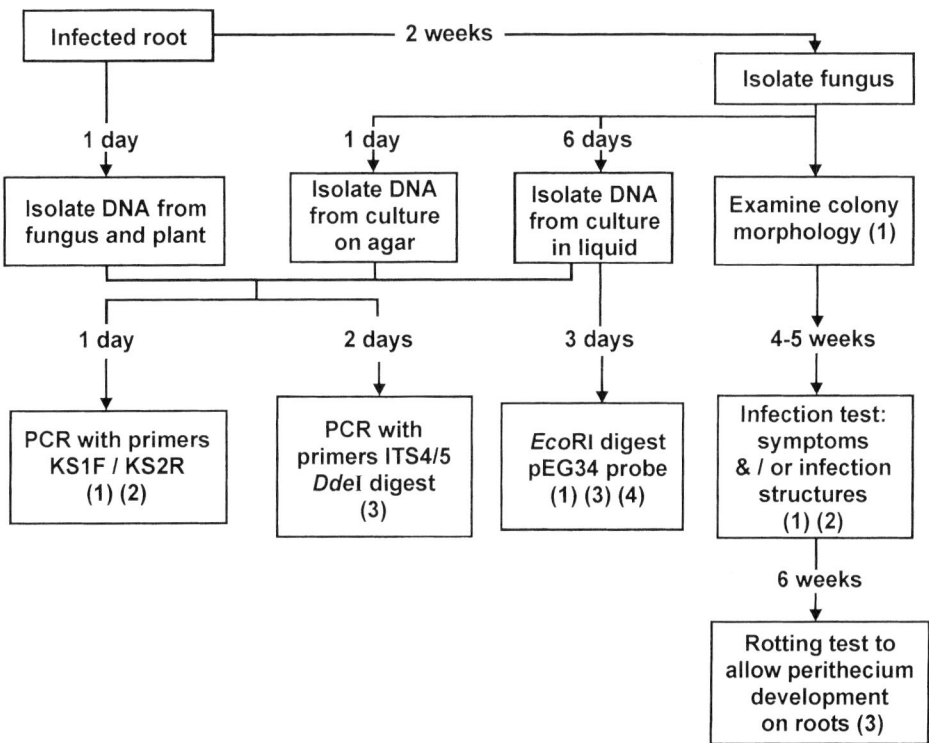

Fig. 5.22. Stages in the identification of *Gaeumannomyces–Phialophora* isolates in use at Rothamsted.
The numbered tests confirm (or give good evidence for) the following: fungi in the *Gaeumannomyces–Phialophora* complex (1); pathogenic varieties (2); varieties and species (3); and types within varieties (4).

Gaeumannomyces–Phialophora is shown in Fig. 5.22. The DNA methods have revealed clear differences among 'typical' isolates of *P. graminicola* and varieties of *G. graminis*. However, some isolates appear to have atypical or intermediate properties and often these isolates had already been observed to differ in some way when investigated by conventional techniques. An example of this is a group of Australian isolates that will attack oats, but which on the basis of ascospore length is more typical of Ggt than Gga (Yeates, 1986; Yeates and Parker, 1986a; Yeates *et al.*, 1986). Several of these isolates have now been analysed by various methods: RFLP analysis of ribosomal DNA, rDNA sequencing, avenacin resistance and avenacinase production (Ward and Gray, 1992; Ward and Akrofi, 1994; Tan *et al.*, 1994; Bryan *et al.*, 1995). They gave results indicating that either they were closest to Gga, or they were different from either Ggt or Gga.

There are reports of isolates with lobed hyphopodia causing take-all symptoms on wheat. Nilsson (1972) reported a lobed hyphopodial strain from Sweden that was virulent on wheat, oats, rye, barley and maize. Roy *et al.* (1982) found lobed hyphopodial isolates on soya beans in the midwest of the USA that produced typical symptoms of take-all on wheat in glasshouse tests. Recent DNA analysis of a lobed hyphopodial isolate from soya bean revealed that it had some characteristics of Ggt isolates and some of Ggg (Ward and Akrofi, 1994).

Molecular methods have shown correlations between particular RFLP types and either host plant preference (in terms of pathogenicity), or the host species of origin. The former situation is exemplified by rye-adapted vs. non-rye-adapted isolates of Ggt (O'Dell et al., 1992), the latter by isolates of Gga from oats and grasses (Ward and Gray, 1992; see also 5.6), isolates of Ggg from different grass hosts (Elliott et al., 1993) and isolates of Ggt from different cereal hosts (Bateman et al., 1997).

5.5.3 Serology

Diagnostic kits based on serological methods such as ELISA (enzyme-linked immunosorbent assay) are being used increasingly to detect plant pathogenic fungi (Fox, 1993; Hansen and Wick, 1993; Schots et al., 1994). Whilst these serological tests are simple to use, it is sometimes difficult to obtain a suitable highly specific antibody for the tests. The prospects improved with the development in 1975 of hybridoma technology to produce monoclonal antibodies, but specific antibodies for Ggt have not yet been made. However, polyclonal antibodies have been used to study fungi in the *Gaeumannomyces–Phialophora* complex, but progress has not matched that achieved by DNA methods. A close relationship amongst varieties of *G. graminis* was inferred from each giving a similar serological reaction (Abbott and Holland, 1975). The serological reactions of *Phialophora* sp. (lobed hyphopodia) from Britain and Ggg were found to be identical (Walker, 1981). At Rothamsted, ELISA tests using polyclonal antiserum to a Ggt isolate gave less conclusive results, although they indicated a closer relationship between the varieties of *G. graminis* than between *G. graminis* and other species (D. Hornby and R.F. White, unpublished data). El-Nashaar et al. (1986) used ELISA to detect Ggt on homogenized roots of inoculated wheat seedlings; absorbances were proportional to the amount of inoculum of Ggt applied. However, the polyclonal antisera used also reacted with several unrelated fungi and a more specific antibody would be needed to study the fungus in field conditions. A monoclonal antibody raised against surface antigens from mycelium of Ggt and believed to recognize an extracellular polyphenol oxidase has recently been identified (Thornton et al., 1997). This monoclonal antibody may have the potential for detecting Ggt both in soil and in infected plant material.

5.5.4 Protein electrophoresis

Analyses of soluble proteins by electrophoretic methods have indicated a closer relationship between Ggt and Gga than between either of these and Ggg (Abbott and Holland, 1975). Similar protein patterns were found for different isolates of Ggt, but isolates of *Phialophora* spp. with similar colony morphology in culture produced patterns which differed from those of Ggt to different extents (Maas et al., 1990). *Gaeumannomyces graminis* var. *maydis* was characterized by morphology, biology and soluble protein profiles (Yao et al., 1992; Yao, 1993); electrophoresis showed it to differ from all the other varieties of *G. graminis*.

5.5.5 Prospects and implications

In addition to the traditional techniques for identification based on morphology, a variety of new methods are being developed that should allow more rapid and reliable identification and detection of *Gaeumannomyces–Phialophora* complex fungi and give a better understanding of the relationships among them (Table 5.3). Advances in DNA-based molecular techniques have become particularly important during the past decade and a variety of molecular methods are now available to aid identification and detection. Conventional pathogenicity testing and perithecial tests take weeks to complete and may be inconclusive, e.g. if perithecia do not develop. Molecular techniques, in contrast, can give an answer within a few days and can be performed on spores or mycelium, or any other part of the fungus. Some of these techniques can also be used on infected roots or infested soil and recently a DNA probe has been used to quantify Ggt in roots and soil (Herdina *et al.*, 1996). The molecular methods have also helped to clarify the position of unusual or intermediate isolates. As well as discriminating between species and varieties, molecular methods can also discriminate between isolates and populations (see 5.6) – something that is very difficult or impossible to do using traditional methods. However, there is still much to be done in the areas of simpler tests, tests suitable for processing large numbers of isolates and tests that provide quantitative data.

Understanding the taxonomic and ecological relationships amongst the fungi of the *Gaeumannomyces–Phialophora* complex should greatly benefit our understanding of take-all biology. A better DNA method than those already described for investigating taxonomic relationships is comparison of ribosomal DNA sequences (White *et al.*, 1990; Bruns *et al.*, 1991). DNA sequences of a number of isolates of *G. graminis* and related *Phialophora* fungi have now been determined and analysed (An *et al.*, 1993; Clyne *et al.*, 1993; Xue *et al.*, 1993; Bryan *et al.*, 1995). As more isolates from different sources are examined by the methods now available, more variants within *G. graminis* varieties are being found. The existence of a few such variants in the past has not seriously upset the familiar separation into distinct varieties, but an accumulating number of variants is giving credence to the view that there is a continuum or gradation of types within the species. Because of this, it is likely that more and more isolates will emerge that are not accommodated in the conventional descriptions of the varieties. The evidence supporting this view includes the following.

- Ascospore sizes overlap for typical Ggt and Gga isolates.
- Oat-attacking isolates with short ascospores from Australia have DNA reactions more characteristic of Gga (which typically has long spores).
- Ggt isolates (short spores) with DNA characteristics of both Ggt and Gga have been found in fields growing continuous wheat at Rothamsted (G.L. Bateman and E. Ward, unpublished data).
- An isolate with lobed hyphopodia from soya beans, which produces take-all symptoms on wheat, gave the same results as Ggt isolates using one DNA test and the same results as Ggg isolates using another DNA test.
- A considerable range of DNA characteristics has been found in isolates provisionally or definitively identified as Ggg (E. Ward and G.L. Bateman, unpub-

Table 5.3. Methods for identifying and characterizing isolates of *Gaeumannomyces graminis*.

Method	References	Diagnostic features	Comments
Non-molecular methods			
Morphology	Cunningham (1981), Walker (1981)	Colony morphology; perithecia and ascospore dimensions	Some isolates fail to produce perithecia
Pathogenicity on cereal roots	Asher (1981)	Host specificity, vascular discoloration	The varieties have different pathogenicities on different hosts
Infection structures	Deacon (1981), Skou 1981), Walker (1981)	Hyphopodia on coleoptiles (lobed for Ggg), runner hyphae, vesicle size and shape in roots (Ggg only)	Assessed in pathogenicity tests; some features are not fully diagnostic
Selective media (agar containing L-DOPA)	Juhnke et al. (1984), Elliott (1991b)	Dark coloration on the medium	Avirulent *Phialophora* spp. that infect cereal roots also produce the dark colour
Selective media (oat-extract agar)	Holden and Ashby (1981), Yeates and Parker (1986a), Bateman et al. (1992)	Allows identification of Gga, selects against Ggt	Not always reliable
Selective media (agar containing avenacin)	Crombie et al. (1986b), Osbourn et al. (1991)	Ggt isolates inhibited by 5–10 µg ml^{-1} avenacin A-1; Gga isolates still grow at 50 µg ml^{-1}	Reliable
DNA methods			
Specific probe pMSU315	Henson (1989), Henson (1992), Bateman et al. (1992), Ward and Bateman (1994)	Derived from mitochondrial DNA; recognizes DNA only from *G. graminis* and related *Phialophora* spp.	No discrimination between Ggt and Gga, nor between Ggg and Pg[a]
Ribosomal probes from other fungi or plants	O'Dell et al. (1992), Tan et al. (1994)	Different species and varieties produce different patterns of DNA bands	Not specific for *G. graminis*, can be used for other fungi[a]
Repetitive DNA probes from *G. graminis*	O'Dell et al. (1992)	*G. graminis* isolates from different hosts produce different DNA band patterns	[a]
Specific *G. graminis* probe GggMR1	Ward and Gray (1992), Ward and Bateman (1994), Ward (1994)	Derived from mitochondrial ribosomal DNA using PCR; relatively specific for *G. graminis* and *P. graminicola*	Discriminates between Ggt, Gga, Ggg and Pg, and some isolates within varieties[a]
Specific probe pG158	Herdina et al. (1996)	Highly repetitive DNA sequence, specific for Ggt and Gga	Has been used for quantification of fungus in infected roots and soil[a]

Method	References	Description
PCR with specific nested primers KS1F + KS2R and KS4F + KS5R	Schesser et al. (1991), Henson (1992), Henson et al. (1993), Elliott et al. (1993), Ophel-Keller et al. (1995)	Two-stage nested PCR which amplifies a band only with G. graminis and closely related fungi	No discrimination between varieties but some discrimination between Ggg in different grasses[b]; can detect the fungus in soil[c]
PCR with specific primers KS1F + KS2R	Ward (1994, 1995)	Single stage, G. graminis-specific PCR	Modification of the previous method; there is a size difference between Ggt/Gga and Ggg[b]
PCR with specific primers pGa1 and pGa2 (Gga) or pGt1 and pGt2 (Ggt)	Bryan et al. (1995)	Single stage, Gga-specific and Ggt-specific PCRs for nuclear rDNA genes (high copy number)	Simple and sensitive methods for discriminating between Ggt and Gga; has been developed as a quantitative competitive PCR technique (see Section 5.5.2)[b]
PCR with specific primers GG1 and GG2	Goodwin et al. (1995)	PCR specific for Ggg and Gga	Simple method for discriminating between the varieties (Ggg and Gga) pathogenic on turf grasses and other fungi (including Ggt, Magnaporthe poae, M. rhizophila, Leptosphaeria korrae and Phialophora graminicola)[b]
PCR-RFLP of ribosomal DNA	Ward and Akrofi (1994), Ward (1994)	Amplification by PCR followed by digestion with restriction enzymes	Discriminates between Ggt, Gga, Ggg, Pg and other fungi; general method – applicable to many different fungal species[b]
Random amplified polymorphic DNA analysis (RAPDs)	Fouly et al. (1996b), Wetzel et al. (1996)	Amplification by PCR using short random primers	Discriminates between Gaeumannomyces spp. and vars; reproducibility of results can be a problem[b]
Sequencing of ribosomal DNA	An et al. (1993), Bryan et al. (1995), Bunting et al. (1996), Tan and Wong (1996)	Amplification by PCR followed by DNA sequencing	Gives detailed taxonomic information about strains; costly and can be technically difficult[b]

Ggt, G. graminis var. tritici; Gga, G. graminis var. avenae; Ggg, G. graminis var. graminis; Pg, Phialophora graminicola.
[a]Requires purified DNA; [b]small quantities of impure DNA can be used; [c]detection in soil requires purified DNA.

lished data). These characteristics are often associated with isolates from different host plants (e.g. wheat, maize or grasses), different continents or even different fields in the same locality. Many isolates that are atypical (i.e produce atypical banding patterns with probe GggMR1, which is usually specific for the *Gaeumannomyces–Phialophora* complex) have failed to produce perithecia as supporting evidence for their identity.

5.6 Population Studies

Earlier work using collections of isolates of Ggt, usually in attempts to relate particular characteristics to virulence, or to stages in take-all epidemics, is referred to in Hornby (1979) and Asher and Shipton (1981). A summary of some of the observations appears in Table 5.1. More recent work on occurrence of R-types and N-types is mentioned in Section 5.2. Population studies, perhaps involving DNA analysis or mutants (see 5.7.1), may help to resolve questions concerning the present distribution, importance and potential of Gga, raised in Section 5.2.

In recent years, collections of isolates of Ggt made at Rothamsted have been studied using DNA characteristics (RFLPs) determined by the mitochondrial ribosomal DNA probe pEG34 (see 5.5.2 and Fig. 5.20). The findings suggest an association between different genotypes of the fungus and different host species, especially wheat and barley (Ward and Gray, 1992; Bateman *et al.*, 1997). For some time, disparate pieces of evidence have been accumulating to support the view that changes occur in the population of Ggt during monoculture of infected cereals (Table 5.1). The evidence for changes in population structure in relation to stages of the polyetic epidemic is reasonably convincing. The influence of these changes, if any, on the disease is unknown, but they may have some value as indicators of disease progress. Differences in disease in different cereal sequences in the same field (Hornby and Gutteridge, 1995) were associated weakly with differences in the predominant RFLP type in isolates of Ggt from soil (Fig. 5.23; Bateman *et al.*, 1997), which raises the possibility that the RFLP type may also be related to the stage of the epidemic. The association might have appeared stronger had the sampling been on a disease patch basis, rather than on a plot basis. Assessing the take-all risk more precisely than is possible at present might help farmers to formulate better management plans, especially if a seed treatment fungicide becomes available for controlling take-all. However, the procedures now used would need considerable development and simplification before becoming practicable as a forecasting tool. Analysis of pooled data from several wheat crops showed that RFLP type T2, usually most frequent in second wheats (and therefore possibly an indicator of severe disease to come), produced fewer perithecia in seedling infection tests. The implications of this are not clear, particularly as perithecial production is traditionally associated with pathogenicity (Holden and Hornby, 1981).

Similar preliminary studies of Gga have also indicated an association of different genotypes with different hosts. First findings were that isolates from grasses were entirely of the A1 RFLP type (see 5.5.2) and those from oats were entirely A2

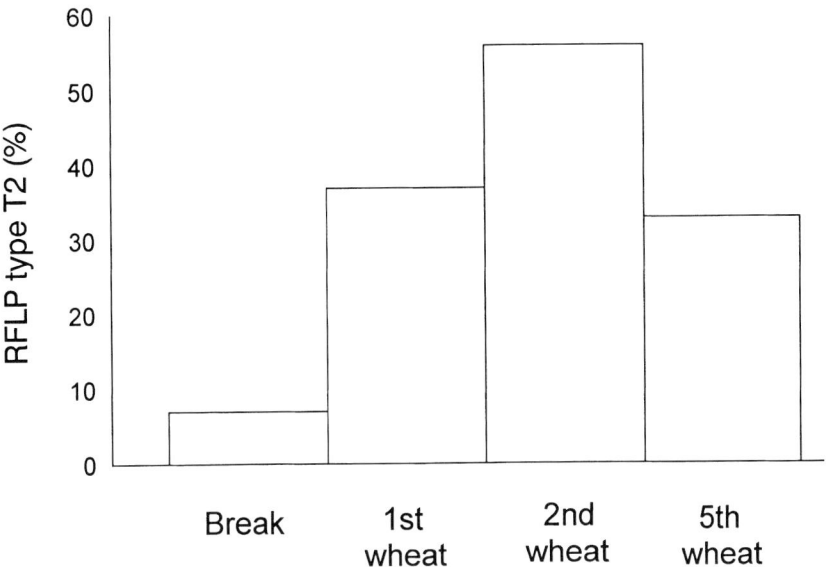

Fig. 5.23. Percentages of RFLP type T2 (as a percentage of T1 + T2 types) in samples of Ggt from plots having grown different numbers of consecutive winter wheat crops. The samples (25–44 per crop sequence) were obtained by seedling bioassay of the soils, which came from West Barnfield, Rothamsted, where the severity of take-all peaked in the third wheat crop.

type (Ward and Gray, 1992), but subsequent work has shown that the A1 type also predominates in some oat crops (G.L. Bateman, unpublished data).

Using isozymes, RFLP analysis, probes and PCR to study genetic diversity between and within varieties of *G. graminis* in the cereal belt of southern Australia, Harvey (1993) found that Ggt and Gga were widespread and often coexisted, but Ggg was not detected in areas of intensive cereal production. In general, populations dominated by Gga (from oat-growing areas) showed less genetic variation than those dominated by Ggt (found where wheat and barley were grown).

5.7 Genetics of *G. graminis*

5.7.1 The need for mutants

The establishment of collections of mutant isolates of *G. graminis* and related fungi is an important step towards classical genetic analysis of these fungi. If mutants are available then mutations can be mapped and linkage groups can be established, loci involved in pathogenicity can be identified, and studies of sexual crossing, heterokaryon formation and parasexuality will be greatly helped. Mutations can also be used as markers to follow isolates in population studies. Furthermore, the advent of molecular biology now offers the opportunity of identifying DNA encoding functions of interest by introducing a 'library' of wild-type DNA into a mutant and looking for DNA which can repair the defect of that mutant.

5.7.2 Development of techniques for mutagenesis

Genetic analysis of G. graminis has been hampered because of difficulties in producing the large numbers of germinable propagules required for screening for spontaneous or induced mutations. G. graminis is haploid during vegetative growth and can produce two types of conidia (phialospores) – small sickle-shaped microconidia and larger conidia which are straight or slightly curved. The former are produced in abundance, but do not germinate in culture. The larger conidia are reported to germinate (Deacon and Henry, 1978) and potentially represent ideal targets for mutagenesis, because they are generally uninucleate. However, several laboratories have been unable to produce and/or germinate them in the numbers required for such experiments.

Isolates of all three varieties of G. graminis have been made to produce ascospores in the laboratory with differing degrees of success (Cunningham, 1981; Holden and Hornby, 1981; Speakman, 1982, 1984b). An easy, reliable method involving inoculated, autoclaved wheat leaves in test tubes containing a low nutrient agar can give mature perithecia in 2–4 weeks, rather than the longer incubation times generally required for the other methods (Pilgeram and Henson, 1992). However, none of these methods produces ascospores in numbers large enough for mutagenesis experiments. Even if prolific ascospore production could be achieved, the spores may be unsuitable for these purposes because they generally have five to nine cells, each containing a nucleus.

In the first successful mutagenesis experiment on G. graminis (Rochefrette et al., 1979), hyphal fragments treated with the chemical mutagen nitrosoguanidine (NTG) produced mutants with resistance to the fungicides benomyl and carboxin. There were no auxotrophic mutants and this was attributed to the probable multinucleate nature of the starting material. Blanch et al. (1981) isolated one auxotrophic mutant of Ggt (requiring supplementation with p-aminobenzoic acid) after UV irradiation of macerated hyphal fragments. A method for NTG mutagenesis of protoplasts (Rochefrette et al., 1979), which generated auxotrophs, was more successful than mutagenesis of hyphal fragments, probably because approximately three-quarters of viable protoplasts were uninucleate and the remainder had two nuclei (Bowyer et al., 1992). This protoplast mutagenesis technique was adapted to utilize 4-nitroquinolene oxide (NQO) as a mutagen (Bowyer et al., 1992). NTG and NQO have similar mutagenic properties, but NQO is safer to use and can be neutralized with sodium thiosulphate prior to disposal. The technique using NQO has yielded auxotrophic, pigmentation and morphological mutants of both Ggt and Gga, and mutants with resistance to a range of inhibitors.

5.7.3 Sexual crosses

G. graminis is homothallic (White, 1939) and therefore a single, genetically homogeneous isolate can produce perithecia. Consequently, the proportion of crossed perithecia may be small and so naturally occurring or introduced markers are required to distinguish crossed perithecia from selfed ones. Because the ascospores of G. graminis are intertwined in the ascus, ordered tetrad analysis is impossible and genetic analysis must proceed by analysis of random meiotic products, or of unordered tetrads (Asher, 1981). In intra-isolate crosses involving auxotrophic mutants and/or antibiotic resistant transformants or mutants of a single isolate, the percentage of crossed perithecia was generally 10–25% (Pilgeram and Henson, 1992), but in one instance it was as high as 70% (Musker, 1994). The progeny derived from crosses of a weakly pathogenic auxotrophic mutant with each of six different, fully pathogenic wild-type isolates included hybrid perithecia with three of the isolates at frequencies of 4%, 27% and 50% of the total number of perithecia analysed (Blanch *et al.*, 1981). This indicates that sexual recombination can take place between different isolates. Pathogenicity, growth rate and pigmentation all seem to be under the control of several genes with no good linkage between pigmentation and pathogenicity (Blanch *et al.*, 1981). In separate experiments, crosses between marked strains of two different Ggt isolates produced 15–25% hybrid perithecia (Musker, 1994), but no conclusions were drawn about the number of genes that determine pathogenicity or growth rate, because of the influence of one auxotrophic marker on the phenotypes. The generation of more markers in laboratory isolates will help in learning more about the extent and limits of inter-isolate crossing and about the heritability of pathogenicity and other properties.

Attempts to make forced heterokaryons of *G. graminis* by pairing isolates on selective medium produced stable heterokaryons only with mutants of the same isolate (Musker, 1994). Other pairings involving different isolates of the same variety or of isolates of Ggt and Gga were unsuccessful, which may reflect vegetative incompatibility between isolates. In laboratory experiments most isolates are found to be vegetatively incompatible with other isolates, unless they were isolated from the same field at the same time and consequently may be clonally related. Analysis of the progeny of a cross between two vegetatively incompatible isolates indicated that more than six genes were involved in determining incompatibility (Musker, 1994). Because numerous vegetative incompatibility groups must exist within *G. graminis* (Asher, 1981; Jamil *et al.*, 1984; Jamil and Buck, 1991; Musker, 1994), parasexuality is unlikely to play an important role in the exchange of genetic information between isolates.

5.7.4 *Molecular biology of G. graminis*

The development of pulsed field gel electrophoresis (PFGE) has made possible the separation of fungal chromosomes on an agarose gel to obtain a molecular karyotype. There is considerable variation in the number and size of chromosomes in some fungi pathogenic on plants, even amongst isolates of the same species (Mills and McCluskey, 1990). However, isolates of *G. graminis* (including Ggt, Australian short-spored, oat-attacking isolates and Gga) were found to have identical karyotypes. Four bands, ranging in size from 3 to 5 megabases, were clearly visible, which is consistent with cytological evidence (Jones, 1926) indicating the existence of four chromosomes. The total genome size estimated from the sum of these bands is approximately 17 megabases (P. Bowyer, unpublished results), which conflicts with the 29-megabase estimate of the sequence complexity of *G. graminis* based on DNA reassociation kinetics (McFadden and Buck, 1983). However, the amount of error associated with estimates of sequence complexity may be large and it is also possible that one or more of the bands observed after PFGE may have been doublets. Double-stranded linear DNA plasmids have been detected in isolates of *G. graminis* (Honeyman and Currier, 1986; Henson and Caesar-Tonthat, 1995), but so far no function or phenotype has been assigned to these.

Ggt, Gga and Ggg have now been transformed with DNA encoding resistance to various antibiotics (Henson *et al.*, 1988; Pilgeram and Henson, 1990; Bowyer *et al.*, 1995), but the frequency of transformation was low (normally one to five transformants per microgram of transforming DNA per 10^7 protoplasts) and was inadequate for identification of DNA clones from genomic libraries by complementation of mutants. Also, the transforming DNA is not always stable in subsequent cell divisions (Pilgeram *et al.*, 1993). However, transformation has been used to generate specific mutations in a gene of interest by gene disruption (Bowyer *et al.*, 1995; see 5.8.3) and experiments of this nature can be achieved with a relatively small number of transformants.

The production of hyphopodia was affected in a number of Ggg transformants after introduction of non-homologous plasmids conferring drug resistance (Epstein *et al.*, 1994). Characterization of the DNA sequences flanking the site of insertion of the transforming DNA should allow the isolation of genes required for formation of hyphopodia.

Transformation systems currently used for *G. graminis* involve the expression of antibiotic resistance genes from heterologous organisms. Cloning of genes from *G. graminis* encoding biosynthetic enzymes such as ornithine carbamoyltransferase and orotidine-5'-phosphate decarboxylase (enzymes required for arginine and uridine metabolism, respectively) may allow development of more effective transformation procedures, in which appropriate mutants are complemented by transformation with DNA from the homologous fungus. A simple and rapid strategy to identify such genes has been used (Bowyer *et al.*, 1994) and involved cotransformation of a fungal replicating vector pHELP1 and linear *G. graminis* DNA into mutants of *Aspergillus nidulans*. This resulted in the identification of *G. graminis* DNA that can complement *argB* and *pyrG* mutants of *A. nidulans*.

5.8 Pathogenicity

5.8.1 *Host specificity*

There is no evidence for specialization of Ggt on different wheat cultivars and resistance to Ggt has not been identified in the *Triticum* germplasm (Scott, 1981). (Section 3.2 discusses differences amongst cultivars, e.g. in immunization response or in efficiency of using Mn, that affect take-all in different parts of the world.) Consequently the classical gene-for-gene interactions that have been described for so many plant–fungus interactions involving pathogen avirulence genes and host resistance genes (De Wit, 1992) have not been identified for Ggt and wheat. There are, however, examples of specialization of isolates of *G. graminis* for different host species:

- Some isolates of Ggt (R-type isolates) cause more root blackening on rye than do others (N-type isolates). It is not clear from field experiments whether R-type isolates can cause substantial yield losses on rye (Hollins and Scott, 1990; and Section 5.2).
- Differential interaction of Ggt isolates with a range of grasses has been reported by Turner (1953).
- Oats are resistant to attack by Ggt isolates in the UK and this resistance has been attributed to the presence of a pre-formed inhibitor of fungal growth, avenacin, which is present in the roots (Maizel *et al.*, 1964). Gga isolates, however, can overcome the toxic effects of avenacin (see 5.8.3).
- Some Australian isolates of Ggt attack oats (see 5.2, 5.10); it is suggested that the oat-attacking Ggt is a specific race.

5.8.2 *Factors implicated in pathogenicity*

The hyphae of *G. graminis* penetrate plant cell walls during infection and enzymes that degrade cell walls are implicated as key determinants of pathogenicity. The literature on cellulolytic and pectinolytic enzyme activity by *G. graminis* was summarized by Sivasithamparam and Parker (1981). Despite 'intriguing' indications that cellulase and pectinase activities were involved in pathogenicity and competitive saprophytic ability, there has been little subsequent work. Activities of polygalacturonase (PG), pectin methyl esterase (PME), carboxymethylcellulase, C_1 and C_x cellulose enzymes and peptidase, but not pectin transeliminase (PTE), had all been detected amongst isolates of *G. graminis* and *Phialophora* spp. C_x cellulase and PME appeared to be adaptive enzymes. However, there are discrepancies amongst these reports, which may reflect differences between the isolates used, the way in which they were grown and/or the methods and conditions used for enzyme assays.

Subsequent work (Martyniuk, 1988) which found PTE, but not PME, in Ggt is the opposite of an earlier finding (Weste, 1970); and evidence for a positive correlation between level of PG and pathogenicity in Ggt contradicts the absence of correlation reported by Pearson (1974). Whilst evidence for a link between pectinolytic and/or cellulolytic enzyme activity and pathogenicity continues to be contradictory, consideration should be given to additional factors such as differences in growth rate between isolates which may be complicating the issue. Cellulytic activity was detected in inoculated wheat roots as visual symptoms appeared, but not in healthy wheat seedlings, in a study to isolate and characterize cellulolytic enzymes produced *in vitro* and *in vivo* by *G. graminis* (Dori *et al.*, 1995).

The most stringent test to assess the role of such enzymes is to generate mutants defective in the ability to produce a single enzyme of interest. Comparisons of the ability of these mutants and of the wild-type fungus to infect plants will give a clear indication of the importance of individual enzymes in pathogenesis. Dori *et al.* (1992, 1993) have gone some way towards achieving this by purifying endopolygalacturonase activity from Ggt. They demonstrated that a single enzyme is induced when the fungus is cultured in media containing pectin and that an enzyme with the same isoelectric point is the only detectable activity in infected plants. It should now be possible to use the techniques of reverse genetics and gene disruption for a formal analysis of the importance of this enzyme in infection.

Cell walls of monocotyledonous plants have relatively more arabinoxylan and relatively less cellulose and pectic polysaccharide compared with dicotyledonous plants, and the major enzymes produced by the cereal pathogens *Rhizoctonia cerealis*, *Fusarium culmorum* and *Pseudocercosporella herpotrichoides* are arabanase, xylanase and laminarinase (Cooper *et al.*, 1988). This is in contrast to pathogens of dicotyledonous plants, where endopectic enzymes predominate. In studies of xylanases produced by *G. graminis*, two major activities from an isolate of Gga were purified (Southerton *et al.*, 1993). The enzymes had very similar physicochemical properties and identical N-terminal amino acid sequences and so it is possible that they may be encoded by a single gene. Genetic approaches can now be used to test the role of these xylanases in pathogenesis.

It is well known that isolates of Ggt can quickly lose their pathogenicity when maintained on artificial media (see 5.3.2). Fewer than half of 111 isolates of Ggt remained pathogenic after storage on agar medium for 9 months, whether or not they were subcultured at 10-day intervals or kept at 12°C or 24°C (Naiki and Cook, 1983a). Loss of pathogenicity was slightly less frequent in monoascosporic isolates. Mycoviruses were not thought to be involved in this loss of pathogenicity (see 5.9) and it was suggested that the heterokaryon contained a proportion of nuclei that lacked virulence genes and/or carried genes that suppressed virulence genes or cytoplasmic determinants of virulence. In this way, culturing on agar could favour a shift in genetic determinants towards an inability to cause disease.

5.8.3 *Specificity for oats: avenacin and avenacinase*

Oats are resistant to attack by Ggt but not by Gga. The antifungal triterpenoid compound avenacin is found in the roots of most oat species and has been implicated as a pre-formed determinant of resistance to attack by avenacin-sensitive fungi such as Ggt (Osbourn *et al.*, 1994a). The chemistry, particularly the structure and identification, of avenacin received much attention in the 1980s (Begley *et al.*, 1986; Crombie and Crombie, 1986; Crombie, *et al.*, 1986a). It was found to be a mixture of four compounds, avenacins A-1, A-2, B-1 and B-2. Avenacin A-1 was the most abundant in young oat roots, as well as the most fungitoxic, and was considered to be the main resistance factor described in earlier work (Crombie and Crombie, 1986). This compound has a characteristic blue fluorescence when illuminated by UV light and is known to be the major fluorescent component of young oat roots. UV microscopy has shown it to be localized in the epidermal cell layer of such roots, presenting an effective antifungal barrier to the rhizosphere (Osbourn *et al.*, 1994b)(Fig. 5.24).

Isolates of Gga can attack oats and are more resistant than Ggt to avenacin in agar plate tests (Crombie *et al.*, 1986b; Osbourn *et al.*, 1991). The early view (Turner, 1961) that Gga isolates are resistant to avenacin because they produce an enzyme, avenacinase, which can deglycosylate avenacin to less toxic forms is now confirmed (Crombie *et al.*, 1986b; Osbourn *et al.*, 1991). The later work concluded that the ability to infect oats is associated with resistance to avenacin

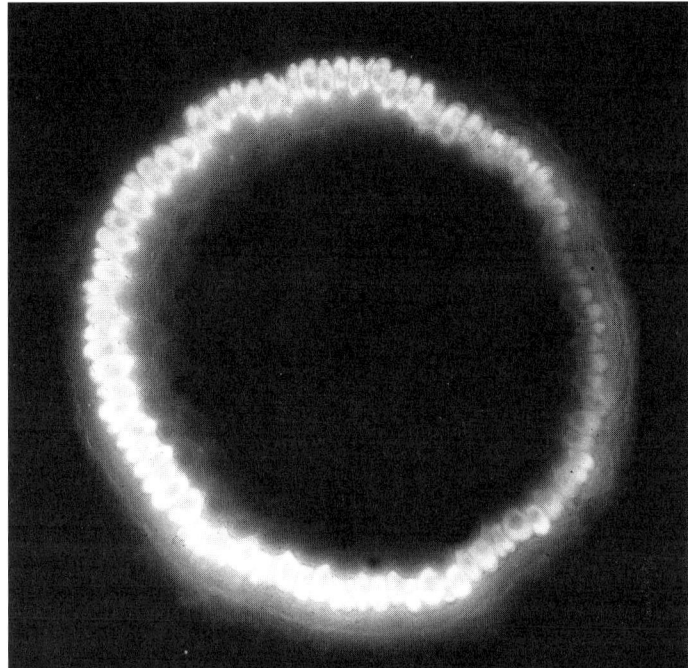

Fig. 5.24. Localization of avenacin A-1 in roots, shown by fluorescence under UV light.

in vitro and with presence of avenacinase activity. The degree of pathogenicity to oats was not directly correlated with levels of avenacinase in culture, suggesting that once the epidermal cell layer of the root has been penetrated, the extent of subsequent infection may depend on other factors. Osbourn *et al.* (1991) purified avenacinase as a single protein species. The DNA encoding the enzyme was isolated by raising polyclonal antiserum and using this to screen a cDNA expression library. Mutants of Gga lacking avenacinase activity were generated using gene disruption techniques and were found to have increased sensitivity to avenacin. In pathogenicity tests these mutants were fully pathogenic to wheat, but had lost their pathogenicity to oats (Bowyer *et al.*, 1995), indicating that avenacinase is required for pathogenicity to oats. In other experiments, the oat species *Avena longiglumis* was shown to lack avenacin and, significantly, to be susceptible to Ggt (Osbourn *et al.*, 1994b). Unfortunately *A. longiglumis* does not hybridize readily with other diploid oat species that produce avenacins, making it difficult to test whether the ability to synthesize avenacins and resistance to Ggt co-segregate. Recently, mutants of the diploid oat species *A. strigosa* lacking avenacin A-1 have been isolated following sodium azide mutagenesis (A.E. Osbourn, R.E. Melton and M.J. Daniels, unpublished data). The mutants show increased susceptibility to isolates of both Gga and Ggt and also to infection by other root-infecting fungi such as *Fusarium avenaceum*, *F. culmorum* and *F. graminearum*. These observations are consistent with a role for avenacin A-1 as a determinant of resistance of oats to a range of phytopathogenic fungi, although further genetic analysis is required to test whether the absence of this saponin and increased susceptibility are causally related. Avenacinase was recently found to be closely related to tomatinase, another enzyme that detoxifies host plant saponins and which is produced by the tomato-leaf infecting fungus *Septoria lycopersici* (Osbourn *et al.*, 1995; Osbourn, 1996).

5.9 Viruses of *G. graminis*

Viruses in *G. graminis* have been reviewed by Rawlinson and Buck (1981) and Buck (1986, 1987). They were discovered by French workers (Lapierre *et al.*, 1970; Lemaire *et al.*, 1970) and since then much work has been done to characterize them and to determine whether their presence is associated with hypovirulence in Ggt and/or with TAD. The viruses do not have an extracellular phase and are probably transmitted by vegetative (somatic hyphal) fusion or by mating. They are sometimes, but not always, eliminated during ascospore formation.

G. graminis viruses are isometric and have double-stranded RNA (dsRNA) genomes. There is much variation in the properties of these viruses and they have been classified into five groups depending on the physical properties of the particle, the capsid polypeptide, serology, and the numbers and sizes of the dsRNAs they contain (Buck, 1986). There may be several viruses present in one isolate of the fungus. Similar viruses have been observed in different vegetative compatibility (v-c) groups of Ggt, three varieties of *G. graminis* and *P. graminicola* (McGinty *et al.*, 1981; Jamil *et al.*, 1984; Jamil and Buck, 1991). This suggests that either different v-c groups and species are not barriers to virus transmission, or the viruses were associated with these fungi before the groups evolved from one another. In China, viruses in *G. graminis* have been studied in relation to v-c groups (Liang *et al.*, 1989) and a molecular assay (dot blot hybridization) was used to demonstrate homology between viruses in *G. graminis* and distantly related fungi (Liang and Chen, 1990).

The distribution of *G. graminis* viruses is complex; Ggt isolates from a single field can have varying combinations of several different viruses. This has created difficulties in understanding the relationships among virus, host and pathogenicity. Since workers realized that further characterization of the viruses, by structure, serology and dsRNA analysis, would be necessary to remedy this situation (Rawlinson and Buck, 1981), there have been relatively few studies (Romanos *et al.*, 1981; Jamil *et al.*, 1984; Buck, 1986, 1987; Jamil and Buck, 1990, 1991, 1992). There have been no published studies of detailed molecular characterization of the viral RNAs, although techniques are now available to achieve this. Available information suggests that viruses in general do not appear to modify pathogenicity, but specific viral RNAs (or combinations of RNAs) may have a role in this. However, without further work, the precise effects of viruses will not be understood and any potential they may have for biological control of Ggt is unlikely to be realized.

Research on chemical inhibitors produced by isolates of Ggt (see 4.3.1) was sometimes associated with that on fungal viruses during the 1970s and 1980s (Buck, 1986). The inhibitors, low molecular weight substances known as Q-factors, occurred at low pH (3.5–5.0) (Romanos *et al.*, 1980). They inhibited the growth of the isolates producing them and also other fungi. The incidence of Q-factors (14% of field isolates of Ggt grown on PDA) was shown not to be associated with the presence of viruses (McGinty *et al.*, 1984). No role for Q-factors has been proved, although it has been hypothesized that they may be a defence mechanism when conditions are unfavourable for growth, or that they may be associated with TAD or other forms of suppressiveness. The former suggestion is consistent with an observed tendency for increased self-inhibition when cultures are kept in culture (Naiki and Cook, 1983b). As with viruses, research interest in Q-factors has dwindled.

5.10 Regional Approaches

Ggt is commonly found in association with other members of the *Gaeumannomyces–Phialophora* complex in the UK. The present status of Gga on cereals in the UK is uncertain (see 5.2). There is no evidence for the presence of virulent oat-attacking isolates of Ggt in the UK, although take-all lesions are occasionally found on oat roots early in the season (but seem rarely to develop further) and some Ggt isolates clearly infect oat seedlings in pot tests. *P. graminicola* is common in the UK and is thought to be a major cause of the delayed onset of take-all in sequences of cereals following grass leys. *Phialophora* sp. (lobed hyphopodia), although by no means rare, is apparently less common in the UK. Tests on pot plants, however, suggest that it has even greater potential for suppressing take-all than *P. graminicola*, as well as an ability to stimulate root production. However, in some wheat-growing areas where take-all is prevalent (for example, the dry, upland areas of northwestern USA), *Phialophora* spp. appear not to be important. There, other pathogens are important components of a root disease complex and consequently the selective media in which these fungi produce pigmentation in the presence of L-DOPA are used routinely to confirm the presence of Ggt.

The concept of Ggt, which does not attack oats, and Gga, which does, has been weakened by the discovery in Australia of oat-attacking isolates of Ggt. In the UK, oats serve as a non-susceptible break, but in Australia they are a carrier. In the UK, Gga traditionally predominated where oats were grown (Wales, Scotland and Cumberland: Garrett and Dennis, 1943), but it has been reported only infrequently in the last 25 years. Gga was thought to be absent in Australian cereal crops, a situation that may have resulted from annual grasses in the alternating pastures and the climate. However, population studies using DNA techniques are now claimed as indicating that both Ggt and Gga are widespread in cereal-growing regions, often in mixed populations (Harvey, 1993).

The problem of identification of Ggt on wheat prompted useful cooperation on approaches using DNA methods. Methods devised in Montana (Henson, 1989; Schesser *et al.*, 1991) provided a basis for some of the work at Rothamsted. Subsequently, research in the USA has concentrated on turf grass pathogens in the *Gaeumannomyces–Phialophora* complex that are of less direct interest in Europe (e.g. Henson *et al.*, 1993; Wetzel *et al.*, 1996). Table 5.3 summarizes methods that have been used or are becoming available for the identification of take-all fungi.

The Disease: Field Techniques

6.1 Introduction

Field studies of plant diseases caused by soil-borne organisms present particular problems because infection usually occurs below the soil surface and is not easily assessed and, typically, the distribution of disease is patchy. Take-all is proving to be no exception as more is learned of the complexities of its behaviour and interactions in different parts of the world. This chapter concerns mostly experimental approaches to handling and studying the disease in field situations and complements those parts of Chapter 5 dealing with problems of experimentation with the pathogen. It starts, however, with a consideration of methods used by field pathologists and advisers to assess, monitor or survey take-all in non-experimental fields.

6.2 Assessing, Monitoring and Surveying Disease in Non-experimental Fields

Traditional methods of assessing take-all in whole fields, such as measuring disease severity on root systems and measuring host responses (Clarkson and Polley, 1981b), are still in use today and many are described in Section 6.3.3. Where field assessments are required for management or advisory purposes, the sampling methods tend to be basic (see also 6.3.2 for other advice on sampling large areas). For instance, in Northern Ireland, as a guide to the likely disease levels in a crop, 10–20 clumps of plants containing at least 10 shoots per clump are taken, preferably using a trowel, along a diagonal of the field. Fields larger than 4 acres (1.6 ha) are divided into two and separate samples are taken from each half (P.C. Mercer, personal communication).

Following the first survey of cereal diseases (foliar diseases in spring barley) in England and Wales in 1967, ADAS surveys of wheat diseases in England and Wales have been done annually since 1970, except for 1983/84 (Polley and Thomas, 1991). Take-all was not included from 1970 to 1975 and eyespot was first included in 1975 (King, 1977). From 1977 to 1979 the frequencies of slight, moderate and severe take-all were estimated using a single plant assessment method on samples of 500–600 plants from each of four crops in each of those 3 years (Polley and Clarkson, 1980). Each plant was allocated to one of four infection categories: healthy, slight, moderate or severe, representing none, < 25%, 25–75% or > 75% of the root system infected, respectively. These categories were compared with measurements of ears (numbers) and grains (numbers and weights) from the same plants and a disease–yield relationship was established. This was adjusted to account for the splitting of plants during sampling and washing and applied to the survey data (Table 6.1).

In 1985 and 1987/88, take-all was assessed on the appearance of the crop, using a 0–4 assessment key (Polley and Thomas, 1991) (Table 6.2): 0 = no take-all seen; 1 = a scatter of plants showing premature ripening; 2 = occasional small patches (less than 5 m across) showing premature ripening and/or stunting (less than 1% of field area affected); 3 = many small, or a few large, areas affected (1% to 10% of field area); 4 = many large areas affected (more than 10% of field area).

Although checks were made for obviously blackened roots to make sure take-all was associated with the above symptoms, no attempt was made to assess take-all on plants with no above-ground symptoms. This key has been used in subsequent winter wheat surveys to date (Polley and Slough, 1993; Polley et al.,

Table 6.1. Yield losses in winter wheat in England and Wales attributed to take-all in 1977–1979.

	Take-all (%)			% Yield loss nationally
Year	Slight	Moderate (x_1)	Severe (x_2)	($y = 0.156x_1 + 0.531x_2$)
1977	8.9	1.6	1.3	0.94
1978	29.8	8.1	2.4	2.55
1979	24.9	7.1	3.1	2.76

Based on ADAS Survey data and a disease–yield relationship established by single plant assessment (Polley and Clarkson, 1980).

Table 6.2. Assessment of take-all in England and Wales in 1985, 1987 and 1988, using a field key.

	% Fields affected at GS 73–75		
Take-all category[a]	1985	1987	1988
1 (Least take-all)	13.7	24.9	16.8
2	7.0	22.8	5.4
3	4.3	13.2	6.1
4 (Most take-all)	4.7	9.1	0.7
% Crops with take-all	29.7	70.0	29.0

[a]More details of these categories are in the text.
From Polley and Thomas (1991).

1994, 1995, 1996). As an example of the sort of information obtained, the 1992 survey revealed that Riband was the most popular cultivar (39.3%) and of the 345 crops assessed for take-all, 12.5% had take-all patches and the occurrence of the disease was correlated with previous cropping and sowing date. More of the data are in Tables 1.2 to 1.4 and 1.13 to 1.16.

Assessments based on terms such as slight, moderate and severe, or on numerical classifications (e.g. as above; or the disease severity assay, DSA, with a 0–4 scale: Cotterill *et al.*, 1988), have featured prominently, and still do, in take-all research, but are considered inappropriate or of little use in quantitative analysis of epidemics (Campbell and Madden, 1990).

'Whiteheads' is an imprecise term, commonly applied to bleached, prematurely ripened ears. It is also applied to symptoms ranging from dead, bleached inflorescences, through bleached, unfilled ears that are often seen on single plants scattered throughout the crop and not necessarily caused by take-all, to stunted plants in groups with small but not bleached ears, ripening early. Thus, prematurely ripened plants in disease patches, which contrast with the surrounding crop but are not obviously bleached, may be referred to as whiteheads. In Australia, severely infected plants that are unable to fill grain are referred to as 'haydie', a name used as an alternative to take-all (Wallwork, 1996). Whiteheads are not unique to take-all. For example, *Pseudocercosporella herpotrichoides*, *Fusarium culmorum* and *Ceratobasidium* sp. caused whiteheads in wheat in Germany in the absence of Ggt (Heyland and Kühnhold, 1984). Similarly, in the UK, whiteheads are usually seen at the end of June to July and may arise from diseases other than take-all, most commonly the stem base diseases eyespot *(P. herpotrichoides)*, sharp eyespot *(Rhizoctonia cerealis)* and fusarium. With eyespot, lesions less severe than those causing lodging may cause scattered plants throughout the crop to die prematurely and produce shrivelled ears. As with take-all, these ears, or 'whiteheads', may be colonized by secondary moulds in wet weather and turn black (Fitt, 1992).

Another possible source of confusion is cephalosporium stripe disease caused by the fungus *Cephalosporium gramineum* Nisikado & Ikata (= *Hymenula cerealis* Ell. & Ev.), which can significantly reduce crop height (Bruehl, 1957). This is another cereal pathogen with taxonomic problems that plant pathologists have tried to ignore (see also *C. maydis*, 5.4.2): *Cephalosporium* has been consigned to synonymy with *Acremonium*, and *Hymenula* is an illegitimate name. The effect of cephalosporium stripe disease on the height of plants in a first winter wheat crop following a 2-year ryegrass ley is illustrated in Fig. 6.1, which also shows typical symptoms of the disease. This disease causes premature ripening of the crop and when the leaves have lost their characteristic yellow stripe through senescence, aboveground symptoms may be indistinguishable from take-all. Cephalosporium stripe occurred more frequently in winter wheat grown after grass than after other break crops (Slope and Bardner, 1965). It was also more frequent where straw residues were left on the soil surface or incorporated shallowly than where the straw had been burnt (Christian and Miller, 1984). The introduction of set-aside and the banning of straw burning may lead to more situations where conditions would be favourable for this disease.

Visible patches, either of stunted plants or of whiteheads, where they are attributable to take-all, represent an extreme effect of the disease. Under other soil conditions the same amount of root infection may be present on plants that appear normal above ground. This arises when soil conditions favour development and function of surviving roots or replacement roots and/or when disease develops late. Consequently, the proportion of a crop showing whiteheads seems to be a better indicator of yield loss than are TAR or percentage of plants with moderate or severe take-all (Gutteridge *et al.*, 1987; Headrick and Bockus, 1996; see also 6.5.2).

Aerial surveys for take-all are not undertaken on a regular basis. It is difficult to distinguish between cereal diseases on aerial photographs, and soil-borne

Fig. 6.1. Typical stripe symptoms (arrowed) of cephalosporium stripe disease on wheat leaves and the stunted appearance of infected plants in a first winter wheat crop following a 2-year ryegrass ley.

diseases have been neglected generally as subjects for aerial photography (Harris, 1985). However, take-all has fared better than most in that it has received some attention in this area (Blakeman, 1990; and Figs 3.5 to 3.7, 3.14 to 3.17 and 3.19).

Good surveys can be useful in assessing the cost or importance of disease and in analysing relationships of disease and agronomic factors. They need to be repeated at intervals if trends are to be monitored. In the case of take-all in the UK, adequate surveys have been limited to a very few years, none recent; more frequent recent annual surveys that ignored below-ground symptoms probably did not reveal the true effect of the disease nationally. The lack of funds for take-all surveys and consequently the paucity of good survey information soon become obvious to those seeking to establish that take-all is sufficiently important to justify the costs of finding and developing control products.

6.3 Field Experiments

6.3.1 Design

Experimental design is the arrangement of experimental units to optimize error control against a background of environmental variability and practical limitations. In practice, however, experimental designs are more commonly determined by tradition and their suitability is seldom questioned. This is particularly true for the randomized block design and the use of large plots amongst which relative homogeneity of environmental conditions within the blocks cannot be confidently predicted. Hornby *et al.* (1983) demonstrated significant effects of systematic field variation and interactions between field variation and weather on yield of spring barley in years when take-all was slight. The decision to adopt one particular experimental design may involve a number of specific considerations, e.g. the number of treatments, replication, plot shape, plot size and block size (Gilligan, 1990b). Possible designs range from the simple complete randomized block to more complex incomplete block, nested and fractional replication designs. The use of any particular design should be determined by the number of treatments and the scale of important environmental heterogeneity (if known).

Treatment effects are tested against the residual variation amongst the experimental units, which may include imprecise plot estimates, caused by inadequate sampling techniques (see 6.3.2) and lack of environmental uniformity amongst the experimental units. Simple blocking will not account for interactions between environmental differences and treatments, which are indicated by a high residual mean square.

Treatments that are particularly sensitive to inoculum density and variation in the soil environment, or that have small effects, may be tested in micro-plots in incomplete block designs to minimize within-block variability (see later in this section). Variability arising from a patchy distribution of natural inoculum may be decreased or avoided by using artificial inoculum sources in experiments (see 6.4). Alternatively, estimates of inoculum density and distribution before treatment application may be used as covariates, presupposing an unlikely linear relationship between inoculum density and disease that may be resolved by logarithmic transformations. Such covariates may also be used to allow for variable initial inoculum densities and variable rates of disease increase in estimating and describing disease progress. It may, however, be more appropriate to use covariates pertaining to systematic field variation in soil fertility, soil type and soil moisture (Hornby *et al.*, 1983), hitherto unexplored in relation to take-all. Whilst it remains to be demonstrated that traditional blocking techniques can successfully and reliably account for a significant amount of the environmental heterogeneity observed in take-all experiments, it may be rewarding to explore the use of nearest-neighbour models in take-all research, despite the unreliability of nearest-neighbour analysis in mitigating the effects of inter-plot interference (Ainsley *et al.*, 1995).

Field experiments on cereals are frequently laid out in plots the same width as the seed drill (often 3 m) and several times this in length. This area is suitable for combine harvesting to determine grain yields per plot. However, the unreliability of take-all epidemics and the distribution of the disease in patches means that average disease levels may be too slight to achieve a result (e.g. to show the effects of putative control treatments) or that disease occurs at greatly different intensities in different plots, leading to ambiguous results. Sufficient replication of plots and of sites to overcome these problems is usually impracticable when such large plots are used.

Novel procedures for field experiments have therefore been sought to test candidate biological and chemical control treatments against natural infections of take-all. For example, Hornby *et al.* (1989) used small plots (31 cm × 37 cm) grouped in two adjacent complementary clusters of four (allowing six treatments plus two control plots) in a single replicate 'block-pair' (2 m × 1 m). It was argued that the area of each block-pair was sufficiently small that the likelihood of uniform disease among the plants occurring within it would be greater than in larger areas. Consequently, the precision of the experiments was expected to increase. In each year many block-pairs were placed on each of two sites, one on clay-loam at Rothamsted and one on sandy loam at Woburn. Some fields were used in successive years, but none of the block-pair positions were re-used. An additional advantage of this approach is that only small amounts of test materials are required. The disease assessment data for small plots (as opposed to block-sized areas) consistently had a smaller variance in the crop where take-all was at a maximum (i.e. in the third cereal), but small plots were less effective as a means of decreasing variability after the time of peak disease (i.e. in fourth or fifth cereals). Comparing the variability of small plots within blocks, blocks within block-pairs and block-pairs scattered throughout the experimental sites in different seasons revealed evidence of changing disease patterns and suggested more than one scale of disease pattern (summarized in Fig. 6.2). This scheme fits sites at Woburn better than Rothamsted.

Disease uniformity tests were used to provide further information on the optimum size of sampling unit (D. Hornby and G.L. Bateman, unpublished data). These tests were situated as replicate groups of 16 × 8 small plots, each group occupying an area of 4 m × 2 m, in third and fourth wheats at Rothamsted and

Woburn. Take-all was assessed on all plants in these plots. Differently sized sampling units were obtained by a progressive doubling of the plot unit and the variance for each size was calculated. Preliminary analyses showed that in the third wheat the variance was greatest amongst 2 m × 2 m sampling units, which may have been close to the average size of the take-all patches. In fourth wheats, variances decreased with increasing size of sampling unit up to 2 m × 1 m and

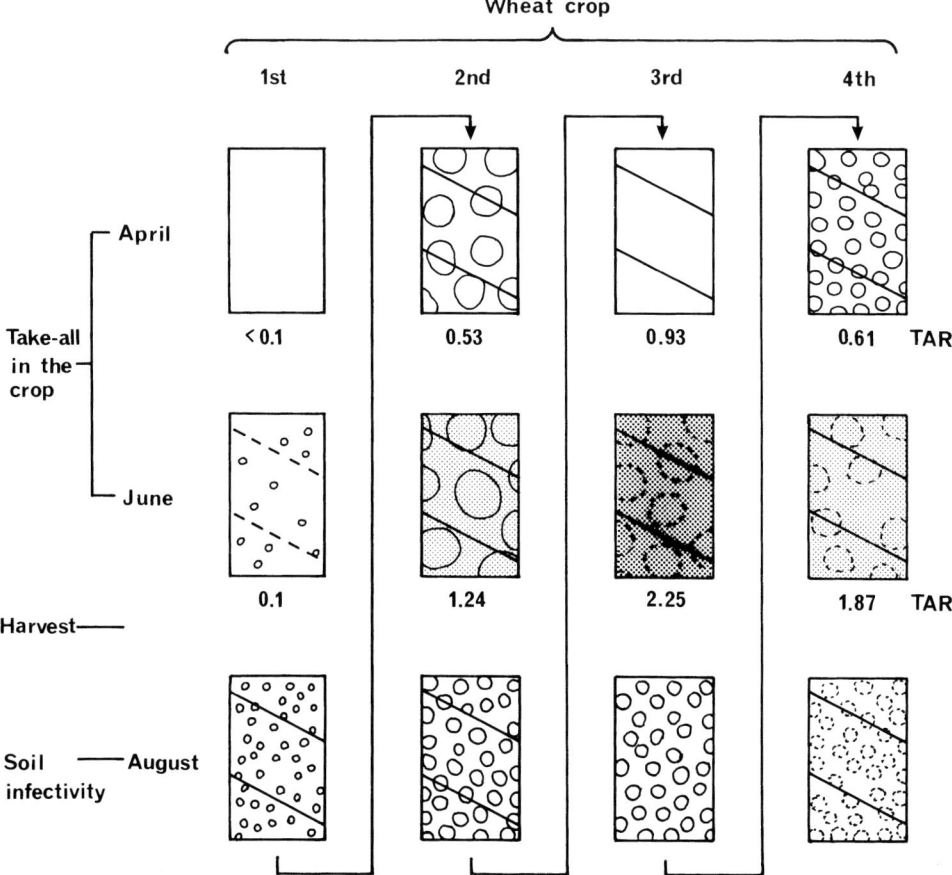

Fig. 6.2. A scheme which incorporates the changing patterns of take-all suggested by data from successive wheat crops at Woburn during 1982–1986.
The rectangles represent a field at three sampling times in four consecutive crops. Medium-scale and small-scale patterns attributed to disease foci, spread from foci and changes in disease intensity are represented by circles. These occur on a background of a large-scale pattern (represented as the three major divisions of the rectangle) which is probably a site factor and is nearly always detected.

◯ = strong pattern; ◌ = weak pattern
⌐▼ = discontinuity in hosts; i.e. 'between-crop' period.
TAR, the take-all rating, has been divided by 100, giving a range 0–3:
☐ = < 1 (slight); ▒ = 1– < 2 (moderate); ■ = 2–3 (severe).
From Hornby *et al.* (1989).

thereafter increased steadily for bigger areas. This suggests both smaller-scale and larger-scale patchiness than that in the more severely diseased third crop. In these experiments, the scale of patchiness seemed to be more closely related to the number of consecutive cereals in the rotation than to severity of take-all in the crop.

In general, therefore, sampling all the plants from very small plots gave rise to less variable results than sampling areas that were four (where block-pairs were scattered throughout a site) or 16 (disease uniformity tests) times larger, especially in crops in which take-all was severe. In longer runs of consecutive cereals, where TAD was assumed to be operating, sampling units up to eight times larger tended to produce the smallest variances. Nonetheless, all these areas are much less than conventional plot areas.

6.3.2 Sampling

Sampling strategy is concerned with the choice of sampling unit (e.g. single tillers, 10 cm lengths of plant row), the number of sampling units (frequently determined by the availability of resources, in particular the time available for disease assessment) and the sampling procedure (e.g. random or systematic). When the distribution of disease is patchy it is desirable that as much as possible of the area is sampled. More accurate assessments of soil-borne diseases result, therefore, from taking large numbers of small samples rather than from taking small numbers of large samples from a given area (Gilligan, 1982). However, complicated sampling strategies involving numerous random sampling positions are often impracticable. Simpler systematic procedures are resorted to when relatively few samples are taken from large areas. An example of this is the 'W' pattern, which is preferred to other patterns (such as parallel diagonal transects) because it achieves a better representation of the area sampled (Delp *et al.*, 1986).

Guidance on numbers of samples is frequently sought by fieldsmen and advisers both for trials aimed specifically at take-all and for other relevant trials where plot size and experimental design are determined by factors other than phytopathological requirements. For monitoring combinable plot trials, ADAS plant pathologists take 30 plants from 0.01 ha plots where there is sixfold replication, but only ten plants per plot on very large trials where factorial designs ensure much higher levels of replication for individual treatments. There is currently no simple general guidance against which to check the adequacy of such choices in advance. In estimating the number of sampling units for a desired level of confidence, it is advisable that some functional relationship between the variance and the mean is used to obtain reliable sample sizes (Gilligan, 1990b). Not all field experiments require the same degree of precision for estimating treatment differences; alternatively, sampling strategy may be based on management costs or decision-making criteria (Gilligan, 1990b).

6.3.3 Assessing disease

Image analysis systems exist for root measurement and cope well with measurements of length, width, projected area, surface area, volume, tip counts, fork counts, etc. Unfortunately, quick and efficient assessment of root discoloration on

The variables most often used in studies of take-all at Rothamsted are listed in Table 6.3. Once plant samples have been prepared, the data are collected using the sort of visual inspection and counting or estimation methods discussed in Section 5.4.1. These data are used to calculate totals, proportions, variances and/or intensity ratings (e.g. TAR). In designed experiments the data provide input for more detailed statistical analysis using Genstat (Payne *et al.*, 1993). Where percentages are derived from proportions, percentage data (p) are usually analysed using the logit transformation. To accommodate 0% and 100%, p is calculated as:

$$p = (\text{number of items infected} + 0.5) \times 100/(\text{total number of items} + 1)$$

where items would be plants or roots. Further, a constant (0.5) may be included to make the scale similar to a probit scale:

$$\text{logit} = 0.5 \times \log_e[p/(100-p)]$$

Where percentages only are available, a similar correction can be made in the following form:

$$p = 100 \times (p \times 10 + 0.5)/1001$$

If it is wished to transform means back to percentages after the statistical analysis, Genstat provides the function ILOGIT, which gives the inverse of the logit transformation.

Disease intensity is taken to be disease incidence × mean disease severity throughout this book. This definition and the use, sometimes, in take-all research of number of roots infected to indicate disease severity may differ from definitions elsewhere. Nutter *et al.* (1991), for example, defined disease severity as the area of a sampling unit infected and as such it was considered as a measure of disease intensity – a general term for the amount of disease in a population. In field experiments, measures of disease incidence and severity have included the mean number of infected roots (seminal, nodal or both) per plant, the percentage of plants infected, the percentage of roots infected (seminal, nodal or both) and disease

Table 6.3. Plant and disease variables most often recorded in take-all studies at Rothamsted.

Variables	Units
Plant	
Emergence counts	no. m^{-1}
Plants	no.
Shoots/straws	no.
Height	cm
Seminal roots	no.
Nodal roots	no.
Yield (85% DM)	t ha^{-1}
Disease	
Infected plants	no.
Infected seminal roots	no.
Infected nodal roots	no.

Table 6.4. Derived variates or estimates used in the assessment of take-all in the UK and France and some yield relationships.

Variables	Examples		References
	Crop	Details	
Plants infected (%)	SW	RES, 0.4% loss in yield for each 1% of disease.	
Plants with moderate or severe take-all (%)	WW	Infection categories: healthy, slight, moderate or severe. In survey of England and Wales the % loss in yield estimated as in Table 6.1.	Polley and Clarkson (1980)
			Gutteridge et al. (1987)
Tillers infected (%)	WW	Results probably refer to % shoots[a], 0.35% loss in yield for each 1% of disease. Feekes growth stage 10.5, N range 64–102 units, average incidence of take-all, 23.2% (incidence based on tillers in moderate and severe categories only).	Rosser and Chadburn (1968)
		0.6% loss in yield for each 1% of disease.	Slope and Etheridge (1971)
Shoots infected (%)			
Roots infected (%)	SW	RES, 0.44% loss of yield for each 1% of roots infected.	
Whiteheads (%)			Gutteridge et al. (1987)
Area of prematurely ripened shoots (%)	SW	Experiment at RES, suggested relationship very dependent on soil conditions.	
Above-ground symptoms category	WW	0–4, see Table 6.2 and associated text.	Polley and Thomas (1991)
Take-all rating (TAR)	WW		Dyke and Slope (1978)
		1.4 t ha^{-1} loss per 100 TAR units (max TAR = 300).	Gutteridge et al. (1987)
Take-all, or disease, index (TI or DI)		Below are examples of how these phrases have been applied to several different scales for assessing take-all.	
		Individual straws and roots graded: < 25% (1), 25–50% (2), 50–75% (3), > 75% (4) diseased roots, respectively. Disease index = aggregate as a % of maximum possible grade.	Cunningham (1985)
		Maximum TI = 100. For each 10% increase in TI recorded during Nov–Jan, the yield of WW was decreased by 1.1 t ha^{-1}; a 0.56 t ha^{-1} decrease occurred for each 10% increase in TI during grain filling.	R.W. Clare, I. Ap Dewi and D.J. Yarham (unpublished data).
		Plants assigned to five disease severity classes, 0–4 (0, 1–10%, 11–30%, 31–60%, 61–100% of the root system with take-all lesions), or sometimes to four classes, 0–3 (0, 1–30%, 31–60% and 61–100%) for the assessment of field plants. DI calculated by multiplying the number of plants in each class by the class weight, summing and dividing by the total number of plants.	Lucas et al. (1988, 1994)

SW, spring wheat; WW, winter wheat; RES, Rothamsted Experimental Station.
[a]Sometimes it is not clear that a distinction has been made between the main shoot and the tillers.

severity ratings, based on the proportion of the root system blackened (either scored by visual assessment or calculated from the percentage of roots infected) (Table 6.4). Different measurements differ in usefulness and costliness (Werker and Gilligan, 1990) and some disease variables require care in interpretation. For example, DPCs based on percentage of roots infected may show a decrease in disease at a time during which average number of infected roots per plant is in fact increasing. This arises when rapid growth of new roots outpaces disease development, often in spring for take-all of winter wheat. Therefore, sequential samples of the number of roots infected give a more realistic impression of the amount of disease in the system and how it is changing, whereas the likely effect of that disease may be deduced better from data for the percentage of roots infected. Graphical representations and different interpretations of the different measures of disease are discussed in Sections 2.5.4 and 2.6.3.

An index of root blackening is commonly used in experimental work, disease being scored according to the proportion of the recovered root system that is blackened. The infection categories of healthy, slight, moderate or severe are explained in Section 6.2; they are described in the MAFF disease assessment keys (Anon., 1976) and illustrated in Plate 4.11 by wheat plants sampled at about the time of anthesis. An adaptation of the MAFF key that has more examples of root systems in each of the categories (Fig. 6.3) has been in use for some years at Rothamsted, where the categories are often referred to in shorthand form by numbers, so that 0 = no infection, 1 = slight, 2 = moderate, 3 = severe. These are the same numbers that are used as the weights in calculating a weighted take-all rating (TAR) with a maximum value of 300 (Dyke and Slope, 1978), i.e. TAR = percentage of plants with slight infection + 2(% moderate) + 3(% severe). In statistical analyses, the TAR has been used as a disease variable without transformation (e.g. Gutteridge *et al.*, 1987; Bateman *et al.*, 1990a).

The advantage of such an assessment is speed. Also, there is a relatively close relationship between proportions of plants in the moderate or severe categories and grain yield, provided that assessments are made between anthesis and ripening (but see 6.5.1 and 6.5.2). A disadvantage of a TAR is that it is not ideally suited to statistical analysis. Linear or logarithmic disease assessment scales with equal divisions assist the decision process and provide data that are more easily handled in parametric statistical analyses (Campbell and Neher, 1994). Logarithmic scales take into account the Weber–Fechner law, that visual acuity is proportional to the log of the intensity of the stimulus and is poorest in the 40–60% range of disease intensity. The Horsfall–Barratt logarithmic scale, originally designed for foliar diseases, has 12 classes, but can be adapted for root diseases (Campbell and Neher, 1994). These classes are: 0, > 0–3, > 3–6, > 6–12, > 12–25, > 25–50, > 50–75, > 75–87, > 87–94, > 94–97, > 97–< 100, 100% diseased tissue. A system which in effect groups the 2nd to 4th and the 9th to 12th classes of this scale was used in scoring a series of field experiments at Rothamsted (D. Hornby and G.L. Bateman, unpublished data) in which large amounts of sample material prohibited root counting (see below). The ratings used were 0 = no infection, 1 = > 0–12.5%, 2 = > 12.5–25%, 3 = > 25–50%, 4 = > 50–75%, 5 = > 75–87.5% and 6 = > 87.5–100%. Using standard drawings for reference, samples were handled

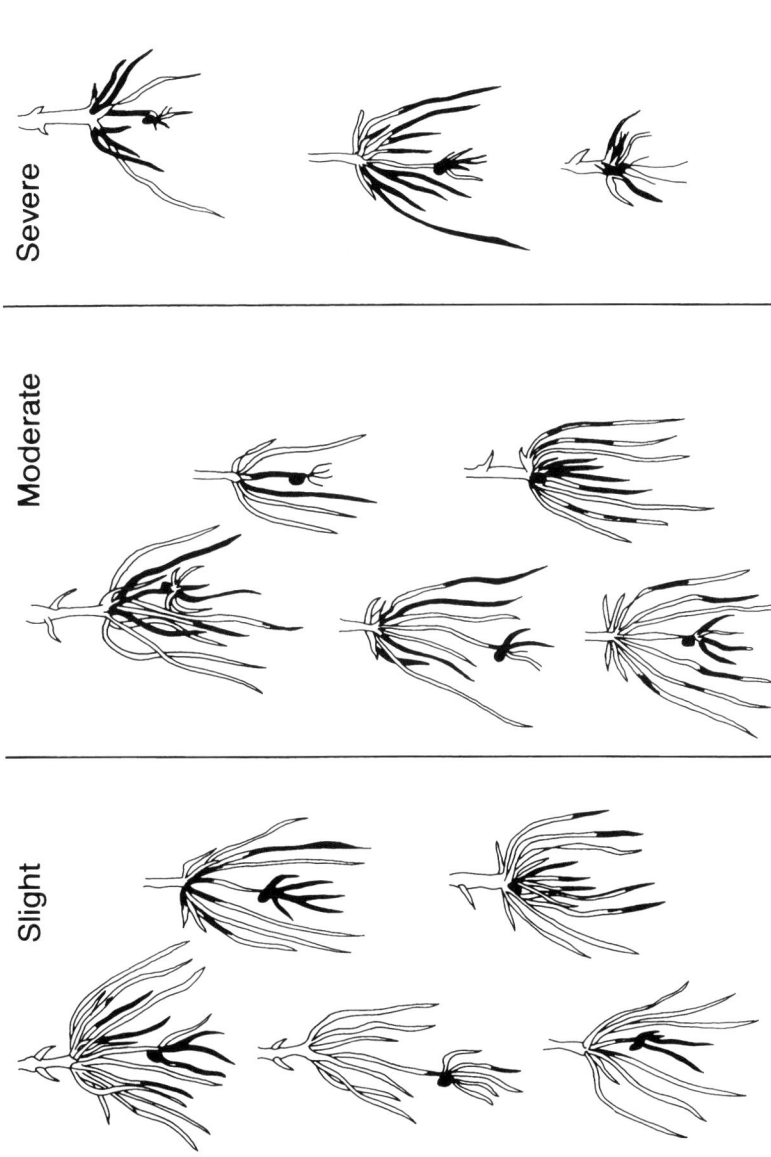

Fig. 6.3. A pictorial take-all assessment key in use at Rothamsted for allocating plants to infection categories. Experience has shown that the contrast between white (healthy) and black (diseased) roots on the key is improved by colouring the healthy roots yellow. Adaptation of an earlier MAFF key (Anon., 1976; Clarkson and Polley, 1981b).

almost as quickly as with the more commonly used 0–3 rating scale discussed above. Other research groups have divided up one or more of the severity categories to produce 0–4 or 0–5 scales. In the middle range of the scale, there is no evidence that observers can readily distinguish finer differences (Campbell and Neher, 1994).

Most information and greatest precision can be derived from counting root axes on each plant in the sample, determining the number of healthy roots and the number of diseased roots. At the time of sampling it is not usually known what the functional state of diseased roots is: discoloured roots may have ceased to function, or they may be more or less dysfunctional. Counting seminal and nodal roots separately can have advantages such as indicating the timing or location of the effects of a control agent. With a fungicide, for instance, decreased infection on seminal roots only would suggest that the fungicide was acting early and close to the seed, whereas decreased nodal root infection may indicate a fungicide with greater persistence and, if it was applied to the seed, that it is mobile. It is usual to calculate the percentage of roots infected from root counts (e.g. Hornby *et al.*, 1989), which, as already indicated above, integrates disease increase and root growth. In disease assessment there is a balance or trade-off between sample size and method of assessment: are large samples assessed crudely better than small samples assessed in greater detail? In this context, it is interesting that TARs, derived by allocating data for percentage of roots infected to the four infection categories of the 0–3 scale, correlated reasonably well with TARs based on simpler assessments of the proportion of the root system blackened (Fig. 6.4). Whereas the TAR is a measure of disease intensity in the crop, $(300 - TAR)/3$ may also be an indication of the percentage of roots remaining functional.

6.3.4 *Patterns of disease*

Diseases continue to be a major and costly factor in field crop production. Many, particularly those caused by soil-borne plant pathogens, often manifest themselves as patches of poorly growing and/or discoloured plants in more or less complex patterns within the crop. The information in these patches is rarely exploited in farm management systems or used efficiently in national disease surveys and yield predictions.

There are a number of methods for studying the spatial distribution of soil-borne inoculum and root disease (Campbell and Benson, 1994). Ggt inoculum is usually assessed by bioassay of the soil and, depending on the subsequent treatment of soil cores, vertical and horizontal distribution profiles or maps are possible (Hornby, 1981). These take time to create and can rapidly become out of date. Knowledge of take-all distribution requires assessment of root disease on plants that have been dug up, or assessment of the distribution of prematurely ripened patches when these occur. Assessments made on a per plant basis present sampling choices that range from all plants in given areas of crop, through defined lengths of plant rows to isolated plants. Line transects, in which all winter wheat plants were assessed for take-all and measured for growth, were used in an ecological study of patch development in crops at Rothamsted and Woburn (Hornby, 1994;

D. Hornby and R.J. Gutteridge, unpublished data). Line transects of 5 m had been used to study effects of late infestations of take-all on yield, yield components and photosynthetic potential of winter wheat cv. Virtue (Green and Ivins, 1984). The line transect offered a convenient way of ensuring that samples were taken within and outside patches of disease.

Mapping of take-all in fields generally has been limited in scale and/or detail, showing usually only the extent of patches of visibly diseased plants once a year (Hornby, 1994). This is crude, in that usually only two levels of disease severity

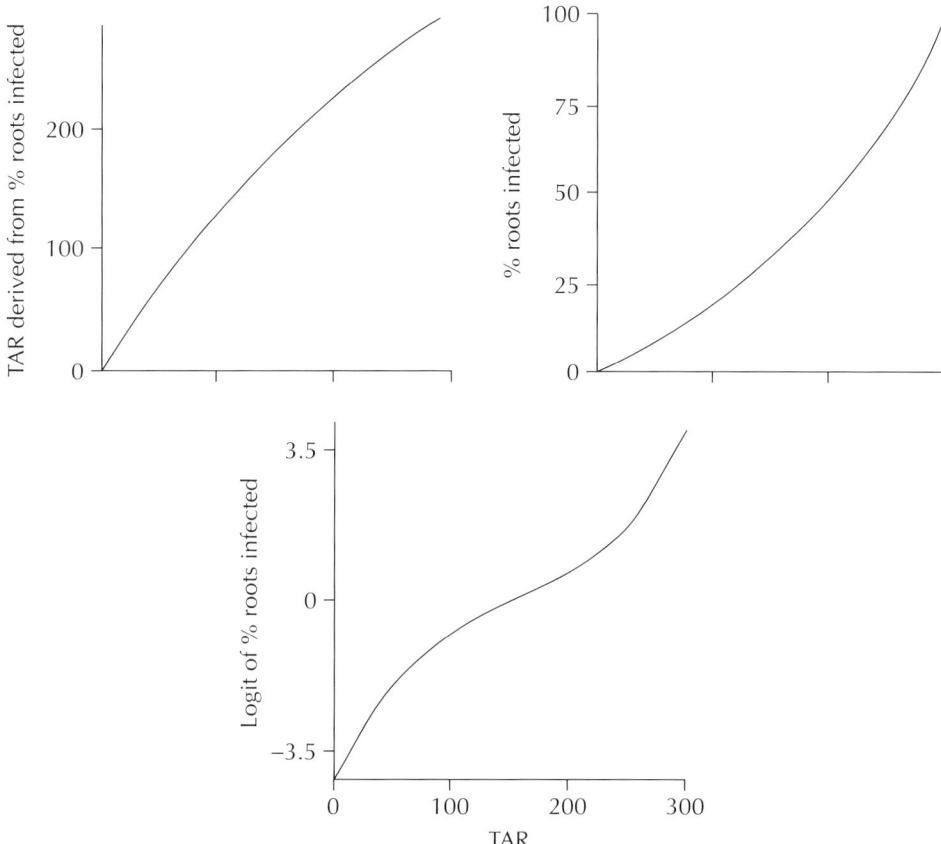

Fig. 6.4. Diagrammatic representation of the relationships between methods of assessing take-all infection of wheat.

Y-axes are based on root counts expressed as percentage roots infected. The x-axis is the conventional take-all rating (TAR) – the sum of the percentages of diseased plants allocated to each of four disease severity categories on the basis of how much of the root system was scored as discoloured. The disease categories are weighted: 0, healthy; 1, slight; 2, moderate; and 3, severe. Therefore the maximum value of the TAR is 300.

The first graph shows TARs on the y-axis which have been derived differently by using data of percentage of roots infected per plant to allocate each plant to one of the four infection categories above. The conventional TAR and the derived TAR, which is unlikely to be used in practice, are not the same. The two following graphs show a more commonly used measure of disease and one of its transformations. The values and ratings plotted are on a per plot basis and each graph summarizes nine graphs, each with 80 points.

(inside and outside patches) are distinguished and there is an assumption that the patchiness of growth is due to take-all. In fact, patchiness often results from differences in soil conditions, where poor soil conditions exacerbate the effects of root disease to produce the symptoms seen above ground. Aerial photography has been used to provide visual records of case histories (see Sections 3.3.3, 3.5.3, 3.5.5 and 3.6) and to collect information on take-all patches, especially in eastern England (Blakeman, 1990). The photographs can be especially useful in identifying areas of poor growth in crops affected by take-all, where soil structure is poor and where remedies are needed, or where cultural practices have exacerbated the disease (e.g. Catt et al., 1986).

Indications of scales of patchiness (or the approximate size of patches) may be useful for determining optimum plot sizes for testing chemical and biological control agents (Hornby et al., 1989). Information about patterns of take-all at Rothamsted was obtained by applying four procedures to disease data collected in four different ways (Table 6.5). Also, root disease data from small contiguous plots (Table 6.5, data source 2) were used to map disease distribution and to compare the variation of different sites and different areas of crop in a single site (see 6.3.1). The same data were analysed geostatistically using the variogram to express spatial structure (Webster and Oliver, 1990). The analysis revealed no structure in two fields, but in a third the data were spatially correlated and it was possible to map the patchiness by kriging (Fig. 6.5). Take-all was almost certainly spatially correlated in the other two fields, but sampling was too sparse to reveal it. Kriging, originally developed for mining and petroleum engineering, is proving to be of increasing value in many branches of ecology (for an introduction for biologists, see Webster, 1996) and with denser sampling it should be profitable for studying the distribution of take-all. Denser sampling was also a recommendation in a geostatistical study of another soil-borne pathogen, *Phytophthora nicotianae* var. *parasitica*, and the disease pineapple heart rot. The sampling used was, however, sufficient to reveal differences between the spatial patterns of inoculum and of disease incidence (Chellemi et al., 1988).

Table 6.5. Procedures and sources of data used in studying patterns of take-all disease at Rothamsted.

Procedures	Data from:
Variance ratio comparisons	1. Many subsamples in conventional large plots in experiments for comparing treatment effects.
	2. Small plots in balanced incomplete block designs paired to form complete replicates scattered throughout the site for comparing treatment effects (Hornby et al., 1989).
Geostatistical methods (Fig. 6.5)	3. Small plots in grids (root disease uniformity studies) for observing changes in variance as the plots are grouped into progressively larger blocks.
Mapping (Fig. 2.21)	
Spatial aggregation analyses	4. Transects and area samples (Fig. 2.21) based on visible patches (Hornby, 1994).

6.3.5 Inoculum

Conventional methods of estimating inoculum in field soil in the UK have been described in detail (Hornby, 1981). In general, these involve taking soil samples, often as soil cores, and growing wheat seedlings in them. These seedling infection tests, or soil bioassays, are usually done under controlled conditions. The cores may be used undisturbed, or subjected to different kinds of preparation. The amount of seedling infection in such tests has been a standard way of measuring soil infectivity for decades. Cores split longitudinally can be used to examine the vertical distribution of inoculum in soil, as shown in the photograph of a 'living histogram' (Fig. 6.6). Mixing and preparing a series of soil dilutions with a suitable diluent (e.g. sand) for seedling infection tests enables inoculum density estimates to be made from quantal data using the maximum likelihood estimate (most probable number method). More precise estimates of inoculum density may be achieved by testing infested plant debris extracted from the soil by wet sieving and elutriation. These conventional types of soil bioassay and soil infectivity estimates are referred to throughout the book (Sections 2.4.2, 2.5.4, 2.5.5, 2.6.8, 3.5.1, 4.3.3, 4.3.4 and 4.5.1; Figs 2.4, 2.5, 3.11, 4.2 to 4.6 and 5.23). They are labour intensive and time-consuming, but continue to be the mainstay of natural inoculum studies (e.g. Cotterill and Sivasithamparam, 1987, 1988b; Cotterill *et al.*, 1988). Serological methods (see 5.5.3), DNA methods (see 5.5.2) and other modern approaches have yet to produce rapid and effective substitutes.

An example of a large study, based on the conventional bioassay of soil, involved the determination of the most probable number of infectious fragments of plant residues per volume of soil (λ), as well as soil infectivity, on 115 occasions during 16 years of continuous spring barley at Woburn, Bedfordshire (Hornby and Henden, 1986). Annual changes in λ and differences in λ in short and long soil core samples were studied. Infected root counts obtained from plants in dilution series could be used to indicate soil infectivity, provided the counts were given in terms of a unit volume of neat soil. However, correlation between soil infectivity and λ varied with time, with a suggestion that a linear relationship was more pronounced during May–July than during November–June. Sampling between crop rows underestimated soil infectivity, but this horizontal discrepancy ceased to exist once crowns had been ploughed down.

Soil infectivity was also greater within rows than between rows in Western Australia (Cotterill and Sivasithamparam, 1987) and the distribution of inoculum after direct drilling (more infective inoculum in the top 3.3 cm), shallow cultivation and deep cultivation (both with inoculum more evenly distributed down to 10 cm) was similar to that in the UK (Cotterill and Sivasithamparam, 1988b). In the Australian work, take-all was least with direct drilling, but the inoculum density in soil after this was similar to that after deep cultivation. A clear regional difference is that the reduction in inoculum that occurs after harvest in the UK (Fig. 2.5) is delayed by the absence of soil moisture during hot, dry summers in Western Australia (Cotterill and Sivasithamparam, 1987). Also, in southern Australia the grass component of clover-ley pastures in pasture/wheat rotations is important for the carry-over of Ggt (Cotterill *et al.*, 1988). However, only one

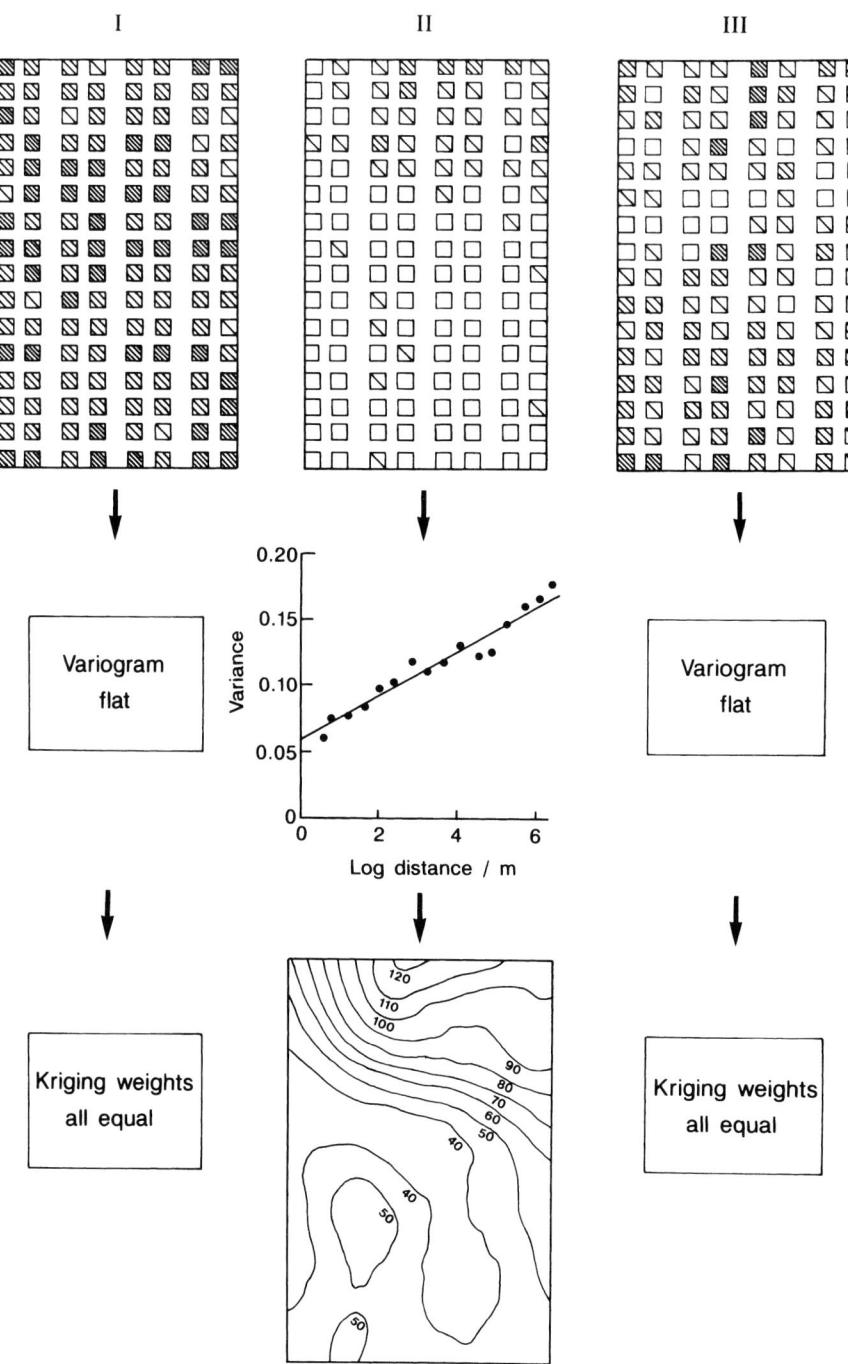

Fig. 6.5 (opposite). Plans of three field experiments showing take-all intensity (top), the results of tests for spatial dependence (middle) and, where spatial dependence was detected, a map (bottom) of kriged estimates of take-all to represent the data in a rational and optimal way.

Fields
I: Summerdells field, Rothamsted, third winter wheat crop, 1987.
II: Long Hoos field, Rothamsted, fourth winter wheat crop, 1986.
III: White Horse field, Woburn, fourth winter wheat crop, 1986.

Assessment of take-all
The plans are of areas containing 128 small plots, each 33 cm × 37 cm. Although the areas between the plots also contained wheat, take-all was assessed only on wheat within the plots (all plants). Each plant was allotted to one of seven weighted infection categories (see 6.3.3) and take-all intensity per plot was recorded as the sum of the percentages of plants in each category multiplied by the category weight (maximum score = 600). For clarity the scores are grouped on the plans as:
☐ 0 to < 100; ▨ 100 to < 200; ▨ 200 to < 300; ▨ 300 to 600.

Variograms
Those for fields I and III were flat, indicating no spatial dependence in the data, and are therefore omitted. The variogram for take-all intensity in field II shows spatial dependence (i.e. strong correlation between disease assessments in adjacent plots). It has a step of 0.4 m; experimental values are plotted as points and the best fitting linear model is shown as a solid line.

Kriging
This is used to predict unknown values of take-all intensity at particular places from the measurements above, without bias and with minimum variance.

(2,4-D amine + propyzamide) of three herbicide treatments tested decreased inoculum density in the field. The effect was short-lived and no effect of herbicide on inoculum was detected in the second year.

Much effort has centred on artifical inoculum of Ggt and its effectiveness in pot work, or, as discussed further in the next section, in field work. Fragments of oat grains colonized by Ggt were slightly more infectious than fragments of colonized crown from the field, which, in turn, were better than colonized roots from plants grown in a glasshouse (Wilkinson *et al.*, 1985b). The smallest fragments from field soil that caused lesions were in the range 0.5–1 mm (this agrees with Australian work reported by Cotterill and Sivasithamparam, 1988b).

Sand grains colonized by hyphae of Ggt provide a uniform, low-nutrient inoculum that is easily mixed with soil (MacNish *et al.*, 1986). Using this inoculum, disease incidence (measured as the length of vascular discoloration of roots) was proportional to the inoculum density in the range 23–225 units per 100 cm^3 of soil for up to 48 days at 10°C, but for only 31 days at 20°C. At greater concentrations the relationship became non-proportional and the reversal of the curve was attributed to antagonism, rather than lack of infection sites. Ggt was able to grow into soil from these low-nutrient inoculum sources.

6.4 Artificial Inoculation

6.4.1 Background

Artificial infestation of experimental plots and fields with Ggt has been done usually with the intention of creating predictable and uniform disease for testing fungicides, or the resistance of host cultivars to take-all (Kollmorgen, 1985). Although

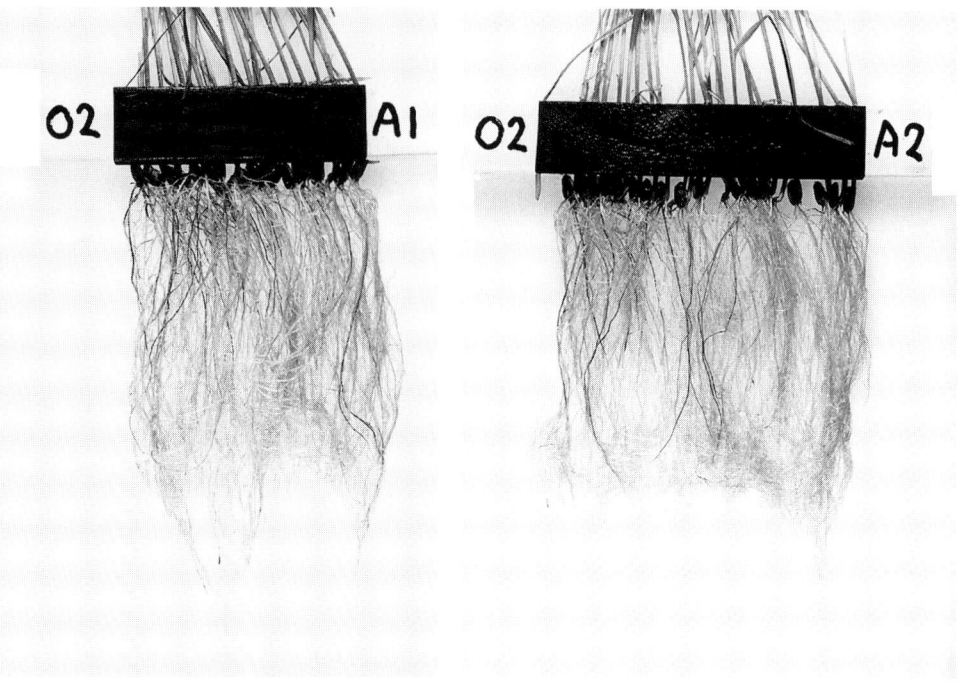

Fig. 6.6. A 'living histogram' of soil infectivity down the profile of a silty clay loam.

The location was plot 02 of the ploughed treatment in a cultivation experiment after 6 years of continuous spring barley at the National Institute of Agricultural Engineering, Silsoe, Bedfordshire. A cylindrical soil core, A (7.5 cm diameter × 28 cm), had been extracted in early November 1976, before cultivation. After cutting the core radially into three sections (A1, 0–8 cm; A2, 8–18 cm; A3, 18–28 cm) and slicing each section longitudinally, one half of each section had been laid with the flat surface in contact with a moist bed of sand and oriented so that the uppermost radial surface was to the left. Wheat seeds had been sown along the top of the curved surface of each section. After 4 weeks at 19°C, the shoots had been taped into position between two wooden strips and the soil washed away from the roots. The amount of root blackening indicates that soil infectivity was most in the upper part of the profile.

stubble and roots infected naturally with Ggt have been transferred from one field to another to achieve artificial infestation (Prew, 1980), artificially produced inoculum has been a more popular choice. Typical preparations comprise the fungus on agar media (cf. pathogenicity testing, 5.4.1), or on sterilized seed, meal or straw, sometimes with an inert carrier such as vermiculite. Partially dried vermiculite–maizemeal inoculum mixed with wheat seed has been found to be suitable for application through a variety of seed drills (T.W. Hollins, personal communication). Substrates other than agar are often dried before use. Dead oat grains, usually killed by autoclaving and subsequently colonized by Ggt, have been a very popular and convenient form of inoculum. Simon *et al.* (1987b) considered them less realistic than smaller grains in the field, because of their larger food reserves. Smaller grains such as ryegrass and millet seed colonized by Ggt were used to manipulate the number of uniformly sized propagules in soil. Millet seed produced consistently more severe disease than the ryegrass, because of its larger size and weight (5.3 mg vs. 1.8 mg). These seeds are small enough to use without grinding and contain sufficient food reserves to support infective Ggt for at least 4 weeks in moist soil (Simon and Rovira, 1985). However, the effects of these grains as inocula are not always as expected. Comparisons of concentrations in the range 0.09–0.75% (w/w) of colonized grains of oats, millet and ryegrass showed similar percentages of roots infected on wheat plants grown in infested sand (Hornby and Davis, 1985). Whereas all treatments decreased plant height and weight, ryegrass did so the least and the colonized ryegrass treatment produced more roots than a check treatment with autoclaved ryegrass alone. Another complication is that in some situations colonized oat or wheat grains may be eaten by mice (T.W. Hollins, personal communication).

The requisites of artificially produced inoculum include virulent isolates of the pathogen and consideration of the importance of the nutrient status of the medium and the tendency for staling products to be formed in nutrient-rich media (Cunningham, 1981). The medium should not have a great influence on the performance of the host plant, either beneficially by providing abundant nutrients, or unfavourably through toxins. The inoculum may need to be in a form suitable for application by farm machinery; this is one of the attractions of colonized oat grains.

Colonized oat grains were used to investigate inoculum placement at different distances from the seed furrow on the same day as sowing (Bockus, 1991). Inoculum placed above or below the seed caused significant yield reductions, but there were no yield decreases when inoculum was placed at distances ≥ 7.6 cm to the side of the seed. The greatest incidence of whiteheads and the lowest yields occurred when inoculum was placed with the seed. Therefore, in tests of putative biological or chemical control treatments, care should be taken not to place artificially produced inoculum so as to increase the likelihood of a favourable result from the candidate treatment, unless the objective is preliminary screening or the investigation of relationships between infection sites and effects of treatments (e.g. Garcia and Mathre, 1987). Close proximity of inoculum and seed treatment fungicides may result in good disease control in experiments, but the performance of the seed treatment against natural infection cannot be predicted reliably from this. An exception may be where such inoculation closely simulates natural infection, i.e. where warm conditions allow rapid infection of spring-sown wheat (see 4.2.2).

Artificially produced inoculum of Ggt has also been used in the field to investigate the relationship between amounts of take-all and grain yield (see 6.5.2) and for infection studies and epidemiological research. However, artificially produced inoculum can cause earlier disease as well as retardation of growth, shoot yellowing and wilting not present in naturally infected plants in the same locality (Jensen and Jørgensen, 1973). In Australia, survival of Ggt in naturally infested soil was greater than survival of Ggt on artificially colonized straw in soil (MacNish, 1976). It remains to be shown whether artificial inoculum is similar to natural inoculum in sensitivity to temperature, particularly in relation to thermal inactivation (Bockus *et al.*, 1994b). Also in Australia, disease in a second wheat crop after an artificially inoculated first wheat crop was distributed more evenly than in one that followed a non-inoculated crop (Cotterill and Sivasithamparam, 1989a). Much of the infection in the crop after the inoculated crop would have arisen from infected root remains (i.e. natural inoculum) and this use of artificially produced inoculum to create more uniformly infested sites for experimentation in the following year may overcome some of the objections to experiments on artificially inoculated crops (Cotterill and Sivasithamparam, 1989b).

6.4.2 Latest research in the UK

In the last decade attempts have been made at Rothamsted to create epidemics of take-all in wheat artificially by applying colonized oat-grain inoculum. The objectives of the research were the following.

- *To establish relationships between different amounts of disease and grain yield.* See Section 6.5.2.
- *To compare natural and artificially created epidemics.* The use of inoculated plots for studying disease or testing potential control agents may be misleading if the resulting epidemics behaved differently to natural ones.
- *To assess the evenness of infection after inoculation.* One expectation from applying inoculum artificially is that uniformly severe disease will occur; this is of benefit in testing control materials. If, in a cereal sequence, inoculation also affected the evenness of disease in the year after application, when most of the inoculum would be in the more natural form of colonized residues of the previous crop, it would have added advantages as an experimental technique.

Introducing inoculum into soil generally produced uniform, severe infection in spring wheat (Hornby and Bateman, 1990) and in spring barley, but only in the year in which it was applied (Hornby *et al.*, 1981; Fig. 6.7). In later experiments on winter wheat, the results were less reliable (Bateman and Hornby, 1995). Shallow inoculum, placed by combined drilling with the wheat seed (Fig. 6.8), produced more severe infection than similar amounts of inoculum applied in drills (Fig. 6.9) and then incorporated more deeply by ploughing-in or mixing in by rotary harrow before sowing. The wider horizontal and vertical distribution of inoculum in the deeper applications resulted in much less inoculum close to the wheat seed than in the shallow application.

None of these methods of application invariably produced infection that was severe enough to cause obvious stunting or premature ripening of winter wheat. For instance, inoculation in autumn 1988, although clearly producing an effect, resulted in less severe take-all in 1989 than expected, perhaps because of warm, dry conditions in spring and summer. This explanation is supported by disease levels in plots without artificial inoculum in the same experiment; there was very little take-all in a first crop in 1988/89, and also in a second wheat in 1990, another year with a warm, dry summer. In contrast, inoculum applied to winter wheat on a different site in autumn 1991 produced severe and damaging take-all in 1992. Figure 6.10 shows year-to-year disease progress, as percentage of roots infected at the time of grain filling, for plots receiving the most effective applied-inoculum treatment and for plots with natural inoculum. One possibility is that artificial inoculation in 1988 did not induce TAD and that the peak in 1992 in plots that had received artificial inoculum arose through the natural epidemic superimposing itself on the waning artificial one. However, the amounts of disease

Fig. 6.7. The effect of introduced artificial inoculum of Ggt on spring barley in plots on sandy loam, Woburn, Bedfordshire, 16 July 1979.
The crop in the plot on the left is an 11th consecutive spring barley crop and the one in the plot to the right is a break crop of field beans. The plot in the centre has a second spring barley crop, which is severely diseased after the introduction of artificial inoculum of Ggt in the form of colonized whole oats placed close to the barley grain at sowing. In untreated spring barley crops, there was never more than 4% of roots infected with take-all, even in barley sequences following a 2-year break.

in artificially and naturally infested soils in 1992 were similar; they were also large, indicating favourable conditions for take-all in that year. This suggests an alternative possibility that TAD had developed in the artificially inoculated plots and that the peak in 1992 was a secondary peak caused by favourable weather. The fact that, by 1993, disease in the plots that had received artificial inoculum was less than in the naturally infested plots would support the view that TAD was happening sooner in those plots, because they had had more disease in previous years. Soil infectivity, determined by bioassay (not shown), decreased progressively after inoculation and did not show a peak in the fourth year, as did disease on the plants; in contrast, soil infectivity in plots that had not been inoculated increased progressively during the experiment.

Comparison of the disease data from the natural and artificially created epidemics in 1990 and 1992 supported the view that greater uniformity of disease can be achieved by adding artificial inoculum to the soil. This probably contributed to the observation that in 1992, when the natural epidemic seemed to have reached a peak at an average take-all severity of moderate, patches of prematurely ripened plants appeared only in the naturally infested crop. Such 'hot' spots were absent in the inoculated plots (which had no patches), even though their average take-all severity was moderate also. Although disease was less in a crop following a first crop that has been exposed to introduced inoculum, this did not result in continued disease suppression in subsequent susceptible crops (Bateman and Hornby, 1995), so the question of whether artificial inoculation of first wheat crops can induce TAD is still unanswered.

Fig. 6.8. Equipment used for introducing laboratory-produced inoculum of Ggt (colonized, dead oat grains) into soil in large-scale field experiments: a direct drill with two hoppers for applying the inoculum in seed furrows along with wheat seed.

6.5 Disease–Yield Relationships

Quantifying effects of disease on yield is complicated by the effects of the environment on the host. Estimates of yield loss from take-all on a national scale have been made for England and Wales (see Table 1.11 and Section 1.2.1), but their reliability is doubtful because of the absence of detailed survey data and the problems of associating take-all with yield loss in individual fields (see below). For example, observations show that years which favoured take-all often also favoured high yields nationally. This section describes attempts to determine losses attributable to take-all and outlines some of the difficulties that beset such endeavours.

6.5.1 *Theoretical background*

Disease–yield relationships are concerned with the quantification of loss in yield (dependent variable) per unit increase in disease (independent variable). They may be simple, with the rate of yield loss constant over a range of disease intensities estimated by a single disease variable (single point models), or complex, with several explanatory variables such as estimates of disease, host growth and environmental factors (multiple regression models) (Table 6.6). A number of independent variables may show correlation with yield and, more problematically, amongst themselves. Therefore, choice of disease variables, timing of the measurements and the statistical methods for selecting a minimum number of variables to explain the maximum variability in yield are important. However, multiple

Fig. 6.9. Equipment used for introducing laboratory-produced inoculum of Ggt (colonized, dead oat grains) in large-scale field experiments: a fertilizer drill used for applying inoculum at a depth below that at which the seed will be drilled.

regression models are multidimensional and so are difficult to visualize and interpret. They are frequently concerned solely with explaining as much of the variation in yield as possible. The complex interactions, including the effect of disease on host growth, the response by the host to disease and how these are mediated by the environment, are undoubtedly reasons that reliable and repeatable disease–yield relationships are difficult to obtain. Yield is considered elsewhere in the context of the disease cycle (2.2), disease–environment interactions (2.3.1), temperature and soil moisture (2.3.3), practical problems related to DPC studies (2.5.4) and modelling host growth and predicting yield (2.6.6).

Models for disease–yield relationships

Disease–yield relationship models have been reviewed by James (1974), James and Teng (1979), Teng (1985), Teng and Johnson (1988) and Campbell and Madden (1990). These include single point models (Table 6.6), which have been applied to describe the effects of take-all on grain yield: estimates of the incidence of take-all were made only once, usually towards the end of the growing season (Slope, 1967; Slope and Etheridge, 1971; Shipton, 1975). All single point models are based on the principle that a straight line adequately describes an epidemic and that the

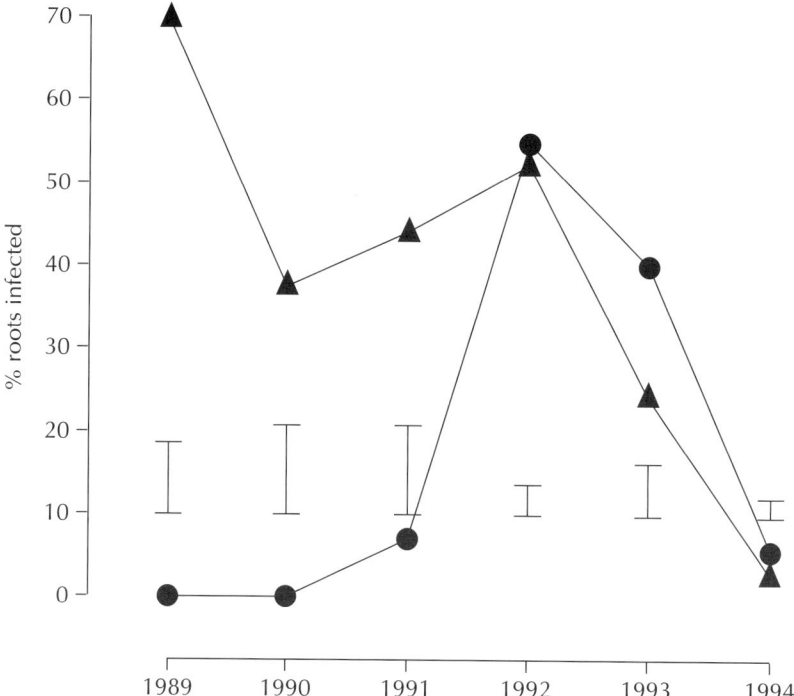

Fig. 6.10. Take-all in July (GS 77–83) in six consecutive wheat crops following a break crop (1988).

Natural inoculum (●). Artificial inoculum in the form of colonized oat grains (▲), applied at the time of drilling the seed in autumn 1988. Vertical bars show SEDs; analyses of logit-transformed percentages showed comparable differences.

straight line is defined by a point defined by two coordinates and a slope. A single point model may also be a critical point model (Table 6.6), for example describing yield in terms of the time of onset of the epidemic and assuming that the subsequent rate of disease increase is predictable. Three critical point models (critical time, critical level and disease-free period) and their use in predicting crop loss from disease severity were discussed in detail by Zadoks and Schein (1979). Unfortunately all DPCs (disease progress curves) going through the same (critical) point do not follow the same course and therefore do not result in the same yield loss. Consequently, the models are valid only within the area and epoch in which they were developed. There are other problems. The critical time is usually a physiological time or state that lasts only a few days. The estimated loss is the difference between final yield of a healthy crop and the yield already available on the day that the critical level is reached. Data to support the assumption that yield production gradually stops when disease severity reaches a critical level are limited. A really disease-free period for most epidemics hardly exists, so the 'onset' of disease tends to become an arbitrary perception threshold. In take-all, the time of onset of the epidemic is usually far more difficult to determine than the time at which a certain critical level of disease (measured as incidence and/or severity) occurs.

Table 6.6. Some disease–yield relationship models.

Description	Equation	Symbols
Single point model (also known as a critical point model when a certain stage that has the best statistical relationship to yield is used).	$Y = B_0 + B_1 \cdot D + e$	Y is yield. B_0 and B_1 are constants. D is disease.[a] e is a measure of error.
Multiple point model (or multiple regression model).	$Y = B_0 + B_1 \cdot D_1 + B_2 \cdot D_2 + \ldots + B_n \cdot D_n + e$	$D_1, D_2 \ldots D_n$ are estimates of disease[a] at times 1, 2...n. $B_1, B_2 \ldots B_n$ are the associated parameters. e is a measure of error.
Area under the disease progress curve (AUDPC) or integral model.	Can be estimated using the midpoint rule method (Campbell and Madden, 1990): $$\text{AUDPC} = \sum_{i}^{n-1} \left(\frac{y_i + y_{i+1}}{2} \right)(t_{i+1} - t_i)$$	AUDPC is in percent-days or proportion-days. n is the number of assessment times. i is the order of the inter-assessment interval. y is disease intensity. t is time. The product of the first and second terms is an estimate of the inter-assessment area.

[a] D could be any summary statistic of the DPC, e.g. disease onset, rate of increase, AUDPC, disease at the end of the season, or at some critical time. In multiple regression these can be used in combination, or in conjunction with other variables affecting yield (e.g. sowing date).

Efforts to improve the relationship have concentrated on disease measurement and experimental techniques for generating different amounts of disease. Shipton (1975) found that the incidence of severely diseased plants correlated better with yield than did the incidence of infected plants. Other methods of disease assessment may give even better correlations: see 6.2.) Although they are popular, it is questionable whether single, end-of-season estimates accurately reflect the effect of disease at a specific time and the cumulative effect of disease as the crop develops (Teng and Johnson, 1988). The shape of the DPC may be particularly important in epidemics of long duration and where the diseased portions of the plant are distinct from those taken for yield (Teng and Gaunt, 1981), as in take-all. The progress of take-all can vary markedly with cropping history or with season and may be variable within seasons (Werker *et al.*, 1991).

Multiple point models (Table 6.6) utilize two or more disease variables that describe selected aspects of the course of an epidemic. For example, relationships between yield loss and disease estimated at several points in time throughout the growing season have been recommended for long-duration epidemics with variable infection rates (James, 1974). Decreasing the time interval between successive observations is likely to lead to increased correlation amongst the independent variables, but using the difference between successive observations, or disease variables that summarize the relationship between the successive observations (e.g. the mean and average rate of disease increase), may overcome this. High correlations amongst the independent variables, as well as with yield loss, may therefore point to future models that would need fewer samplings.

Whilst the derivative of the DPC provides an estimate of the rate of disease increase, the integral – the area under the disease progress curve (AUDPC) – provides an estimate of the total amount of disease the host has been exposed to throughout the growing season (Table 6.6). It takes no account of the shape of the DPC and gives equal weight to disease occurrences at all stages during the growing season. Combining the AUDPC with the rate of disease progress or partitioning the AUDPC into 'critical' stages in relation to host growth are refinements. Alternatively, the AUDPC may be described in terms of both disease and time (Shaw and Royle, 1987), but such surface response models require much data (Teng, 1985). Where yield measurements of winter wheat were made after sequential assessments of take-all throughout seasons (Christen, 1992), different critical point models using a regression approach were compared to test their ability to explain and predict subsequent changes in grain yield. It proved impossible to establish constant thresholds for disease based on a single assessment during the growing season. The AUDPC gave the best results showing that, as the disease developed slowly, accumulated effects were more important than a single critical stage. The same conclusion was drawn from a parallel study of wheat powdery mildew.

Simple models that predict a constant rate of yield loss negate the complex interactions between host, disease and environment. In artificially inoculated micro-plots in Kansas, USA, linear and quadratic trends between yield loss and inoculum density of Ggt were significant in each of the 5 years of the experiment (Bockus *et al.*, 1994b). Disease tolerance (surplus root function), stimulation of new roots and compensation by non-diseased tillers result in a range of disease intensities for which there is little or no loss in yield, otherwise termed minimum thresholds (Teng, 1985). There is evidence for a threshold in the take-all results obtained in the UK by Prew *et al.* (1986), although experiments in Germany suggested that it was not possible to establish a single threshold (Christen, 1992). When disease is less than the minimum threshold, there may be a small but significant amount of yield loss before compensation occurs, or there may be yield enhancement as in other systems (Burleigh *et al.*, 1972; Gaunt, 1980; Teng, 1985). Where the TAR was used at Rothamsted, fitting split lines to yield–take-all data often accounted for slightly more of the variance than the linear regression model (Bateman *et al.*, 1990a). In the split-line model the steeper slope after the breakpoint showed that a unit of disease in the moderate-to-severe range had a greater effect on yield than a unit in the slight disease category (i.e. TAR < 100). Relationships between root infection indices and grain yield were sought using experimental data spanning 10 years in Czechoslovakia (Herman and Dovrtèl, 1991). The relationship was not linear and was most variable at lower levels of disease, when yield sometimes increased with increasing disease. The effects also varied according to the growth stage (during the ripening period) at which the assessments were made. Non-linear models can explicitly incorporate parameters to estimate both lower and upper threshold levels (Seinhorst, 1965; Madden *et al.*, 1981; Madden, 1983), but evidence of an upper threshold above which no further loss occurs is weak. If root function is destroyed completely prior to grain filling, then 100% yield loss may be expected, but a similar disease severity achieved during grain filling is not likely to cause 100% loss (see also Section 1.2.1 and Figs 1.12 and 1.13). In addition, what is perceived as 100% diseased roots may not mean the total absence of a functional root system.

Factors that are known to affect yield independently of disease (e.g. weather, soil type, date of sowing and soil fertility) need quantifying. If the dependent variable is represented by the absolute yield, then the disease–yield relationship will vary: this may be overcome by normalizing actual yield with respect to potential, or disease-free, yield. Circumstances may arise in which yield is adequately defined in terms of selected agronomic and environmental factors without reference to disease (A.R. Werker and C.A. Gilligan, personal communication). An early attempt to quantify the relative effects of environmental factors and crop damage explained 78% of variation in wheat yields in Canada over a 10-year period (Sallans, 1948).

Statistical considerations
In developing a disease–yield model there may be many independent variables to consider, including some that have a significant overlap of information. Procedures for 'picking out' significant explanatory variables or 'dropping' least significant variables have been devised (Neter and Wasserman, 1974). If the number of parameters is too large to test all possible models, it is possible to resort to stepwise regression techniques. These are powerful and relatively simple, but no attention is paid to the biological interpretation of the order or combination of variables used to give rise to a 'best' model. To overcome some of these problems an order of priority in the addition of variables, based on biological, practical or other criteria,

may be used instead. Multiple regression models assume that the explanatory variables exert independent effects on yield and that correlation between variables has no significant effect on the validity of using such regression techniques (Teng, 1985). More sophisticated multivariate techniques (e.g. principal components analysis and canonical correlation) have been used to select variables amongst which correlation is minimized. These have predicted yields of wheat using 12 explanatory variables and explained 84% of the variation in yield (Stynes, 1980).

6.5.2 *Practical approaches to assessing losses in grain yield and quality*

Relationships between naturally occurring take-all and grain yield have been explored in a number of ways, using: (i) survey or experimental data involving different fields, locations and/or treatments (e.g. Rosser and Chadburn, 1968; Polley and Clarkson, 1980; Headrick and Bockus, 1996); (ii) data for 1 year from one field with experimental plots treated differently to influence disease (e.g. Nilsson, 1969; Prew *et al.*, 1986; Gutteridge *et al.*, 1987), including fumigation to eliminate inoculum from some plots (e.g. Roth *et al.*, 1984; Folwell *et al.*, 1991); (iii) data for 1 year from a field with take-all patches (MacNish and Dodman, 1973).

None of these was entirely satisfactory and in each case it can be argued that disease was confounded with one or more other factors. Conditions that contribute to poor yield, such as waterlogging, stress caused by drought, or nutrient deficiency, can also increase the effects of root infection. ADAS workers have on occasions observed areas of fields where, because poor soil conditions have exacerbated the effects of root loss, the expression of symptoms was most severe, yet the percentage of roots infected was less than elsewhere in the field.

Yield loss on farms is usually ascribed to take-all only when growth is visibly poor, or more commonly when patches of prematurely ripened plants develop. ADAS cereal disease surveys assess such patches to estimate the severity of take-all nationally (see 6.2). Incidence of whiteheads (see 6.2) in experimental plots often relates better than root infection to yield (Gutteridge *et al.*, 1987), but yield loss is also likely to occur when disease is severe but conditions are such that whiteheads do not become visible. This means that a farmer may often have no indication of the extent of yield loss from take-all, or even of the severity of the disease in a field. Conversely, some effects of take-all on grain quality can be anticipated when whiteheads are present: grains in the early-ripened patches are small and often have mould growth that causes discoloration of the flour.

Effects on grain quality occur most directly as a result of decreased grain-filling. Thousand-grain weight, total yield, tillering and the number of grains per ear were decreased by the most severe infection in artificially inoculated plots of wheat, although 1000-grain weight was sometimes increased in less severely infected plots compared with healthy plots (cf. 'minimum threshold' in 6.5.1) (Manners and Myers, 1981). In a different experiment, using a single assessment at GS 59, triadimenol plus fuberidazole seed treatment decreased the percentage of winter wheat plants infected and increased 1000-grain weight and specific weight (Jones, 1987).

Take-all can affect crop maturation, as is seen in patches where severe take-all has led to premature ripening (and yield loss). The need for continued experimentation to assess the relationship between disease, crop maturation and Hagberg falling number has been pointed out (Stevens *et al.*, 1988). When a prematurely ripened patch stands during a period of wet weather before the crop is harvested, there may be increased α-amylase activity that can decrease the Hagberg falling number below a level at which, for wheat, the flour is suitable for making bread.

There is as yet no consensus as to when assessments of take-all should be done to estimate yield loss. Although disease assessments during anthesis and grain-filling are generally accepted as relating best to yields (e.g. Rosser and Chadburn, 1968), this was not so in a series of ADAS trials between 1981 and 1983. In plots where no nitrogen was added, the correlation between yield and take-all was best at GS 31; where 120 kg or 200 kg N ha^{-1} were applied, there was no correlation between disease and yield (L.V. Vaidyanathan, personal communication). Another ADAS survey (Polley and Clarkson, 1980) relied on assessments made just before harvest, and in artificially inoculated spring wheat relationships between take-all and yield were strong up to GS 85 (Bateman and Hornby, 1989, and unpublished data). Late assessments of cereal root and stem base disease symptoms, when diagnosis is difficult, may sometimes be strongly associated with yield, but doubts as to the cause and the limited practical value of such intelligence must be set against this.

Assessing losses by artificially applied inoculum
Artificial inoculum has been used to explore the disease–yield relationship. In Australia and the USA, the approach successfully achieved different amounts of disease and yield after application of different amounts of inoculum (Rovira, 1978; Rothrock, 1987). However, such differences may sometimes result from differences in disease incidence rather than from different severities of infection on individual plants. This remains one of the major problems in determining yield–loss relationships, because plants without disease compensate. It also raises questions about the relationships between single plants and their yields and field assessments of crops and their yields. In experiments on winter wheat in Chile, varying densities of oat-grain inoculum were used to produce sigmoid curves relating inoculum concentration to yield and to percentage of whiteheads, but there was no attempt to relate amount of root disease to yield (Madariaga and Mellado, 1985). In Kansas, USA, different amounts of oat-grain inoculum were applied in the furrow with seed in single-row plots to establish relationships between amounts of take-all and yield (Bockus *et al.*, 1994b). Mixed inoculations with *Cephalosporium gramineum* were also studied and relatively simple response-surface models generated to describe the effect of the interaction of the two pathogens on grain yield. The interaction between the two pathogens was negative, indicating competition, but the antagonism was not thought to be of importance to the epidemiology or management of the diseases in farmers' fields. Bearing in mind the observations in Section 6.4.1, it needs to be established in such procedures that the different amounts of oat grain do not have different effects on plant growth, which are independent of the effects of the pathogen.

The relationship between take-all severity and yield in spring wheat was investigated at Rothamsted using different methods of applying inoculum (Bateman and Hornby, 1989; Hornby and Bateman, 1990; Section 6.4.2). The procedures followed were intended to achieve uniform infection within field plots, but different severities among treatments, by using a sequence of cultivations that allowed inoculum to be placed at different depths. In this experiment, incidence (percentage of plants infected) and severity (percentage of roots infected) of take-all were related similarly to yield, whilst the correlation between incidence of whiteheads and yield was less good.

If all the main axes of a cereal root system were blocked proximally and became non-functional sufficiently early in growth because of take-all, then little or no grain would result. Because it is unusual for all plants in a crop to be so severely diseased, this extreme is rarely experienced in UK agriculture. Some published disease–yield relationships are given in Table 6.4 and these suggest that even were 100% of roots to be infected, yield losses would be partial only. For example, regression lines in the Rothamsted spring wheat experiment revealed that for each 1% increase in roots infected there was a decrease in yield of 0.44%. In an artificial inoculation experiment in Ireland, yield losses of 50% and 24% were recorded for similar levels of severe take-all in spring wheat and spring barley, respectively (Cunningham *et al.*, 1968). The colour of flour from this experiment was poor (caused by mould growth on prematurely ripened ears) and 1000-grain weight and bushel weight (formerly used as the equivalent of specific or hectolitre weight) were decreased by disease in both crops. Take-all also caused an increase in the protein content (smaller grains contain relatively more protein as the carbohydrate content decreases) and in diastase (barley grains only). In wheat, gluten was increased slightly, but not sufficiently to affect baking; there was little or no effect on α-amylase activity and protein quality and dough characteristics were unaffected.

6.5.3 *Recent research in the UK*

Grain yield

Oat breaks at different stages in sequences of wheat crops provide an experimental means of generating different amounts of take-all without affecting the nutritional status of the soil. In experiments at Rothamsted during the 1980s losses from take-all were 2.2 t ha^{-1} (3-year mean) for winter wheat (Prew *et al.*, 1986) and 1.0 t ha^{-1} (6-year mean) for winter barley (unpublished in full; 3-year means in Jenkyn *et al.*, 1992b) when compared with crops grown after oats. In years when take-all was severe the losses were 3.0 t ha^{-1} for wheat and 1.5 t ha^{-1} for barley.

Subsequently, interactions between agronomy, disease and yield were investigated in a series of factorial field experiments on winter wheat (Rothamsted, 1986–1989) (see 4.4). Large variations in amounts of disease enabled relationships to be determined between amounts of take-all infection and grain yield (Fig. 6.11) and quality (Fig. 6.12). The main factor affecting take-all was sowing date (September vs. October), and grain yield was affected additionally by plant density, a consequence of sowing date and seed rate.

In 1986, disease was slight to moderate and regressions of TAR in June on yield were not significant. In 1987, disease was moderate and accounted for 17% and 9% of variation in yields in early-sown and later-sown plots, respectively. In 1988, some plots were severely infected and the association between TAR and yield in early-sown plots was stronger (64% of the variance accounted for) than in the less severely infected, later-sown plots (14% of the variance accounted for). Disease levels were similar in 1989, but only 38% of the variance in yield was accounted for by disease. These regressions could be improved by accounting for a variable

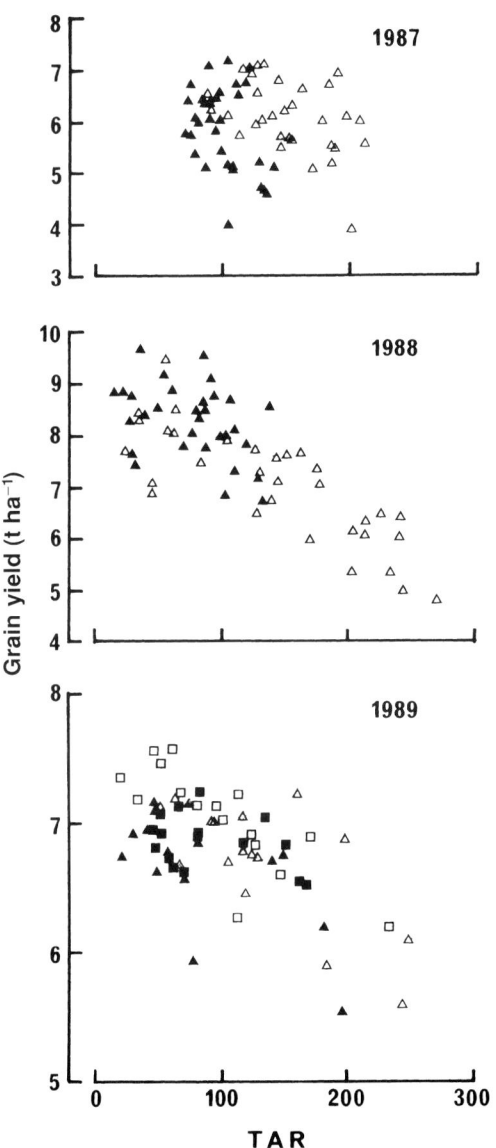

Fig. 6.11. Relationships between take-all rating (TAR) in June and grain yield in three factorial experiments at Rothamsted.
1987 and 1988: seed rate 200 kg ha^{-1}, sown September (△), sown October (▲).
1989: seed rate 100 kg ha^{-1}, sown September (▲), sown October (■).

yield response to increasing TAR, which was negligible at TARs less than 100. In each of the last 2 years of the experiments, yields of the most severely diseased plots were 35–45% less than those of least severely diseased plots, although because of the different treatments not all of this difference can be assumed to result directly from take-all.

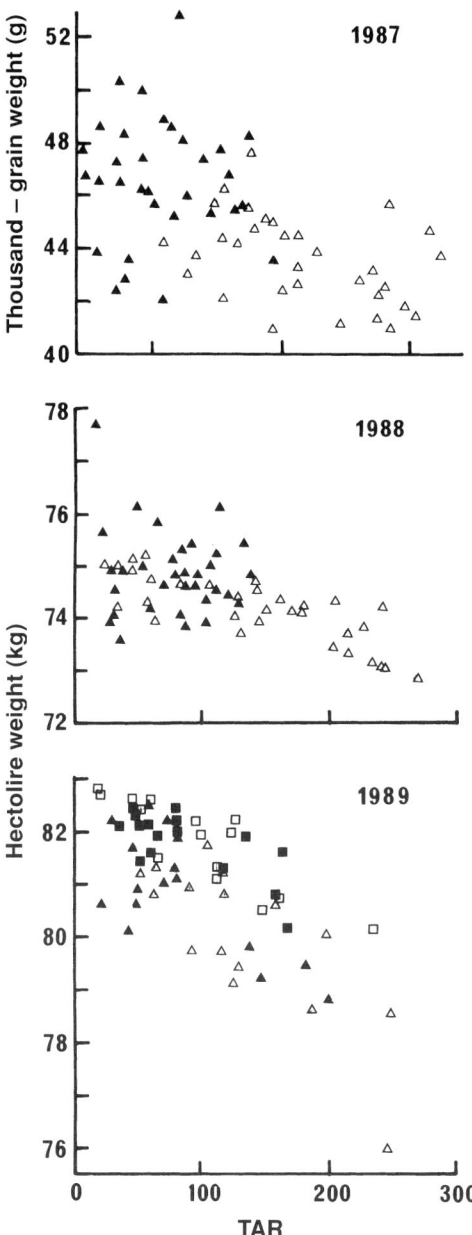

Fig. 6.12. Relationships between take-all rating (TAR) in June and 1000-grain weight or hectolitre weight in three factorial experiments at Rothamsted. For key, see Figure 6.11.

ADAS work in the years 1983–1986 revealed significant relationships between TI, assessed either during November–January or at harvest, and yield of winter wheat (R.W. Clare, I. Ap Dewi and D.J. Yarham, unpublished data also referred to in Table 6.4). The amount of yield loss per unit of TI was about twice as much for November–January assessments as for assessments at harvest.

The Rothamsted results and those of others emphasize some of the experimental requirements for studying yield loss in relation to take-all severity. These are the following.

- A small experimental area with a wide range of disease severities. Ways of achieving this have been discussed: applying artificial inoculum, experiments over several years using phased-in crop sequences, utilizing natural variation (patchiness).
- Avoidance of confounding factors on the site, e.g. different sowing dates, other diseases.
- Seasons in which conditions otherwise favour good yields. This is perhaps a contentious point, but it serves to emphasize that such seasons also favour take-all. Because responses at different sites, and probably also at the same site in different seasons, will not be identical, a more realistic approach would abandon this requirement and consider losses in potential yield (mentioned briefly in 6.5.1, 'Models for disease–yield relationships'; yield levels and associated constraints are discussed in Campbell and Madden, 1990, and Cook and Veseth, 1991) for sites and seasons.

Grain quality
The main effect of take-all on grain quality in the Rothamsted factorial experiments was decreased grain size. This was evident in the negative correlation between TAR and specific weight of grain (Fig. 6.12) in 2 years (1988 and 1989), which paralleled the relationship between TAR and grain yield in those years (Fig. 6.11). In the year with most disease (1987, average disease level in the moderate category), it was also evident in 1000-grain weights, which decreased with increasing TAR. This relationship was better than that between yield and disease in that year. As with yields, specific weights were not correlated with TAR in the year with least disease (1986). There was no clear relationship between take-all and percentage N in the grain, but there was some negative correlation between take-all and Hagberg falling number in one year (1988) when the Hagberg falling numbers were acceptably high. In the following year, warm dry weather continued throughout the ripening and harvesting period and Hagberg falling numbers were excessively high. These were not associated with take-all disease, which was only slight–moderate.

Increasingly, there are indications of similar effects of disease on winter barley (Jenkyn *et al.*, 1992a,b). In an experiment at Rothamsted take-all affected yield, particularly when disease was in the moderate to severe disease range (Fig. 6.13), and 1000-grain weight. Amounts of nitrogen per grain from barley crops with much take-all and from crops with little take-all were similar, although crops with little take-all produced more and larger grains and these were of better malting quality (Jenkyn *et al.*, 1991a). The regression of each of several qualitative and quantitative yield variables on TAR was investigated for winter wheat, winter triticale and winter barley in each of 7 consecutive years at Rothamsted. A summary in Table 6.7 shows that relationships were detected most often in wheat, that the pattern of relationships was different amongst the cereals and that some unexpected associations occurred in barley and triticale at low levels of disease.

6.6 Comparisons of Regional Approaches to Field Experimentation

Yield responses in wheat to soil fumigation, used over some 15 years in the Pacific Northwest of the USA as a research tool (Cook, 1994), were attributed mostly to the control of take-all where the crop was grown frequently. This approach led to important discoveries, especially about the antagonistic components of the soil microflora. In Germany, annual fumigation of soil with metham-sodium established that take-all was a major contributor to losses in rye yields and that rye may be more susceptible to the disease than is often thought to be the case (Roth *et al.*, 1984; Section 3.2.1). Fumigation has been used much less in the UK, where the emphasis has been on research on epidemics developing and declining naturally under normal farming conditions. Also, applying inoculum artificially is standard practice in many places – for example, for testing chemical controls in North

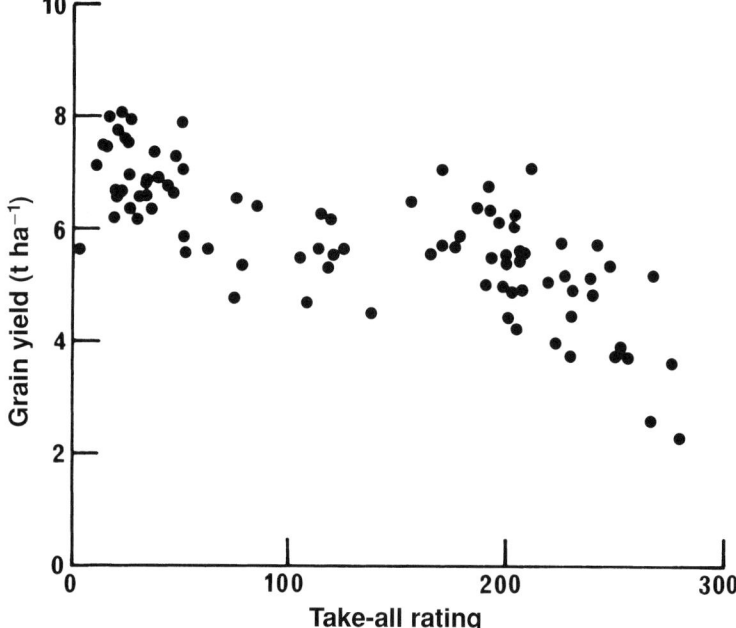

Fig. 6.13. Relationship between take-all rating (TAR) in June and grain yield in a winter barley experiment at Rothamsted, 1987.
Data provided by J.F. Jenkyn.

Table 6.7. Variables related to cereal grain quality or quantity and their association with TARs in regression analyses. The data sets analysed came from three different winter cereals, each grown in 7 consecutive years in a rotation experiment at Rothamsted. Consequently, each year represents the pooled results for several different runs of cereals, e.g. 1992 includes results from first, second, third and fifth consecutive crops.

	Winter wheat					Winter triticale					Winter barley				
	TAR range (maximum	% Grain (mm)			Yield	TAR range (maximum	% Grain (mm)			Yield	TAR range (maximum	% Grain (mm)			Yield
Year	0–300)	kg hl^{-1}	1–2.2	2.2–3.5	TGW	(t ha^{-1})	0–300)	kg hl^{-1}	1–2.2	2.2–3.5	TGW	(t ha^{-1})	0–300)	kg hl^{-1}	1–2.2	2.2–3.5	TGW	(t ha^{-1})
1989	0–142	–					0–41					–	0–48					–
1990	0–270	•				•	0–92					•	0–114					•
1991	15–300	•				•	0–243					•	5–274					
1992	63–286					•	4–272					•	36–280			•		
1993	41–295	•				•	29–200			•		•	16–210			•		
1994	11–137	•				•	0–97			•		•	0–88					•
1995	5–261	•					0–46			•		•	3–179			•		•

• Variance ratio probability ≤ 0.05 (i.e. null hypothesis of no relationship rejected).
– Not tested; TGW, 1000-grain weight.
Additionally, for wheat, % N (grain) was • in 1990, 1993 and 1994 and – in 1995; Hagberg falling number was – in 1989 and • in 1990.

America. This, too, has been less common in the UK, where only recently (as described above) has detailed evaluation of inoculation procedures and their effects been undertaken. Much of that detail has rarely been considered elsewhere.

Recently, take-all/yield loss relationships have been studied in Kansas, USA, using the new benzamide fungicide MON 41100 to control the disease (Headrick and Bockus, 1996). Yield losses of 5–6% for every 10% incidence of whiteheads occurred consistently over three seasons. Only about 3.5% yield loss occurred for every 10% of whiteheads in an experiment on nitrogen fertilizers in the UK (Gutteridge *et al.*, 1987).

The Future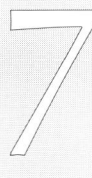

7.1 The Uncertainties

7.1.1 The cereal production system

At a time when the future of arable farming remains unclear, a prognosis for take-all in Britain is difficult. The size of the disease problem seems to be inextricably bound up with the proportion of cereals in rotations and it would be expected to change with any significant move away from current levels of intensive production of wheat and barley. A move towards more first wheats (i.e. less intensive production) has been in progress since the latter half of the 1980s. At the time of writing, winter wheat prices have been falling, prompting further debate about increasing the proportion of break crops in rotations at the expense of second wheats (Johnson, 1997). However, if intensification were to increase because of a large loss of agricultural land, or increased specialization in production, such intensive areas would be likely to experience an increase in the take-all problem. Paradoxically, a greater commitment to growing cereals continuously might ultimately see a decrease in damage caused by take-all, because of the creation of more widespread natural biological control (TAD). Current and proposed changes in agricultural practice, other than those dispensing with rotations that put cereals at risk, seem unlikely to create many new opportunities for managing take-all.

Growing a significant proportion of the nation's wheat in rotations where there is a high risk of take-all currently seems unavoidable, unless there is a major change in the way the grain requirements are met. The implications of less specialization and less intensive cereal production are not clear in this respect. The number of acceptable break crops is limited, the future of global wheat production in what appears to be a period of unprecedented climatic change is uncertain and a market turnround could result from the discovery of an industrial use for wheat or one of the major alternative crops. The implications and feasibility of replacing wheat with barley at periods of high risk from disease are not well understood and will depend on the relative profitability of the two crops. At the time of writing, barley could be as profitable as, or more profitable than, wheat on light soils where a good malting sample can be obtained. On heavy soils, where barley is likely to be of feed quality only, it is unlikely to become a profitable replacement, even for a feed wheat, let alone wheat grown for milling. Neither is such a move likely to afford any protection against a severe outbreak of disease on a return to wheat (see 3.2.1).

Table 3.3 shows how some aspects of the system have changed in the space of 3 years. Of course, not all the changes that will affect wheat production in the future can be foreseen. In 1996 Britain faced a food crisis arising out of the perceived risk of BSE (bovine spongiform encephalopathy) to consumers of beef and the consequent beef export ban imposed by the EU is still in effect at the time of writing. There appear to be implications for cereals. A reduction in cattle numbers that might follow the proposed slaughter policy, or diminished demand, could mean less barley required for cattle compounds and, as a consequence, a bigger surplus of barley. If, as a result, pig and poultry production were to expand, the demand for feed wheat would grow (Clark, 1996). This potential influence on wheat production was not foreseen and any real effects it may have had remain to be evaluated.

There are changes occurring in the environment and these could have repercussions for take-all. Sulphur deficiencies in cereals in the UK are rising as atmospheric deposits decrease and although the main concern is with wheat quality there may be implications for take-all. It was predicted that 25% of the UK's wheat crop would be susceptible to sulphur deficiency in 1996 (Blake, 1996b).

Just so much can be done in the laboratory and then practical ideas need to be tested in the field. Field work is time-consuming and reliable results are rarely obtained in the space of one season's experimentation. Herein lie the reasons behind much of the confusion and controversy that have surrounded take-all research. Trying to produce timely answers to current questions is almost impossible. Frequently, by the time adequate experiments have produced reliable results, circumstances have changed or the topic is no longer current and interest in the hard-won results has waned or evaporated. Attempts to assess the interaction between set-aside and take-all exemplify this point. A challenge for the future is to find ways out of this dilemma. Breakthroughs may come from developments in fields such as molecular biology and biotechnology. Other, more conventional routes could be taken deliberately now. Developing long-term core research would provide a strong basis against which to consider ephemera. More research could be brought into the realms of computer modelling, but this will require a sufficient grasp of the system to model it realistically (this point was discussed in relation to PatchMaker in 2.6.7).

7.1.2 Climate change

The effects of weather on take-all and the importance of take-all in different regions of the world where cereals are grown were discussed in Chapters 2 and 3. Considerable attention has been paid among scientists and in the media in recent years to the prospect of global warming. Some of the effects of global warming on both crops and disease might be deduced from what is known about their environmental requirements. Climatic change scenarios indicate that wheat yields will increase considerably in northern Europe, but not in southern Europe (Harrison, 1996). A programme of disease monitoring in eastern Germany between 1977 and 1990 provided data for regression analyses to calculate the consequences of hypothetical climate changes (Jahn *et al.*, 1996). The effects on take-all (Table 7.1) were calculated as being much less than on some other cereal diseases such as rusts, although it is not stated how disease incidence was measured. The calculated values for take-all and other diseases were claimed as valid only for the continentally influenced inland climate where the measurements were made. Such calculated effects might be different for the UK, because the present climate is different from that in eastern Germany. However, one assumption is that the climate in the UK will also become warmer and drier (although cooling may occur if the Gulf Stream fails), and so take-all may decrease. Hornby (1990a) suggested that a weak trend towards wetter soils might have contributed to increased disease at a site at Woburn in the 1980s.

It is inevitable that climatic change will affect the disease in some way, but it is not possible to predict how. Several possible scenarios come readily to mind. An increase in mean temperature might favour soil organisms antagonistic to Ggt, thus reducing the incidence of take-all. Alternatively, were winters to become milder, early infection by Ggt might become more of a problem. If winters became wetter, heavy soils could be waterlogged more frequently and the resultant poor rooting would lead to plants less able to withstand root loss caused by disease. Increased soil moisture early in the season might also favour infection. An early onset of summer drought could check infection at a time when plants were still able to replace diseased roots and so lead to decreased disease severity. Perhaps the worst scenario for increasingly serious losses from take-all would be a change to wet weather persisting later into the spring, followed by drought conditions.

Table 7.1. Factors by which 'infestation' by take-all is expected to decrease for the given climate changes.

Increase in temperature		Decrease in rainfall		1°C increase in temperature plus 30% decrease in rainfall
1°C	2°C	30%	60%	
0.82	0.69	0.65	0.35	0.52

Effects of climate change on take-all based on data collected in eastern Germany with reference to weather data from the meteorological station in Potsdam. Each change in conditions is an average over a whole year.
Source: Jahn *et al.* (1996), in which temperatures were given in kelvin units.

7.1.3 *Unforeseen developments and applications*

As plant pathology evolves and benefits from application or adaptation of new developments in other disciplines, new vistas open up and the means become available to re-examine old ones critically. Some of these may not be in the mainstream of expected or predicted developments. The two examples that follow perhaps exemplify this.

New approaches to chemical control

Chemical control of take-all has been a long time coming, for reasons discussed in Section 4.2. However, the promising seed treatment fungicides now under development, at least one of which makes use of novel fungicide chemistry, may not be the only solutions to this problem. An approach that seems only recently to have received consideration in relation to take-all concerns the use of elicitors of systemic acquired resistance (SAR) in the host plant. Cross-protection, or immunization, is one form of resistance that has never come to the forefront of UK research on take-all. Now, using molecular biological techniques to study cross-protected and non-protected plants, it may be possible to detect resistance-related substances which could be exploited.

In nature, a local infection with a necrogenic pathogen triggers the release of an SAR signal, resulting in the protection of remote parts of the plant from further infection. Salicylic acid is required for transmitting the SAR signal systemically and SAR is usually associated with an increase in pathogenesis-related (PR) proteins. The first accounts of SAR concerned resistance to viruses induced by the same or different viruses (Ross, 1961), by derivatives of salicylic acid such as aspirin (White, 1979), or by other chemicals (Kessmann *et al.*, 1994; Lyon *et al.*, 1996). The subject was reviewed extensively in the last two papers cited.

A screening programme by Ciba-Geigy led to the identification of small molecules that activate the SAR response. Of these, benzo[1,2,3]thiadiazole-7-carbothioic acid S-methyl ester (CGA 245704) has been developed for commercial use (Ruess *et al.*, 1995, 1996). It seems that this compound can replace salicylic acid without the need for prior infection (Kessmann *et al.*, 1996). It is possible, therefore, that a wide range of diseases could be subject to some degree of control by this means.

Seah *et al.* (1996) compared the effects of applying salicylic acid to wheat seedlings by different methods (foliar dip, root drench or pre-germination soak) on take-all arising from inoculum of Ggt grown on infested ryegrass seeds. No treatment induced resistance in wheat roots. The authors pointed to evidence that some degree of host resistance, absent in the wheat–take-all pathosystem, is required for SAR to operate. In research at Rothamsted on infecting barley roots with Ggt (Schlichter, 1997), there was little evidence for gene expression being induced at high levels in heavily infected roots except in the case of the gene encoding for a cysteine proteinase. (A similar gene was induced in wheat leaves after chemical induction of SAR; Görlach *et al.*, 1996.) There was very little expression of this gene in uninfected roots, and most in the most severely diseased roots. Some expression in roots with only very slight disease was thought to suggest systemic expression limited to the roots, because there was no expression in leaves, even after powdery mildew infection. These results raise the question of whether there are fundamental differences in SAR in roots and shoots.

Expression of the gene, though weaker, was also detected after infection of barley roots with non-pathogenic *Polymyxa graminis* and after wounding. Nothing is known about how this proteinase might be involved in a defence reaction. If it is, it might operate directly on Ggt at the site of infection, breaking down fungal proteins that release elicitors that trigger defence responses. Alternatively, it might provide free amino acids for the synthesis of other defence-related proteins. Also, if it were induced by non-pathogenic *Phialophora* spp., and was effective as suggested above, it could be involved in their ability to protect against take-all. Such questions are the basis of continuing work and the subject of SAR is clearly one for future research.

There may also be chemicals with yet other novel modes of action that might also be sought, such as those that alter the plant structurally to increase resistance (see 4.2.3).

Viruses and virulence
Although viruses, or dsRNA, have been associated with reduced virulence in some pathogenic fungi, this has not been demonstrated for viruses found in the take-all fungi. Interest in this peaked in the 1970s and early 1980s (Buck, 1986) and has now declined to a negligible level. However, it is possible that only certain individual dsRNAs or specific combinations of them have an effect on virulence, but this has not been studied thoroughly. Much of the previous work was done when analytical techniques for studying dsRNAs were laborious, but new molecular methods are easier. For instance, the presence of dsRNA particles once required electron microscopy following ultracentrifugation to concentrate large mycelial samples. Particles of different sizes are now easily detected by gel electrophoresis of relatively crude DNA samples from small pieces of mycelium and can be identified further by DNA probes.

Precision farming
Although precision farming (also known as computer-aided farming systems, or CAFS) is a relatively new phenomenon for arable farmers, yield-mapping has already proved to be an invaluable management aid in determining future input strategies (Fenton, 1996). Opportunity may arise in the future to bring together expertise in remote-sensing plant diseases (using infrared aerial photography), yield-mapping, and plant disease assessment and epidemiology in order to improve the control and management of crops and their diseases and the prediction of yield shortfalls. Using take-all as a model it would be possible to apply the new technology in the form of combine harvesters equipped with the Global Positioning System and yield loggers on diseased fields. Developing and exploring this system might establish the extent to which damaging disease extends beyond visible patches, and the strength of correlations between disease maps produced by remote sensing and yield maps. The possibility of future rapid diagnosis techniques for use in the field, particularly ones that could be automated, and control treatments for take-all would open up further avenues. This is speculation for take-all, but already research on semi-automated weed-mapping and computer-controlled herbicide application has been undertaken (Rew *et al.*, 1996). The potential reduction in herbicide usage as a result of patch spraying varied with patch morphology and infestation level.

7.2 Resources

Behind the perceived intractability of take-all is a complex problem, which has not yielded to inadequate and fragmentary investigation. Experience has shown repeatedly that isolated exercises are inefficient and unlikely to succeed. The formation of the IACR/ADAS/Universities Cereal Root Pathology Group in 1988 (details in the Preface) was recognition that in the face of dwindling resources it is important to pool experience and effort to achieve progress. It is still true, but now more difficult to achieve with current attitudes to agricultural research, that more attention should be given to coordination of research and collaboration within the UK, and further afield if possible. Such endeavours would need careful targeting at what are identified as the major practical and research issues by a consensus of those concerned.

The build-up of take-all and development of TAD in experimental sites takes several years. The long-term field experiment has a vital role to play in take-all research and should not be underestimated. Experiments with phased-in sequences of cereals, that provide comparisons of different runs of susceptible cereals each year, are extremely valuable scientifically but relatively expensive and uncommon. Recently in the UK, some well planned, well documented long-term experiments on take-all have been discontinued, or data collection has been terminated as a result of cut-backs. Unfortunately, such important resources, once discontinued, are not readily replaced.

7.3 Research

In the HGCA Review (Hornby and Bateman, 1991), the forerunner of this book, recommendations for research were made and classified as high, medium or low priority for achieving objectives. The objectives were grouped in four major research areas:

- Establishing the importance of take-all.
- Improving forecasting and risk assessment.
- Understanding take-all biology.
- Controlling take-all.

Much of what was recommended for initiation and expansion of projects, and the appropriate organizations for undertaking the work, has been rendered of historical interest only by the passage of time and changes in funding. However, the objectives and priority ratings largely remain valid and are given below with explanatory text. Forecasting, risk assessment and control obviously have more immediate appeal to the farmer. However, it cannot be stressed too strongly that work in all the areas is required if the complex problem of take-all is to be understood sufficiently to ensure critical research and to improve chances of significant progress in controlling the disease. Some topics such as resistance are long-running themes, others such as hypovirulence and TAD, self-inhibitors, differences in decline and non-decline populations of Ggt recur periodically. Finding new members of the *Gaeumannomyces–Phialophora* complex has exhibited a regional shift from being essentially UK orientated before Asher and Shipton (1981) to involving subsequently researchers as far afield as China, South Africa and the USA.

7.3.1 Establishing the importance of take-all

Surveys (high priority)
There has been only one extensive survey in the UK in which root symptoms of take-all have been assessed (see 6.2) and that was undertaken in the late 1970s in years when levels of take-all were not generally high. Although patchiness has since been assessed in national wheat disease surveys, it is too crude an approach for establishing the national importance of the disease, its importance on different soil types at different pHs and under different climatic conditions, and the effects of rotational position and agronomic practices (e.g. cultivation, cultivar and sowing date).

Continuous winter wheat on a sandy clay loam in North Yorkshire seemed not to be affected by soil-borne diseases in the years 1979–1985 (Hodgeson *et al.*, 1989). Such reports draw attention to the perennial question of localities and soil types in Britain which appear not to suffer from take-all. Usually the reports contain insufficient information about disease, previous cropping and typical crop sequences for the district. Consequently, the disease situation remains in doubt and if little or no disease is the norm, clues as to the explanation are few.

A survey of take-all should use an agreed procedure based on the best information available. This may require a pilot study to optimize sampling and data collection strategy, which in turn might also provide the guidance on sampling take-all sites that is so badly needed (see 6.3.2). The simplest survey would also need to record take-all and yields in crops throughout the country in a series of contrasting seasons. The possibility of linking such a survey to the existing ADAS cereal disease survey should be considered. A structured national survey over a number of years to provide qualitative and quantitative information about take-all should include supporting agronomic and weather data and the weather should be recorded during all phases of cultivation (e.g. harvest of the previous crop, sowing) and stages of crop growth (e.g. seminal root development, crown root development, cessation of root growth and grain filling).

The absence of fully reliable survey data to quantify the extent of take-all and the losses in yield that it causes has been keenly felt by those making application for funding in the public sector research, or contemplating the development of commercial control products. Agricultural chemical companies must inevitably be reluctant to make the enormous investment necessary to develop a fungicide to solve a problem whose extent is uncertain. Even so, such companies have been as unwilling as the public funding bodies to finance surveys.

Disease–yield relationship (high priority)
This relationship is not well understood and is usually confounded by other factors. Many empirical disease–yield relationships lack repeatability and have poor

predictive powers. There is a need for alternative mechanistic approaches (Teng, 1985) using crop growth models and simulators incorporating the effects of disease on root and shoot function. In considering alternative models the relationship between the effects of disease on the yield of individual plants and on plant populations may need to be determined in advance. Quantification of the spatial distribution of disease in relation to yield loss is now of interest (Ferrandino, 1989; Hughes, 1990), notably amongst nematologists (Seinhorst, 1973; Noe and Barker, 1985). The use of artificial infestation to create artificial epidemics may be a useful exploratory tool, with the potential to eliminate patchiness and, if the inoculum is uniform in size and infectivity, to achieve reproducible levels of disease. Many factors will influence the effect of a given level of disease on yield through their effects on the vigour of the root system. Consequently, information will be needed for a range of soil types over a number of years and agronomic factors such as nitrogen, fungicide (seed treatment, etc.) and method of cultivation. Much of this could be obtained from monitoring trials laid down to explore effects of these factors on disease incidence. Other approaches would be: (i) the use of large numbers of small plots (half of each to be destructively sampled to assess disease and half harvested) scattered throughout areas of interest; and (ii) harvesting the grain from single tillers with known levels of disease.

Diagnosis (medium priority)

Take-all research continues to be hampered because of the time taken to achieve reliable diagnoses by traditional methods. Because of the similarities between Ggt and some of its close but less pathogenic relatives, it is also quite possible that where Koch's postulates are not carried out fungi may be wrongly identified. New diagnostic methods are much needed and amongst the most promising areas of research in need of support is the use of serological and molecular biological techniques to identify and quantify the pathogen. Where there has been progress (see 5.5.2–5.5.5), it is now necessary to explore how such techniques can be developed for routine use in the surveys suggested above.

Economic evaluation (high priority)

Information on the effect of take-all on the profitability of various rotations on different soil types is needed for the proper evaluation of the financial constraints the disease imposes on the flexibility of cropping. A paper exercise using information from a national survey and work on yield loss prediction would indicate for a range of soil types the profitability of second to fourth wheat crops compared with a range of break crops. Data from the ADAS Arable Crop Recording System should form the basis of another desk study to evaluate the effects of soil type and agricultural practices on the comparative yields of first and subsequent wheats, which by inference would indicate the importance of take-all. An economic survey of intensive cereals in the UK should also consider whether the present intensity is unavoidable, or whether adjustments in rotations could be made to minimize take-all.

7.3.2 Improving forecasting and risk assessment

Data storage and availability (high priority)

Much information on take-all in field experiments, trials and surveys, collected annually, has not been generally available. Some of it is idiosyncratic and would have benefited from conforming to some general standard for the collection of take-all data. To overcome these inefficiencies the establishment of a central database and data archive was proposed. A pilot study, carried out by Cambridge University and IACR-Rothamsted under the AFRC Linked Research Group Scheme a few years ago, was based primarily on a long-running Rothamsted cereal experiment. It proved very useful (Werker *et al.*, 1991) and such a project would be ideal on a national basis, helping to systematize the collection of data and providing a powerful tool for re-examining data and trying out new hypotheses.

Forecasting (medium priority)

The question of forecasting take-all is a vexed one of long standing. Although changes in the cereal production system have contributed to the difficulties of forecasting, it is the relationship between weather and the disease that is at the centre of the problem. In the 1980s and early 1990s more long runs of reliable take-all data began to be available in the UK. (Regrettably, two major sources of such data have since been terminated in cut-backs at Rothamsted.) These data need investigating in relation to detailed meteorological data which are available for many localities. Initial analyses revealed long-term trends in the occurrence of take-all and such understanding is vital to predicting national losses over time. Much work on weather relationships is necessary before the state of explaining or forecasting disease trends within crops advances beyond the rudimentary. However, the importance of this work is likely to increase because of: (i) the debate about climatic change and its implications for take-all and the profitability of wheat under UK conditions; (ii) the possibility that in the near future farmers will be faced with making decisions about buying seed treated with a fungicide (see 7.3.4, '*Fungicides*'), which may be expensive, to control take-all; (iii) the possibility of new methods of assessing risk in subsequent crops, such as one based on observations of DNA types of Ggt in field populations (see 5.6).

Agronomic and edaphic factors (high priority)

Further research is required on the relative importance and interactions of agronomic and edaphic factors in the development of take-all. Of the few factors that both influence take-all and are under the farmer's control, sowing date is the most important. The use of cultivars least sensitive to sowing date (Sylvester-Bradley and Scott, 1990), including rapid-developing, autumn-sown spring wheat cultivars, which allow further delay in sowing date without yield or quality penalties, needs to be investigated in situations where the risk from take-all is high. Information restricted to a list of sowing dates, cultivars and yields limits interpretation, and trials including other factors are needed to decide on a cultivar's suitability for early or late sowing (Sylvester-Bradley and Scott, 1990). Trials with second and third wheats on sites prone to take-all would, with sufficient disease monitoring, allow interpretation of sowing date × cultivar interactions in the presence of significant take-all.

Further information is required about the relationship between water tension (in the soil and in the plant), take-all development and root growth. Because the major effect of root loss due to take-all is the restriction of water uptake by the host, differences in the ability of different cultivars to withstand moisture stress may be important. More information is required also about the long-term effects of straw incorporation for a range of soil types and on the possible effects of recently introduced herbicides on crop susceptibility to take-all.

7.3.3 *Understanding take-all biology*

Epidemiology (high priority)
There is much in the epidemiology of take-all that remains unknown. This hampers the proper interpretation of most field work on the disease and the realistic extension of laboratory and glasshouse studies to the field. Consequently, epidemiology must not be overlooked or under-resourced. Further work (including work on artificially created epidemics) is essential for progress in predicting epidemics, better risk assessment and proper testing of putative controls. Work on models for spatial and temporal development of disease will aid understanding.

Field work methodology (medium priority)
Take-all occurs naturally in patches. This heterogeneity has meant that disease in many traditional large plot experiments has been so variable that the precision of the test of any treatments has been low. Various ways of tackling this problem, such as many small plots or more even artificial infestation of soil, have been tried. However, new designs and procedures for field work are still required. Manipulation of, and dependence on, natural inoculum in field work have drawbacks. There is, for instance, no certainty that epidemics will develop and, if they do, the relationship between disease and loss of yield often is not clear-cut. There is doubt that artificially produced inoculum produces epidemics which simulate those occurring naturally and this could be misleading, e.g. when using artificial inoculum to test BCAs. The development of disease patches, naturally and artificially created, needs further study in order that the best use can be made of field sites for experimentation.

Gaeumannomyces–Phialophora *complex (medium priority)*
Over the last 25 years the number of fungi in this complex has steadily increased and several outstanding and intriguing taxonomic problems remain. Some members are pathogenic, potential BCAs, or may be mistaken for the pathogen, whereas others do not appear to be pathogenic or to complicate the pathology of the take-all fungi. Clarifying the relationships amongst these fungi will benefit from the continued development of serological and molecular biological diagnostic techniques. The following example illustrates this approach in action at Rothamsted.

Five of six isolates of *Phialophora*-like fungi from maize plants (two from India and two from Egypt, in culture for 11–13 years, provisionally identified as

Cephalosporium and subsequently as *Phialophora* – J.M. Waller, personal communication – and two from China, identified respectively as Ggg and *G. graminis* var. *maydis*) had *Gaeumannomyces*-like colonies on PDA. On maize roots the same five produced dark runner hyphae and swollen terminal vesicles that were smaller than those of typical wheat isolates of Ggg produced on wheat roots. One of the Egyptian isolates was the exception which showed none of these characteristics and had probably degenerated in culture. All six isolates failed to produce perithecia when infected roots were rotted down in test tubes and all reacted positively in a PCR test diagnostic for *G. graminis*, the Indian and Egyptian isolates appearing identical to Rothamsted Ggg standards. Using a mitochondrial DNA probe (pEG34), all isolates produced a pattern similar to Ggg, except for *G. graminis* var. *maydis* which produced a pattern like typical British Ggt. Restriction digestion of PCR-amplified ribosomal DNA showed the four isolates from India and Egypt to be the same (as in the other tests), but with patterns unlike those found previously for *G. graminis* varieties. In this test the Chinese isolates (Ggg and *G. graminis* var. *maydis*) reacted like British Ggt. An unusual 'Ggg' isolate from soya bean and others from Australia had reacted similarly.

Ecology of pathogens and antagonists (medium priority)
Ecological studies of the pathogens and potential BCAs on the root surface and in the soil are needed. Related practical issues are: (i) delivery of BCAs to the root zone; (ii) enhancing the natural suppressiveness of field soil by such procedures as incorporating organic matter; and (iii) the use of root-stimulating bacteria to alleviate the effects of disease even though not suppressing infection.

7.3.4 Controlling take-all

Rotations (medium priority)
This topic has been the subject of much research, but many questions remain to be answered and older work needs to be repeated under the conditions of modern husbandry. More information is required on the following.

- Effects of different break crops on the survival and subsequent build up of Ggt.
- Effects of introducing different break crops into long runs of wheat in which TAD is well established.
- Effects of different break crops on the yield of first and second wheats.
- Effects of manipulating the husbandry of a preceding wheat (e.g. delaying the sowing date) on the effectiveness of a break crop in decreasing disease incidence in a following wheat.
- Effects of less susceptible cereals such as triticale and rye on the development of TAD in cereal sequences and the effectiveness of any TAD that develops in protecting subsequent crops of wheat.
- Effects of the sowing dates of sequential wheats on the development of TAD.
- Effects of the various systems of management of set-aside on take-all.
- Effects on take-all of catch crops introduced to decrease the leaching of nitrogen.

It would be prudent to consider long-term experiments aimed at accumulating information about the effects of potential future changes in wheat production, such as lower inputs and organic systems. Cuts in subsidies and changes in consumer demands may cause control strategies to shift in the long term.

Natural biological control phenomena (high priority)
Detailed assessment of biological and other factors implicated in TAD in one site may help to establish which control mechanisms predominate during TAD, and when. Following variations in soil suppressiveness through seasons should help to establish when the phenomenon is strongest. Research on how to exploit TAD should also examine how early sowing affects the phenomenon. Such insights are relevant to, and possibly essential to, the development of effective uses of biological control for soil-borne diseases of field crops.

The hypothesis that interactions between different suppressive (natural biological control) phenomena occur (see 4.3.2) has received little or no attention from take-all researchers. Figure 7.1 is a simple scheme showing how suppression caused by *Phialophora graminicola*, a fungus that can develop under grass leys, may delay the onset of TAD. There is a small amount of circumstantial evidence of interactions between the two phenomena other than delayed onset of TAD in experiments done at Rothamsted (Gutteridge and Slope, 1978). In one experiment the take-all peak after a grass ley was decreased, as well as delayed (see 4.3.2),

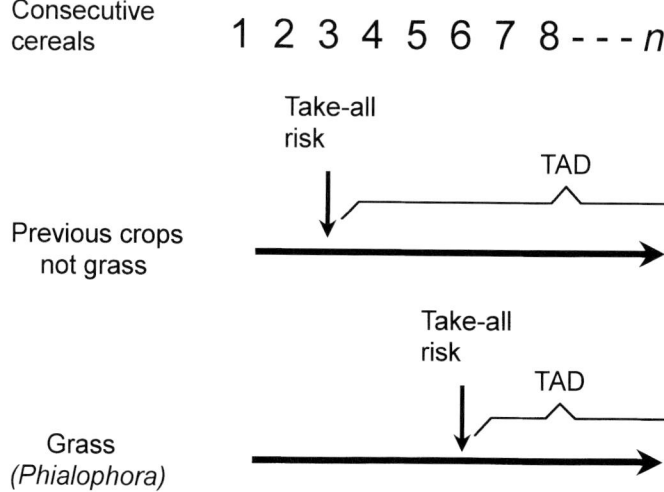

Fig. 7.1. A scheme showing two biological control phenomena that protect against take-all.

In the upper cereal sequence, TAD occurs in consecutive crops after a severe attack of take-all, usually resulting in an unprofitable crop, which is the price paid for achieving the protection that TAD subsequently affords. The phenomenon is strongest in consecutive crops of wheat. In the lower sequence, '*Phialophora*' delays the onset of severe take-all through the cross-protection activity of the fungus, *P. graminicola*, carrying over from previous grass. It is possible to have both phenomena in one cereal sequence (as shown in the sequence after grass), but apart from the delayed onset of TAD little is known about the nature or magnitude of the interaction.

which on the face of it seems to indicate that subsequent TAD would be less effective, because TAD tends to be most effective when it follows the greatest peaks of severe take-all (see 4.3.1). However, in another experiment in pots, the addition of *P. graminicola* to different soils resulted in more infections by *P. graminicola* of plants grown in continuous wheat soil showing TAD, than of plants grown in soils from shorter sequences. This greater colonization of roots in continuous wheat soil also occurred in the field at a lower level of infection. It is possible, therefore, that the greater population of *P. graminicola* could compensate for the possible decreased effectiveness of the subsequent TAD. If these ideas are valid, it might be possible to grow sequences of wheat crops with less penalty (yield loss) than is usually incurred at peak take-all, without losing the advantage of a subsequent, effective TAD. Perhaps this has already been done unknowingly in the past.

Introduced BCAs (medium priority)
Although the introduction of specific BCAs has met with little success in the UK, the search for new ones and the assessment of combinations and formulations and their evaluation under a range of conditions should continue. The prospects for their integration with other control methods should be explored. The view that the search for BCAs should be largely the province of agrochemical companies is proving unrealistic. Without public awareness of, and education in, the nature of biological control, the industry is not going to develop BCAs it knows it cannot use (Hemming and Houghton, 1993). Also, companies are unlikely to undertake the wide-ranging studies that still need to be done (see the relevant sections of 4.3.) and which are likely to materialize only with continued involvement of the public sector. Indeed, in the early 1990s, projects of the Monsanto Co. involving field tests of genetically engineered and non-engineered pseudomonads in the USA and Australia fell victim to economic recession and a realignment of strategy towards biotechnological applications in the pharmaceutical area (Hemming and Houghton, 1993). Those experiments had been primarily to evaluate new marked strains of pseudomonads having the potential to suppress take-all. Although in earlier experiments wheat roots had been well colonized by the bacteria, there had been no significant suppression of the disease. From the industry's point of view, further development was required to achieve a product for the biological control of take-all. High-resolution image analysis for more rapid quantitative analysis of disease and quality control in the manufacture of a product (e.g. microencapsulated bacteria) were needed (Hemming and Houghton, 1993). Modifications of conventional aminopeptidase profiling were yielding sensitive and extremely rapid methods of differentiating and monitoring recombinant organisms. Such technological developments need to be accompanied by scientifically sound and politically stable regulatory and registration policies. Figure 7.2 attempts to illustrate some aspects of the situation that has created a biological control credibility gap that continues to discourage industry from a greater involvement.

Resistance (low priority)
Public sector research in the UK in this area seems to have been discontinued from the time the Plant Breeding Institute became part of the AFRC Institute of Plant

Science Research in 1985, prior to its commercial activities being privatized in 1988. However, it would be prudent to continue to look for resistance to take-all in new wheat genotypes produced by traditional (breeding) methods. Since the British view is that there is currently little of interest to work on, this is an activity awaiting developments. The development of genetically engineered cultivars with resistance to disease would offer a new and exciting method of control which, should it materialize, would need evaluating in the field.

Before 2000 it is likely that hybrid wheats, developed predominantly from using chemical hybridizing agents, will enter the registration system at an increasing rate, particularly in countries such as France (Edwards and Dorlencourt, 1994). One prediction is that they will be in the hands of growers in the UK by 1998 (Abel, 1995). In France and Spain hybrid wheats have yielded 12–17% more than

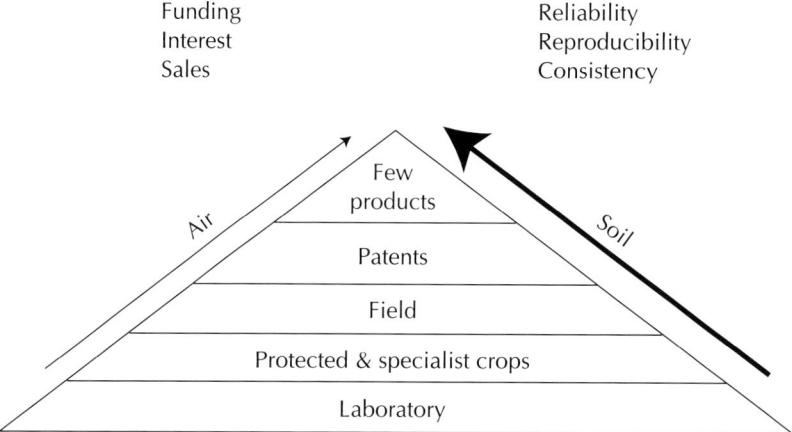

Fig. 7.2. A scheme indicating relative achievement in seeking BCAs at different levels of environmental complexity and commercial reality and some of the current problems.

In the triangle the consecutive layers indicate decreasing likelihood of success at the different levels of biological control endeavour, i.e. successful results are most likely in laboratory studies. Principles and mechanisms of biological control probably do not differ greatly between soil-borne and air-borne diseases, but soil-borne diseases receive the most attention and are the better known, e.g. bacteria to control crown gall, mushroom blotch or take-all. A multitude of laboratory studies, by and large, feed the optimism for biological control. Many concern antagonism on agar plates, or in simple systems in pots, which, although attractive in that they provide promising results for publication, rarely work when applied in the field. Sophisticated techniques to elucidate the mechanisms of antagonism between organisms, where that antagonism has been demonstrated in these simple systems, may have little relevance to more natural systems. Moving from laboratory studies to controlled environments, glasshouses and finally to fields presents a reality of increasing complexity. Successful examples of introduced biological control and management of suppression are negatively related to that sequence. Also, many interactions that are reported to decrease disease do so only at low or moderate levels of disease. At the apex are a few tens of products (Faull and Powell, 1995; Van Driesche and Bellows, 1996) based on a lesser number of organisms, available to growers for the biological control of plant diseases. These form a minute proportion of the biological products market and are mostly for protected and specialist crops. Lack of interest and support from industry for potential products that would lack regular sales, problems of reliability of current BCAs and reproducibility and consistency of the biological control effect are some reasons for emphasizing the flatness of this triangle.

leading check cultivars (Edwards and Dorlencourt, 1994). The current paucity of incentives to grow high quality wheat and the current grain prices may initially limit their use to the better growing regions under intensive management. Since hybrids provide 'a more flexible and rapid means of responding to a new pest than conventional breeding procedures', the implications for take-all need to be assessed.

Fungicides (medium priority)
Intimations of new seed treatments (see 4.2.2) have raised the profile of this topic since its previous low priority rating in Hornby and Bateman (1991). In July 1996, the CRPG visited a site in Norfolk to see a large strip trial of a benzamide fungicide, MON 65500, which when judged on crop patchiness appeared to be decreasing take-all. (Work involving another formulation, MON 41100, is described in 4.2.2, '*Seed treatment*') Following current comprehensive efficacy trials in several European countries, Monsanto expects a new benzamide fungicide seed treatment for take-all to be available to growers in 2–3 years (P. O'Reilly, personal communication).

Clearly, if such chemicals fulfil their promise, it would represent a giant step forward in the control of take-all. Should they achieve partial control only, particularly in the later stages of the crop, then the farmers' reaction to still seeing symptoms after treatment would become a factor in their acceptance, and the possibility of large amounts of soil-borne inoculum following a treated crop would need to be considered. However, until there is authoritative release of more detailed information, it would be premature to dismiss the statement that no fungicide has given consistent, satisfactory and economic control of take-all. Certain triazoles (particularly triadimenol seed treatment) have sometimes suppressed infection, have occasionally seemed to increase disease severity and have sometimes increased the yield from severely infected plants without any obvious effect on root infection. The physicochemical properties of the ideal soil-applied fungicide for take-all have been determined in government-funded research, but synthesizing and developing new fungicides of this type would be more appropriately in the province of the agrochemical companies. Field trials by others will always be required if and when promising fungicides and delivery systems emerge. This has already been happening with the new seed treatments, but the details are not generally available at this stage, presumably for commercial reasons. The value of a treatment cannot be assessed on the basis of one or a few seasons of trials, and its performance (should it come onto the market for controlling take-all) should be monitored throughout its period of use – more particularly as the development of fungicide resistance by the pathogen cannot be precluded.

Integrated control (medium priority)
It is unlikely that any single material or technique will ever give a satisfactory and consistent control of take-all. Successful disease control strategies will almost certainly depend on combining treatments that favour root development and treatments that act against the pathogen. Some work has already been done by IACR, ADAS and INRA to identify best 'packages' based on current treatments and

procedures, but such work should be ongoing as long as new factors are identified and promising new treatments emerge.

Attempts to get integrated control by combining the methods that are currently the best would include re-examining the use of fertilizers, such as ammonium sulphate, to control take-all. Non-nitrate sources of nitrogen, nitrification inhibitors, chloride-containing fertilizers and soil-applied manganese have decreased take-all in Australia and USA, but not in experiments in the UK. They and the currently available phloem-mobile formulations of phosphate for foliar application need more careful evaluation over a range of UK soil types. It will be necessary to look more closely at effects of different sources, timings and rates of nitrogen under UK conditions. Some outstanding questions are as follows.

- What is the range of UK soils on which the use of ammonium sulphate is likely to be beneficial?
- Would the use of a urea-based product with a nitrification inhibitor offer a possible alternative?
- Can nitrogen rates be adjusted to take account of take-all risks?
- Are disease levels in late winter or spring a guide to making such adjustments?
- By how much should our standard recommendations be increased if take-all levels are high?

7.4 Summary

Take-all represents a major challenge in plant pathology: (i) because of the losses it causes and the constraints it imposes on rotational practices; (ii) because, being caused by a soil-borne, root-infecting fungus, it has not responded reliably to conventionally applied fungicides; and (iii) because no important cultivar resistance exists. Much research on many fronts over many years has contributed to take-all becoming an important model for the study of soil-borne diseases generally. Research from abroad on the use of bacteria as biological control agents and on particular forms of nitrogen-containing fertilizers and chloride-containing fertilizers to control take-all has been much publicized, but such treatments have not performed well in British farming conditions. Regional differences need to be recognized and understood when evaluating and applying research findings from elsewhere.

The geographical distribution of take-all is determined largely by climate, as are the places where its host cereals are grown. Its incidence and severity are determined principally by the proportion of susceptible crops in rotations and by edaphic factors such as soil type and moisture content. Other choices and practices over which the farmer has some control (e.g. sowing date, methods of cultivation, application of fertilizers and the nutritional status of the soil) also affect the disease but usually to a lesser extent. Because damaging take-all does not normally occur in a susceptible crop (wheat, barley, rye or triticale) that is grown after a non-susceptible crop in a rotation, the disease could mostly be avoided, but there are often economic reasons for such rotations not being practised. Therefore the risk of take-all imposes constraints on the husbandry practices of farmers growing sequences of cereals.

Potential additions to the scant armoury of control measures are fungicides and BCAs. Commercial chemical seed treatments are undergoing field trials, but suitable fungicidal compounds for use in the complex soils found in the UK are not available at the time of writing. Biological control using resident microorganisms in the soil is already practised, knowingly or unknowingly, on many farms where soil suppressiveness is exploited. This is done either by inducing TAD in long sequences of cereal crops, using fields known not to favour take-all, or by inducing the build-up of antagonistic *Phialophora* spp. under preceding grass crops. On the other hand, BCAs suitable for application to control the disease in high-risk situations are still not available, despite much research. Problems that have to be resolved to remedy this include the unreliability of BCAs because of their environmental requirements, poor root colonization, lack of persistence and stability, and the question of regional adaptation. Whilst the importance of delayed sowing was paramount and single fungicide or fertilizer treatments were more or less effective, combinations of agronomic treatments rarely achieved better control than the best of the constituent treatments in the UK.

Control strategies are unlikely to be implemented successfully without the ability to assess risk, which depends on a fundamental understanding of the epidemiology of the disease. Modelling has come to the fore in attempts to explain the spread of infection and rates of epidemic development. Detailed information from long-running field experiments, a scarce and valuable resource that has suffered in recent rounds of cut-backs in agricultural research in the UK, is useful for validating models and provoking ideas. With the aid of computers, large sets of data from such experiments have been used to create images of epidemics which show trends in disease over decades and make seasonal comparisons clearer.

The ability to identify the pathogen and related fungi is also fundamental to the study, and ultimately the control, of take-all. Modern serological and molecular biological techniques have been used to study the taxonomic relationships among these fungi. DNA methods in particular are beginning to prove useful for accurate diagnosis of taxa. Quantification of infection using these approaches is developing more slowly and the challenge will be to develop such techniques in ways that can be applied in ecological and epidemiological research and in disease monitoring in the field.

Bioassays aside, much pot work with take-all has limited value in explaining the disease in crops. However, good field experiments are costly in time and money and those that rely on naturally occurring inoculum are often inconclusive because of the patchy distribution and unpredictability of the disease. Appropriate experimental designs and the use of artificially produced inoculum are tools in helping to overcome these problems. Artificial infestation of field soil can also be used to evaluate losses in grain yield and reductions in grain quality without many of the confounding factors relating to crop sequences and soil conditions. More results of this type and more national and regional surveys would lead to a better understanding of the importance of the disease in cereal production and provide data for analysing disease–yield relationships and testing mathematical models for yield loss.

Take-all has changed in the UK in the last two or three decades, presumably in response to changing farming practices and weather trends. Continued research is required to understand these changes so as to achieve a more reliable assessment of risk. Perhaps, soon, the long wait for BCAs and/or the even longer wait for fungicides that are effective, economic and environmentally safe will be rewarded. This would increase the options beyond rotation for farmers faced with the risk of damaging take-all. Assessing risk and applying control measures efficiently require knowledge of the complex nature of the disease that has still to be acquired. Acquiring that knowledge involves continued epidemiological, ecological and biological research that would benefit from more coordinated and less fragmentary effort than hitherto.

References

The letters and numbers in parentheses at the end of each reference indicate the chapter(s) in which it is cited, e.g. (C5) refers to Chapter 5; (C0) refers to the Introduction.

Abbott, L.K. and Holland, A.A. (1975) Electrophoretic patterns of soluble proteins and isoenzymes of *Gaeumannomyces graminis*. *Australian Journal of Botany* 23, 1–12. (C5)

Abd-el Moity, T.H., Papavizas, G.C. and Shatla, M.N. (1982) Induction of new isolates of *Trichoderma harzianum* tolerant to fungicides and their experimental use for control of white rot of onion. *Phytopathology* 72, 396–400. (C4)

Abel, C. (1994) Simple system to produce the highest quality. *Farmers Weekly* 12 August, 58. (C3)

Abel, C. (1995) Higher yielding hybrid wheats here by 1998. *Farmers Weekly* 20 January, 55–56. (C4, C7)

Abel, C. (1996) Check for lack of manganese. *Farmers Weekly* 15 March, 63 . (C3)

Adamatzky, A. (1994) *Identification of Cellular Automata*. Taylor & Francis, London. (C2)

ADAS (1989) Crop Intelligence Report, Eastern Region. CI89/27, 7 July 1989. (C1)

ADAS (1992a) Crop Intelligence Report, Eastern Region. CI9206, 14 February 1992. (C3)

ADAS (1992b) Autumn manganese for cereals. *Crop Action 11 92/93*, 12 November, 5. (C3)

Addiscott, T.M., Whitmore, A.P. and Powlson, D.S. (1991) *Farming, Fertilizers and the Nitrate Problem*. CAB International, Wallingford. (C1)

Agricultural and Food Research Council (1988) *Keeping the Balance. A Rothamsted Guide. Soil and Fertiliser Nitrogen*. IACR Rothamsted Experimental Station, 4 pp. (C1)

Ahl, P., Voisard, C. and Défago, G. (1986) Iron-bound siderophores, cyanide, and antibiotics involved in suppression of *Thielaviopsis basicola* by a *Pseudomonas fluorescens* strain. *Journal of Phytopathology* 116, 121–134. (C4)

Ainsley, A.E., Dyke, G.V. and Jenkyn, J.F. (1995) Inter-plot interference and nearest-neighbour analysis of field experiments. *Journal of Agricultural Science, Cambridge* 125, 1–9. (C6)

Allan, J. (1996) How continuous wheat can pay. *Farmers Weekly* 1 March, Arable Supplement, p. S10. (C3)

Almassi, F., Ghisalberti, E.L., Narbey, M.J. and Sivasithamparam, K. (1991) New antibiotics from strains of *Trichoderma harzianum*. *Journal of Natural Products* 54, 396–402. (C4)

Altman, J. (1985) Impact of herbicides on plant diseases. In: Parker, C.A., Rovira, A.D., Moore, K.J., Wong, P.T.W. and Kollmorgen, J.F. (eds) *Ecology and Management of Soilborne Plant Diseases*. APS Press, St Paul, Minnesota, pp. 227–231. (C3)

Amein, T.A.M. (1988) [Occurrence of soil fungi in wheat cultivation and their effects on *Gaeumannomyces graminis* var. *tritici* in vitro conditions.] *Roczniki Nauk Rolniczych, Seria E* 18, 153–160. (C4)

An, Z.-Q., Siegel, M.R., Hollin, W., Tsai, H.-F., Schmidt, D. and Schardl, C.L. (1993) Relationships among non-*Acremonium* sp. fungal endophytes in five grass species. *Applied and Environmental Microbiology* 59, 1540–1548. (C5)

Andrade, O. (1995) *Fungicida al suelo aplicado al momento de la siembra. Nueva alternativa para el control de enfermedades en cereales.* Boletin Technico No. 170, Instituto de Investigaciones Agropecuarias, Centro Regional de Investigacion Carillanca, Temuco, Chile, 28 pp. (C4)

Andrade, O.A., Mathre, D.E. and Sands, D.C. (1994a) Natural suppression of take-all disease of wheat in Montana soils. *Plant and Soil* 164, 9–18. (C2)

Andrade, O.A., Mathre, D.E. and Sands, D.C. (1994b) Suppression of *Gaeumannomyces graminis* var. *tritici* in Montana soils and its transferability between soils. *Soil Biology and Biochemistry* 26, 397–402. (C2)

Angus, J.F., Gardner, P.A., Kirkegaard, J.A. and Desmarchelier, J.M. (1994) Biofumigation: isothiocyanates released from *Brassica* roots inhibit growth of the take-all fungus. *Plant and Soil* 162, 107–112. (C3)

Anon. (1960) France. In: Schlömer, F.C., Weinhues, F., Kopetz, L.M., Coïc, Y., Lichnowsky, L. and Sakoff, A.N. (eds) *Progressive Wheat Production*. Centre d'Étude de l'Azote, Genève, pp. 142–147. (C1)

Anon. (1976) *Manual of Plant Growth Stages and Disease Assessment Keys*. Ministry of Agriculture, Fisheries and Food, Harpenden. (C6)

Anon. (1981) En bref... dans les régions. *Perspectives Agricoles* 52, 4–5. (C1)

Anon. (1984) Techniques culturales en 1981, blé tendre d'hiver. *Productions Végétales et Forestières* 25, 157–162. (C1)

Anon. (1987) *Annual Review of Agriculture 1987*. HMSO, London. (C3)

Anon. (1988a) *Agriculture in the United Kingdom: 1988*. HMSO, London. (C3)

Anon. (1988b) Techniques culturales en 1986, blé tendre d'hiver. *Productions Végétales et Forestières* 34, 110. (C1)

Anon. (1992a) Post-CAP reform crop plan survey. *Farmers Weekly* 4 December, 49. (C3)

Anon. (1992b) Low cost – but manganese has to be targeted. *Farmers Weekly* 13 November, 33. (C3)

Anon. (1994a) *Agriculture. Annuaire Statistique 1994*. Office des Publications des Communautés Européennes, Luxembourg. (C1)

Anon. (1994b) *The Digest of Agricultural Census Statistics. United Kingdom. 1993*. HMSO, London. (C1)

Anon. (1994c) *Victorian Institute for Dryland Research, Horsham, Biennial Report for 1992/4*, pp. 21–25. (C3)

Anon. (1995a) Grandes cultures: estimations des productions en 1er février 1995. *Agreste, Conjoncture* 1 (3), 4pp. Service Central des Enquêtes et Etudes Statistiques (SCEES), Ministère de l'Agriculture et de la Pêche, Paris, 4 pp. (C1)

Anon. (1995b) L'utilisation du territoire en 1994. *Agreste, Données Chiffrées – Agriculture* 66, 15–31. (C1)

Anon. (1995c) Falling S prompts a reformulation. *Farmers Weekly*, 26 May, 59, (C1)

Anon. (1997) European trials could bring take-all control. *Farmers Weekly* 7 March, 71. (C4)

Asher, M.J.C. (1972) Effect of *Ophiobolus graminis* infection on the growth of wheat and barley. *Annals of Applied Biology* 70, 215–223. (C2)

Asher, M.J.C. (1980) Variation in pathogenicity and cultural characters in *Gaeumannomyces graminis* var. *tritici*. *Transactions of the British Mycological Society* 75, 213–220. (C2)

Asher, M.J.C. (1981) Pathogenic variation. In: Asher, M.J.C. and Shipton, P.J. (eds) *Biology and Control of Take-all*. Academic Press, London, pp. 199–218. (C4, C5)

Asher, M.J.C. and Shipton, P.J. (eds) (1981) *Biology and Control of Take-all*. Academic Press, London. (C0, C1, C2, C5, C7)

Augustin, C. (1992) Verfahren zur Herstellung eines Myzelgranulats. East Germany, Basic Patent, DD 298038, 6 February 1992. (C4)

Augustin, C., Elsner, G. and Weirauch, A. (1990) Verfahren zur Einschränkung von *Gaeumannomyces graminis* var. *tritici* bei Weizen mittels einer Präinokulation von *Phialophora graminicola*. East Germany, Basic Patent, DD 278487, 9 May 1990. (C4)

Augustin, C., Elsner, G. and Weirauch, A. (1991a) Verfahren zur Einschränkung von *Gaeumannomyces graminis* var. *tritici* bei Weizen mittels einer Präinokulation von *Phialophora* sp. (lobed hyphopodia). East Germany, Basic Patent, DD 287193, 21 February 1991. (C4)

Augustin, C., Elsner, G. and Weirauch, A. (1991b) Verfahren zur Einschränkung der Schadwirkung von *Gaeumannomyces graminis* var. *tritici* bei Weizen mittels einer Präinokulation von avirulenten Stämmen des Erregers. East Germany, Basic Patent, DD 293252, 29 August 1991. (C4)

Ballinger, D.J. and Kollmorgen, J.F. (1986a) Glasshouse and field evaluation of benomyl and triadimefon applied at seeding to control take-all in wheat. *Plant Pathology* 35, 61–66. (C4)

Ballinger, D.J. and Kollmorgen, J.F. (1986b) Control of take-all of wheat in the field with benzimidazole and triazole fungicides applied at seeding. *Plant Pathology* 35, 67–73. (C4)

Ballinger, D.J. and Kollmorgen, J.F. (1988) Effect of triazole coated superphosphate, applied at sowing, on take-all and yield of wheat. *Australian Journal of Experimental Agriculture* 28, 635–638. (C1, C4)

Bateman, D.F. (1978) The dynamic nature of disease. In: Horsfall, J.G. and Cowling, E.B. (eds) *Plant Disease: an Advanced Treatise*. Academic Press, New York, pp. 53–83. (C2)

Bateman, G.L. (1980) Prospects for fungicidal control of take-all of wheat. *Annals of Applied Biology* 96, 275–282. (C4)

Bateman, G.L. (1981) Effects of soil application of benomyl against take-all (*Gaeumannomyces graminis* var. *tritici*) and footrot diseases of wheat. *Zeitschrift für Pflanzenkrankheiten und Pflanzenschutz* 88, 249–255. (C4)

Bateman, G.L. (1982) Formulations of soil-applied fungicides for controlling take-all (*Gaeumannomyces graminis* var. *tritici*) in experiments with pot-grown wheat. *Zeitschrift für Pflanzenkrankheiten und Pflanzenschutz* 89, 480–486. (C4)

Bateman, G.L. (1984a) Effects of surfactants on the performance of soil-applied fungicides against take-all (*Gaeumannomyces graminis* var. *tritici*) in wheat. *Zeitschrift für Pflanzenkrankheiten und Pflanzenschutz* 91, 345–353. (C4)

Bateman, G.L. (1984b) Soil-applied fungicides for controlling take-all in field experiments with winter wheat. *Annals of Applied Biology* 104, 459–465. (C4)

Bateman, G.L. (1985) The effects of distribution of two soil-incorporated fungicides on control of take-all (*Gaeumannomyces graminis* var. *tritici*) in wheat. *Zeitschrift für Pflanzenkrankheiten und Pflanzenschutz* 92, 194–203. (C4)

Bateman, G.L. (1986) Effects of triadimenol-containing seed treatment on winter wheat infected with take-all. *Zeitschrift für Pflanzenkrankheiten und Pflanzenschutz* 93, 404–414. (C2, C4)

Bateman, G.L. (1989) Progress in research on the control of take-all in cereals with fungicides: a review. *Crop Protection* 8, 75–81. (C4)

Bateman, G.L. and Hornby, D. (1989) Artificial epidemics to study the effects of take-all on yield. *Aspects of Applied Biology* 22, 435–436. (C6)

Bateman, G.L. and Hornby, D. (1995) Take-all of cereals: challenges of field experimentation on a patch-forming disease caused by a soil-borne pathogen. *Aspects of Applied Biology* 43, 31–38. (C4, C6)

Bateman, G.L. and Nicholls, P.H. (1982) Experiments on soil drenching with fungicides against take-all in wheat. *Annals of Applied Biology* 100, 297–303. (C4)

Bateman, G.L., Hornby, D. and Gutteridge, R.J. (1990a) Effects of take-all on some aspects of grain quality of winter wheat. *Aspects of Applied Biology* 25, 339–348. (C1, C6)

Bateman, G.L., Nicholls, P.H. and Chamberlain, K. (1990b) The effectiveness of eleven sterol biosynthesis-inhibiting fungicides against the take-all fungus, *Gaeumannomyces graminis* var. *tritici*, in relation to their physical properties. *Pesticide Science* 29, 109–122. (C4)

Bateman, G.L., Ward, E. and Antoniw, J.F. (1992) Identification of *Gaeumannomyces graminis* var. *tritici* and *G. graminis* var. *avenae* using a DNA probe and non-molecular methods. *Mycological Research* 96, 737–742. (C5)

Bateman, G.L., Hornby, D., Payne, R.W. and Nicholls, P.H. (1994) Evaluation of fungicides applied to soil to control naturally-occurring take-all using a balanced-incomplete-block design and very small plots. *Annals of Applied Biology* 124, 241–251. (C4)

Bateman, G.L., Ward, E., Hornby, D. and Gutteridge, R.J. (1997) Comparisons of isolates of the take-all fungus, *Gaeumannomyces graminis* var. *tritici*, from different cereal sequences using DNA probes and non-molecular methods. *Soil Biology and Biochemistry* 29, 1225–1232. (C5)

Batey, T. (1971) Manganese and boron deficiency. In: Anon. (ed.) *Trace Elements in Soils and Crops*. HMSO, London, pp. 137–149. (C3)

Becker, J.O. and Schwinn, F.J. (1993) Control of soil-borne pathogens with living bacteria and fungi: status and outlook. *Pesticide Science* 37, 355–363. (C4)

Begley, M.J., Crombie, L., Crombie, W.M.L. and Whiting, D.A. (1986) The isolation of avenacins A-1, A-2, B-1 and B-2, chemical defences against cereal 'take-all' disease. Structure of their 'aglycones', the avenestergenins, and their anhydro dimers. *Journal of the Chemical Society. Perkin Transactions I*, 1905–1915. (C5)

Bernatzky, R. (1988) Restriction fragment length polymorphisms. In: Gelvin, S.B. and Schilperoort, R.A. (eds) *Plant Molecular Biology Manual*. Kluwer, Dordrecht, pp. C2:1–18. (C5)

Berry, R.J. (1993) A voyage around Julian Huxley. *New Scientist* 140, 46–47. (C0)

Bingham, J., Law, C. and Miller, T. (1991) *Wheat Yesterday, Today and Tomorrow*. Plant Breeding International, Cambridge. (C0, C1)

Blake, A. (1989) "Cover" yourself with the right crop for set-aside. *Farmer's Weekly*, 19 May, 48. (C1)

Blake, A. (1996a) Home-saved seed just as good as bought-in. *Farmers Weekly*, 11 October, 61. (C4)
Blake, A. (1996b) Quarter at risk of sulphur deficiency. *Farmers Weekly* 12 April, 52. (C7)
Blakeman, R.H. (1990) The identification of crop disease and stress by aerial photography. In: Steven, M.D. and Clark, J.A. (eds) *Applications of Remote Sensing in Agriculture*. Butterworths, London, pp. 229–254. (C6)
Blanch, P.A., Asher, M.J.C. and Burnett, J.H. (1981) Inheritance of pathogenicity and cultural characters in *Gaeumannomyces graminis* var. *tritici*. *Transactions of the British Mycological Society* 77, 391–399. (C5)
Board of Agriculture and Fisheries (1915) 'White-Heads' or 'Take-all' of wheat and oats (*Ophiobolus graminis*, Sacc.). Leaflets Nos 273 to 300. Leaflet No. 273, 4 pp., dated July 1913. (C1)
Bockus, W.W. (1983) Effects of fall infection by *Gaeumannomyces graminis* var. *tritici* and triadimenol seed treatment on severity of take-all in winter wheat. *Phytopathology* 73, 540–543. (C4)
Bockus, W.W. (1991) Effect of position of inoculum of *Gaeumannomyces graminis* var. *tritici* relative to the seed on yield of winter wheat. *Phytopathology* 81, 1241. (C6)
Bockus, W.W., Davis, M.A. and Norman, B.L. (1994a) Effect of soil shading by surface residues during summer fallow on take-all of winter wheat. *Plant Disease* 78, 50–54. (C2)
Bockus, W.W., Davis, M.A. and Todd, T.C. (1994b) Grain yield responses of winter wheat coinoculated with *Cephalosporium gramineum* and *Gaeumannomyces graminis* var. *tritici*. *Plant Disease* 78, 11–14. (C6)
Boesewinkel, H.J. (1976) Storage of fungal cultures in water. *Transactions of the British Mycological Society* 66, 183–185. (C5)
Boggini, G. (1986) Durum wheat in the centre-north. Crop intensification of durum wheat in central and northern Italy. *Informatore Agrario* 43, 29–32. (C1)
Bowerman, P. (1989) Reviewing the rotation. *Boxworth – 1989*. Experimental Husbandry Farm, Ministry of Agriculture, Fisheries and Food, ADAS, Boxworth, Cambridge, 3–4. (C1, C3)
Bowyer, P., Musker, R., Osbourn, A.E., Clarke, B., Caten, C. and Daniels, M.J. (1992) Production of mutants of *Gaeumannomyces graminis* var. *tritici* and var. *avenae* by 4-nitroquinolene oxide treatment of protoplasts. *Fungal Genetics Newsletter* 39, 13–15. (C5)
Bowyer, P., Osbourn, A.E. and Daniels, M.J. (1994) An 'instant gene bank' method for heterologous gene cloning: complementation of two *Aspergillus nidulans* mutants with *Gaeumannomyces graminis* DNA. *Molecular and General Genetics* 242, 448–454. (C5)
Bowyer, P., Clarke, B.R., Lunness, P., Daniels, M.J. and Osbourn, A.E. (1995) Host range of a plant pathogenic fungus determined by a saponin detoxifying enzyme. *Science* 267, 371–373. (C5)
Boyeldieu, J. (1980) *Les Cultures Céréalières*. Hachette, Paris. (C1)
Brassett, P.R. and Gilligan, C.A. (1988) A model for primary and secondary infection in botanical epidemics. *Zeitschrift für Pflanzenkrankheiten und Pflanzenschutz* 95, 352–360. (C2)
Brassett, P.R. and Gilligan, C.A. (1989) Fitting of simple models for field disease progress data for the take-all fungus. *Plant Pathology* 38, 397–407. (C2)
Brassett, P.R. and Gilligan, C.A. (1990) Effects of self-sown wheat on levels of the take-all disease on seedlings of winter wheat grown in a model system. *Journal of Phytopathology* 129, 46–57. (C2, C4)
Brennan, J.P. and Murray, G.M. (1988) Australian wheat diseases: assessing their economic importance. *Agricultural Sciences* 7, 26–36. (C1)
Brennan, R.F. (1988) Effect of phosphorus deficiency in wheat on the infection of roots by *Gaeumannomyces graminis* var. *tritici*. *Australian Journal of Agricultural Research* 39, 541–546. (C3)
Brennan, R.F. (1989a) Effect of copper application on take-all severity and grain yield of wheat in field experiments near Esperance, Western Australia. *Australian Journal of Experimental Agriculture* 31, 255–258. (C3)
Brennan, R.F. (1989b) Effect of nitrogen and phosphorus deficiency in wheat on the infection of roots by *Gaeumannomyces graminis* var. *tritici*. *Australian Journal of Agricultural Research* 40, 489–495. (C3)
Brennan, R.F. (1989c) Effect of superphosphate and superphosphate plus flutriafol on yield and take-all of wheat. *Australian Journal of Experimental Agriculture* 29, 247–252. (C1, C4)
Brennan, R.F. (1992) The effect of zinc fertilizer on take-all and the grain yield of wheat grown on zinc-deficient soils of the Esperance region, Western Australia. *Fertilizer Research* 31, 215–219. (C3)
Brennan, R.F. (1995) Effect of levels of take-all and phosphorus fertiliser on the dry matter and grain yield of wheat. *Journal of Plant Nutrition* 18, 1159–1176. (C3)
Brisbane, P.G. and Rovira, A.D. (1988) Mechanisms of inhibition of *Gaeumannomyces graminis* var. *tritici* by fluorescent pseudomonads. *Plant Pathology* 37, 104–111. (C4)
Britton, D.K. (1969) *Cereals in the United Kingdom*. Pergamon Press, Oxford. (C1)

Brooks, D.H. and Dawson, M.G. (1968) Influence of direct-drilling of winter wheat on incidence of take-all and eyespot. *Annals of Applied Biology* 61, 57–64. (C3)

Brown, L.R., Durning, A.T., Flavin, C., French, H.F., Lenssen, N., Lowe, M.D., Misch, A., Postel, S., Renner, M., Starke, L., Weber, P. and Young, J.E. (1994) *State of the World 1994. A Worldwatch Report on Progress Toward a Sustainable Society*. W.W. Norton & Company, New York. (C1)

Brown, M.E. and Hornby, D. (1987) Effects of nitrate and ammonium on wheat roots in gnotobiotic culture: amino acids, cortical cell death and take-all (caused by *Gaeumannomyces graminis* var. *tritici*). *Soil Biology and Biochemistry* 19, 567–573. (C2, C3)

Brown, M.E., Hornby, D. and Pearson, V. (1973) Microbial populations and nitrogen in soil growing consecutive cereal crops infected with take-all. *Journal of Soil Science* 24, 296–310. (C2)

Brown, P.D. and Morra, M.J. (1995) Glucosinolate-containing plant tissues as bioherbicides. *Journal of Agricultural and Food Chemistry* 43, 3070–3074. (C3)

Bruehl, G.W. (1957) Cephalosporium stripe disease of wheat. *Phytopathology* 47, 641–649. (C6)

Bruns, T.D., White, T.J. and Taylor, J.W. (1991) Fungal molecular systematics. *Annual Review of Ecology and Systematics* 22, 525–564. (C5)

Bryan, G.T., Daniels, M.J. and Osbourn, A.E. (1995) Comparison of fungi within the *Gaeumannomyces–Phialophora* complex by analysis of ribosomal DNA sequences. *Applied and Environmental Microbiology* 61, 681–689. (C5)

Buck, K.W. (1986) Viruses of the wheat take-all fungus, *Gaeumannomyces graminis* var. *tritici*. In: Buck, K.W. (ed.) *Fungal Viruses*. CRC Press, Boca Raton, Florida, pp. 221–236. (C2, C5, C7)

Buck, K.W. (1987) Viruses of plant pathogenic fungi. In: Day, P.R. and Jellis, G.J. (eds) *Genetics and Plant Pathogenesis*. Blackwell Scientific Publications, Oxford, pp. 111–126. (C5)

Buddin, W. and Garrett, S.D. (1941) Seasonal occurrence of take-all disease of wheat. *Annals of Applied Biology* 28, 74. (C1)

Bunting, T.E., Plumley, K.A., Clarke, B.E. and Hillman, B.I. (1996) Identification of *Magnaporthe poae* by PCR and examination of its relationship to other fungi by analysis of their nuclear rDNA ITS-1 regions. *Phytopathology* 86, 398–404. (C5)

Burleigh, J.R., Roelfs, A.P. and Eversmeyer, M.G. (1972) Estimating damage to wheat caused by *Puccinia recondita tritici*. *Phytopathology* 62, 944–946. (C6)

Butler, E.J. and Jones, S.G. (1961) *Plant Pathology*. Macmillan, London. (C3)

Butler, F.C. (1961) Root and foot rot diseases of wheat. *Department of Agriculture N.S.W. Science Bulletin* 77, 98 pp. (C1)

Campbell, C.L. and Benson, D.M. (1994) Spatial aspects of the development of root disease epidemics. In: Campbell, C.L. and Benson, D.M. (eds) *Epidemiology and Management of Root Diseases*. Springer-Verlag, Berlin, pp. 195–243. (C2, C6)

Campbell, C.L. and Madden, L.V. (1990) *Introduction to Plant Disease Epidemiology*. John Wiley & Sons, New York. (C2, C6)

Campbell, C.L. and Neher, D.A. (1994) Estimating disease severity and incidence. In: Campbell, C.L. and Benson, D.M. (eds) *Epidemiology and Management of Root Diseases*. Springer-Verlag, Berlin, pp. 117–147. (C6)

Campbell, R. and Faull, J.L. (1979) Biological control of *Gaeumannomyces graminis*: field trials and the ultrastructure of the interaction between the fungus and a successful antagonist bacterium. In: Schippers, B. and Gams, W. (eds) *Soil-borne Plant Pathogens*. Academic Press, London, pp. 603–609. (C4)

Cannon, P.F. (1994) The newly recognised family *Magnaporthaceae* and its interrelationships. *Systema Ascomycetum* 13, 25–42. (C5)

Capper, A.L. and Campbell, R. (1986) The effect of artificially inoculated antagonistic bacteria on the prevalence of take-all disease of wheat in field experiments. *Journal of Applied Bacteriology* 60, 155–160. (C4)

Capper, A.L. and Higgins, K.P. (1993) Application of *Pseudomonas fluorescens* isolates to wheat as potential biological control agents against take-all. *Plant Pathology* 42, 560–567. (C4)

Catt, J.A., Gutteridge, R.J. and Slope, D.B. (1986) Take-all distribution and soil type on Chalky Boulder Clay. *Journal of Agricultural Science, Cambridge* 106, 61–66. (C2, C6)

Cavelier, N. (1989) Protection intégrée contre les maladies du pied des céréales. In: *L'Europe et des Maladies Cryptogamiques du Blé*. Record of meeting organized by Schering SA, Palais des Congres, Versailles, pp. 107–108. (C4)

Cavelier, N. and Lucas, P. (1985) Efficacité du triadimenol sur le piétin échaudage et sur d'autres maladies du pied des céréales. In: *Fungicides for Crop Protection. 100 years of progress*, British Crop Protection Council Monograph, 31. BCPC, Croydon, pp. 347–350. (C4)

Central Statistical Office (1935–1995) *Annual Abstract of Statistics*. HMSO, London. (C1)

Chakraborty, S. (1983) Population dynamics of amoebae in soils suppressive and nonsuppressive to wheat take-all. *Soil Biology and Biochemistry* 15, 661–664. (C4)

Chakraborty, S. and Old, K.M. (1982) Mycophagous soil amoeba: interactions with three plant pathogenic fungi. *Soil Biology and Biochemistry* 12, 247–255. (C2, C4)

Chakraborty, S. and Warcup, J.H. (1983) Soil amoebae and saprophytic survival of *Gaeumannomyces graminis* var. *tritici* in a suppressive pasture soil. *Soil Biology and Biochemistry* 15, 181–185. (C2, C4)

Chakraborty, S. and Warcup, J.H. (1984) Population dynamics of amoebae in soils suppressive and non-suppressive to wheat take-all. *Soil Biology and Biochemistry* 16, 197–199. (C2, C4)

Chakraborty, S. and Warcup, J.H. (1985) Reduction of take-all by mycophagous amoebas in pot bioassays. In: Parker, C.A., Rovira, A.D., Moore, K.J., Wong, P.T.W. and Kollmorgen, J.F. (eds) *Ecology and Management of Soil-borne Plant Pathogens*. APS Press, St Paul, Minnesota, pp. 107–109. (C4)

Chakraborty, S., Old, K.M. and Warcup, J.H. (1983) Amoebae from a take-all suppressive soil which feed on *Gaeumannomyces graminis* var. *tritici* and other soil fungi. *Soil Biology and Biochemistry* 15, 17–24. (C2, C4)

Chalmers, A.G., Church, B.M., Dyer, C.J. and Leech, P.K. (1991) *Survey of Fertiliser Practice: Fertiliser Use on Farm Crops in England and Wales 1990*. ADAS/FMA, London. (C1)

Chamberlain, K., Bateman, G.L. and Nicholls, P.H. (1991) Volatile analogues of penconazole and their activity against the take-all fungus *Gaeumannomyces graminis* var. *tritici*. *Pesticide Science* 31, 185–196. (C4)

Chambers, S.C. and Flentje, N.T. (1967) Studies on variation with *Ophiobolus graminis*. *Australian Journal of Biological Science* 20, 941–951. (C2)

Chamen, W.C.T., Watts, C.W., Leede, P.R. and Longstaff, D.J. (1992) Assessment of a wide span vehicle (gantry), and soil and cereal crop responses to its use in a zero traffic regime. *Soil and Tillage Research* 24, 359–380. (C3)

Chapman, G.P. (1996) *The Biology of Grasses*. CAB International, Wallingford. (C4)

Charigkapakorn, N. and Sivasithamparam, K. (1987) Changes in the composition and population of fluorescent pseudomonads on wheat roots inoculated with successive generations of root-piece inoculum of the take-all fungus. *Phytopathology* 77, 1002–1007. (C4)

Chellemi, D.O., Rohrbach, K.G., Yost, R.S. and Sonoda, R.M. (1988) Analysis of the spatial pattern of plant pathogens and diseased plants using geostatistics. *Phytopathology* 78, 221–226. (C6)

Chen, C., Collins, D.J. and Morgan-Jones, G. (1996) Fungi associated with root rot of winter wheat in Alabama. *Journal of Phytopathology* 144, 193–196. (C2)

Cherfas, J. (1990) Genes unlimited. *New Scientist* 126, 29–33. (C5)

Christen, O. (1992) Yield-loss-relationships in winter wheat affected by the interaction between fungal disease and husbandry practices – a comparison of methods. In: *Proceedings of the Second Congress of the European Society for Agronomy*. European Society for Agronomy, Wellesbourne, pp. 50–53. (C6)

Christensen, N.W. and Brett, M. (1985) Chloride and liming effects on soil nitrogen form and take-all of wheat. *Agronomy Journal* 77, 157–163. (C3, C4)

Christensen, N.W., Powelson, R.L. and Brett, M. (1987) Epidemiology of wheat take-all as influenced by soil pH and temporal changes in inorganic soil N. *Plant and Soil* 98, 221–230. (C2)

Christensen, N.W., Brett, M.A., Hart, J.W. and Weller, D.M. (1990) Disease dynamics and yield of wheat as affected by take-all, N sources and fluorescent *Pseudomonas*. In: *Transactions of the 14th International Congress of Soil Science, Kyoto, Japan, August 1990*, Vol. III. International Society of Soil Science, pp. 10–15. (C3)

Christian, D.G. and Miller, D.P. (1984) Cephalosporium stripe in winter wheat grown after different methods of straw disposal. *Plant Pathology* 33, 605–606. (C6)

Chu Chou, M. and Hornby, D. (1972) Attempts to explain the nature of 'decline' of take-all. *Rothamsted Experimental Station Report for 1971. Part 1*, p. 141. (C2)

Church, B.M. (1979) *Survey of Fertilizer Practice. Fertilizer Use on Farm Crops in England & Wales. 1978. With Comparative Metric Data from 1970*. (SS/SAF/29). Agricultural Development and Advisory Service, Rothamsted Experimental Station and Fertiliser Manufacturers Association, 8 pp. + tables. (C3)

Church, B.M. and Leech, P.K. (1982) *Survey of Fertilizer Practice. Fertilizer Use on Farm Crops in England & Wales. 1981*. (SS/CH/8). Agricultural Development and Advisory Service, Rothamsted Experimental Station and Fertiliser Manufacturers Association, 8 pp. + tables. (C3)

Clare, R.W., Ap Dewi, I. and Madge, W.E.R. (1986) The effect of autumn nitrogen, insecticide and fungicide on winter wheat sown at two dates and with three levels of spring nitrogen. In: *Proceedings of the 1986 British*

Crop Protection Conference – Pests and Diseases, Vol. 1. The British Crop Protection Council, Thornton Heath, pp. 165–172. (C3)

Clark, P. (1996) BSE – good and bad for grain trade. *Farmers Weekly* 5 April, 22. (C7)

Clarke, J.M. and Cooper, F.B. (1992) Vegetation changes and weed levels in set-aside and subsequent crops. In: Clarke, J. (ed.) *Set-aside*. BCPC, Farnham, pp. 103–110. (C3)

Clarkson, J.D.S. and Polley, R.W. (1981a) Assessment of losses caused by stem-base and root diseases in cereals. In: *Proceedings of the 1981 British Crop Protection Conference – Pests and Diseases*. BCPC Publications, Croydon, pp. 223–231. (C1)

Clarkson, J.D.S. and Polley, R.W. (1981b) Diagnosis, assessment, crop-loss appraisal and forecasting. In: Asher, M.J.C. and Shipton, P.J. (eds) *Biology and Control of Take-all*. Academic Press, London, pp. 251–269. (C2, C6)

Cleal, R.A.E. (1993) Effect of growth regulators on the grain yield and quality of triticale and wheat grown as a second cereal on light soil. *Aspects of Applied Biology* 36, 281–286. (C3)

Clyne, P., Henson, J., White, T.J. and Bowman, B.H. (1993) Highly variable pattern of insertion sequences in 18S ribosomal RNA genes of the plant pathogenic fungal genus *Gaeumannomyces*. *Abstracts of the General Meeting of the American Society for Microbiology* 93, 296. (C5)

Coïc, Y. (1950) Influence de l'époque d'apport des engrais azotés sur les composantes du rendement du blé. *Comptes Rendus des Séances de l'Academie d'Agriculture de France* 36, 231–234. (C1)

Colbach, N. (1995) Modélisation de l'influence des systèmes de culture sur les maladies du pied et des racines du blé tendre d'hiver. Thèse, Institut National Agronomique Paris-Grignon. (C0, C2, C3)

Colbach, N. and Huet, P. (1995) Modelling the frequency and severity of root and foot diseases in winter wheat monocultures. *European Journal of Agronomy* 4, 217–227. (C2)

Colbach, N., Lucas, P. and Cavelier, N. (1994) Influence des successions culturales sur les maladies du pied et des racines du blé d'hiver. *Agronomie (Paris)* 14, 525–540. (C2)

Colbach, N., Maurin, N. and Huet, P. (1996) Influence of cropping system on foot rot of winter wheat in France. *Crop Protection* 15, 295–305. (C0, C2)

Colbach, N., Lucas, P. and Meynard, J.M. (1997) Influence of crop management on take-all development and infection cycles on winter wheat. *Phytopathology* 87, 26–32. (C2)

Conner, R.L. and Kuzyk, A.D. (1990) Evaluation of seed-treatment fungicides for control of take-all in soft white spring wheat. *Canadian Journal of Plant Pathology* 12, 213–216. (C4)

Conner, R.L., MacDonald, M.D. and Whelan, E.D.P. (1988) Evaluation of take-all resistance in wheat-alien amphiploid and chromosome substitution line. *Genome* 30, 597–602. (C3)

Cook, R.J. (1981) The effect of soil reaction and physical conditions. In: Asher, M.J.C. and Shipton, P.J. (eds) *Biology and Control of Take-all*. Academic Press, London, pp. 343–352. (C2)

Cook, R.J. (1992a) A customized approach to biological control of wheat root diseases. In: Tjamos, E.C., Papavizas, G.C. and Cook, R.J. (eds) *Biological Control of Plant Diseases: Progress and Challenges for the Future*. Plenum Press, New York and London, pp. 211–222. (C2, C3, C4)

Cook, R.J. (1992b) Wheat root health management and environmental concern. *Canadian Journal of Plant Pathology* 14, 76–85. (C2)

Cook, R.J. (1993) Making greater use of introduced microorganisms for the biological control of plant pathogens. *Annual Review of Phytopathology* 31, 53–80. (C2)

Cook, R.J. (1994) Problems and progress in the biological control of wheat take-all. *Plant Pathology* 43, 429–437. (C2, C4, C6)

Cook, R.J. (1995) Getting to the root of wheat take-all. *Agricultural Research, Washington DC* August, 2. (C4)

Cook, R.J. and Naiki, T. (1982) Virulence of *Gaeumannomyces graminis* var. *tritici* from fields under short-term and long-term wheat cultivation in the Pacific Northwest, USA. *Plant Pathology* 31, 201–208. (C2)

Cook, R.J. and Reis, E. (1981) Cultural control of soil-borne pathogens of wheat in the Pacific North-West of the U.S.A. In: Jenkyn, J.F. and Plumb, R.T. (eds) *Strategies for the Control of Cereal Disease*. Blackwell Scientific Publications, Oxford, pp. 167–177. (C3)

Cook, R.J. and Rovira, A.D. (1976) The role of bacteria in the biological control of *Gaeumannomyces graminis* by suppressive soils. *Soil Biology and Biochemistry* 8, 269–273. (C4)

Cook, R.J. and Veseth, R.J. (1991) *Wheat Health Management*. APS Press, St Paul, Minnesota. (C0, C1, C3, C4, C6)

Cook, R.J., Papendick, R.W. and Griffin D.M. (1972) Growth of two root rot fungi as affected by osmotic and matric water potential. *Proceedings of the Soil Science Society of America* 36, 78–82. (C2)

Cook, R.J., Wilkinson, H.T. and Alldredge, J.R. (1986) Evidence that microorganisms in suppressive soil associated with wheat take-all decline do not limit the number of lesions produced by *Gaeumannomyces graminis* var. *tritici*. *Phytopathology* 76, 342–345. (C2)

Cook, R.J., Weller, D.M., Kovacevich, P., Drahos, D., Hemming, B., Barnes, G. and Pierson, E.L. (1990) Establishment, monitoring and termination of field tests with genetically altered bacteria applied to wheat for biological control of take-all. In: Mackenzie, D.R. and Henry, S.C. (eds) *Biological Monitoring of Genetically Engineered Plants and Microbes*. Agricultural Research Institute, Bethesda, Maryland, pp. 177–187. (C4)

Cook, R.J., Polley, R.W. and Thomas, M.R. (1991) Disease-induced losses in winter wheat in England and Wales in 1985–89. *Crop Protection* 10, 504–508. (C1)

Cook, R.J., Ownley, B.H. and Rasmussen, P. (1992) Nitrates left in the soil profile after harvest of wheat and barley relate to severity of root diseases. *Phytopathology* 82, 1113. (C2)

Cook, R.J., Thomashow, L.S., Weller, D.M., Fujimoto, D., Mazzola, M., Bangera, G. and Kim, D.-S. (1995) Molecular mechanisms of defense by rhizobacteria against root disease. *Proceedings of the National Academy of Sciences of the USA* 92, 4197–4201. (C4)

Cooper, R.M., Longman, D., Campbell, A., Henry, M. and Lees, P.E. (1988) Enzymatic adaptation of cereal pathogens to the monocotyledonous cell wall. *Physiological and Molecular Plant Pathology* 32, 33–47. (C5)

Cortazar, S.R. (1989) Factores que influyeron el comportamiento de los trigos en la estacion experimental La Platina en los anos 1965 a 1986. *Agricultura Tecnica (Chile)* 49, 193–201. (C1)

Cortazar, S.R., Ramirez, A., Moreno, M.O. and Hacke, E.E. (1987) Analysis of the behaviour of wheat cultivars at La Platina Experimental Station, Santiago, Chile years 1982, 1983 and 1984. *Agricultura Tecnica (Chile)* 47, 57–62. (C1)

Cotterill, P.J. (1991) Biological mode of action of soil-applied flutriafol in controlling take-all of wheat. *Soil Biology and Biochemistry* 23, 323–329. (C4)

Cotterill, P.J. and Sivasithamparam, K. (1987) Take-all inoculum: its occurrence and distribution within a wheat crop and a fallow. *Soil Biology and Biochemistry* 19, 221–222. (C6)

Cotterill, P.J. and Sivasithamparam, K. (1988a) Effect of sowing date on take-all of wheat in Western Australia. *Phytophylactica* 20, 11–14. (C1)

Cotterill, P.J. and Sivasithamparam, K. (1988b) The effect of tillage practices on distribution, size, infectivity and propagule number of the take-all *Gaeumannomyces graminis* var. *tritici*. *Soil and Tillage Research* 11, 183–195. (C6)

Cotterill, P.J. and Sivasithamparam, K. (1989a) An autecological study of the take-all fungus (*Gaeumannomyces graminis* var. *tritici*) in Western Australia. *Australian Journal of Agricultural Research* 40, 229–240. (C1, C2, C6)

Cotterill, P.J. and Sivasithamparam, K. (1989b) The effect of addition to field soil of inoculum of *Gaeumannomyces graminis* var. *tritici* on take-all of wheat. *Journal of Phytopathology* 124, 119–122. (C6)

Cotterill, P.J., De'ath, A.G., Thorn, C.W. and Sivasithamparam, K. (1988) Effect of certain herbicide treatments on pasture composition and inoculum of the take-all fungus. *Plant and Soil* 105, 153–162. (C6)

Cotterill, P.J., Ballinger, D.J. and McLean, L.K. (1992) Evaluation of screening methods for fungicides to control take-all of wheat. *Australasian Plant Pathology* 21, 83–87. (C4)

Coventry, D.R., Brooke, H.D., Kollmorgen, J.F. and Ballinger, D.J. (1989) Increases in wheat yield on limed soil after reduction of take-all by fungicide application and crop rotation. *Australian Journal of Experimental Agriculture* 29, 85–89. (C4)

Cowan, M.C. (1978) Lignification in wheat roots parasitised by *Gaeumannomyces graminis* and *Phialophora radicicola*. *Annals of Applied Biology* 89, 101. (C4)

Crombie, W.M.L. and Crombie, L. (1986) Distribution of avenacins A-1, A-2, B-1 and B-2 in oat roots: their fungicidal activity towards the 'take-all' fungus. *Phytochemistry* 25, 2069–2073. (C5)

Crombie, L., Crombie, W.M.L. and Whiting, D.A. (1986a) Structures of the oat root resistance factors to 'take-all' disease, avenacins A-1, A-2, B-1 and B-2 and their companion substances. *Journal of the Chemical Society, Perkin Transactions I*, 1917–1927. (C5)

Crombie, W.M.L., Crombie, L., Green, J.B. and Lucas, J.A. (1986b) Pathogenicity of the take-all fungus to oats: its relationship to the concentration and detoxification of the four avenacins. *Phytochemistry* 25, 2075–2083. (C5)

Cunningham, P.C. (1975) Some consequences of cereal monoculture on *Gaeumannomyces graminis* (Sacc.) Arx & Olivier and the take-all disease. *EPPO Bulletin* 5, 297–317. (C2)

Cunningham, P.C. (1981) Isolation and culture. In: Asher, M.J.C. and Shipton, P.J. (eds) *Biology and Control of Take-all*. Academic Press, London, pp. 103–123. (C5, C6)

Cunningham, P.C. (1985) Characteristics of trends in diseases caused by soilborne pathogens with spring barley monoculture. In: Parker, C.A., Rovira, A.D., Moore, K.J., Wong, P.T.W. and Kollmorgen, J.F. (eds) *Ecology and Management of Soilborne Plant Pathogens*. American Phytopathological Society, St Paul, Minnesota, pp. 9–13. (C2, C3, C6)

Cunningham, P.C., Spillane, P.A., Foreman, B.T. and Conliffe, D. (1968) Effects of infection by *Ophiobolus graminis* Sacc. on grain yields, baking characteristics of wheat and quality of malting barley. *Irish Journal of Agricultural Research* 7, 183–193. (C6)

Daamen, R.A. (1990) Surveys of cereal diseases and pests in the Netherlands. 1. Weather and winter wheat cropping during 1974–1986. *Netherlands Journal of Plant Pathology* 96, 227–236. (C1)

Daamen, R.A. and Stol, W. (1990) Surveys of cereal diseases and pests in the Netherlands. 2. Stem-base diseases of winter wheat. *Netherlands Journal of Plant Pathology* 96, 251–260. (C1)

Darbyshire, J.F., Davidson, M.S., Scott, N.M., Sparling, G.P. and Shipton, P.J. (1979) Ammonium and nitrate in the rhizosphere of spring barley, *Hordeum vulgare* L. and take-all disease. *Soil Biology and Biochemistry* 11, 453–485. (C3)

Datnoff, L.E., Elliott, M.L. and Jones, D.B. (1993) Black sheath rot caused by *Gaeumannomyces graminis* var. *graminis* on rice in Florida. *Plant Disease* 77, 210. (C5)

Deacon, J.W. (1974) Further studies on *Phialophora radicicola* and *Gaeumannomyces graminis* on roots and stem bases of grasses and cereals. *Transactions of the British Mycological Society* 63, 307–327. (C5)

Deacon, J.W. (1981) Ecological relationships with other fungi: competition and hyperparasites. In: Asher, M.J.C. and Shipton, P.J. (eds) *Biology and Control of Take-All*. Academic Press, London, pp. 75–102. (C4, C5)

Deacon, J.W. and Berry, L.A. (1993) Biocontrol of soil-borne plant pathogens: concepts and their application. *Pesticide Science* 37, 417–426. (C4)

Deacon, J.W. and Henry, C.M. (1978) Studies on virulence of the take-all fungus *Gaeumannomyces graminis*, with reference to methodology. *Annals of Applied Biology* 89, 401–409. (C5)

Deacon, J.W. and Henry, C.M. (1981) Death of the root cortex of winter wheat in field conditions; effects of break crops and possible implications for the take-all fungus and its biological control agent, *Phialophora radicicola* var. *graminicola*. *Journal of Agricultural Research, Cambridge* 96, 579–585. (C2)

Deacon, J.W. and Scott, D.B. (1983) *Phialophora zeicola* sp. nov. and its role in the root rot–stalk rot complex of maize. *Transactions of the British Mycological Society* 81, 247–262. (C5)

Défago, G., Berling, C.H., Burger, U., Haas, D., Kahr, G., Keel, C., Voisard, C., Wirthner, P. and Wüthrich, B. (1990) Suppression of black root rot of tobacco and other root diseases by strains of *Pseudomonas fluorescens*: potential applications and mechanisms. In: Hornby, D. (ed.) *Biological Control of Soil-borne Plant Pathogens*. CAB International, Wallingford, pp. 93–108. (C4)

Dejong, D., Kurtbøke, D.I., Shankar, M. and Sivasithamparam, K. (1993) Effect of placement of inoculum in soil on infectivity and disease protection ability of a sterile red fungus. *Soil Biology and Biochemistry* 25, 1641–1647. (C4)

Delp, B.R., Stowell, L.J. and Marois, J.J. (1986) Evaluation of field sampling techniques for estimation of disease incidence. *Phytopathology* 76, 1299–1305. (C6)

Dennis, R.W.G. and Spooner, B.M. (1992) The fungi of North Hoy, Orkney – I. *Persoonia* 14, 493–507. (C5)

Descalzo, R.C., Punja, Z.K., Lévesque, C.A. and Rahe, J.E. (1996) Identification and role of *Pythium* species as glyphosate synergists on bean (*Phaseolus vulgaris*) grown in different soils. *Mycological Research* 100, 971–978. (C3)

Dewan, M.M. and Sivasithamparam, K. (1988) A plant-growth-promoting sterile fungus from wheat and rye-grass roots with a potential for suppressing take-all. *Transactions of the British Mycological Society* 91, 687–692. (C4)

De Wit, P.J.G.M. (1992) Molecular characterization of gene-for-gene systems in plant–fungus interactions and the application of avirulence genes in control of plant pathogens. *Annual Review of Phytopathology* 30, 391–418. (C5)

Diehl, J.A., Oliveira, M.A.R.D., Igarashi, S., Reis, E.M., Mehta, Y.R. and Gomes, L.S. (1984) Levantamento da ocorrência de doenças radiculares do trigo no Paraná. *Fitopatologia Brasileira* 9, 179–188. (C1)

Dommergues, Y. and Mangenot, F. (eds) (1970) *Ecologie Microbienne du Sol*. Masson, Paris. (C3)

Dori, S., Hershenhorn, J., Solel, Z. and Barash, I. (1992) Characterization of an endopolygalacturonase associated with take-all disease of wheat. *Physiological and Molecular Plant Pathology* 40, 203–210. (C5)

Dori, S., Solel, Z. and Barash, I. (1993) Cell-wall-degrading enzymes associated with take-all disease of wheat. *Phytoparasitica* 21, 143. (C5)

Dori, S., Solel, Z. and Barash, I. (1995) Cell wall-degrading enzymes produced by *Gaeumannomyces graminis* var. *tritici in vitro* and *in vivo*. *Physiological and Molecular Plant Pathology* 46, 189–198. (C5)

Doube, B.M., Stephens, P.M., Davoren, C.W. and Ryder, M.H. (1995) Interactions between earthworms, beneficial soil microorganisms and root pathogens. *Abstracts of the 7th International Symposium on Microbial Ecology (Sao Paulo, 1 September 1995)*, International Committee on Microbial Ecology, p. 142. (C4)

Duffy, B.K. and Weller, D.M. (1994) A semiselective and diagnostic medium for *Gaeumannomyces graminis* var. *tritici*. *Phytopathology* 84, 1407–1415. (C5)

Duffy, B.K. and Weller, D.M. (1995) Use of *Gaeumannomyces graminis* var. *graminis* alone and in combination with fluorescent *Pseudomonas* spp. to suppress take-all of wheat. *Plant Disease* 79, 907–911. (C4)

Dunlop, R.W., Simon, A., Sivasithamparam, K. and Ghisalberti, E.L. (1989) An antibiotic from *Trichoderma koningii* active against soilborne plant pathogens. *Journal of Natural Products* 52, 67–74. (C2, C4)

Dyke, G.V. (1991) *John Bennet Lawes: the Record of his Genius*. Research Studies Press, Taunton. (C1)

Dyke, G.V. and Slope, D.B. (1978) Effects of previous legume and oat crops on grain yield and take-all in spring barley. *Journal of Agricultural Science, Cambridge* 91, 443–451. (C6)

Eastwood, R.F., Kollmorgen, J.F. and Hannah, M. (1993) *Triticum tauschii* reaction to the take-all fungus *Gaeumannomyces graminis* var. *tritici*. *Australian Journal of Agricultural Research* 44, 745–754. (C3)

Eastwood, R.F., Kollmorgen, J.F., Hannah, M. and Williams, W.M. (1994) Reaction of somaclonal variants of wheat to the take-all fungus (*Gaeumannomyces graminis* var. *tritici*). *Plant Pathology* 43, 644–650. (C3)

Ebbels, D.L. (1969) Effects of soil fumigation on disease incidence, growth and yield of spring wheat. *Annals of Applied Biology* 63, 81–93. (C4)

Edwards, I.B. and Dorlencourt, G. (1994) Hybrid wheat – current status and future role in European wheat production. *Melhoramento* 33, 87–104. (C7)

Elliott, M.L. (1989) An improved selective medium for isolation of *Gaeumannomyces*-like fungi. *Phytopathology* 79, 1177. (C5)

Elliott, M.L. (1991a) A selective medium for *Gaeumannomyces*-like fungi. *Plant Disease* 75, 1075. (C5)

Elliott, M.L. (1991b) Determination of an etiological agent of bermudagrass decline. *Phytopathology* 81, 1380–1384. (C5)

Elliott, M.L., Des Jardin, E.A. and Henson, J.M. (1993) Use of a polymerase chain reaction assay to aid in identification of *Gaeumannomyces graminis* var. *graminis* from different grass hosts. *Phytopathology* 83, 414–418. (C5)

El-Nashaar, H.M., Moore, L.W. and George, R.A. (1986) Enzyme-linked immunosorbent assay quantification of initial infection of wheat by *Gaeumannomyces graminis* var. *tritici* as moderated by biocontrol agents. *Phytopathology* 76, 1319–1322. (C5)

Epstein, L., Kaur, S., Goins, T., Kwon, Y.H. and Henson, J.M. (1994) Production of hyphopodia by wild-type and three transformants of *Gaeumannomyces graminis* var. *graminis*. *Mycologia* 86, 72–81. (C5)

Eriksson, O.E. and Hawksworth, D.L. (1993) Outline of the ascomycetes – 1993. *Systema Ascomycetum* 12, 51–257. (C5)

Evans, I., Demulder, J., Maurice, D., Penney, D. and Solberg, E. (1992) Take-all and other diseases of wheat affected by copper deficiency and herbicide application. *Canadian Journal of Plant Pathology* 14, 241–242. (C3)

Evans, I.R., Solberg, E.D., Penney, D.C. and Maurice, D. (1995) Copper, a key element for disease control in wheat and barley crops in Alberta. *Phytopathology* 85, 1125. (C3)

Faull, J.L. and Powell, K.A. (1995) Biological control agents. In: Godfrey, C.R.A. (ed.) *Agrochemicals from Natural Products*. Marcel Dekker, New York, pp. 369–393. (C7)

Feest, A. and Campbell, R. (1986) The microbiology of soils under successive wheat crops in relation to take-all disease. *FEMS Microbiology Ecology* 38, 99–112. (C4)

Fenton, J.P. (1996) Precision farming – a hands-on perspective. In: *Brighton Crop Protection Conference – Pests and Diseases 1996: Proceedings*. British Crop Protection Council, Farnham, pp. 1105–1112. (C7)

Ferrandino, F.J. (1989) A distribution free method for estimating the effect of aggregated plant damage on crop yield. *Phytopathology* 79, 1229–1232. (C7)

Fiddian, W.E.H. (1973) The changing pattern of cereal growing. *Annals of Applied Biology* 75, 123–128. (C1)

Fitt, B.D.L. (1992) Eyespot of cereals. In: Singh, U.S., Mukhopadhyay, A.N., Kumar, J. and Chaube, H.S. (eds) *Plant Diseases of International Importance*. Prentice-Hall, Englewood Cliffs, New Jersey, pp. 336–354. (C6)

Fitt, B.D.L. and Hornby, D. (1978) Effects of root-infecting fungi on wheat transport processes and growth. *Physiological Plant Pathology* 13, 335–346. (C4)

Folwell, R.J., Cook, R.J., Heim, M.N. and Moore, D.L. (1991) Economic significance of take-all on winter wheat in the Pacific Northwest USA. *Crop Protection* 10, 391–395. (C1, C6)

Fouly, H.M., Pedersen, W.L., Wilkinson, H.T. and Abd El-Kader, M.M. (1996a) Wheat root rotting fungi in the 'old' and 'new' agricultural lands of Egypt. *Plant Disease* 80, 1298–1300. (C1)

Fouly, H.M., Wilkinson, H.T. and Domier, L.L. (1996b) Use of random amplified polymorphic DNA (RAPD) for identification of *Gaeumannomyces* species. *Soil Biology and Biochemistry* 28, 703–710. (C5)

Fox, R.T.V. (1993) Immunological techniques for identification. In: Fox, R.T.V. (ed.) *Principles of Diagnostic Techniques in Plant Pathology*. CAB International, Wallingford, pp. 129–151. (C5)

Foxroberts, J.J. and Deacon, J.W. (1988) Biological control of take-all fungi. Europe, Patent, 279676, 24 August 1988. (C4)

Fraselle, J. and Schiffers, B. (1982) L'enrobage des semences en tant que vecteur phytosanitaire pour une protection à long terme. *Mededelingen van de Faculteit Landbouwwetenschappen der Rijksuniversiteit te Gent* 47, 665–673. (C4)

French, C.S. (1985) *Computer Science*. DP Publications, Eastleigh. (C2)

Gair, R., Jenkins, J.E.E. and Lester, E. (1983) *Cereal Pests and Diseases*. Farming Press, Ipswich. (C0, C1)

Garcia, C. and Mathre, D.E. (1987) Factors affecting control of take-all of spring wheat by seed treatment with sterol biosynthesis-inhibiting fungicides. *Plant Disease* 71, 743–746. (C6)

Gardner, W.K. and Flynn, A. (1988) The effect of gypsum on copper nutrition of wheat grown in marginally deficient soil. *Journal of Plant Nutrition* 11, 475–493. (C3)

Garrett, S.D. (1937) The take-all or whiteheads disease of wheat and barley, and its control. *The Journal of the Royal Agricultural Society of England* 98, 24–34. (C1)

Garrett, S.D. (1938) Soil conditions and the take-all disease of wheat. III. Decomposition of the resting mycelium of *Ophiobolus graminis* in infected wheat stubble buried in the soil. *Annals of Applied Biology* 25, 742–766. (C2)

Garrett, S.D. (1950) The control of take-all under intensive cereal cultivation. *Agriculture, London* 56, 514–516. (C1)

Garrett, S.D. (1956) *Biology of Root-infecting Fungi*. Cambridge University Press, Cambridge. (C2)

Garrett, S.D. (1981) Introduction. In: Asher, M.J.C. and Shipton, P.J. (eds) *Biology and Control of Take-all*. Academic Press, London, pp. 1–11. (C2, C3, C5)

Garrett, S.D. and Buddin, W. (1947) Control of take-all under the Chamberlain System of intensive barley growing. *Agriculture, London* 54, 425–426. (C3)

Garrett, S.D. and Dennis, R.W.G. (1943) Note on the occurrence of *Ophiobolus graminis* Sacc. var. *avenae* E.M. Turner in Scotland in 1942. *Transactions of the British Mycological Society* 26, 146–147. (C5)

Garthwaite, D.G., Thomas, M.R. and Hart, M. (1995) *Pesticide Usage Survey Report 127: Arable Farm Crops in Great Britain 1994*. MAFF Publications, London, 97 pp. (C4)

Gaunt, R.E. (1980) Physiological basis of yield loss. In: Teng, P.S. and Krupa, S.V. (eds) *Crop Loss Assessment*. Minnesota Agricultural Experimental Station, St Paul, Minnesota, pp. 98–111. (C6)

Geddens, R.M., Appleby, A.P. and Powelson, R.L. (1990) Effects of herbicides on take-all disease (*Gaeumannomyces graminis*) in winter wheat (*Triticum aestivum*). *Weed Technology* 4, 478–481. (C4)

Ghisalberti, E.L., Narbey, M.J., Dewan, M.M. and Sivasithamparam, K. (1990) Variability among strains of *Trichoderma harzianum* in their ability to reduce take-all and to produce pyrones. *Plant and Soil* 121, 287–291. (C4)

Gilligan, C.A. (1982) Size and shape of sampling units for estimating incidence of sharp eyespot, *Rhizoctonia cerealis*, in plots of wheat. *Journal of Agricultural Science, Cambridge* 99, 461–464. (C6)

Gilligan, C.A. (1985a) Construction of temporal models. III. Disease progress of soil-borne pathogens. In: Gilligan, C.A. (ed.) *Mathematical Modelling of Crop Disease*. Academic Press, London, pp. 67–102. (C2)

Gilligan, C.A. (1985b) Probability models for host infection by soil-borne fungi. *Phytopathology* 75, 61–67. (C2)

Gilligan, C.A. (1990a) Comparison of disease progress curves. *New Phytologist* 115, 223–242. (C2)

Gilligan, C.A. (1990b) Mathematical modeling and analysis of soilborne pathogens. In: Kranz, J. (ed.) *Epidemiology of Plant Diseases: Mathematical Analysis and Modeling*. Springer-Verlag, Heidelberg, pp. 96–142. (C2, C6)

Gilligan, C.A. (1994a) Temporal aspects of the development of root disease epidemics. In: Campbell, C.L. and Benson, D.M. (eds) *Epidemiology and Management of Root Diseases*. Springer-Verlag, Berlin, pp. 148–194. (C2)

Gilligan, C.A. (1994b) The dynamics of infection by the take-all fungus on seminal roots of wheat: sensitivity analysis of a stochastic simulation model. *New Phytologist* 128, 539–553. (C2)

Gilligan, C.A. and Brassett, P.R. (1990) Modelling and estimation of the relative potential for infection of winter wheat by inoculum of *Gaeumannomyces graminis* derived from propagules and infected roots. *Journal of Phytopathology* 129, 58–68. (C2)

Gilligan, C.A. and Simons, S.A. (1987) Inoculum efficiency and pathozone width for two host–parasite systems. *New Phytologist* 107, 549–566. (C2)

Gilligan, C.A., Brassett, P.R. and Campbell, A. (1994) Modelling of early infection of cereal roots by the take-all fungus: a detailed mechanistic simulator. *New Phytologist* 128, 515–537. (C2)

Glenn, O.F. and Parker, C.A. (1988) Growth and infectivity of *Gaeumannomyces graminis* var. *tritici* in soil. *Soil Biology and Biochemistry* 20, 575–576. (C2)

Glenn, O.F. and Sivasithamparam, K. (1991) The influence of soil pH on the saprophytic growth in soil of the take-all fungus *Gaeumannomyces graminis* var. *tritici*. *Australian Journal of Soil Research* 29, 627–634. (C2)

Glenn, O.F., Parker, C.A. and Sivasithamparam, K. (1988) Use of ^{14}C-labelled wheat tissue to demonstrate saprophytic growth of *Gaeumannomyces graminis* var. *tritici* in soil. *Transactions of the British Mycological Society* 90, 545–550. (C2, C4)

Goodwin, P.H., Hsiang, T., Xue, B.G. and Liu, H.W. (1995) Differentiation of *Gaeumannomyces graminis* from other turf-grass fungi by amplification with primers from ribosomal internal transcribed spacers. *Plant Pathology* 44, 384–391. (C5)

Görlach, J., Volrath, S., Knauf-Beiter, G., Hengy, G., Beckhove, U., Kogel, K.H., Oostendorp, M., Staub, T., Ward, E., Kessmann, H. and Ryals, J. (1996) Benzothiadiazole, a novel class of inducers of systemic acquired resistance, activates gene expression and disease resistance in wheat. *The Plant Cell* 8, 629–643. (C7)

Graham, J.H. and Menge, J.A. (1982) Influence of vesicular-arbuscular mycorrhizae and soil phosphorus on take-all disease of wheat. *Phytopathology* 72, 95–98. (C3)

Graham, R.D. (1983) Effects of nutrient stress on susceptibility of plants to disease with particular reference to trace elements. *Advances in Botanical Research* 10, 221–276. (C3)

Graham, R.D. (1991) Micronutrients and disease resistance and tolerance in plants. In: Welch, R.M. (ed.) *Micronutrients in Agriculture*. Soil Science Society of America, Madison, Wisconsin, pp. 329–370. (C3)

Graham, R.D. and Rovira, A.D. (1984) A role for manganese in the resistance of wheat plants to take-all. *Plant and Soil* 78, 441–444. (C3)

Graneto, M.J., Phillion, D.P., Pratt, J.K. and Wong, S.C. (1994) Fungicides for the control of take-all diseases of plants. Canada, Patent, 2121917, 7 October 1994. (C4)

Green, C.F. and Ivins, J.D. (1984) Late infestations of take-all (*Gaeumannomyces graminis* var. *tritici*) on winter wheat (*Triticum aestivum* cv. Virtue): yield, yield components and photosynthetic potential. *Field Crops Research* 8, 199–206. (C1, C6)

Green, D. II, Fry, J., Pair, J. and Tisserat, N. (1992) Pathogenicity of fungi associated with a patch disease of zoysiagrass in Kansas. *Phytopathology* 82, 1123. (C4)

Greenacre, M.J. (1988) *Theory and Applications of Correspondence Analysis*. Academic Press, London. (C2)

Gregory, P.H. (1951) The fungi of Hertfordshire. *Transactions of the Hertfordshire Natural History Society and Field Club* 23, 135–208. (C1)

Griffiths, R.L. (1933) 'Take-all.' Incidence and control on the lighter soils of the mallee. *Journal of the Department of Agriculture of South Australia* 36, 774–778. (C3)

Grose, M.J. (1989) Nitrogen form and growth of *Gaeumannomyces graminis* var. *tritici* in soil. *Mycological Research* 93, 112–114. (C4)

Grose, M.J., Parker, C.A. and Sivasithamparam, K. (1984) Growth of *Gaeumannomyces graminis* var. *tritici* in soil: effects of temperature and water potential. *Soil Biology and Biochemistry* 16, 211–216. (C2, C4)

Grose, M.J., Gilligan, C.A., Spencer, D. and Goddard, B.V.D. (1996) Spatial heterogeneity of soil water around single roots: use of CT-scanning to predict fungal growth in the rhizosphere. *New Phytologist* 133, 261–272. (C2)

Gudauskas, R.T., Hagan, A.K., Morgan-Jones, G. and Williams, C. (1984) Take-all disease of wheat moves into Alabama. *Alabama Agricultural Experiment Station, Auburn University Highlights of Agricultural Research* 31, 16. (C2)

Gurusiddaiah, S., Jois, Y.H.R., Weller, D.M. and Cook, R.J. (1986a) Phenazine antibiotic from *Pseudomonas fluorescens*. US Patent Application, US 817374, 6 June 1986. (C4)

Gurusiddaiah, S., Weller, D.M., Saker, A. and Cook, R.J. (1986b) Characterisation of an antibiotic produced by a strain of *Pseudomonas fluorescens* inhibitory to *Gaeumannomyces graminis* var. *tritici* and *Pythium* spp. *Antimicrobial Agents and Chemotherapy* 29, 488–495. (C4)

Gutteridge, R.J. and Slope, D.B. (1978) Effect of inoculating soils with *Phialophora radicicola* var. *graminicola* on take-all disease of wheat. *Plant Pathology* 27, 131–135. (C4, C7)

Gutteridge, R.J., Bateman, G.L. and Hornby, D. (1987) Comparison of the effects of spring applications of ammonium chloride and other nitrogen fertilizers on take-all in winter wheat. *Journal of Agricultural Science, Cambridge* 108, 567–572. (C1, C3, C6)

Gutteridge, R.J., Hornby, D., Hollins, T.W. and Prew, R.D. (1993) Take-all in autumn-sown wheat, barley, triticale and rye grown with high and low inputs. *Plant Pathology* 42, 425–431. (C3)

Gutteridge, R.J., Jenkyn, J.F. and Poulton, P.R. (1996) Occurrence of severe take-all in winter wheat after many years of growing spring barley, and effects of soil phosphate. *Aspects of Applied Biology* 47, 453–458. (C3)

Hansen, M.A. and Wick, R.L. (1993) Plant disease diagnosis. In: Andrews, J.H. and Tommerup, I.C. (eds) *Advances in Plant Pathology*. Academic Press, San Diego, pp. 65–126. (C5)

Harris, J.R. (1985) Use of aerial photography for assessing soilborne disease. In: Parker, C.A., Rovira, A.D., Moore, K.J., Wong, P.T.W. and Kollmorgen, J.F. (eds) *Ecology and Management of Soil-borne Plant Pathogens*. APS Press, St Paul, Minnesota, pp. 27–30. (C6)

Harrison, P.A. (1996) Modelling the effects of climate change on wheat productivity in Europe. *Aspects of Applied Biology* 45, 41–48. (C7)

Harvey, P. (1993) Genetic diversity among populations of the take-all fungus *Gaeumannomyces graminis*. PhD Dissertation, The University of Adelaide, Australia. (C5)

Headrick, J.M. and Bockus, W.W. (1996) Advances in the development of take-all–yield loss relationships using the novel fungicide MON 41100. *Phytopathology* 86 (Suppl.), S27–S28. (C4, C6)

Heim, M., Folwell, R.J., Cook, R.J. and Kirpes, D.J. (1986) Economic benefits and costs of biological control of take-all to the Pacific Northwest wheat industry. *Research Bulletin* 0988, Agriculture Research Center, College of Agriculture and Home Economics, Washington State University, Pullman, 15 pp. (C1, C4)

Hemming, B.C. and Houghton, J.M. (1993) Influence of biotechnology on biocontrol of take-all disease of wheat. In: Chet, I. (ed.) *Biotechnology in Plant Disease Control*. Wiley-Liss, New York, pp. 15–38. (C4, C7)

Henson, J.M. (1989) DNA probe for identification of the take-all fungus, *Gaeumannomyces graminis*. *Applied and Environmental Microbiology* 55, 284–288. (C5)

Henson, J.M. (1992) DNA hybridisation and polymerase chain reaction (PCR) tests for the identification of *Gaeumannomyces*, *Phialophora* and *Magnaporthe* isolates. *Mycological Research* 96, 629–636. (C5)

Henson, J.M. and Caesar-Tonthat, C. (1995) Mitochondrial plasmids of the *Gaeumannomyces–Phialophora* complex and their detection by primed, in situ fluorescence labeling. *Experimental Mycology* 19, 263–274. (C5)

Henson, J.M. and French, R. (1993) The polymerase chain reaction and plant disease diagnosis. *Annual Review of Phytopathology* 31, 81–109. (C5)

Henson, J.M., Blake, N.K. and Pilgeram, A.L. (1988) Transformation of *Gaeumannomyces graminis* to benomyl resistance. *Current Genetics* 14, 113–117. (C5)

Henson, J.M., Goins, T., Grey, W., Mathre, D.E. and Elliott, M.L. (1993) Use of polymerase chain reaction to detect *Gaeumannomyces graminis* DNA in plants grown in artificially and naturally infested field soil. *Phytopathology* 83, 283–287. (C5)

Herdina, Harvey, P. and Ophel-Keller, K. (1996) Quantification of *Gaeumannomyces graminis* var. *tritici* in infected roots and soil using slot-blot hybridization. *Mycological Research* 100, 962–970. (C5)

Heritage, A.D., Rovira, A.D., Bowen, G.D. and Correll, R.L. (1989) Influence of soil water on the growth of *Gaeumannomyces graminis* var. *tritici* in soil: use of a mathematical model. *Soil Biology and Biochemistry* 21, 729–732. (C1, C2)

Herman, M. and Dovrtěl, J. (1991) [Relations between winter wheat yields and *Gaeumannomyces graminis* infection]. *Sborník ÚVTIZ, Ochrana Rostlin* 27, 53–62. (C1, C6)

Heyland, K.-U. and Kühnhold, J. (1984) Fusskrankheitsbefall und dessen Einfluss auf die Ertragsbildung von Winterweizen in extremen Getriedefruchtfolgen. *Zeitschrift für Pflanzenkrankheiten und Pflanzenschutz* 91, 354–370. (C6)

HMSO (1988) *Agricultural Statistics United Kingdom 1986*. HMSO, London. (C1)

HMSO (1989a) *Agricultural Statistics United Kingdom 1987*. HMSO, London. (C1)

HMSO (1989b) *CSO Annual Abstract of Statistics 1989 Edition*. HMSO, London. (C1)

HMSO (1990) *Agricultural Statistics United Kingdom 1988*. HMSO, London. (C1)

Hodgeson, D.R., Hipps, N.A. and Braim, M.A. (1989) Direct drilling compared with ploughing for winter wheat grown continuously and the effects of subsoiling. *Soil Use and Management* 5, 189–194. (C7)

Holden, M. and Ashby, M. (1981) Growth on oat-seedling agar of isolates of *Gaeumannomyces graminis* and some *Phialophora* species from cereal roots. *Transactions of the British Mycological Society* 77, 543–547. (C5)

Holden, M. and Hornby, D. (1981) Methods of producing perithecia of *Gaeumannomyces graminis* and their application to related fungi. *Transactions of the British Mycological Society* 77, 107–118. (C2, C5)

Hollins, T.W. and Scott, P.R. (1990) Pathogenicity of *Gaeumannomyces graminis* isolates to wheat and rye seedlings. *Plant Pathology* 39, 269–273. (C5)

Hollins, T.W., Scott, P.R. and Gregory, R.S. (1986) The relative resistance of wheat, rye and triticale to take-all caused by *Gaeumannomyces graminis*. *Plant Pathology* 35, 93–100. (C3, C5)

Home-Grown Cereals Authority (1995) *Cereal Statistics 1994*. Market Information, Home-Grown Cereals Authority, London. (C1)

Honeyman, A.L. and Currier, T.C. (1986) Isolation and characterization of linear DNA elements from the mitochondria of *Gaeumannomyces graminis*. *Applied and Environmental Microbiology* 52, 924–929. (C5)

Hoog, G.S. de and Guarro, J. (eds) (1995) *Atlas of Clinical Fungi*. CBS, Baarn. (C5)

Hornby, D. (1975) Inoculum of the take-all fungus: nature, measurement, distribution and survival. *EPPO Bulletin* 5, 319–333. (C2)

Hornby, D. (1978a) *Gaeumannomyces–Phialophora* complex: an early isolate of Prr. *Rothamsted Experimental Station Report for 1977, Part 1*, 216. (C1)

Hornby, D. (1978b) The problems of trying to forecast take-all. In: Scott, P.R. and Bainbridge, A. (eds) *Plant Disease Epidemiology*. Blackwell Scientific Publications, Oxford, pp. 151–158. (C1, C2, C4)

Hornby, D. (1979) Take-all decline: a theorist's paradise. In: Schippers, B. and Gams, W. (eds) *Soil-borne Plant Pathogens*. Academic Press, London, pp. 133–156. (C2, C5)

Hornby, D. (1981) Inoculum. In: Asher, M.J.C. and Shipton, P.J. (eds) *Biology and Control of Take-all*. Academic Press, London, pp. 271–293. (C2, C5, C6)

Hornby, D. (1982) A wheat root pathologist in South America. *ARC News*, Autumn 1982, p. 22. (C1)

Hornby, D. (1983) Suppressive soils. *Annual Review of Phytopathology* 21, 65–85. (C2)

Hornby, D. (1985) Soil nutrients and take-all. *Outlook on Agriculture* 14, 122–128. (C1, C3, C4)

Hornby, D. (1987) Field testing putative biological controls of take-all: rationale and results. *EPPO Bulletin* 17, 615–623. (C4)

Hornby, D. (1989) Take-all. *Institute of Arable Crops Research. Report for 1988*, 111–112. (C4)

Hornby, D. (1990a) Diseases. *Institute of Arable Crops Research. Report for 1989*, 65. (C7)

Hornby, D. (1990b) Root diseases. In: Lynch, J.M. (ed.) *The Rhizosphere*. John Wiley & Sons, Chichester, pp. 233–258. (C1, C2)

Hornby, D. (1992a) New information about take-all decline and its relevance to research on the control of take-all by biological control agents. In: Tjamos, E.C., Papavizas, G.C. and Cook, R.J. (eds) *Biological Control of Plant Diseases: Progress and Challenges for the Future*. Plenum Press, New York and London, pp. 95–98. (C2)

Hornby, D. (1992b) The changing face of take-all: evidence, causes and implications. In: McCracken, A.R. and Mercer, P.C. (eds) *Disease Management in Relation to Changing Agricultural Practice*. DANI/Queens University, Belfast, pp. 58–64. (C1)

Hornby, D. (1994) Aspects of the autecology of the take-all fungus. In: Blakeman, J.P. and Williams, B. (eds) *Ecology of Plant Pathogens*. CAB International, Wallingford, pp. 209–226. (C1, C2, C6)

Hornby, D. and Bateman, G.L. (1990) Artificial infestation of soil with *Gaeumannomyces graminis* var. *tritici* to study the relationship between take-all and wheat yields in field experiments. *Soil Use and Management* 6, 209–217. (C4, C6)

Hornby, D. and Bateman, G.L. (eds) (1991) *Take-all Disease of Cereals*. Home-Grown Cereals Authority, London. (C0, C1, C7)

Hornby, D. and Bateman, G.L. (1997) Potential use of plant root pathogens as bioindicators of soil health. In: Pankhurst, C.E., Doube, B.M. and Gupta, V.V.S.R. (eds) *Biological Indicators of Soil Health*. CAB International, Wallingford, pp. 179–200. (C1)

Hornby, D. and Davis, C.M.P. (1985) Assessing soil suppressiveness. *Rothamsted Report for 1984, Part 1*, 128–129. (C6)

Hornby, D. and Fitt, B.D.L. (1981) Effects of root-infecting fungi on structure and function of cereal roots. In: Ayres, P.G. (ed.) *Effects of Disease on the Physiology of the Growing Plant*. Cambridge University Press, pp. 101–130. (C2, C4)

Hornby, D. and Goring, C.A.I. (1972) Effects of ammonium and nitrate nutrition on take-all disease of wheat in pots. *Annals of Applied Biology* 70, 225–231. (C3)

Hornby, D. and Gutteridge, R.J. (1988) Known and supposed factors affecting take-all in winter wheat. In: *Abstracts of the 5th International Congress of Plant Pathology, Kyoto, 1988.* No. 446, p. 191. (C2)

Hornby, D. and Gutteridge, R.J. (1995) The natural biological control phenomenon of take-all decline in different sequences of cereals. In: Mańka, M. (ed.) *Environmental Biotic Factors in Integrated Plant Disease Control.* European Foundation for Plant Pathology and The Polish Phytopathological Society, Poznań, pp. 53–60. (C2, C3, C4, C5)

Hornby, D. and Henden, D.R. (1986) Epidemics of take-all during 16 years of continuous spring barley. *Annals of Applied Biology* 108, 251–264. (C1, C2, C3, C4, C6)

Hornby, D., Slope, D.B., Gutteridge, R.J. and Sivanesan, A. (1977) *Gaeumannomyces cylindrosporus*, a new ascomycete from cereal roots. *Transactions of the British Mycological Society* 69, 22–25. (C5)

Hornby, D., Henden, D.R. and Den Toom, A. (1979) Take-all disease. *Rothamsted Experimental Station Report for 1978, Part 1*, pp. 213–215. (C5)

Hornby, D., Henden, D.R. and Bedford, I. (1981) Studies of a site with little take-all. *Rothamsted Experimental Station Report for 1980, Part 1*, p. 184. (C6)

Hornby, D., Henden, D.R. and Catt, J.A. (1983) Some causes of yield variation in an intensive spring barley experiment at Woburn, 1972–8. *Journal of Agricultural Science, Cambridge* 100, 175–189. (C6)

Hornby, D., Bateman, G.L., Payne, R.W., Brown, M.E. and Henden, D.R. (1989) An experimental design and procedures for testing putative controls against naturally-occurring take-all in the field. *Annals of Applied Biology* 115, 195–208. (C2, C6)

Hornby, D., Bateman, G.L., Gutteridge, R.J., Lucas, P., Montfort, F. and Cavelier, A. (1990a) Experiments in England and France on fertilisers, fungicides and agronomic practices to decrease take-all. *Brighton Crop Protection Conference – Pests and Diseases 1990: Proceedings*, Vol. 2, 771–776. (C3, C4)

Hornby, D., Gutteridge, R. and Parsonage, R. (1990b) *Phialophora* spp. in relation to take-all disease of cereals. *Institute of Arable Crops Research Report for 1990*, 73–74. (C4, C5)

Hornby, D., Bateman, G.L., Payne, R.W., Brown, M.E., Henden, D.R. and Campbell, R. (1993) Field tests of bacteria and soil-applied fungicides as control agents for take-all in winter wheat. *Annals of Applied Biology* 122, 253–270. (C4)

Hossain, I. and Schlosser, E. (1986) Selective identification of *Gaeumannomyces graminis* on wheat plants with foot rot disease. *Mededelingen van de Faculteit Landbouwwetenschappen der Rijksuniversiteit te Gent* 51, 597–601. (C5)

Howie, W.J., Cook, R.J. and Weller, D.M. (1987) Effects of soil matric potential and cell motility on wheat root colonization by fluorescent pseudomonads suppressive to take-all. *Phytopathology* 77, 286–292. (C4)

Huber, D.M. (1981a) Incidence and severity of take-all of wheat in Indiana. *Plant Disease* 65, 734–737. (C1)

Huber, D.M. (1981b) The role of nutrients and chemicals. In: Asher, M.J.C. and Shipton, P.J. (eds) *Biology and Control of Take-all*. Academic Press, London, pp. 317–342. (C1, C3, C4)

Huber, D.M. (1990) Fertilizers and soil-borne diseases. *Soil Use and Management* 6, 168–173. (C4)

Huber, D.M. and McCay-Buis, T.S. (1993) A multiple component analysis of the take-all disease of cereals. *Plant Disease* 77, 437–447. (C1, C2, C3)

Huber, D.M., Painter, C.C., McKay, H.C. and Petersen, D.L. (1968) Effect of nitrogen fertilization on take-all of winter wheat. *Phytopathology* 58, 1470–1478. (C4)

Hughes, G. (1990) Characterizing crop response to patchy pathogen attack. *Plant Pathology* 39, 2–4. (C7)

Huisman, O.J. (1982) Interrelations of root growth dynamics to epidemiology of root-invading fungi. *Annual Review of Phytopathology* 20, 303–327. (C2)

Hunt, L.A. and Pararajasingham, S. (1995) CROPSIM-WHEAT: a model describing the growth and development of wheat. *Canadian Journal of Plant Science* 75, 619–632. (C2)

Jackson, N. and Landschoot, P.J. (1984) *Gaeumannomyces cylindrosporus* associated with diseased turfgrass in Rhode Island. *Phytopathology* 76, 654. (C4)

Jadot, R., Frankinet, M., Raepsaet, J. and Grevy, L. (1982) La lutte biologique contre *Gaeumannomyces graminis* var. *tritici*, agent du piétin échaudage. 2. Évolution du potentiel infectieux après introduction de souches d'agressivités différentes et effet sur le rendement. *Bulletin des Recherches Agronomiques de Gembloux* 17, 345–362. (C4)

Jahn, M., Kluge, E. and Enzian, S. (1996) Influence of climate diversity on fungal diseases of field crops – evaluation of long-term monitoring data. *Aspects of Applied Biology* 45, 247–252. (C7)

James, W.C. (1974) Assessment of plant diseases and losses. *Annual Review of Phytopathology* 12, 27–48. (C6)

James, W.C. and Teng, P.S. (1979) The quantification of production constraints associated with plant diseases. *Applied Biology* 4, 201–267. (C6)

Jamil, N. and Buck, K.W. (1990) Virion-associated RNA polymerase in group III virus from *Gaeumannomyces graminis*. *Karachi University Journal of Science* 18, 173–179. (C5)

Jamil, N. and Buck, K.W. (1991) Effect of vegetative incompatability on double stranded RNA and mycovirus transmission in *Gaeumannomyces graminis* var. *tritici*. *Pakistan Journal of Botany* 23, 160–164. (C5)

Jamil, N. and Buck, K.W. (1992) Isolation and characterization of some viruses associated with the wheat pathogen *Gaeumannomyces graminis* var. *tritici*. *Pakistan Journal of Botany* 24, 187–196. (C5)

Jamil, N., Buck, K.W. and Carlile, M.J. (1984) Sequence relationship between virus double stranded RNA from isolates of *Gaeumannomyces graminis* in different vegetative compatibility groups. *Journal of General Virology* 65, 1741–1747. (C5)

Jeger, M.J. (1987) The influence of root growth and inoculum density on the dynamics of root disease epidemics: theoretical analysis. *New Phytologist* 107, 459–478. (C2)

Jeger, M.J. and Lyda, S.D. (1986) Epidemics of Phymatotrichum root rot (*Phymatotrichum omnivorum*) in cotton: environmental correlates of final incidence and forecasting criteria. *Annals of Applied Biology* 109, 523–534. (C1)

Jenkyn, J.F., Gutteridge, R.J. and Thomas, M.R. (1988) Effects of straw incorporation and cultivations on cereal diseases. *Aspects of Applied Biology* 17, 181–189. (C3)

Jenkyn, J.F., Gutteridge, R.J. Darby, R.J. and Carreck, N. (1991a) Interactions of N with diseases. *Institute of Arable Crops Research Report for 1990*, 42–43. (C6)

Jenkyn, J.F., Gutteridge, R.J. and Todd, A.D. (1991b) Effects of fungicides, applied in autumn, and a growth regulator, applied in spring, on the growth and yield of winter barley grown on contrasting soil types. *Journal of Agricultural Science, Cambridge* 117, 287–297. (C4)

Jenkyn, J.F., Carter, N., Darby, R.J., Gutteridge, R.J., Mullen, L.A., Plumb, R.T., Ross, G.J.S., Todd, A.D., Widdowson, F.V. and Wood, D.W. (1992a) Effects of seven factors on the growth and yield of winter barley grown as a third consecutive take-all susceptible crop and of growing the barley after oats or a fallow. *Journal of Agricultural Science, Cambridge* 119, 303–333. (C4, C6)

Jenkyn, J.F., Gutteridge, R.J. and Todd, A.D. (1992b) Effects of sowing winter barley on different dates in autumn on the severity of take-all (*Gaeumannomyces graminis* var. *tritici*) in those and the subsequent crops. *Journal of Agricultural Science, Cambridge* 119, 19–25. (C3, C6)

Jenkyn, J.F., Gutteridge, R.J., Gladders, P. and Yarham, D.J. (1996) Effects of cultivating one-year rotational set-aside at different times on take-all disease of winter wheat. *Aspects of Applied Biology* 47, 449–452. (C3)

Jensen, H.P. and Jørgensen, J.H. (1973) Reactions of five cereal species to the take-all fungus (*Gaeumannomyces graminis*) in the field. *Phytopathologische Zeitschrift* 78, 193–203. (C6)

Johnson, G. (1997) Why wheat is losing its gloss. *Crops* 1 February, 30–31. (C7)

Johnston, A.E. (1994) The Rothamsted Classical Experiments. In: Leigh, R.A. and Johnston, A.E. (eds) *Long-term Experiments in Agricultural and Ecological Sciences*. CAB International, Wallingford, pp. 9–37. (C1, C4)

Jones, D.G. and Clifford, B.C. (1978) *Cereal Diseases: their Pathology and Control*. BASF United Kingdom Ltd, Ipswich. (C0)

Jones, D.R. (1987) The effects of early-season fungicides on quality of winter wheat. *Aspects of Applied Biology* 15, 395–401. (C4, C6)

Jones, S.G. (1926) The development of the perithecium of *Ophiobolus graminis* Sacc. *Annals of Botany* XL, 607–629. (C5)

Jouan, B. and Lemaire, J.-M. (1976) Product for prevention against ophiobolus take-all in cereals and the like. United States, Patent, 3,987,165, 19 Oct 1976. (C4)

Juhnke, M.E., Mathre, D.E. and Sands, D.C. (1984) A selective medium for *Gaeumannomyces graminis* var. *tritici*. *Plant Disease* 68, 233–236. (C5)

Keel, C., Voisard, C., Haas, D. and Défago, G. (1990) Role of 2,4-diacetylphloroglucinol in disease suppression by a strain of *Pseudomonas fluorescens*. *Phytopathology* 80, 1024. (C4)

Kent, N.L. and Evers, A.D. (1994) *Technology of Cereals. An Introduction for Students of Food Science and Agriculture*. Pergamon (Elsevier Science), Oxford. (C0, C1, C3)

Kessmann, H., Staub, T., Hofmann, C., Maetzke, T., Herzog, J., Ward, E., Uknes, S. and Ryals, J. (1994) Induction of systemic acquired disease resistance in plants by chemicals. *Annual Review of Phytopathology* 32, 439–459. (C7)

Kessmann, H., Oostendorp, M., Staub, T., Goerlach, J., Friedrich, L., Lawton, K. and Ryals, J. (1996) CGA 245704: mode of action of a new plant activator. *Proceedings of the 1996 Brighton Crop Protection Conference – Pests and Diseases* 3, 961–966. (C7)

King, J.E. (1977) Survey of foliar and stem base diseases and take-all of winter wheat in England and Wales. MAFF ADAS 'Open' Conference of advisory Plant Pathologists PP/0/405, 9 pp. (C1, C6)

Kirk, J.J. and Deacon, J.W. (1987) Control of the take-all fungus by *Microdochium bolleyi*, and interactions involving *M. bolleyi*, *Phialophora graminicola* and *Periconia macrospinosa* on cereal roots. *Plant and Soil* 98, 231–237. (C4)

Kirkegaard, J.A., Wong, P.T.W. and Desmarchelier, J.M. (1996) *In vitro* suppression of fungal root pathogens of cereals by *Brassica* tissues. *Plant Pathology* 45, 593–603. (C3)

Kjaergaard, T. (1995) Agricultural development and nitrogen supply from an historical point of view. In: Kristensen, L. (ed.) *Nitrogen Leaching in Ecological Agriculture*. AB Academic Publishers, Bicester, pp. 3–14. (C1)

Klepper, B. and Rickman, R.W. (1990) Modeling crop root growth and function. *Advances in Agronomy* 44, 113–132. (C2)

Kohlmeyer, J., Volkmann-Kohlmeyer, B. and Ericksson, O.E. (1995) Fungi on *Juncus roemerianus*. 4. New marine Ascomycetes. *Mycologia* 87, 532–542. (C5)

Kollmorgen, J.F. (ed.) (1985) Proceedings of the First International Workshop on Take-all of Cereals. In: Parker, C.A., Rovira, A.D., Moore, K.J. and Wong, P.T.W. (eds) *Ecology and Management of Soilborne Plant Pathogens*. The American Phytopathological Society, St Paul, Minnesota, pp. 289–351. (C0, C1, C3, C6)

Kranz, J. (1974) *Epidemics of Plant Diseases: Mathematical Analysis and Modelling*. Springer-Verlag, Berlin. (C2)

Kranz, J. and Rotem, J. (1988) Preface. In: Kranz, J. and Rotem, J. (eds) *Experimental Techniques in Plant Disease Epidemiology*. Springer-Verlag, Heidelberg, 2 pp. (C2)

Lagneau, C., Dandois, J., Bastin, V., Poncelet, J. and Maraite, H. (1986) Évolution et sévérité des maladies sur froment d'hiver et escourgeon en 1985. *Mededelingen van de Faculteit Landbouwwetenschappen der Rijksuniversiteit te Gent* 51, 603–616. (C1)

Landschoot, P.J. (1993) Taxonomy and biology of ectotrophic root-infecting fungi associated with patch diseases of turfgrasses. In: Clarke, B.B. and Gould, A.B. (eds) *Turfgrass Patch Diseases caused by Ectotrophic Root-infecting Fungi*. American Phytopathological Society, St Paul, Minnesota, pp. 41–71. (C5)

Landschoot, P.J. and Jackson, N. (1989) *Gaeumannomyces incrustans* sp. nov., a root-infecting hyphopodiate fungus from grass roots in the United States. *Mycological Research* 93, 55–58. (C5)

Lapierre, H., Lemaire, J.-M., Jouan, B. and Molin, G. (1970) Mise en évidence de particules virales associées à une perte de pathogénicité chez le piétin-échaudage des céréales, *Ophiobolus graminis* Sacc. *Comptes Rendus Hebdomadaires des Séances de l'Académie des Sciences* 271, 1833–1836. (C5)

Large, E.C. (1954) Growth stages in cereals. Illustrations of the Feekes' scale. *Plant Pathology* 3, 128–129. (C0)

Lawes, J.B. and Gilbert, J.H. (1870) 'Take-all'. *Journal of the Royal Horticultural Society* 2, lxxxvi–lxxxviii. (C1)

Leech, P.K. and Chalmers, A.G. (1985) *Survey of Fertiliser Practice. Fertiliser Use on Farm Crops in England & Wales. 1984.* (SS/CH/21). Agricultural Development and Advisory Service, Rothamsted Experimental Station and Fertiliser Manufacturers Association. 13 pp. + tables. (C3)

Leisinger, T. and Margraff, R. (1979) Secondary metabolites of the fluorescent pseudomonads. *Microbiological Reviews* 43, 422–442. (C4)

Lemaire, J.-M. and Coppenet, M. (1968) Influence de la succession céréalière sur les fluctuations de la gravité du piétin-échaudage (*Ophiobolus graminis* Sacc.). *Annales des Épiphyties* 19, 589–599. (C2)

Lemaire, J.-M. and Ponchet, J. (1963) *Phialophora radicicola* Cain, forme conidienne du *Linocarpon cariceti* B. et Br. *Comptes Rendus des Séances de l'Académie d'Agriculture de France* 49, 1067–1069. (C1)

Lemaire, J.-M., Lapierre, H., Jouan, B. and Bertrand, G. (1970) Découverte de particules virales chez certaines souches d'*Ophiobolus graminis*, agent du piétin-échaudage des céréales: consequences agronomique prévisibles. *Comptes Rendus Hebdomadaires des Séances de l'Académie des Sciences* 56, 1134–1138. (C5)

Lemaire, J.-M., Carpentier, F., Dalle, J.F. and Doussinault, G. (1979a) Lutte biologique contre le piétin-échaudage des céréales. Modifications physiologiques chez le blé inoculé par une souche atténuée d'*Ophiobolus graminis*. 2. Changement de la teneur en chlorophylle. *Annales de Phytopathologie* 11, 193–197. (C3, C4)

Lemaire, J.-M., Carpentier, F., Dalle, J.F., Doussinault, G. and Perraton, B. (1979b) Lutte biologique contre le piétin-échaudage des céréales, modifications physiologiques chez le blé inoculé par une souche atténuée d'*Ophiobolus graminis*. 1. Précocité accrue aux premiers stades du blé. *Comptes Rendus des Séances de l'Académie d'Agriculture de France* 65, 766–772. (C3, C4)

Lemaire, J.-M., Doussinault, G., Lucas, P., Perraton, B. and Messager, A. (1982) Possibilités de sélection pour l'aptitude à la prémunition dans le cas du piétin-échaudage des céréales, *Gaeumannomyces graminis*. *Cryptogamie, Mycologie* 3, 347–359. (C3, C4)

Lemerle, D., Tang Hong Yuan, Murray, G.M. and Morris, S. (1996) Survey of weeds and diseases in cereal crops in the southern wheat belt of New South Wales. *Australian Journal of Experimental Agriculture* 36, 545–554. (C1, C3)

Lennartson, E.K.M. (1990) The effect of green manuring and mixed species cropping on take-all disease of wheat. In: *Crop Protection in Organic and Low Input Agriculture, British Crop Protection Council Monograph 45*, BCPC, Farnham, Surrey, 219. (C4)

Lewis, T. (1994) Commitment to long-term agricultural research: a message for science, sponsors and industry. *Brighton Crop Protection Conference – Pests and Diseases 1994: Proceedings* 1, 3–20. (C1)

Liang, P.-Y. and Chen, K.-Y. (1990) Homology between *Gaeumannomyces graminis* virus and other mycoviruses by dot blot hybridization. *Chinese Journal of Virology* 6, 245–249. (C5)

Liang, P.-Y., Chou, S.-M. and Chen, K.-Y. (1989) Properties of double stranded RNA viruses from different strains of *Gaeumannomyces graminis* in different vegetative compatible groups. *Virologica Sinica* 4, 68–75. (C5)

Linde-Laursen, I., Jensen, H.P. and Jørgensen, J.H. (1973) Resistance of *Triticale*, *Aegilops* and *Haynaldia* species to the take-all fungus, *Gaeumannomyces graminis*. *Zeitschrift für Pflanzenzüchtung* 70, 200–213. (C3)

Long, E. (1990) Straw incorporation. Plan to beat the ban. *Crops* 7, 20. (C1)

Lucas, P. and Nignon, M. (1986) Influence du type de sol et de ses composantes physicochimiques sur les relations entre une variété de blé (*Triticum aestivum* L. var. Rescler) et deux souches, agressive et hypoagressive de *Gaeumannomyces graminis* (Sacc.) von Arx & Olivier var. *tritici* Walker. *Plant and Soil* 97, 105–117. (C4)

Lucas, P. and Sarniguet, A. (1990) Soil receptivity to take-all: influence of some cultural practices and soil chemical characteristics. *Symbiosis* 9, 51–58. (C3)

Lucas, P., Lemaire, J.-M., Doussinault, G., Perraton, B., Tivoli, B. and Carpentier, F. (1986) Lutte biologique contre *Gaeumannomyces graminis* Sacc. von Arx & Olivier var. *tritici* (Walker), agent du piétin-échaudage par l'utilisation d'une souche hypoaggressive du parasite. Résultats, perspectives. In: *Les rotations céréalières intensives. Dix années d'études concertées INRA–ONIC–ITCF, 1973–1983*. INRA, Paris, pp. 113–125. (C4)

Lucas, P., Montfort, F., Cavelier, N. and Cavelier, M. (1988) Attempts to combine different methods of control of foot-and-root diseases of winter wheat. In: Cavalloro, R. and Sunderland, K.D. (eds) *Integrated Crop Protection on Cereals*. A.A. Balkema, Rotterdam/Brookfield, pp. 237–249. (C4, C6)

Lucas, P., Capron, G. and Guillerm, A.-Y. (1994) Intérêt d'un fongicide expérimental (MON41100) pour l'étude de la nuisibilité du piétin-échaudage en culture de blé d'hiver. Conséquences sur l'utilisation de l'azote du sol par les plantes. In: *Annales, 4. Conférence Internationale sur les Maladies des Plantes*. ANPP (Association Nationale de Protection des Plantes), Paris, pp. 111–119. (C1, C2, C4, C6)

Lyon, G.D., Forrest, R.S. and Newton, A.C. (1996) SAR – the potential to immunise plants against infection. *Proceedings of the 1996 Brighton Crop Protection Conference – Pests and Diseases* 3, 939–946. (C7)

Maas, E.M.C. and Kotzé, J.M. (1987) *Trichoderma harzianum* and *Trichoderma polysporum* as biological control agents of take-all of wheat in the greenhouse. *Phytophylactica* 19, 365–368. (C4)

Maas, E.M.C., Bezuidenhout, J.J., Kotzé, J.M. and Grimbeek, R.J. (1989) A quantitative method for the simultaneous assessment of take-all of wheat in field plots and the isolation of the pathogen. *Phytophylactica* 21, 171–174. (C4)

Maas, E.M., Van Zyl, E., Steyn, P.L. and Kotzé, J.M. (1990) Comparison of soluble proteins of *Gaeumannomyces graminis* var. *tritici* and *Phialophora* spp. by polyacrylamide gel electrophoresis. *Mycological Research* 94, 78–82. (C5)

Macdonald, A.J., Powlson, D.S., Poulton, P.R. and Jenkinson, D.S. (1989) Unused fertiliser nitrogen in arable soils – its contribution to nitrate leaching. *Journal of the Science of Food and Agriculture* 46, 407–419. (C2)

Macdonald, A.J., Poulton, P.R., Powlson, D.S. and Purkis, C. (1991) Soil-applied fertiliser N. *AFRC Institute of Arable Crops Research Report for 1990*, 41. (C2)

Macdonald, A.J., Poulton, P.R., Powlson, D.S. and Jenkinson, D.S. (1997) Effects of season, soil type and cropping on recoveries, residues and losses of ^{15}N-labelled fertilizer applied to arable crops in spring. *Journal of Agricultural Science, Cambridge* 129, 125–154. (C2)

MacDonald, J.D. (1994) The soil environment. In: Campbell, C.L. and Benson, D.M. (eds) *Epidemiology and Management of Root Diseases*. Springer-Verlag, Berlin, pp. 82–116. (C2)

MacDuff, J.H. (1989) Growth and nutrient uptake in response to root temperature. *Aspects of Applied Biology* 22, 23–32. (C2)

MacKenzie, D. (1995) *Farmers Weekly*, 26 May, 59. (C1)

MacLeod, W.J., MacNish, G.C. and Thorn, C.W. (1993) Manipulation of ley pastures with herbicides to control take-all. *Australian Journal of Agricultural Research* 44, 1235–1244. (C4)

MacNish, G.C. (1976) Survival of *Gaeumannomyces graminis* var. *tritici* in artificially colonized straws buried in naturally infested soil. *Australian Journal of Biological Science* 29, 163–174. (C6)

MacNish, G.C. and Dodman, R.L. (1973) Relation between incidence of *Gaeumannomyces graminis* var. *tritici* and grain yield. *Australian Journal of Biological Science* 26, 1289–1299. (C6)

MacNish, G.C. and Speijers, J. (1982) The use of ammonium fertilizers to reduce the severity of take-all (*Gaeumannomyces graminis* var. *tritici*) on wheat in Western Australia. *Annals of Applied Biology* 100, 83–90. (C3)

MacNish, G.C., Liddle, J.M. and Powelson, R.L. (1986) Studies on the use of high-nutrient and low-nutrient inoculum for infection of wheat by *Gaeumannomyces graminis* var. *tritici*. *Phytopathology* 76, 815–819. (C5, C6)

MacNish, G.C. and Nicholas, D.A. (1987) Some effects of field history on the relationship between grass production in subterranean clover pasture, grain yield and take-all (*Gaeumannomyces graminis* var. *tritici*) in a subsequent crop of wheat at Bannister, Western Australia. *Australian Journal of Agricultural Research* 38, 1011–1018. (C3)

Madariaga, B.R. and Mellado, Z.M. (1985) Efecto de concentraciones de inoculo del hongo *Gaeumannomyces graminis* var. *tritici* sobre caracteristicas agronomicas de un trigo (*Triticum aestivum* L.). [Effect of inoculum concentration of *Gaeumannomyces graminis* var. *tritici* on agronomic features of a wheat (*Triticum aestivum* L.).] *Agricultura Tecnica (Chile)* 45, 15–20. (C6)

Madden, L.V. (1983) Measuring and modeling crop losses at the field level. *Phytopathology* 73, 1591–1596. (C6)

Madden, L.V. (1986) Statistical analysis and comparison of disease progress curves. In: Leonard, K.J. and Fry, W.E. (eds) *Plant Disease Epidemiology: Population Dynamics and Management*. Macmillan, New York, pp. 55–84. (C2)

Madden, L.V., Pennypacker, S.P., Antle, C.E. and Kingsolver, C.H. (1981) A loss model for crops. *Phytopathology* 71, 685–689. (C6)

Madge, W.E.R. (1987) Winter wheat: improving the yield of second crops grown on an organic fen soil. *Research and Development in Agriculture* 4, 37–41. (C3)

Maizel, J.V., Burkhardt, H.J. and Mitchell, H.K. (1964) Avenacin, an antimicrobial substance isolated from *Avena sativa*. I. Isolation and antimicrobial activity. *Biochemistry* 3, 424–431. (C5)

Manners, J.G. and Myers, A. (1981) Effects on host growth and physiology. In: Asher, M.J.C. and Shipton, P.J. (eds) *Biology and Control of Take-all*. Academic Press, London, pp. 237–248. (C1, C6)

Marschner, P., Ascher, J.S. and Graham, R.D. (1991) Effect of manganese-reducing rhizosphere bacteria on the growth of *Gaeumannomyces graminis* var. *tritici* and on manganese uptake by wheat *Triticum aestivum* L. *Biology and Fertility of Soils* 12, 33–38. (C3)

Martyniuk, S. (1987) The occurrence of *Phialophora*-like fungi related to *Gaeumannomyces graminis* under various grass species and some characteristics of these fungi. *EPPO Bulletin* 17, 609–614. (C5)

Martyniuk, S. (1988) Pectolytic enzymes activity and pathogenicity of *Gaeumannomyces graminis* var. *tritici* and related Phialophora-like fungi. *Plant and Soil* 107, 19–23. (C5)

Martyniuk, S. and Myśków, W. (1984) Control of the take-all fungus by *Phialophora* sp. (lobed hyphopodia) in microplot experiments with wheat. *Zentralblatt für Mikrobiologie* 139, 575–579. (C4)

Massee, G. (1912) 'White-heads' or 'take-all' of wheat and oats. *Bulletin of Miscellaneous Information. Royal Botanic Gardens, Kew*, 435–439. (C1)

Mathre, D.E. (1992) *Gaeumannomyces*. In: Singleton, L.L., Mihail, J.D. and Rush, C.M. (eds) *Methods for Research on Soilborne Phytopathogenic Fungi*. American Phytopathological Society, St Paul, Minnesota, pp. 60–63. (C5)

Mathre, D.E., Johnson, R.H. and Engel, R. (1986) Effect of seed treatment with triadimenol on severity of take-all of spring wheat caused by *Gaeumannomyces graminis* var. *tritici*. *Plant Disease* 70, 749–751. (C4)

Mattingley, G.E.G., Slope, D.B. and Gutteridge, R.J. (1980) Effects of phosphate and potassium on take-all and yield of winter wheat. *Report of Rothamsted Experimental Station for 1979* 1, 227–229. (C3)

Maurhofer, M., Sacherer, P., Keel, C., Haas, D. and Défago, G. (1994) Role of some metabolites produced by *Pseudomonas fluorescens* strain CHAO in the suppression of different plant diseases. In: Ryder, M.H., Stephens, P.M. and Bowen, G.D. (eds) *Improving Plant Productivity with Rhizosphere Bacteria*. CSIRO Division of Soils, Glen Osmond, South Australia, pp. 117–119. (C4)

McCarty, L.B. and Lucas, L.T. (1989) *Gaeumannomyces graminis* associated with spring dead spot of bermudagrass in the Southeastern United States. *Plant Disease* 73, 659–661. (C5)

McCay-Buis, T.S., Huber, D.M., Graham, R.D., Phillips, J.D. and Miskin, K.E. (1995) Manganese seed content and take-all of cereals. *Journal of Plant Nutrition* 18, 1711–1721. (C3)

McClelland, M. and Welsh, J. (1995) DNA Fingerprinting using arbitrarily primed PCR. In: Dieffenbach, C.W. and Dveksler, G.S. (eds) *PCR Primer. A Laboratory Manual*. Cold Spring Harbor, New York, pp. 203–211. (C5)

McFadden, J.J.P. and Buck, K.W. (1983) Sequence complexities of the nuclear and mitochondrial genomes of the take-all fungus, *Gaeumannomyces graminis* var. *tritici*. *Journal of General Microbiology* 129, 3515–3517. (C5)

McGinty, R.M., Buck, K.W. and Rawlinson, C.J. (1981) Virus particles and double-stranded RNA in isolates of *Phialophora* species with lobed hyphopodia, *Phialophora graminicola* and *Gaeumannomyces graminis* var. *graminis*. *Phytopathologische Zeitschrift* 102, 153–162. (C5)

McGinty, R.M., McFadden, J.J.P., Rawlinson, C.J. and Buck, K.W. (1984) Widespread inhibitor production in culture by isolates of *Gaeumannomyces graminis* var. *tritici*. *Transactions of the British Mycological Society* 82, 429–434. (C5)

McGrath, S.P. and Hellon, P.W. (1986) Micronutrient concentrations in wheat grain. *Rothamsted Report for 1985* 1, 172. (C3)

McGrath, S.P., Zhao, F.J., Withers, P.J.A., Sinclair, A.H. and Evans, E.J. (1995) *Sulphur Nutrition of Cereals in Britain: Yield Responses and Prediction of Likely Deficiency*. Home-Grown Cereals Authority, London. (C3)

Mekwatanakarn, P. and Sivasithamparam, K. (1987) Effect of certain herbicides on soil microbial populations and their influence on saprophytic growth in soil and pathogenicity of take-all fungus. *Biology and Fertility of Soils* 5, 174–180. (C3)

Mepham, T.B., Tucker, G.A. and Wiseman, J. (eds) (1995) *Issues in Agricultural Bioethics*. Nottingham University Press, Nottingham. (C0)

Meynard, J.M. (1985) Construction d'itinéraires techniques pour la conduite du blé d'hiver. Docteur-Ingénieur Thesis, INA-PG, Paris. (C1)

Mielke, H. (1974) Untersuchungen über die Anfälligkeit verschiedener Getreide-arten gegen den Erreger der Schwarzbeinigkeit, *Ophiobolus graminis* Sacc. *Mitteilungen aus der Biologischen Bundesanstalt für Land- und Forstwirtschaft, Berlin-Dahlem* 160, 61 pp. (C3)

Mielke, H. (1983) Schwarzbeinigkeit bei Getriede. Information zum Integrierten Pflanzenschutz. *Nachrichtenblatt des Deutschen Pflanzenschutzdienstes (Stuttgart)* 35, 143. (C1)

Mielke, H. (1988) Zum integrierten Pflanzenschutz gegen Fusskrankheiten und parasitäre Auswinterung des Getreides. *Gesunde Pflanzen* 40, 362–367. (C1, C2, C3, C4)

Mielke, H. (1992) Untersuchungen zum Befall der Gerste durch *Gaeumannomyces graminis* (Sacc.) von Arx & Olivier var. *tritici* Walker unter Berücksichtigung der Arten- und Sortenanfälligkeit [sic]. *Mitteilungen aus der Biologischen Bundesanstalt für Land- und Forstwirtschaft, Berlin-Dahlem* 276, 1–74. (C3)

Mielke, H. (1995) Schwarzbeinigkeit im Weizenbau. *Nachrichtenblatt des Deutschen Pflanzenschutzdienstes (Stuttgart)* 47, 177–180. (C2)

Mills, D. and McCluskey, K. (1990) Electrophoretic karyotypes of fungi: the new cytology. *Molecular Plant–Microbe Interactions* 3, 351–357. (C5)

Ministry of Agriculture, Fisheries and Food (1980) *Fertilizers for Cereals*. Leaflet 553, Agricultural Development and Advisory Service, 8 pp. (C3, C4)

Ministry of Agriculture, Fisheries and Food (1994) *Fertiliser Recommendations for Agricultural and Horticultural Crops*. Reference Book 209, HMSO, London. (C3)

Misaghi, I.J., Stowell, L.J., Grogan, R.G. and Spearman, L.C. (1982) Fungistatic activity of water-soluble fluorescent pigments of fluorescent pseudomonads. *Phytopathology* 72, 33–36. (C4)

Mögling, R. (1984) Befall von Triticale durch Schwarzbeinigkeit (*Gaeumannomyces graminis* Walk.). *Nachrichtenblatt für den Pflanzenschutz in der DDR* 38, 23–24. (C3)

Mögling, R. (1987) Verhalten von Sommergerstensortenmischungen gegenüber *Gaeumannomyces graminis* Sacc. *Nachrichtenblatt für den Pflanzenschutz in der DDR* 41, 152. (C3)

Mögling, R., Honermeier, B. and Gawlik, G. (1988) Befall von Wintertriticale durch *Pseudocercosporella herpotrichoides* und *Gaeumannomyces graminis*. *Nachrichtenblatt für den Pflanzenschutz in der DDR* 42, 68–70. (C3)

Monckton, A. (1996) *Farmer's Weekly*, 12 January, 46–47. (C2)

Moore, K.J. and Cook, R.J. (1984) Increased take-all of wheat with direct drilling in the Pacific Northwest. *Phytopathology* 74, 1044–1049. (C3)

Moore, W.C. (1948) Take-all of cereals in England and the epidemic of 1948. *Agriculture, London* 55, 383–385. (C1, C2)

Moule, G. (1995) Crop health. In: Soffe, R.J. (ed.) *Primrose McConnell's The Agricultural Notebook*. Blackwell Science, Oxford, pp. 252–320. (C4)

Mróz, A., Martyniuk, S. and Kuś, J. (1994) Response of winter wheat to seed applied microorganisms. *Phytopathologia Polonica* 7, 15–20. (C4)

Müller, A. (1873) 'The Take-all (the Corn Disease of Australia) Scientifically Considered'. By Dr Carl Mücke. Notices of Books. *The Gardeners' Chronicle and Agricultural Gazette*. 19 July, p. 1004. (C1)

Mullis, K.B. (1990) The unusual origin of the polymerase chain reaction. *Scientific American* 262, 36–43. (C5)

Murphy, M.C. (1989) *Report on Farming in the Eastern Counties of England, 1987/88*. Department of Land Economy, University of Cambridge, Cambridge, 188 pp. (C1)

Murray, G.M., Scott, B.J., Hochmann, Z. and Butler, B.J. (1987) Failure of liming to increase grain yield of wheat and triticale in acid soils may be due to the associated increase in the incidence of take-all (*Gaeumannomyces graminis* var. *tritici*). *Australian Journal of Experimental Agriculture* 27, 411–418. (C1)

Musker, R. (1994) Genetics and population studies of *Gaeumannomyces graminis*. PhD Thesis, University of Birmingham. (C5)

Mussared, D. (1996) Finding ways to beat take-all. *Rural Research* 170(3), 7–9. (C1, C2)

Naiki, T. and Cook, R.J. (1983a) Factors in loss of pathogenicity in *Gaeumannomyces graminis* var. *tritici* to cause take-all. *Phytopathology* 73, 1652–1656. (C5)

Naiki, T. and Cook, R.J. (1983b) Relationship between production of a self-inhibitor and inability of *Gaeumannomyces graminis* var. *tritici* to cause take-all. *Phytopathology* 73, 1657–1660. (C5)

Narita, Y. and Suzui, T. (1991) Influence of a sterile dark mycelial fungus on take-all of wheat. *Annals of the Phytopathological Society of Japan* 57, 301–305. (C2, C4)

National Institute of Agricultural Botany (1996) *Cereal Variety Handbook. NIAB Recommended Lists of Cereals 1996*. NIAB, Cambridge. (C1)

Nayudu, M., Groom, K.A.E., Fernance, J., Wong, P.T.W. and Turnbull, K. (1994) The genetic nature of biological control of the take-all fungal pathogen by *Pseudomonas*. In: Ryder, M.H., Stephens, P.M. and Bowen, G.D. (eds) *Improving Plant Productivity with Rhizosphere Bacteria*. CSIRO Division of Soils, Glen Osmond, South Australia, pp. 122–124. (C4)

Neate, S.M. (1988) Effect of tillage on disease of cereals caused by *Gaeumannomyces graminis*, *Rhizoctonia solani* and *Heterodera avenae*: a review. *Plant Protection Quarterly* 3, 5–7. (C3)

Neate, S.M. (1994) Soil and crop management practices that affect root diseases of crop plants. In: Pankhurst, C.E., Doube, B.M., Gupta, V.V.S.R. and Grace, P.R. (eds) *Soil Biota. Management in Sustainable Farming Systems*. CSIRO, East Melbourne, pp. 96–106. (C3)

Neter, J. and Wasserman, W. (1974) *Applied Linear Statistical Models*. Richard D. Urwin, Homewood, Illinois. (C6)

Newby, H. (1988) *Country Life. A Social History of Rural England*. Cardinal, Sphere Books, Penguin Group, London. (C1)

New South Wales Department of Agriculture (1987) *Plant disease survey NSW 1986–87. 56th Annual Report*. Plant Pathology Branch, Biological and Chemical Research Institute, Rydalmere, NSW. (C1)

Nilsson, H.E. (1969) Studies of root and foot rot diseases of cereals and grasses. I. On resistance to *Ophiobolus graminis* Sacc. *Landbrukshögskolans Annaler* 35, 275–807. (C1, C6)

Nilsson, H.E. (1972) The occurrence of lobed hyphopodia on an isolate of the take-all fungus, '*Ophiobolus graminis* Sacc.' on wheat in Sweden. *Swedish Journal of Agricultural Research* 2, 105–118. (C5)

Nilsson, H.E. (1973a) Influence of herbicides on take-all and eyespot disease of winter wheat in a field trial. *Swedish Journal of Agricultural Research* 3, 115–118. (C3)

Nilsson, H.E. (1973b) Influence of the herbicide mecoprop on *Gaeumannomyces graminis* and the take-all disease in spring wheat. *Swedish Journal of Agricultural Research* 3, 105–113. (C3)

Nilsson, H.E. (1973c) Varietal differences in resistance to take-all disease of winter wheat. *Swedish Journal of Agricultural Research* 3, 89–93. (C3)

Nix, J. (1993) *Farm Management Pocketbook*. Department of Agricultural Economics, Wye College, Ashford. (C3)

Nix, J. (1996) *Farm Management Pocketbook*. Wye College Press, Ashford. (C3)

Noe, J.P. and Barker, K.R. (1985) Overestimation of yield loss of tobacco caused by aggregated spatial pattern of *Meloidogyne incognita*. *Journal of Nematology* 17, 245–251. (C7)

Novotný, J. and Herman, M. (1981) [Effect of soil cultivation on incidence of take-all (*Gaeumannomyces graminis*) on winter wheat.] *Sborník ÚVTIZ, Ochrana Rostlin* 17, 151–156. (C3)

Nutter, F.W. Jr, Teng, P.S. and Shokes, F.M. (1991) Disease assessment terms and concepts. *Plant Disease* 75, 1187–1188. (C6)

OCDE/OECD (1991) *Environmental Indicators*, OCDE/OECD, Paris. (C4)

O'Dell, M., Flavell, R.B. and Hollins, T.W. (1992) The classification of isolates of *Gaeumannomyces graminis* from wheat, rye and oats using restriction fragment length polymorphisms in families of repeated DNA sequences. *Plant Pathology* 41, 554–562. (C5)

Oerke, E.-C., Dehne, H.-W., Schönbeck, F. and Weber, A. (1994) *Crop Production and Crop Protection*. Elsevier, Amsterdam. (C1)

Old, K.M. and Patrick, Z.A. (1979) Giant soil amoebae, potential biocontrol agents. In: Schippers, B. and Gams, W. (eds) *Soil-borne Plant Pathogens*. Academic Press, London, pp. 617–628. (C4)

Oliver, R. (1993) Nucleic acid-based methods for detection and identification. In: Fox, R.T.V. (ed.) *Principles of Diagnostic Techniques in Plant Pathology*. CAB International, Wallingford, pp. 153–169. (C5)

Ophel-Keller, K., Engel, B. and Heinrich, K. (1995) Specific detection of *Gaeumannomyces graminis* in soil using polymerase chain reaction. *Mycological Research* 99, 1385–1390. (C5)

Osbourn, A. (1996) Saponins and plant defence – a soap story. *Trends in Plant Science* 1, 4–9. (C5)

Osbourn, A.E., Clarke, B.R., Dow, J.M. and Daniels, M.J. (1991) Partial characterization of avenacinase from *Gaeumannomyces graminis* var. *avenae*. *Physiological and Molecular Plant Pathology* 38, 301–312. (C5)

Osbourn, A., Bowyer, P., Bryan, G., Lunness, P., Clarke, B. and Daniels, M. (1994a) Detoxification of plant saponins by fungi. In: Daniels, M.J., Downie, J.A. and Osbourn, A.E. (eds) *Advances in Molecular Genetics of Plant–Microbe Interactions*. Kluwer Academic Publishers, Dordrecht, pp. 215–221. (C5)

Osbourn, A.E., Clarke, B.R., Lunness, P., Scott, P.R. and Daniels, M.J. (1994b) An oat species lacking avenacin is susceptible to infection by *Gaeumannomyces graminis* var. *tritici*. *Physiological and Molecular Plant Pathology* 45, 457–467. (C5)

Osbourn, A., Bowyer, P., Lunness, P., Clarke, B. and Daniels, M. (1995) Fungal pathogens of oat roots and tomato leaves employ closely related enzymes to detoxify different host plant saponins. *Molecular Plant–Microbe Interactions* 8, 971–978. (C5)

Ou, S.H. (1972) *Rice Diseases*. Commonwealth Mycological Institute, Kew. (C5)

Ownley, B.H., Duffy, B.K. and Weller, D.M. (1992) Soil factors associated with suppression of take-all by *Trichoderma koningii*. *Phytopathology* 82, 1120. (C4)

Oyanagi, A., Suenaga, K., Kawaguchi, K., Tsuyushige, M., Takada, H., Sato, A. and Eguchi, H. (1990) Varietal differences in resistance to take-all disease in wheat and barley. *Bulletin of the National Agriculture Research Centre, Japan* 18, 19–40. (C3)

Papendick, R.W. and Cook, R.J. (1974) Plant water stress and development of root rot in wheat subjected to different cultural practices. *Phytopathology* 64, 358–363. (C2)

Parry, D.W. (1990) *Plant Pathology in Agriculture*. Cambridge University Press, Cambridge. (C2)

Paul, V. and Schönbeck, F. (1976) Untersuchungen über den Einfluss des Herbizids Diallat auf einige Getreidekrankheiten. *Phytopathologische Zeitschrift* 85, 353–367. (C4)

Paulitz, T.C. (1992) Biological control of damping-off diseases with seed treatments. In: Tjamos, E.C., Papavizas, G.C. and Cook, R.J. (eds) *Biological Control of Plant Diseases. Progress and Challenges for the Future*. Plenum Press, New York and London, pp. 145–156. (C4)

Paveley, N.D., Yarham, D.J., Clare, R. and Capper, A.L. (1992) Take-all disease of winter wheat – field experience in the use of biological control. In: Jensen, D.F., Hockenhull, J. and Fokkema, N.J. (eds) *New Approaches in Biological Control of Soil-borne Diseases*. IOBC/WPRS and EFPP, Copenhagen, pp. 166–168. (C4)

Payne, R.W., Lane, P.W., Digby, P.G.N., Harding, S.A., Leech, P.K., Morgan, G.W., Todd, A.D., Thompson, R., Tunnicliffe Wilson, G., Welham, S.J. and White, R.P. (1993) *Genstat 5 Release 3 Reference Manual*. Clarendon Press, Oxford. (C2, C6)

Pearson, V. (1974) Virulence and cellulolytic enzyme activity of isolates of *Gaeumannomyces graminis*. *Transactions of the British Mycological Society* 63, 199–202. (C5)

Peng, Y., Zhang, Z. and Huang, D. (1994) Effects of Tn5 mutants of *Pseudomonas fluorescens* on biological control of *Gaeumannomyces graminis*. In: Ryder, M.H., Stephens, P.M. and Bowen, G.D. (eds) *Improving Plant Productivity with Rhizosphere Bacteria*. CSIRO Division of Soils, Glen Osmond, South Australia, p. 250. (C4)

Penrose, L. (1987) Thickening and browning of cortical cell walls in seminal roots of wheat seedlings infected with *Gaeumannomyces graminis* var. *tritici*. *Annals of Applied Biology* 110, 463–470. (C3)

Penrose, L.D.J. and Neate, S.M. (1994) Resistance to *Gaeumannomyces graminis* in wheat genotypes grown in field environments and sand culture. *Soil Biology and Biochemistry* 26, 719–726. (C3)

Penrose, L., Rathjen, A.J., Rovira, A. and Warcup, J.H. (1986) Resistance to infection by *Gaeumannomyces graminis* var. *tritici* in wheat. *Biennial Report of the Waite Agricultural Research Institute, 1984–5*, p. 144. (C3)

Philippi, T.E. (1993) Multiple regression: herbivory. In: Scheiner, S.M. and Gurevitch, J. (eds) *Design and Analysis of Ecological Experiments*. Chapman & Hall, New York, pp. 183–210. (C2)

Phillion, D.P., Braccolino, D.S., Graneto, M.J., Phillips, W.G., Van Sant, K.A., Walker, D.M. and Wong, S.C. (1993) Preparation of heterocyclic and aromatic compounds as fungicides for control of take-all disease of plants. Europe, Patent Application, EP 538231, 21 April 1993. (C4)

Phillion, D.P., Graneto, M.J., Pratt, JK and Wong, S.C. (1994) Fungicidal compositions. Hungary, Basic Patent, HU 9400963, 28 June 1994. (C5)

Phillion, D.P., Van Sant, K.A. and Walker, D.M. (1996a) Selected fungicides for the control of take-all disease of plants. United States, Patent, 5,482,974, 9 January 1996. (C4)

Phillion, D., Wong, S.C. and Shortt, B. (1996b) Fungicides for the control of take-all disease of plants. United States, Patent, 5,486,621, 23 January 1996. (C4)

Pilgeram, A.L. and Henson, J.M. (1990) Transformation and cotransformation of *Gaeumannomyces graminis* to phleomycin resistance. *Phytopathology* 80, 1124–1129. (C5)

Pilgeram, A.L. and Henson, J.M. (1992) Sexual crosses of the homothallic fungus *Gaeumannomyces graminis* var. *tritici* based on use of an auxotroph obtained by transformation. *Experimental Mycology* 16, 35–43. (C5)

Pilgeram, A.L., Goins, T. and Henson, J.M. (1993) The fate of integrated DNA in *Gaeumannomyces graminis* transformants. *FEMS (Federation of European Microbiological Societies) Microbiology Letters* 113, 309–314. (C5)

Polley, R.W. and Clarkson, J.D.S. (1980) Take-all severity and yield in winter wheat: relationship established using a single plant assessment method. *Plant Pathology* 29, 110–116. (C1, C6)

Polley, R.W. and Slough, J.E. (1993) *Survey of Winter Wheat Diseases in England and Wales 1992*. Central Science Laboratory, MAFF and ADAS, 44 pp. (C6)

Polley, R.W. and Thomas, M.R. (1991) Surveys of diseases of winter wheat in England and Wales 1976–1988. *Annals of Applied Biology* 119, 1–20. (C1, C2, C6)

Polley, R.W., Slough, J.E. and Jones, D.R. (1994) *Winter Wheat Disease Survey Report 1993*. Central Science Laboratory, MAFF and ADAS, 44 pp. (C1, C6)

Polley, R.W., Slough, J.E. and Jones, D.R. (1995) *Winter Wheat Disease Survey Report 1994*. Central Science Laboratory, MAFF and ADAS, 44 pp. (C1, C6)

Polley, R.W., Slough, J.E. and Jones, D.R. (1996) *Winter Wheat Disease Survey Report 1995*. Central Science Laboratory, MAFF and ADAS, 44 pp. (C1, C6)

Ponchet, J. (1962) Étude des facteurs qui conditionnent le développement du piétin-échaudage: *Linocarpon cariceti* B. et Br. *Annales des Épiphyties* 13, 151–165. (C1)

Ponchet, J. and Coppenet, M. (1957) Le problème du piétin-échaudage des céréales dans les sols bretons. *Phytiatrie–Phytopharmacie* 6, 157–164. (C1)

Ponchet, J. and Coppenet, M. (1962a) Influence de divers facteurs culturaux sur le développement du piétin-échaudage *Linocarpon cariceti* B. et Br. *Annales des Épiphyties* 13, 285–291. (C1)

Ponchet, J. and Coppenet, M. (1962b) Influence de la fumure minérale sur le développement du piétin-échaudage, *Linocarpon cariceti* B. et Br. *Annales des Épiphyties* 13, 277–283. (C1)

Ponomareva, G.Y. (1965) [Effect of *Trichoderma lignorum* on the development of root rots of winter wheat in the foothills of Stavropol]. In: Païkin, D.M. (ed.) *[Chemical and Biological Methods of Plant Protection]*. Kolos, Leningrad, pp. 182–186. (C4)

Pope, A.M.S. and Jackson, R.M. (1973) Effects of wheat field soil on inocula of *Gaeumannomyces graminis* (Sacc.) Arx and Oliver var. *tritici* J.Walker in relation to take-all decline. *Soil Biology and Biochemistry* 5, 881–890. (C4)

Porter, J.R. (1993) AFRCWHEAT2: a model of the growth and development of wheat incorporating responses to water and nitrogen. *European Journal of Agronomy* 2, 69–82. (C2)

Porter, J.R., Jamieson, P.D. and Wilson, D.R. (1993) Comparison of the wheat simulation models AFRCWHEAT2, CERES-Wheat and SWHEAT for non-limiting conditions of crop growth. *Field Crops Research* 33, 131–157. (C2)

Powelson, R.L., Jackson, T.L. and Christensen, N.W. (1985) Enhanced suppression of take-all root rot of wheat with chloride fertilizer. In: Parker, C.A., Rovira, A.D., Moore, K.J. and Wong, P.T.W. (eds) *Ecology and Management of Soilborne Plant Pathogens*. American Phytopathological Society, St Paul, Minnesota, pp. 246–251. (C3)

Prade, K. and Trolldenier, G. (1990) Incidence of *Gaeumannomyces graminis* var. *tritici* and potassium deficiency increase rhizospheric denitrification. *Plant and Soil* 124, 141–142. (C3)

Prew, R.D. (1980) Studies on the spread of *Gaeumannomyces graminis* var. *tritici* in wheat. II. The effects of cultivations. *Annals of Applied Biology* 94, 397–404. (C6)

Prew, R.D. and Dyke, G.V. (1979) Experiments comparing 'break crops' as a preparation for winter wheat followed by spring barley. *Journal of Agricultural Science, Cambridge* 92, 189–201. (C1)

Prew, R.D., Beane, J., Carter, N., Church, B.M., Dewar, A.M., Lacey, J., Penny, A., Plumb, R.T., Thorne, G.N. and Todd, A.D. (1986) Some factors affecting the growth and yield of winter wheat grown as a third cereal with much or negligible take-all. *Journal of Agricultural Science, Cambridge* 107, 639–671. (C2, C3, C4, C6)

Prew, R.D., Ashby, J.E., Bacon, E.T.G., Christian, D.G., Gutteridge, R.J., Jenkyn, J.F., Powell, W. and Todd, A.D. (1995) Effects of incorporating or burning straw, and of different cultivation systems, on winter wheat grown on two soil types, 1985–91. *Journal of Agricultural Science, Cambridge* 124, 173–194. (C4)

Prillieux, E.E. and Delacroix, G. (1890) La maladie du pied du blé, causée par l'*Ophiobolus graminis*, Sacc. *Bulletin de la Société Mycologique de France* 6, 110–113. (C1)

Raepsaet, J. and Defosse, L. (1982) La lutte biologique contre *Gaeumannomyces graminis* var. *tritici*, agent du piétin échaudage du froment. I. Essais orientatifs d'introduction de souches d'agressivités différentes. *Bulletin des Recherches Agronomiques de Gembloux* 17, 263–276. (C4)

Raskin, I. (1992) Role of salicylic acid in plants. *Annual Review of Plant Physiology and Plant Molecular Biology* 43, 439–463. (C4)

Rawlinson, C.J. and Buck, K.W. (1981) Viruses in *Gaeumannomyces graminis* and *Phialophora*. In: Asher, M.J.C. and Shipton, P.J. (eds) *Biology and Control of Take-all*. Academic Press, London, pp. 151–172. (C2, C5)

Rawlinson, C.J., Hornby, D., Pearson, V. and Carpenter, J.M. (1973) Virus-like particles in the take-all fungus *Gaeumannomyces graminis*. *Annals of Applied Biology* 74, 197–209. (C2)

Reis, E.M., Cook, R.J. and McNeal, B.L. (1982) Effect of mineral nutrition on take-all of wheat *Triticum aestivum* cultivar Fielder. *Phytopathology* 72, 224–229. (C3)

Reis, E.M., Cook, R.J. and McNeal, B.L. (1983) Elevated pH and associated reduced trace-nutrient availability as factors contributing to take-all of wheat upon soil liming. *Phytopathology* 73, 411–413. (C3)

Rengel, Z., Graham, R.D. and Pedler, J.F. (1993) Manganese nutrition and accumulation of phenolics and lignin as related to differential resistance of wheat genotypes to the take-all fungus. *Plant and Soil* 151, 255–263. (C3)

Rengel, Z., Graham, R.D. and Pedler, J.F. (1994) Time-course of biosynthesis of phenolics and lignin in roots of wheat genotypes differing in manganese efficiency and resistance to take-all fungus. *Annals of Botany* 74, 471–477. (C3)

Rengel, Z., Gutteridge, R.J., Hirsch, P. and Hornby, D. (1996) Plant genotype, micronutrient fertilization and take-all infection influence bacterial populations in the rhizosphere of wheat. *Plant and Soil* 183, 269–277. (C3)

Rew, L.J., Cussans, G.W., Mugglestone, M.A. and Miller, P.C.H. (1996) A technique for mapping the spatial distribution of *Elymus repens*, with estimates of the potential reduction in herbicide usage from patch spraying. *Weed Research* 36, 283–292. (C7)

Richardson, D. (1995) *Farmer's Weekly*, 11 August, 68.

Rickman, R.W. and Klepper, E.L. (1991) Tillering in wheat. In: Hodges, T. (ed.) *Predicting Crop Phenology*. CRC Press, Boca Raton, Florida, pp. 73–83. (C2)

Rochefrette, C., Tivoli, B. and Boissonnet-Menés, M. (1979) Obtention de mutants résistants à des fongicides et de mutants auxotrophes obtenus à partir de protoplastes chez *Ophiobolus graminis* Sacc. Étude de leur agressivité et de leur aptitude à prémunir le blé. *Annales de Phytopathologie* 11, 43–51. (C4, C5)

Romanos, M.A., Rawlinson, C.J., Almond, M.R. and Buck, K.W. (1980) Production of fungal growth inhibitors by isolates of *Gaeumannomyces graminis* var. *tritici*. *Transactions of the British Mycological Society* 74, 79–88. (C4, C5)

Romanos, M.A., Buck, K.W. and Rawlinson, C.J. (1981) A satellite double-stranded RNA in a virus from *Gaeumannomyces graminis*. *Journal of General Virology* 57, 375–386. (C5)

Ross, A.F. (1961) Systemic acquired resistance induced by localized virus infections in plants. *Virology* 14, 340–358. (C7)

Rosser, W.R. and Chadburn, B.L. (1968) Cereal diseases and their effects in intensive wheat cropping in the East Midland region, 1963–65. *Plant Pathology* 17, 51–60. (C1, C6)

Roth, R., Obenauf, U. and Wendland, U. (1984) Ergebnisse zu Winterroggen in Selbstfolge und im Fruchtwechsel. *Archiv für Acker- und Pflanzenbau und Bodenkunde* 28, 749–756. (C3, C6)

Rothrock, C.S. (1987) Susceptibility and yield losses of small grains to take-all. *Phytopathology* 77, 1773. (C6)

Rothrock, C.S. (1988a) Effect of chemical and biological treatments on take-all of winter wheat. *Crop Protection* 7, 20–24. (C4)

Rothrock, C.S. (1988b) Relative susceptibility of small grains to take-all. *Plant Disease* 72, 883–886. (C3)

Rothrock, C.S. (1989) Suppression of take-all on wheat under wheat sorghum double cropping systems. *Phytopathology* 79, 1166. (C2)

Rothrock, C.S. and Cunfer, B.M. (1986) Absence of take-all decline in double-cropped fields. *Soil Biology and Biochemistry* 18, 113–114. (C2)

Rothrock, C.S. and Cunfer, B.M. (1991) Influence of small grain rotations on take-all in a subsequent wheat crop. *Plant Disease* 75, 1050–1052. (C3)

Rouatt, J.W. and Katznelson, H. (1961) A study of the bacteria on the root surface and in the rhizosphere soil of crop plants. *Journal of Applied Bacteriology* 24, 164–171. (C4)

Rovira, A.D. (1978) Manipulation of the level of take-all disease in the field by inoculation. In: *Epidemiology and Crop Loss Assessment*. Proceedings of the Australian Plant Pathology Society Workshop, Lincoln College, August, 1977, pp. 11-1 to 11-4. (C6)

Rovira, A.D. and Wildermuth, G.B. (1981) The nature and mechanisms of suppression. In: Asher, M.J.C. and Shipton, P.J. (eds) *Biology and Control of Take-all*. Academic Press, London, pp. 385–416. (C2, C4)

Rovira, A.D., Graham, R.D. and Ascher, J.S. (1985) Reduction in infection of wheat roots by *Gaeumannomyces graminis* var. *tritici* with application of manganese to soil. In: Parker, C.A., Rovira, A.D., Moore, K.J., Wong, P.T.W. and Kollmorgen, J.F. (eds) *Ecology and Management of Soilborne Plant Pathogens*. APS Press, St Paul, Minnesota, pp. 212–214. (C3)

Rovira, A.D., Elliott, L.F. and Cook, R.J. (1990) The impact of cropping systems on rhizosphere organisms affecting plant health. In: Lynch, J.M. (ed.) *The Rhizosphere*. John Wiley & Sons, Chichester, pp. 389–436. (C3, C4)

Roy, K.W., Abney, T.S., Huber, D.M. and Keeler, R. (1982) Isolation of *Gaeumannomyces graminis* var. *graminis* from soybeans in the midwest. *Plant Disease* 66, 822–825. (C5)

Ruess, W., Kunz, W., Staub, T., Müller, K., Poppinger, N., Speich, J. and Ahl Goy, P. (1995) Plant activator CGA 245704, a new technology for disease management. *European Journal of Plant Pathology* 101, 424. (C7)

Ruess, W., Müller, K., Knauf-Beiter, G., Kunz, W. and Staub, T. (1996) Plant activator CGA 245704: an innovative approach for disease control in cereals and tobacco. *Proceedings of the 1996 Brighton Crop Protection Conference – Pests and Diseases* 1, 53–60. (C7)

Ryder, M.H. and Rovira, A.D. (1993) Biological control of take-all of glasshouse-grown wheat using strains of *Pseudomonas corrugata* isolated from wheat field soil. *Soil Biology and Biochemistry* 25, 311–320. (C4)

Sallans, B.J. (1948) Interrelations of common root rot and other factors with wheat yields in Saskatchewan. *Scientific Agriculture (Ottawa)* 28, 6–20. (C6)

Sambrook, J., Fritsch, E.F. and Maniatis, T. (1989a) Analysis of genomic DNA by Southern hybridization. In: *Molecular Cloning, a Laboratory Manual*, 2nd edn. Cold Spring Harbor Laboratory Press, Cold Spring Harbor, New York, pp. 9.31–9.62. (C5)

Sambrook, J., Fritsch, E.F. and Maniatis, T. (1989b) DNA Sequencing. In: *Molecular Cloning, a Laboratory Manual*, 2nd edn. Cold Spring Harbor Laboratory Press, Cold Spring Harbor, New York, pp. 13.3–13.104. (C5)

Samuel, G. (1937) Whiteheads or take-all in wheat. *The Journal of the Ministry of Agriculture* 44, 231–241. (C1)

Sandford, G.B. and Broadfoot, W.C. (1931) Studies of the effects of other soil-inhabiting microorganisms on the virulence of *Ophiobolus graminis* Sacc. *Scientific Agriculture* 11, 512–528. (C4)

Sarniguet, A. (1990) Réceptivité des sols au piétin-échaudage du blé: influence des rotations et de la fertilisation azotée en relation avec certains facteurs physicochimiques et les peuplements de *Pseudomonas fluorescens*. Thèse, Université de Paris-Sud. (C3)

Sarniguet, A. and Lucas, P. (1991) Evolution of bacterial populations related to decline of take-all patch on turfgrass. *Phytopathology* 81, 1202. (C2)

Sarniguet, A. and Lucas, P. (1992) Evaluation of populations of fluorescent pseudomonads related to decline of take-all patch on turfgrass. *Plant and Soil* 145, 11–15. (C2)

Sarniguet, A. and Lucas, P. (1993) Relationship between pot induced soil suppressiveness to take-all of wheat and populations of fluorescent pseudomonads. *Phytopathology* 83, 1418. (C2)

Sarniguet, A., Lucas, P. and Lucas, M. (1992a) Relationships between take-all, soil conduciveness to the disease, populations of fluorescent pseudomonads and nitrogen fertilizers. *Plant and Soil* 145, 17–27. (C2, C3, C4)

Sarniguet, A., Lucas, P., Lucas, M. and Samson, R. (1992b) Soil conduciveness to take-all of wheat: influence of the nitrogen fertilizers on the structure of populations of fluorescent pseudomonads. *Plant and Soil* 145, 29–36. (C2, C3, C4)

Sayler, G.S. and Layton, A.C. (1990) Environmental applications of nucleic acid hybridization. *Annual Review of Microbiology* 44, 625–648. (C5)

Scarisbrick, D. and Meikle, S. (1993) Cereals: which way now? *Farmer's Weekly*, 29 October, p. 36. (CI)

Schesser, K., Luder, A. and Henson, J.M. (1991) Use of polymerase chain reaction to detect the take-all fungus, *Gaeumannomyces graminis*, in infected wheat plants. *Applied and Environmental Microbiology* 57, 553–556. (C5)

Schlichter, U. (1997) Differential gene expression in winter barley after infection with the fungal root pathogen *Gaeumannomyces graminis* var. *tritici* and the obligate fungal root parasite *Polymyxa graminis*. PhD Dissertation, University of Cologne. (C7)

Schmidli-Sacherer, P., Keel, C. and Défago, G. (1997) The global regulator GacA of *Pseudomonas fluorescens* CHA0 is required for suppression of root diseases in dicotyledons but not in Gramineae. *Plant Pathology* 46, 80–90. (C4)

Schönbeck, F., Dehne, H.-W. and Zimmerman, I. (1977) Untersuchungen über den Einfluss von Diallat auf den Befall von Weizen mit *Fusarium culmorum* und *F. avenaceum*. *Phytopathologische Zeitschrift* 90, 77–86. (C4)

Schots, A., Dewey, F.M. and Oliver, R. (eds) (1994) *Modern Assays for Plant Pathogenic Fungi: Identification, Detection and Quantification*. CAB International, Wallingford. (C5)

Schulze, D.G., McCay-Buis, T., Sutton, S.R. and Huber, D.M. (1995) Manganese oxidation states in *Gaeumannomyces*-infested wheat rhizospheres probed by micro-XANES spectroscopy. *Phytopathology* 85, 990–994. (C3)

Schumann, G.L. and MacDonald, J.D. (1996) *Turfgrass Diseases: Diagnosis and Management*. APS Press, St Paul, Minnesota. (C5)

Scott, D.B. (1989) *Gaeumannomyces graminis* var. *graminis* on Gramineae in South Africa. *Phytophylactica* 21, 251–254. (C5)

Scott, D.B. (1990) *Wheat Diseases in South Africa*. Technical Communication, Department of Agriculture, Republic of South Africa No. 220, v + 62 pp. (C3)

Scott, D.B. and Deacon, J.W. (1983) *Magnaporthe rhizophila* sp. nov. a dark mycelial fungus with a *Phialophora* conidial state from cereal roots in South Africa. *Transactions of the British Mycological Society* 81, 77–82. (C5)

Scott, P.R. (1981) Variation in host susceptibility. In: Asher, M.J.C. and Shipton, P.J. (eds) *Biology and Control of Take-all*. Academic Press, London, pp. 219–236. (C3, C5)

Scott, P.R. (1990) American Phytopathological Society Annual Meeting, Richmond, Virginia, 20–24 August 1989. *British Society for Plant Pathology Newsletter*, no. 16, 29–31. (C4)

Scott, P.R., Hollins, T.H. and Summers, R.W. (1989) Breeding for resistance to two soil-borne diseases of cereals. *Vorträge für Pflanzenzüchtung* 16, 217–230. (C3)

Seah, S., Sivasithamparam, K. and Turner, D.W. (1996) The effect of salicylic acid on resistance in wheat (*Triticum aestivum*) seedling roots against the take-all fungus, *Gaeumannomyces graminis* var. *tritici*. *Australian Journal of Botany* 44, 499–507. (C7)

Seidel, D. and Spaar, D. (1980) Zur Beeinflussung der Ertragsstruktur des Getreides durch einen Schaderregerbefall. *Tagungsbericht, Akademie der Landwirtschaftswissenschaften der DDR* 181, 17–26. (C1)

Seidel, D., Spaar, D. and Wächter, V. (1981) Einfluss eines Befalls durch *Gaeumannomyces graminis* var. *tritici* Walk. auf die Ertragskomponenten und das Wurzelwachstum in Abhängigkeit von den Entwicklungsstadien der Weizenpflanze. *Archiv für Phytopathologie und Pflanzenschutz* 17, 95–104. (C1)

Seinhorst, J.W. (1965) The relation between nematode density and damage to plants. *Nematologica* 11, 137–154. (C6)

Seinhorst, J.W. (1973) The relation between nematode distribution in a field and loss in yield at different average nematode densities. *Nematologica* 19, 421–427. (C7)

Shaw, M.W. and Royle, D.J. (1987) A new method of disease-loss studies used to estimate effects of *Pyrenophora teres* and *Rhynchosporium secalis* on winter barley yields. *Annals of Applied Biology* 110, 247–262. (C6)

Shipton, P.J. (1972) Take-all in spring sown cereals under continuous cultivation: disease progress and decline in relation to crop succession and nitrogen. *Annals of Applied Biology* 71, 33–46. (C1, C4)

Shipton, P.J. (1975) Yield trends during take-all decline in spring barley and wheat grown continuously. *EPPO Bulletin* 5, 363–374. (C1, C6)

Shipton, P.J. (1981) Saprophytic survival between susceptible crops. In: Asher, M.J.C. and Shipton, P.J. (eds) *Biology and Control of Take-all*. Academic Press, London, pp. 295–316. (C2)

Shipton, P.J., Cook, R.J. and Sitton, J.W. (1973) Occurrence and transfer of a biological factor in soil that suppresses take-all of wheat in eastern Washington. *Phytopathology* 63, 511–517. (C4)

Shivanna, M.B., Meera, M.S. and Hyakumachi, M. (1996) Role of colonization of plant growth promoting fungi in the suppression of take-all and common root rot of wheat. *Crop Protection* 15, 497–504. (C4)

Silvertown, J., Holtier, S., Johnson, J. and Dale, P. (1992) Cellular automaton models of interspecific competition for space – the effect of pattern on process. *Journal of Ecology* 80, 527–534. (C2)

Sim, T. IV, Willis, W.G. and Eversmeyer, M.G. (1988) Kansas plant disease survey. *Plant Disease* 72, 832–836. (C1)

Simon, A. (1989) Biological control of take-all of wheat by *Trichoderma koningii* under controlled environmental conditions. *Soil Biology and Biochemistry* 21, 323–326. (C4)

Simon, A. and Rovira, A.D. (1985) New inoculation technique for *Gaeumannomyces graminis* var. *tritici* to measure dose response and resistance in wheat field experiments. In: Parker, C.A., Rovira, A.D., Moore, K.J., Wong, P.T.W. and Kollmorgen, J.F. (eds) *Ecology and Management of Soilborne Plant Pathogens*. APS Press, St Paul, Minnesota, pp. 183–184. (C6)

Simon, A. and Sivasithamparam, K. (1988a) Crop rotation and biological suppression of *Gaeumannomyces graminis* var. *tritici* in soil. *Transactions of the British Mycological Society* 91, 279–286. (C2)

Simon, A. and Sivasithamparam, K. (1988b) Interactions among *Gaeumannomyces graminis* var. *tritici*, *Trichoderma koningii* and soil bacteria. *Canadian Journal of Microbiology* 34, 871–876. (C2)

Simon, A. and Sivasithamparam, K. (1988c) Microbiological differences between soils suppressive and conducive of the saprophytic growth of *Gaeumannomyces graminis* var. *tritici*. *Canadian Journal of Microbiology* 34, 860–864. (C2)

Simon, A. and Sivasithamparam, K. (1988d) The soil environment and the suppression of saprophytic growth of *Gaeumannomyces graminis* var. *tritici*. *Canadian Journal of Microbiology* 34, 865–870. (C2)

Simon, A. and Sivasithamparam, K. (1989) Pathogen suppression: a case study in biological suppression of *Gaeumannomyces graminis* var. *tritici* in soil. *Soil Biology and Biochemistry* 21, 331–338. (C2)

Simon, A., Rovira, A.D. and Foster, R.C. (1987a) Inocula of *Gaeumannomyces graminis* var. *tritici* for field and glasshouse studies. *Soil Biology and Biochemistry* 19, 363–370. (C6)

Simon, A., Sivasithamparam, K. and MacNish, G.C. (1987b) Biological suppression of the saprophytic growth of *Gaeumannomyces graminis* var. *tritici* in soil. *Canadian Journal of Microbiology* 33, 515–519. (C2)

Simon, A., Sivasithamparam, K. and MacNish, G.C. (1988a) Effect of application to soil of nitrogenous fertilizers and lime on biological suppression of *Gaeumannomyces graminis* var. *tritici*. *Transactions of the British Mycological Society* 91, 287–294. (C2)

Simon, A., Dunlop, R.W., Ghisalberti, E.L. and Sivasithamparam, K. (1988b) *Trichoderma koningii* produces a pyrone compound with antibiotic properties. *Soil Biology and Biochemistry* 20, 263–264. (C4)

Sivasithamparam, K. and Parker, C.A. (1979) Bacterial antagonists to the take-all fungus and fluorescent pseudomonads in the rhizosphere of wheat. *Soil Biology and Biochemistry* 11, 161–165. (C4)

Sivasithamparam, K. and Parker, C.A. (1981) Physiology and nutrition in culture. In: Asher, M.J.C. and Shipton, P.J. (eds) *Biology and Control of Take-all*. Academic Press, London, pp. 125–150. (C3, C5)

Skou, J.P. (1981) Morphology and cytology of the infection process. In: Asher, M.J.C. and Shipton, P.J. (eds) *Biology and Control of Take-all*. Academic Press, London, pp. 175–197. (C5)

Slope, D.B. (1966) Getting to grips with disease. *Arable Farmer*, December, pp. 21–23. (C3)

Slope, D.B. (1967) Disease problems of intensive cereal growing. *Annals of Applied Biology* 59, 317–319. (C6)

Slope, D.B. and Bardner, R. (1965) Cephalosporium stripe of wheat and root damage by insects. *Plant Pathology* 14, 184–187. (C6)

Slope, D.B. and Etheridge, J. (1971) Grain yield and incidence of take-all (*Ophiobolus graminis* Sacc.) in wheat grown in different crop sequences. *Annals of Applied Biology* 67, 13–22. (C1, C6)

Slope, D.B. and Gutteridge, R.J. (1979) Epidemiology of take-all. *Rothamsted Report for 1978, Part 1*, 215. (C4)

Slope, D.B. and Gutteridge, R.J. (1982) Cultural variation in the take-all fungus. *Rothamsted Report for 1981, Part 1*, 192. (C2, C4)

Slope, D.B., Gutteridge, R.J. and Henden, D.R. (1977) Epidemiology of take-all. *Rothamsted Report for 1976, Part 1*, 215. (C4)

Slope, D.B., Salt, G.A., Broom, E.W. and Gutteridge, R.G. (1978) Occurrence of *Phialophora radicicola* var. *graminicola* and *Gaeumannomyces graminis* var. *tritici* on roots of wheat in field crops. *Annals of Applied Biology* 88, 239–246. (C5)

Slope, D.B., Prew, R.D., Gutteridge, R.J. and Etheridge, J. (1979) Take-all, *Gaeumannomyces graminis* var. *tritici*, and yield of wheat grown after ley and arable rotations in relation to the occurrence of *Phialophora radicicola* var. *graminicola*. *Journal of Agricultural Science, Cambridge* 93, 377–389. (C2, C4)

Slope, D.B., Gutteridge, R.J. and Swaby, A. (1982) Take-all on wheat after leys. *Rothamsted Report for 1981, Part 1*, 191–192. (C3)

Slope, D.B., Gutteridge, R.J. and Poulton, P.R. (1984) Phosphate manuring and take-all. *Rothamsted Report for 1983, Part 1*, 124. (C3, C6)

Smiley, R.W. (1978) Antagonists of *Gaeumannomyces graminis* from the rhizoplane of wheat in soils fertilized with ammonium- or nitrate-nitrogen. *Soil Biology and Biochemistry* 10, 169–174. (C4)

Smiley, R.W., Ingham, R.E., Uddin, W. and Cook, G.H. (1994) Crop sequences for managing cereal cyst nematode and fungal pathogens of winter wheat. *Plant Disease* 78, 1142–1149. (C4)

Smith, D. (1984) Maintenance of fungi. In: Kirsop, B.E. and Snell, J.J.S. (eds) *Maintenance of Microorganisms*. Academic Press, London, pp. 83–107. (C5)

Smith, H.T.E. (1960) Cereal growing in Cambridgeshire. *Agricultural Progress* 35, 69–81. (C1)

Solel, Z. and Anikster, Y. (1988) Susceptibility of wild emmer wheat from Eastern Galilee to take-all disease. *Phytoparasitica* 16, 76. (C3)

Solel, Z., Ben-Ze'Ev, I.S. and Dori, S. (1990) Features of resistance to take-all disease in cereal species evaluated by laboratory assays. *Journal of Phytopathology* 130, 219–224. (C3)

Southerton, S.G., Osbourn, A.E., Dow, J.M. and Daniels, M.J. (1993) Two xylanases from *Gaeumannomyces graminis* with identical N-terminal amino acid sequences. *Physiological and Molecular Plant Pathology* 42, 97–107. (C5)

Speakman, J.B. (1982) A simple, reliable method of producing perithecia of *Gaeumannomyces graminis* var. *tritici* and its application to isolates of *Phialophora* spp. *Transactions of the British Mycological Society* 79, 350–353. (C5)

Speakman, J.B. (1984a) Control of *Gaeumannomyces graminis* var. *tritici* in wheat by isolates of the *Gaeumannomyces graminis* var. *graminis* / *Phialophora* sp. (lobed hyphopodia) complex under field conditions. *Phytopathologische Zeitschrift* 109, 188–191. (C4)

Speakman, J.B. (1984b) Perithecia of *Gaeumannomyces graminis* var. *graminis* and *G. graminis* var. *tritici* in pure culture. *Transactions of the British Mycological Society* 82, 720–723. (C5)

Speakman, J.B. and Krüger, W. (1984) Control of *Gaeumannomyces graminis* var. *tritici* by a sterile, black mycelial fungus. *Zeitschrift für Pflanzenkrankheiten und Pflanzenschutz* 91, 391–395. (C4)

Stack, R.W. and Thompson, C. (1988) Effect of cropping sequence on take-all and take-all decline in spring durum wheat under irrigation. *Canadian Journal of Plant Pathology* 10, 373 (Abstract). (C2)

Stanners, D. and Bourdeau, P. (eds) (1995) *Europe's Environment. The Dobrís Assessment*. European Environment Agency, Copenhagen. (C1)

Steffan, R.J. and Atlas, R.M. (1991) Polymerase chain reaction: applications in environmental microbiology. *Annual Review of Microbiology* 45, 137–161. (C5)

Steinbrenner, K., Höflich, G. and Sachse, B. (1984) Einfluss des Getreideanteils in der Fruchtfolge auf Ertrag und Befall des Getreides durch *Pseudocercosporella herpotrichoides* (Fron) Deighton und *Gaeumannomyces graminis* (Sacc.) Arx et Olivier sowie auf den Besatz des Bodens mit *Heterodera avenae* Woll., dargestellt an Ergebnissen des Internationalen Fruchtfolge-versuches Dewitz. *Archiv für Acker- und Pflanzenbau und Bodenkunde* 28, 553–559. (C3)

Stelljes, K.B. and Hardin, B. (1995) Tackling wheat take-all. *Agricultural Research, Washington* 43, 4–7. (C0, C1, C4)

Stephens, P.M. and Davoren, C.W. (1995) Influence of soil moisture on the ability of the lumbricid earthworm *Aporrectodea trapezoides* to reduce the disease severity of take-all. *Phytopathology* 85, 1560. (C4)

Stephens, P.M., Davoren, C.W., Doube, B.M. and Ryder, M.H. (1994a) Ability of the lumbricid earthworms *Aporrectodea rosea* and *Aporrectodea trapezoides* to reduce the severity of take-all under greenhouse and field conditions. *Soil Biology and Biochemistry* 26, 1291–1297. (C4)

Stephens, P.M., Davoren, C.W., Ryder, M.H. and Doube, B.M. (1994b) Enhanced growth of 'take-all' affected plants associated with *Pseudomonas corrugata* 2140R and the earthworm *Aporrectodea trapezoides* (Lumbricidae). In: Ryder, M.H., Stephens, P.M. and Bowen, G.D. (eds) *Improving Plant Productivity with Rhizosphere Bacteria*. CSIRO Division of Soils, Glen Osmond, South Australia, pp. 57–63. (C4)

Stevens, D.B., Vaidyanathan, L.V. and Baldwin, J.H. (1988) *Hagberg Falling Number and Breadmaking Quality*. Home-grown Cereals Authority, London. (C6)

Stolp, H. and Gadkari, D. (1981) Nonpathogenic members of the genus *Pseudomonas*. In: Starr, M.P., Stolp, H., Trüper, H.G., Balows, A. and Schlegel, H.G. (eds) *The Prokaryotes: a Handbook on Habitats, Isolation and Identification of Bacteria*. Springer-Verlag, Berlin, pp. 719–741. (C4)

Sturm, K., Fischbeck, G. and Hoffmann, G.M. (1984a) Vergleichende Untersuchungen zur Ertragsbildung in ein- bis mehrjährigen Weizenkulturen und zum Auftreten der Schwarzbeinigkeit (Erreger: *Gaeumannomyces graminis* [Sacc.] Arx et Olivier). *Bayerisches Landwirtschaftliches Jahrbuch* 61, 997–1013. (C1)

Sturm, K., Hoffmann, G.M. and Fischbeck, G. (1984b) Population dynamics of *Gaeumannomyces graminis* in distributed monoculture of wheat. *Zeitschrift für Pflanzenkrankheiten und Pflanzenschutz* 91, 533–542. (C1)

Sturz, A.V. and Bernier, C.C. (1991) Fungal communities in winter wheat roots following crop rotations suppressive and nonsuppressive to take-all. *Canadian Journal of Botany* 69, 39–43. (C4)

Stynes, B.A. (1980) Synoptic methodologies for crop loss assessment. In: Teng, P.S. and Krupa, S.V. (eds) *Crop Loss Assessment*. Minnesota Agricultural Experimental Station, St Paul, pp. 166–175. (C6)

Sylvester-Bradley, R. and Scott, R.K. (1990) *Physiology in the Production and Improvement of Cereals*. Home-grown Cereals Authority, London. (C7)

Tan, M.K. and Wong, P.T.W. (1996) Group I introns in 26S rRNA genes of *Gaeumannomyces graminis* as possible indicators of host specificity of *G. graminis* varieties. *Mycological Research* 100, 337–342. (C5)

Tan, M.K., Wong, P.T.W. and Holley, M.P. (1994) Characterization of nuclear ribosomal DNA (rDNA) in *Gaeumannomyces graminis* and correlation of rDNA variation with *G. graminis* varieties. *Mycological Research* 98, 553–561. (C5)

Teng, P.S. (1985) Construction of predictive models. II. Forecasting crop losses. In: Gilligan, C.A. (ed.) *Mathematical Modelling of Crop Disease*. Academic Press, London, pp. 179–206. (C6, C7)

Teng, P.S. and Gaunt, R.E. (1981) Modeling systems of disease and yield loss in cereals. *Agricultural Systems* 6, 131–154. (C6)

Teng, P.S. and Johnson, K.B. (1988) Analysis of epidemiological components in yield loss assessment. In: Kranz, J. and Rotem, J. (eds) *Experimental Techniques in Plant Disease Epidemiology*. Springer-Verlag, Heidelberg, pp. 179–189. (C6)

Thomashow, L.S. and Weller, D.M. (1988) Role of a phenazine antibiotic from *Pseudomonas fluorescens* in biological control of *Gaeumannomyces graminis* var. *tritici*. *Journal of Bacteriology* 170, 3499–3508. (C4)

Thomashow, L.S. and Weller, D.M. (1990) Application of fluorescent pseudomonads to control root diseases of wheat and some mechanisms of disease suppression. In: Hornby, D. (ed.) *Biological Control of Soil-borne Plant Pathogens*. CAB International, Wallingford, pp. 109–130. (C4)

Thomashow, L.S. and Weller, D.M. (1996) Current concepts in the use of introduced bacteria for biological disease control: mechanisms and antifungal metabolites. In: Stacey, G. and Keen, N.T. (eds) *Plant–Microbe Interactions*. Chapman & Hall, New York, pp. 187–235. (C2, C4)

Thomashow, L.S., Weller, D.M., Bonsall, R.F. and Pierson, L.S. III (1990) Production of the antibiotic phenazine-1-carboxylic acid by fluorescent pseudomonas species in the rhizosphere of wheat. *Applied and Environmental Microbiology* 56, 908–912. (C4)

Thornton, C.R., Dewey, F.M. and Gilligan, C.A. (1997) Production and characterization of a monoclonal antibody raised against antigens from the mycelium of the take-all pathogen *Gaeumannomyces graminis* var. *tritici*; evidence for an extracellular polyphenol oxidase. *Phytopathology* 87, 123–131. (C5)

Tivoli, B., Lemaire, J.-M. and Jouan, B. (1974) Prémunition du blé contre *Ophiobolus graminis* Sacc. par des souches peu agressives du même parasite. *Annales de Phytopathologie* 6, 395–406. (C4)

Tomasini, R.G.A., Jocobsen, L.A. and Ambrosi, I. (1983) [Analysis of 430 wheat fields in the middle plateau of Rio Grande do Sul]. Documentos, Centro Nacional de Pesquisa de Trigo, no. 6, 28. (C1)

Tottman, D.R. and Thompson, D.R. (1978) The influence of herbicides on the incidence of take-all disease (*Gaeumannomyces graminis*) on the roots of winter wheat. In *Proceedings 1978 British Crop Protection Conference – Weeds*, Vol. 2. BCPC, Croydon, 609–616. (C3)

Tottman, D.R., Makepeace, R.J. and Broad, H. (1979) An explanation of the decimal code for the growth of cereals, with illustrations. *Annals of Applied Biology* 93, 221–234. (C0)

Trolldenier, G. (1981) Influence of soil moisture, soil acidity and nitrogen source on take-all of wheat. *Phytopathologische Zeitschrift* 102, 163–177. (C1)

Trolldenier, G. (1982) Influence of potassium nutrition and take-all on wheat yield depending on inoculum density. *Phytopathologische Zeitschrift* 103, 340–348. (C3)

Trolldenier, G. (1985) Effect of varied NPK nutrition and inoculum density on yield losses of wheat caused by take-all. In: Parker, C.A., Rovira, A.D., Moore, K.J., Wong, P.T.W. and Kollmorgen, J.F. (eds) *Ecology and Management of Soilborne Plant Pathogens*. APS Press, St. Paul, Minnesota, pp. 218–220. (C3)

Turner, E.M. (1953) The nature of the resistance of oats to the take-all fungus. *Journal of Experimental Botany* 4, 264–271. (C5)

Turner, E.M. (1961) An enzymic basis for pathogenic specificity in *Ophiobolus graminis*. *Journal of Experimental Botany* 12, 169–175. (C5)

Van Driesche, R.G. and Bellows, T.S.J. (1996) *Biological Control*. Chapman & Hall, New York. (C7)

Wächter, V. (1984) Ertragsbeeinflussung durch *Gaeumannomyces graminis* bei Winterweizen und die Möglichkeit der Toleranzprüfung. *Tagungsbericht, Akademie der Landwirtschaftswissenschaften der DDR*, no. 225, 219–224. (C3)

Wächter, V., Mögling, R. and Seidel, D. (1979) Modellversuch zu den Beziehungen zwischen Wurzelregeneration des Winterweizens und dem Befall durch *Gaeumannomyces graminis var. tritici*. *Wissenschaftliche Zeitschrift der Wilhelm-Pieck-Universität Rostock* 28, 671–674. (C1)

Waggoner, P.E. (1986) Progress curves of foliar diseases: their interpretation and use. In: Leonard, K.J. and Fry, W.E. (eds) *Plant Disease Epidemiology: Population Dynamics and Management*. Macmillan, New York, pp. 3–37. (C2)

Walker, J. (1972) Type studies on *Gaeumannomyces graminis* and related fungi. *Transactions of the British Mycological Society* 58, 427–457. (C1)

Walker, J. (1975) Take-all diseases of Gramineae: a review of recent work. *Review of Plant Pathology* 54, 113–144. (C2)

Walker, J. (1980) *Gaeumannomyces, Linocarpon, Ophiobolus* and several other genera of scolecospored ascomycetes and *Phialophora* conidial states with a note on hyphopodia. *Mycotaxon* 11, 1–129. (C5)

Walker, J. (1981) Taxonomy of take-all fungi and related genera and species. In: Asher, M.J.C. and Shipton, P.J. (eds) *Biology and Control of Take-all*. Academic Press, London, pp. 15–74. (C1, C2, C5)

Wallwork, H. (1989) Screening for resistance to take-all in wheat, triticale and wheat–triticale hybrid lines. *Euphytica* 40, 103–110. (C3)

Wallwork, H. (1996) *Cereal Root and Crown Diseases*. Kondinin Group, Australia. (C4, C6)

Ward, E. (1994) Use of the polymerase chain reaction for identifying plant pathogens. In: Blakeman, J.P. and Williamson, B. (eds) *The Ecology of Plant Pathogens*. CAB International, Wallingford, pp. 143–160. (C5)

Ward, E. (1995) Improved polymerase chain reaction (PCR) detection of *Gaeumannomyces graminis* including a safeguard against false negatives. *European Journal of Plant Pathology* 101, 561–566. (C5)

Ward, E. and Akrofi, A.Y. (1994) Identification of fungi in the *Gaeumannomyces–Phialophora* complex by RFLPs of PCR-amplified ribosomal DNAs. *Mycological Research* 98, 219–224. (C5)

Ward, E. and Bateman, G.L. (1994) Identification of fungi in the *Gaeumannomyces–Phialophora* complex associated with take-all of cereals and grasses using DNA probes. In: Schots, A., Dewey, F.M. and Oliver, R.P. (eds) *Modern Assays for Plant Pathogenic Fungi: Identification, Detection and Quantification*. CAB International, Wallingford, pp. 127–134. (C5)

Ward, E. and Gray, R.M. (1992) Generation of a ribosomal DNA probe by PCR and its use in identification of fungi within the *Gaeumannomyces–Phialophora* complex. *Plant Pathology* 41, 730–736. (C5)

Webster, C.P. and Goulding, W.T. (1995) Effect of one year rotational set-aside on immediate and ensuing nitrogen leaching loss. *Plant and Soil* 177, 203–209. (C2)

Webster, R. (1996) What is kriging? *Aspects of Applied Biology* 46, 57–66. (C6)

Webster, R. and Oliver, M.A. (1990) *Statistical Methods in Soil and Land Resource Survey*. Oxford University Press, Oxford. (C6)

Weed Resistance Action Group (1991) Guidelines for the prevention and control of herbicide-resistant black-grass. UK, 6 pp. (available from ADAS, Cambridge and IACR, Rothamsted). (C1)

Weller, D.M. (1983) Colonization of wheat roots by a fluorescent pseudomonad suppressive to take-all. *Phytopathology* 73, 1548–1553. (C4)

Weller, D.M. (1985) Application of fluorescent pseudomonads to control root diseases. In: Parker, C.A., Rovira, A.D., Moore, K.J., Wong, P.T.W. and Kollmorgen, J.K. (eds) *Ecology and Management of Soilborne Plant Pathogens*. APS Press, St Paul, Minnesota, pp. 137–140. (C4)

Weller, D.M. (1986) Effects of wheat genotype on root colonization by a take-all suppressive strain of *Pseudomonas fluorescens*. *Phytopathology* 76, 1059. (C3)

Weller, D.M. (1988) Biological control of soilborne plant pathogens in the rhizosphere with bacteria. *Annual Review of Phytopathology* 26, 379–407. (C4)

Weller, D.M. and Cook, R.J. (1983) Suppression of take-all of wheat by seed treatment with fluorescent pseudomonads. *Phytopathology* 73, 463–469. (C4)

Weller, D.M., Cook, R.J. and Wilkinson, H.T. (1984) Method for screening bacteria and application thereof for field control of diseases caused by *Gaeumannomyces graminis*. United States, Patent, 4,456,684, 26 June 1984. (C4)

Welsh, J. and McClelland, M. (1990) Fingerprinting genomes using PCR with arbitrary primers. *Nucleic Acids Research* 18, 7213–7218. (C5)

Wendland, U. (1986) Untersuchungen zu den Ursachen für Ertragsdegressionen in mehrjähriger Winterroggenselbstfolge auf Sand-Braunerde. *Archiv für Acker- und Pflanzenbau und Bodenkunde* 30, 781–787. (C3)

Werker, A.R. (1988) The epidemiology of *Gaeumannomyces graminis* var. *tritici* on winter wheat: the effects of selected agronomic factors on the progress and distribution of disease. PhD Dissertation, University of Cambridge. (C2)

Werker, A.R. and Gilligan, C.A. (1990) Analysis of the effects of selected agronomic factors on the dynamics of the take-all disease of wheat in field plots. *Plant Pathology* 39, 161–177. (C2, C3, C4, C6)

Werker, A.R., Gilligan, C.A. and Hornby, D. (1991) Analysis of disease-progress curves for take-all in consecutive crops of winter wheat. *Plant Pathology* 40, 8–24. (C2, C6, C7)

Weste, G. (1970) Extracellular enzyme production by various isolates of *Ophiobolus graminis* and *O. graminis* var. *avenae*. Part 1. Enzymes produced in culture. *Phytopathologische Zeitschrift* 67, 189–204. (C5)

Weston, W.A.R.D. (1944) Take-all or whiteheads of wheat and barley. *Agriculture, London* 51, 226–227. (C2, C3)

Wetzel, H.C. III, Dernoeden, P.H. and Millner, P.D. (1996) Identification of darkly pigmented fungi associated with turfgrass roots by mycelial characteristics and RAPD–PCR. *Plant Disease* 80, 359–364. (C5)

White, N.H. (1939) The sexuality of *Ophiobolus graminis* Sacc. *Journal of the Council for Scientific and Industrial Research, Australia* 12, 209–212. (C5)

White, R.F. (1979) Acetylsalicylic acid (aspirin) induces resistance to tobacco mosaic virus in tobacco. *Virology* 99, 410–412. (C7)

White, T.J., Bruns, T., Lee, S. and Taylor, J. (1990) Amplification and direct sequencing of fungal ribosomal RNA genes for phylogenetics. In: Innis, M.A., Gelfand, D.H., Sninsky, J.J. and White, T.J. (eds) *PCR Protocols. A Guide to Methods and Applications.* Academic Press, San Diego, pp. 315–322. (C5)

Widdowson, F.V., Penny, A., Gutteridge, R.J., Darby, R.J. and Hewitt, M.V. (1985) Tests of amounts and times of application of nitrogen and of sequential sprays of aphicide and fungicides on winter wheat, following either beans or wheat, and the effect of take-all (*Gaeumannomyces graminis* var. *tritici*), on two varieties at Saxmundham, Suffolk 1980–3. *Journal of Agricultural Science, Cambridge* 105, 97–122. (C3)

Wildermuth, G.B. and Rovira, A.D. (1977) Hyphal density as a measure of suppression of *Gaeumannomyces graminis* var. *tritici* on wheat roots. *Soil Biology and Biochemistry* 9, 203–205. (C2)

Wildermuth, G.B., Warcup, J.H. and Rovira, A.D. (1984) Growth of *Gaeumannomyces graminis* var. *tritici* in soil in the presence and absence of wheat roots. *Transactions of the British Mycological Society* 82, 435–441. (C2)

Wilhelm, N.S., Graham, R.D. and Rovira, A.D. (1988) Application of different sources of manganese sulfate decreases take-all (*Gaeumannomyces graminis* var. *tritici*) of wheat grown in a manganese deficient soil. *Australian Journal of Agricultural Research* 39, 1–10. (C3)

Wilhelm, N.S., Graham, R.D. and Rovira, A.D. (1990) Control of Mn status and infection rate by genotype of both host and pathogen in the wheat take-all interaction. *Plant and Soil* 123, 267–275. (C3)

Wilkinson, H.T. and Kane, R.T. (1993) *Gaeumannomyces graminis* var. *graminis* infecting zoysiagrass in Illinois. *Plant Disease* 77, 100. (C4)

Wilkinson, H.T., Alldredge, J.R. and Cook, R.J. (1985a) Estimated distances for infection of wheat roots by *Gaeumannomyces graminis* var. *tritici* in soils suppressive and conducive to take-all. *Phytopathology* 75, 557–559. (C2)

Wilkinson, H.T., Cook, R.J. and Alldredge, J.R. (1985b) Relation of inoculum size and concentration to infection of wheat *Triticum aestivum* roots by *Gaeumannomyces graminis* var. *tritici*. *Phytopathology* 75, 98–103. (C2, C6)

Williams, J.G.K., Kubelik, A.R., Livak, K.J., Rafalski, J.A. and Tingey, S.V. (1990) DNA polymorphisms amplified by arbitrary primers are useful as genetic markers. *Nucleic Acids Research* 18, 6531–6535. (C5)

Wilson, G.L. and Wisniewski, M.E. (1992) Future alternatives to synthetic fungicides for the control of post-harvest diseases. In: Tjamos, E.C., Papavizas, G.C. and Cook, R.J. (eds) *Biological Control of Plant Diseases. Progress and Challenges for the Future.* Plenum Press, New York and London, pp. 133–138. (C4)

Withers, P.J.A. and Sinclair, A.H. (1994) *Sulphur Nutrition of Cereals in the UK: Effects on Yield and Grain Quality.* Home-Grown Cereals Authority, London. (C3)

Wong, P.T.W. (1981) Biological control by cross-protection. In: Asher, M.J.C. and Shipton, P.J. (eds) *Biology and Control of Take-all.* Academic Press, London, pp. 417–431. (C4)

Wong, P.T.W. (1983) Effect of osmotic potential on the growth of *Gaeumannomyces graminis* and *Phialophora* spp. *Annals of Applied Biology* 102, 67–78. (C5)

Wong, P.T.W. (1984) Saprophytic survival of *Gaeumannomyces graminis* and *Phialophora* spp. at various temperature–moisture regimes. *Annals of Applied Biology* 105, 455–461. (C2)

Wong, P.T.W. (1985) Interactions between microbial residents of cereal roots. In: Parker, C.A., Rovira, A.D., Moore, K.J., Wong, P.T.W. and Kollmorgen, J.F. (eds) *Ecology and Management of Soilborne Plant Pathogens.* APS Press, St Paul, Minnesota, pp. 144–147. (C4)

Wong, P.T.W. (1993) Control of take-all disease of cereals. PCT International Application, WO 9322922, 25 November 1993. (C4)

Wong, P.T.W. (1994) Biocontrol of wheat take-all in the field using soil bacteria and fungi. In: Ryder, M.H., Stephens, P.M. and Bowen, G.D. (eds) *Improving Plant Productivity with Rhizosphere Bacteria.* CSIRO Division of Soils, Glen Osmond, South Australia, pp. 24–28. (C4)

Wong, P.T.W. (1995) Biocontrol of take-all using *Phialophora* sp. (lobed hyphopodia). Great Britain, Basic Patent, GB 9502046, 22 March 1995. (C4)

Wong, P.T.W. and Southwell, R.J. (1979) Biological control of take-all in the field using *Gaeumannomyces graminis* var. *graminis* and related fungi. In: Schippers, B. and Gams, W. (eds) *Soil-borne Plant Pathogens.* Academic Press, London, pp. 597–602. (C4)

Wong, P.T.W., Mead, J.A. and Holley, M.P. (1996) Enhanced field control of wheat take-all using cold tolerant isolates of *Gaeumannomyces graminis* var. *graminis* and *Phialophora* sp. (lobed hyphopodia). *Plant Pathology* 45, 285–293. (C4)

Wood, M.J. and Robson, A.D. (1984) Effect of copper deficiency in wheat on infection of roots by *Gaeumannomyces graminis* var. *tritici. Australian Journal of Agricultural Research* 35, 735–742. (C3)

Wüthrich, B. and Défago, G. (1991) Suppression of wheat take-all and black root rot of tobacco by *Pseudomonas fluorescens* strain CHA0: results of field and pot experiments. In: Keel, C., Koller, B. and Défago, G. (eds) *Plant Growth-promoting Rhizobacteria: Progress and Prospects.* Organisation Internationale de Lutte Biologique et Intégrée contre les Animaux et les Plantes Nuisibles, Paris, pp. 17–22. (C4)

Xue, B., Goodwin, P., O'Gorman, D. and Hsiang, T. (1993) Identification of *Leptosphaeria korrae* and *G. graminis* with ribosomal DNA probes. In: *Abstracts of the 6th International Congress of Plant Pathology.* National Research Council Canada, Ottawa, p. 41. (C5)

Yao, J.M. (1993) The discovery and classification of the corn take-all pathogen in China. In: *Abstracts of the 6th International Congress of Plant Pathology.* National Research Council Canada, Ottawa, p. 132. (C5)

Yao, J.M., Wang, Y.C. and Zhu, Y.G. (1992) [A new variety of the pathogen of maize take-all.] *Acta Mycologica Sinica* 11, 99–104. (C5)

Yarham, D.J. (1979) The effect on soil-borne diseases of changes in crop and soil management. In: Schippers, B. and Gams, W. (eds) *Soil-borne Plant Pathogens.* Academic Press, London, pp. 371–383. (C1)

Yarham, D.J. (1981) Practical aspects of epidemiology and control. In: Asher, M.J.C. and Shipton, P.J. (eds) *Biology and Control of Take-all.* Academic Press, London, pp. 353–384. (C1, C2, C3, C4)

Yarham, D.J. (1986) Change and decay – the sociology of cereal foot rots. *Proceedings of the 1986 British Crop Protection Conference – Pests and Diseases,* Vol. 2, 401–410. (C1, C3)

Yarham, D.J. and Gladders, P. (1993) Effect of volunteer plants on crop diseases. *Aspects of Applied Biology* 35, 75–82. (C3)

Yarham, D.J. and Norton, J. (1981) Effects of cultivation methods on disease. In: Jenkyn, J.F. and Plumb, R.T. (eds) *Strategies for the Control of Cereal Disease.* Blackwell Scientific Publications, Oxford, pp. 157–166. (C3)

Yarham, D.J. and Symonds, B.V. (1992) Effect of set-aside on diseases of cereals. In: Clarke, J. (ed.) *Set-Aside.* British Crop Protection Council, Farnham, pp. 41–46. (C3)

Yarham, D.J., Clark, E. and Bell, T.S. (1989) The effect of husbandry on soil-borne pathogens. *Aspects of Applied Biology* 22, 271–278. (C3)

Yeates, J.S. (1986) Ascospore length of Australian isolates of *Gaeumannomyces graminis. Transactions of the British Mycological Society* 86, 131–136. (C5)

Yeates, J.S. and Parker, C.A. (1986a) *In vitro* reaction of Australian isolates of *Gaeumannomyces graminis* to crude oat extracts. *Transactions of the British Mycological Society* 86, 137–144. (C5)

Yeates, J.S. and Parker, C.A. (1986b) Rate of natural senescence of seminal root cortical cells of wheat, barley and oats, with reference to invasions by *Gaeumannomyces graminis*. *Transactions of the British Mycological Society* 86, 683–685. (C3)

Yeates, J.S., Fang, C.S. and Parker, C.A. (1986) Distribution and importance of oat attacking isolates of *Gaeumannomyces graminis* var. *tritici* in Western Australia. *Transactions of the British Mycological Society* 86, 145–152. (C5)

Yoder, O.C., Valent, B. and Chumley, F. (1986) Genetic nomenclature and practice for plant pathogenic fungi. *Phytopathology* 76, 383–385. (C5)

York, K. (1996) Take-all patch in cereals and amenity turf. *Turfgrass Bulletin* (193), 29–30. (C5)

Zadoks, J.C. and Rijsdijk, F.H. (1984) *Agro-ecological Atlas of Cereal Growing in Europe. Vol. III. Atlas of Cereal Diseases and Pests in Europe.* Centre for Agricultural Publishing and Documentation (Pudoc), Wageningen. (C1)

Zadoks, J.C. and Schein, R.D. (1979) *Epidemiology and Plant Disease Management.* Oxford University Press, New York and Oxford. (C2, C6)

Zadoks, J.C., Chang, T.T. and Konzak, C.F. (1974) A decimal code for the growth stages of cereals. *Weed Research* 14, 415–421. (C0)

Zaspel, I. (1992) Studies on the influence of antagonistic rhizosphere bacteria on winter wheat attacked by *Gaeumannomyces graminis* var. *tritici*. In: Jensen, D.F., Hockenhull, J. and Fokkema, N.J. (eds) *New Approaches in Biological Control of Soil-borne Diseases.* IOBC/WPRS and EFPP, Copenhagen, pp. 142–144. (C4)

Zaspel, I., Rudolf, U., Wiemer, M. and Kaske, A. (1989) Verfahren zur Bekämpfung von *Gaeumannomyces graminis* bei Weizen mittels eines *Pseudomonas putida*-Stammes. East Germany, Basic Patent, DD 270653, 9 August 1989. (C4)

Zaspel, I., Wiemer, M. and Rudolf, U. (1990) Verfahren zur Bekämpfung von *Gaeumannomyces graminis* mittels eines *Bacillus subtilis*-Stammes bei Weizen. East Germany, Basic Patent, DD 277008, 21 March 1990. (C4)

Zaspel, I., Wiemer, M. and Obenauf, U. (1991a) Verfahren zur Verhinderung von *Gaeumannomyces graminis* bedingten Schäden bei Weizenjungpflanzen mittels Wachstumsstimulierenden Backterienstammes. East Germany, Basic Patent, DD 291237, 27 June 1991. (C4)

Zaspel, I., Wiemer, M., Obenauf, U. and Kaske, A. (1991b) Verfahren zur Bekämpfung von *Gaeumannomyces graminis* bei Weizen mittels eines antibiotikabildenden *Pseudomonas*-Stammes. East Germany, Basic Patent, DD 291236, 27 June 1991. (C4)

Zi, Y., Gong, M. and Zhang, K. (1991) Preparation of a special fertilizer against *Gaeumannomyces graminis* disease of wheat. China, Basic Patent, CN 1055169, 9 October 1991. (C4)

Zimmermann, A. (1984) Wurzelfäule an Weizen durch *Gaeumannomyces graminis* var. *tritici*-Interactionen zwischen Pflanzendichte und Wasserversorgung bei der Ertragsbildung im Gewächshausversuch. *Zeitschrift für Pflanzenkrankheiten und Pflanzenschutz* 91, 657–668. (C1)

Further Reading

Publication of *Biology and Control of Take-all* (Asher and Shipton, 1981) was a fillip to increased research on the disease. The present book contains an up-to-date account of that subsequent research, although the emphasis tends to be on western Europe and particularly on research of which the contributors have direct experience. Areas that have received less attention concern work carried out elsewhere, often where research priorities have been different, as for example Australia and the USA. Figure I.2 shows the overall rise, and in some cases the decline, of interest in various research areas since 1981. Here, additional references, grouped by the major categories in Fig. I.2, and starting at 1981, are provided for interested readers. Each reference is allotted to one group only and since the groupings are approximate and since some papers could equally well be classified in more than one group, serious searchers are advised to use the groupings here only as a first guide. Some of the papers listed are preliminary reports that may have been superseded by full research papers that may or may not appear in the main text. The main bibliography together with this supplementary bibliography constitute the most complete listing of take-all references since Asher and Shipton (1981).

Pathogen and Related Fungi (includes virus infection)

Buck, K.W. (1984) A new double stranded RNA virus from *Gaeumannomyces graminis*. *Journal of General Virology* 65, 987–990.

Buck, K.W. (1988) Control of plant pathogens with viruses and related agents. *Philosophical Transactions of the Royal Society of London*, Series B, *Biological Sciences* 318, 295–318.

Buck, K.W., McGinty, R.M. and Rawlinson, C.J. (1981) Two serologically unrelated viruses isolated from a *Phialophora* sp. *Journal of General Virology* 55, 235–239.

Buck, K.W., Romanos, M.A., McFadden, J.J.P. and Rawlinson, C.J. (1981) *In vitro* transcription of double-stranded RNA by virion-associated RNA polymerases of viruses from *Gaeumannomyces graminis*. *Journal of General Virology* 57, 157–168.

Campbell, R. (1983) Ultrastructural studies of *Gaeumannomyces graminis* var. *tritici* in the water films on wheat roots and the effect of clay on the interaction between this fungus and bacteria. *Canadian Journal of Microbiology* 29, 39–45.

Huang, L.H. and Luttrell, E.S. (1982) Development of the perithecium in *Gnomonia comari* (Diaporthaceae). *American Journal of Botany* 69, 421–431.

Jamil, N. and Buck, K.W. (1984) Apparently identical viruses from *Gaeumannomyces graminis* var. *tritici* and *Phialophora* sp. (lobed hyphopodia). *Transactions of the British Mycological Society* 83, 519–522.

Jamil, N. and Buck, K.W. (1986) Capsid polypeptides in a group III virus from *Gaeumannomyces graminis* var. *tritici* are related. *Journal of General Virology* 67, 1717–1720.

Jamil, N. and Buck, K.W. (1988) Virion associated RNA polymerase activity in group III virus from *Gaeumannomyces graminis*. *Abstracts of the Annual Meeting of the American Society of Microbiology* 88, 273.

Kackley, K.E., Grybauskas, A.P. and Dernoeden, P.H. (1990) Growth of *Magnaporthe poae* and *Gaeumannomyces incrustans* as affected by temperature-osmotic potential interactions. *Phytopathology* 80, 646–650.

Liang, P., Chen, K. and Liu, H. (1987) Virus particles in sporophore of oyster mushroom *Pleurotus sapidus*. *Chinese Journal of Virology* 3, 369–375.

Martyniuk, S. (1986) [Ecology and properties of the phytopathogen of cereal roots, *Gaeumannomyces graminis* Sacc. Arx and Olivier, and of related fungi of the genus *Phialophora*.] PhD Thesis, Institute of Soil Science and Plant Cultivation Press, R(208), Puławy, Poland.

Martyniuk, S. (1988) [*Ecology and Properties of the Phytopathogen of Cereal Roots,* Gaeumannomyces graminis *Sacc. Arx and Olivier, and of Related Fungi of the Genus* Phialophora.] *Pamiętnik Puławski* No. 87, 187–194.

Martyniuk, S. and Myśków, W. (1983) Das Vorkommen der *Gaeumannomyces graminis*- und *Phialophora radicicola*-Stämme auf Getreidewurzeln von verschiedenen Standorten und die pektinolytische Aktivität dieser Pilze. *Zentralblatt für Mikrobiologie* 138, 465–473.

McFadden, J.J.P., Buck, K.W. and Rawlinson, C.J. (1983) Infrequent transmission of double-stranded RNA virus particles but absence of DNA proviruses in single ascospore cultures of *Gaeumannomyces graminis*. *Journal of General Virology* 64, 927–937.

Osbourn, A.E., Clarke, B.R. and Daniels, M.J. (1989) Molecular pathology of the take-all fungus *Gaeumannomyces graminis*. *The Sainsbury Laboratory Report 1989*, 19–20.

Rothrock, C.S. (1990) Effect of small grains on the inoculum potential of *Gaeumannomyces graminis* var. *tritici*. *Phytopathology* 80, 438.

Schneider, R.W. (1990) Influence of mineral nutrition on fusarium wilt. A proposed mechanism involving cell water relations. In: Ploetz, R.C. (ed.) *Fusarium Wilt of Banana*. APS Press, St Paul, Minnesota, pp. 83–92.

Schreiber, M.T. and Prillwitz, H.G. (1989) Investigations on the antagonism of *Pseudocercosporella anguioides* against *Pseudocercosporella herpotrichoides* var. *acuformis* and *Pseudocercosporella herpotrichoides* var. *herpotrichoides* pathogens of eyespot of cereals. *Zeitschrift für Pflanzenkrankheiten und Pflanzenschutz* 96, 408–427.

Stanway, C.A. and Buck, K.W. (1984) Infection of protoplasts of the wheat take-all fungus, *Gaeumannomyces graminis* var. *tritici*, with double-stranded RNA viruses. *Journal of General Virology* 65, 2061–2065.

Vanetten, H.D., Sandrock, R.W., Wasmann, C.C., Soby, S.D., McCluskey, K. and Wang, P. (1995) Detoxification of phytoanticipins and phytoalexins by phytopathogenic fungi. *Canadian Journal of Botany* 73, S518–S525.

Zhou, Y.L. and Xing, L.J. (1986) [*Mycology. A Textbook for Undergraduates.*] Higher Education Press, Beijing, China.

Soil and rhizosphere

Buyer, J.S. and Sikora, L.J. (1990) Rhizosphere interactions and siderophores. *Plant and Soil* 129, 101–107.

Deacon, J.W. (1996) Ecological implications of recognition events in the pre-infection stages of root pathogens. *New Phytologist* 133, 135–145.

Glenn, O.F., Parker, C.A. and Sivasithamparam, K. (1985) A technique to compare growth in soil of *Gaeumannomyces graminis* var. *tritici* over a range of matric potentials. In: Parker, C.A., Rovira, A.D., Moore, K.J., Wong, P.T.W. and Kollmorgen, J.F. (eds) *Ecology and Management of Soilborne Plant Pathogens*. APS Press, St Paul, Minnesota, pp. 24–26.

Glenn, O.F., Hainsworth, J.M., Parker, C.A. and Sivasithamparam, K. (1987) Influence of matric potential and soil compaction on growth of the take-all fungus through soil. *Transactions of the British Mycological Society* 88, 83–90.

Hannukkala, A. and Koponen, H. (1987) *Microdochium bolleyi* a common inhabitant of barley and wheat roots in Finland. *Karstenia* 27, 31–36.

Kollmorgen, J.F. and Walsgott, D.N. (1984) Saprophytic survival of *Gaeumannomyces graminis* var. *tritici* at various depths of soil. *Transactions of the British Mycological Society* 82, 346–348.

Roseman, T.S. and Huber, D.M. (1989) Influence of *Gaeumannomyces graminis* var. *tritici* on manganese oxidizing bacteria in wheat rhizospheres. *Phytopathology* 79, 1166–1167.

Sivasithamparam, K. and Rowland, I. (1985) Propagule behaviour of the take-all fungus (*Gaeumannomyces graminis* var. *tritici*) at a field site in Western Australia. *Proceedings of the Indian Academy of Sciences (Plant Sciences)* 94, 91–97.

Smith, M.A.L., Spomer, L.A. and McClelland, M.T. (1990) Direct analysis of root zone data in a microculture system. *Plant Cell, Tissue and Organ Culture* 23, 21–26.

Sturz, A.V. and Bernier, C.C. (1987) Survival of cereal root pathogens in the stubble and soil of cereal versus noncereal crops. *Canadian Journal of Plant Pathology* 9, 205–213.

Sung, J.M., Yang, S.S., Lee, E.J. and Park, J.S. (1984) Soil-borne pathogens and its [sic.] damage analysis in nursery box and paddy field. *Research Reports – Office of Rural Development, Suweon, Soil Fertilizer Crop Protection and Mycology* 26, 57–63.

Whipps, J.M. (1997) Interactions between fungi and plant pathogens in soil and the rhizosphere. In: Gange, A.C. and Brown, V.K. (eds) *Multitrophic Interactions in Terrestrial Systems. Symposium of the British Ecological Society, 36*. Blackwell Science, Oxford, pp. 47–63.

Wildermuth, G.B. (1982) Soils suppressive to *Gaeumannomyces graminis* var. *tritici*: effects on other fungi. *Soil Biology and Biochemistry* 14, 561–568.

Wilkinson, H.T. and Cook, R.J. (1981) The effect of size and concentration of inoculum on the infection of wheat by *Gaeumannomyces graminis* var. *tritici* in different soils. *Phytopathology* 71, 265.

Wilkinson, H.T., Alldredge, J.R. and Cook, R.J. (1982) Estimated distances for infection of wheat roots by *Gaeumannomyces graminis* var. *tritici* in take-all suppressive and conducive soils. *Phytopathology* 72, 949.

Wong, P.T.W. and Southwell, R.J. (1987) Saprophytic survival of the take-all fungus and its antagonist, *Gaeumannomyces graminis* var. *graminis* under conventional and no-tillage. *Soil and Tillage Research* 9, 355–362.

In vitro

Arnott, H.J., Roseman, T.S., Graham, R.D. and Huber, D.M. (1991) An experimental study of manganese mineralization in the take-all fungus, *Gaeumannomyces graminis*. *Mycological Society of America Newsletter* 43, 3.

Bockus, W.W. and Norman, B.L. (1990) Heat inactivation of *Gaeumannomyces graminis* var. *tritici*. *Phytopathology* 80, 1024.

Boothby, D. (1981) Enhancement of polygalacturonase activity by potassium phosphate in liquid culture of *Phialophora radicicola* var. *radicicola*. *Transactions of the British Mycological Society* 77, 639–641.

Bowyer, P., Bryan, G., Lunness, P., Clarke, B., Daniels, M. and Osbourn, A. (1994) Saponin-detoxifying enzymes in plant pathogenic fungi. *Abstracts. 4th International Congress of Plant Molecular Biology. Amsterdam, June 19–24*. The International Society for Plant Molecular Biology. Abstract No. 1761.

Brodowsky, I.D., Hamberg, M. and Oliw, E.H. (1992) A linoleic acid (8R)-dioxygenase and hydroperoxide isomerase of the fungus *Gaeumannomyces graminis*. Biosynthesis of (8R)-hydroxylinoleic acid and (7S,8S)-dihydroxylinoleic acid from (8R)-hydroperoxylinoleic acid. *Journal of Biological Chemistry* 267, 14738–14745.

Brodowsky, I.D., Hamberg, M. and Oliw, E.H. (1994) BW A4C and other hydroxamic acids are potent inhibitors of linoleic acid 8R-dioxygenase of the fungus *Gaeumannomyces graminis*. *European Journal of Pharmacology* 254, 43–47.

Brodowsky, I.D. and Oliw, E.H. (1992) Metabolism of 18:2(n-6), 18:3(n-3) 20:4(n-6) and 20:5(n-3) by the fungus *Gaeumannomyces graminis*: identification of metabolites formed by 8-hydroxylation and by w2 and w3 oxygenation. *Biochimica et Biophysica Acta* 1124, 59–65.

Buyer, J.S., Sikora, L.J. and Chaney, R.L. (1989) A new growth medium for the study of siderophore-mediated interactions. *Biology and Fertility of Soils* 8, 97–101.

Dori, S., Solel, Z. and Barash, I. (1988) Detection and characterization of hydroxamate siderophores produced by *Gaeumannomyces graminis* var. *tritici*. *Phytoparasitica* 16, 94.

Dori, S., Solel, Z., Kashman, Y. and Barash, I. (1989) Characterization of hydroxamate siderophores and siderophore-mediated iron in *Gaeumannomyces graminis* var. *tritici*. *Phytopathology* 79, 1136.

Dori, S., Solel, Z., Kashman, Y. and Barash, I. (1990) Characterization of hydroxamate siderophores and siderophore-mediated iron uptake in *Gaeumannomyces graminis* var. *tritici*. *Physiological and Molecular Plant Pathology* 37, 95–106.

Elliott, M.L. (1991) Effect of melanin biosynthesis inhibitors on *Gaeumannomyces* spp. and *Magnaporthe poae*. *Phytopathology* 81, 1168.

Elliott, M.L. (1995) Effect of melanin biosynthesis inhibiting compounds on *Gaeumannomyces* species. *Mycologia* 87, 370–374.
Elliott, M.L. and Chasse, S.A. (1994) Evaluation of non-pathogenic strains of *Gaeumannomyces graminis* var. *graminis* derived by protoplasting and chemical mutagenesis. *Phytopathology* 84, 1085–1086.
Fang, C.S. and Parker, C.A. (1981) An L-drying method for preservation of *Gaeumannomyces graminis* var. *tritici*. *Transactions of the British Mycological Society* 77, 103–106.
Garrett, S.D. (1983) Factors limiting colonization of unsterilized filter paper by cereal root infecting fungi. *Soil Biology and Biochemistry* 15, 101–104.
Garrett, S.D. (1984) Factors controlling growth rate of cellulolytic fungi on sterile filter-paper. *Proceedings of the Indian Academy of Sciences (Plant Sciences)* 93, 189–194.
Hamberg, M., Zhang, L.Y., Brodowsky, I.D. and Oliw, E.H. (1994) Sequential oxygenation of linoleic acid in the fungus *Gaeumannomyces graminis*: stereochemistry of dioxygenase and hydroperoxide isomerase reactions. *Archives of Biochemistry and Biophysics* 309, 77–80.
Henson, J.M., Blake, N.K. and Pilgeram, A. (1988) Transformation of *Gaeumannomyces graminis*. *Journal of Cellular Biochemistry* (Suppl. 12), Part C, 281.
Henson, J.M., Blake, N.K. and Pilgeram, A. (1988) Transformation of *Gaeumannomyces graminis*. *Abstracts of the Annual Meeting of the American Society of Microbiology* 88, 154.
Honeyman, A.L. and Currier, T.C. (1983) The isolation and characterization of two linear DNA elements from *Gaeumannomyces graminis* var. *tritici*, the causitive agent of 'take-all disease' of wheat. *The American Society of Microbiology, Abstracts of the Annual Meeting – 1983*, Abstract H 175, 135.
Osbourn, A.E., Wubben, J.P. and Daniels, M.J. (1997) Saponin detoxification by phytopathogenic fungi. In: Stacey, G. and Keen, N.T. (eds) *Plant–Microbe Interactions*. Chapman & Hall, New York, pp. 99–124.
Su, C., Brodowsky, I.D. and Oliw, E.H. (1995) Studies on linoleic acid 8R-dioxygenase and hydroperoxide isomerase of the fungus *Gaeumannomyces graminis*. *Lipids* 30, 43–50.
Su, C. and Oliw, E.H. (1996) Purification and characterization of linoleate 8-dioxygenase from the fungus *Gaeumannomyces graminis* as a novel hemoprotein. *Journal of Biological Chemistry* 271, 14112–14118.
West, H.M. and Walters, D.R. (1989) Effects of polyamine biosynthesis inhibitors on growth of *Pyrenophora teres*, *Gaeumannomyces graminis*, *Fusarium culmorum* and *Septoria nodorum in vitro*. *Mycological Research* 92, 453–457.
Yarden, O. and Russo, V.E.A. (1995) Changes in chitin deposition accompany runner hypha branching of *Gaeumannomyces graminis* in culture. *Mycological Research* 100, 444–448.
Yu, S.Q., Trione, E.J. and Ching, T.M. (1984) Biochemical determination of the viability of fungal spores and hyphae. *Mycologia* 76, 608–613.

Pathogenicity

Brown, M.E. (1981) Infectivity of lesioned wheat root tissue related to the presence of dehydrogenase enzymes in hyphae of *Gaeumannomyces graminis* var. *tritici* in the lesions. *Soil Biology and Biochemistry* 13, 519–526.
Dewan, M.M. and Sivasithamparam, K. (1990) Differences in the pathogenicity of isolates of the take-all fungus from roots of wheat and ryegrass. *Soil Biology and Biochemistry* 22, 119–122.
Goins, T.L. and Henson, J.M. (1996) Melanization and decreased pathogenicity of *Gaeumannomyces graminis* var. *tritici*. *Phytopathology* 86, S64.
Kol'Nobryts'Kiĭ, M.I. (1985) [Pathogenicity of *Gaeumannomyces graminis* (Sacc.) Arx et Olivier, a causal agent of *Ophiobolus* root rot of winter wheat.] *Ukrains'kiĭ Botanĺchniĭ Zhurnal* 42, 98–100.
Osbourn, A.E. (1990) Molecular approaches towards understanding *Gaeumannomyces graminis* pathogenicity. *Journal of Experimental Botany* 41, 5–6.
Pedler, J.F., Webb, M.J., Buchhorn, S.C. and Graham, R.D. (1996) Manganese-oxidizing ability of isolates of the take-all fungus is correlated with virulence. *Biology and Fertility of Soils* 22, 272–278.
Roseman, T.S., Graham, R.D., Arnott, H.J. and Huber, D.M. (1991) The interaction of temperature with virulence and manganese oxidizing potential in the epidemiology of *Gaeumannomyces graminis*. *Phytopathology* 81, 1215.
Yeates, J.S. (1986) Relationship between colony pigmentation and pathogenicity of *Gaeumannomyces graminis* to oats. *Transactions of the British Mycological Society* 86, 680–682.

Detection (includes identification and quantification)

Detection – unclassified

Innocenti, G. and Govi, G. (1995) [Characterization of take-all isolates from different tillage techniques.] *Micologia Italiana* 24, 25–31.

Detection – conventional

Amelung, D. (1991) A simple method for identification of *Gaeumannomyces graminis* var. *tritici* Walker. In: Beemster, A.B.R., Bollen, G.L., Gerlach, M., Ruissen, B., Schippers, B. and Tempel, A. (eds) *Biotic Interactions and Soil-borne Diseases*. Elsevier, Amsterdam, pp. 412–413.

Cotterill, P.J. and Sivasithamparam, K. (1988) Use of a root assessment tray for the detection of take-all on lateral roots of wheat seedlings. *Plant and Soil* 110, 140–142.

Duffy, B.K. and Weller, D.M. (1991) A semi-selective medium for identification of *Gaeumannomyces graminis* var. *tritici*. *Phytopathology* 81, 1168.

Elliott-Juhnke, M. (1982) Selective differentiation of *Gaeumannomyces graminis* var. *tritici* on a medium utilizing L-dopa. *Phytopathology* 72, 992.

Gams, W. and Van Laar, W. (1982) The use of solacol validamycin as a growth retardant in the isolation of soil fungi. *Netherlands Journal of Plant Pathology* 88, 39–46.

Yeates, J.S. (1986) Failure of *in vitro* growth inhibition by cysteine to differentiate between Australian *Gaeumannomyces graminis* isolates pathogenic and non-pathogenic to oats. *Annals of Applied Biology* 109, 95–100.

Detection – molecular and immunological

Bertioli, D.J., Schlichter, U.H.A., Adams, M.J., Burrows, P.R., Steinbiss, H.-H. and Antoniw, J.F. (1995) An analysis of differential display shows a strong bias towards high copy number mRNAs. *Nucleic Acids Research* 23, 4520–4523.

Fouly, H.M., Domier, L.L., Wilkinson, H.T. and Pedersen, W.L. (1994) Use of RAPD and ribosomal RNA specific primers in PCR to characterize isolated *Gaeumannomyces* species. *Phytopathology* 84, 1145.

Fouly, H.M., Wilkinson, H.T. and Chen, W. (1997) Restriction analysis of internal transcribed spacers and the small subunit gene of ribosomal DNA among four *Gaeumannomyces* species. *Mycologia* 89, 590–597.

Krátká, J., Kyněrová, B., Zemanová, A. and Sýkorová, S. (1997) The diagnosis of *Fusarium culmorum* by polyclonal antibodies – preparation and character of antigens and antibodies. *Ochrana Rostlin* 33, 89–102.

Moore, L.W. and George, R.A. (1985) The use of ELISA for early quantification of *Gaeumannomyces graminis* var. *tritici* associated with winter wheat roots. *Phytopathology* 75, 1363.

O'Dell, M., Hollins, T.W., Flavell, R.B., Simpson, C., Scott, P.R. and Wolfe, M.S. (1987) Variation associated with repeated sequences in *Erysiphe graminis* and *Gaeumannomyces graminis*. *Report of the Plant Breeding Institute for 1986*, p.88.

Pilgeram, A.L. and Henson, J.M. (1990) Transformation and cotransformation of the take-all fungus, *Gaeumannomyces graminis*, to phleomycin resistance. *Phytopathology* 80, 1046.

Rachdawong, S. and Stromberg, E.L. (1997) Development of a molecular diagnostic test for *Gaeumannomyces graminis* var. *tritici*. *Phytopathology* 87, June (Suppl.), S81.

Rachdawong, S., Grabau, E.A., Cramer, C.C., Lacy, G.H. and Stromberg, E.L. (1996) Development of molecular diagnostic test for *Gaeumannomyces graminis* var. *tritici*. *Phytopathology* 86, November (Suppl.), S124.

Salava, J., Hájková, M. and Sommerová, P. (1997) [Characterization of *Fusarium* spp. using RAPD markers]. *Ochrana Rostlin* 32, 143–149.

Sauer, K.M., Hulbert, S.H. and Tisserat, N.A. (1993) Identification of *Ophiosphaerella herpotricha* by cloned DNA probes. *Phytopathology* 83, 97–102.

Schesser, K. and Henson, J.M. (1990) A polymerase chain reaction for detection of the take-all fungus, *Gaeumannomyces graminis*, in infected wheat plants. *Phytopathology* 80, 1073.

Sikora, L.J. and Buyer, J.S. (1989) Detection of ferric pseudobactin using monoclonal antibodies. *Abstracts of the Annual Meeting of the American Society of Microbiology* 89, 295.

Tan, M.K. (1997) Origin and inheritance of group I introns in 26S rRNA genes of *Gaeumannomyces graminis*. *Journal of Molecular Evolution* 44, 637–645.

Disease (includes root infection)

El-Nashaar, H.M., George, R.A. and Moore, L.W. (1986) The effect of sand particles size on infection of wheat roots by *Gaeumannomyces graminis* var. *tritici*. *Phytopathology* 76, 1098–1099.

Kirk, J.J. (1984) Ability of *Gaeumannomyces graminis* var. *tritici* to benefit from senescence of the cereal root cortex during infection. *Transactions of the British Mycological Society* 82, 107–112.

Lewis, S.J. and Deacon, J.W. (1982) Effects of shading and powdery mildew infection on senescence of the root cortex and coleoptile of wheat and barley seedlings, and implications for root- and foot-rot fungi. *Plant and Soil* 69, 401–411.

Potlaichuk, V.I. and Khlopunova, L.B. (1989) [Phytopathogenic species of *Phialophora*.] *Mikologiya i Fitopatologiya* 20, 178–185.

Rengel, Z. and Graham, R.D. (1993) Effects of Mn on accumulation of nutrients in wheat plants infected with the take-all fungus. In: Barrow, N.J. (ed.) *Plant Nutrition from Genetic Engineering to Field Practice*. Kluwer Academic Publishers, Dordrecht, pp. 685–688.

Rothrock, C. (1984) Take-all of wheat. In: Johnson J.W. and Hargrove W.L. (eds) *Small Grain Field Day. Georgia Agriculture Experiment Station Progress Report*, no. 8, 14.

Smiley, R.W., Fowler, M.C. and Reynolds, K.L. (1986) Temperature effects on take-all of cereals caused by *Phialophora graminicola* and *Gaeumannomyces graminis*. *Phytopathology* 76, 923–931.

Věchet, L. (1984) [The effect of supercontamination of the soil by *Gaeumannomyces graminis* var. *tritici* on the health of winter wheat.] *Sborník ÚVTIZ, Ochrana Rostlin* 20, 23–28.

Wang, X.R., Chen, J.X., Bi, J. and Zhang, C.Y. (1994) [Observations on the infection course of take-all in summer sown maize at seedling stage.] *Plant Protection* 20, 9–10.

Wiese, M.V. (1987) *Compendium of Wheat Diseases*. APS Press, St Paul, Minnesota.

Wilhelm, N.S. (1991) Investigations into *Gaeumannomyces graminis* var. *tritici* infection of manganese-deficient wheat. PhD Dissertation, University of Adelaide, Adelaide.

Yao, J.M. and Hu, H.W. (1986) [A brief report of take-all of maize.] *Plant Protection* 12, 21.

Occurrence and surveys

Amir, J., Krikun, J., Orion, D. and Putter, J. (1986) Root pathogens: a major factor in limiting wheat yield in an arid zone environment. *Canadian Journal of Plant Pathology* 8, 347.

Brown Jr, W.M., Perotti, L.E. and Hill, J.P. (1985) Wheat take-all in Colorado high country irrigated spring wheat. *Phytopathology* 75, 1296.

Carmona, M. and Barreto, D. (1995) [Fungal diseases on barley in the province of Buenos Aires (Argentina) in 1991.] *Fitopatologia Brasileira* 20, 509–510.

Cunfer, B. and Rothrock, C.S. (1983) 'Take-all' disease increasing on wheat. *Southeast Farm Press* 10, 20.

Daamen, R.A. (1981) Surveys of diseases and pests of winter wheat in the Netherlands, 1979–1980. *Mededelingen van de Faculteit Landbouwwetenschappen der Rijksuniversiteit te Gent* 46/3, 933–937.

Devaux, A. (1991) Survey of winter wheat diseases in 1990. *Canadian Plant Disease Survey* 71, 82.

El-Meleigi, M.A. (1988) Fungal diseases of spring wheat in central Saudi Arabia. *Crop Protection* 7, 207–209.

Gao, S.R., Wu, X.C., Shi. C.K. and Du, C.X. (1995) [Investigation and study on condition of incidence of 'Take-all disease of wheat'.] *Journal of Shandong Agricultural University* 26, 83–88.

Golzar, H., Forotan, A. and Torabi, M. (1994) Occurrence of take-all disease of wheat in Gorzan and Mazanduran. *Applied Entomology and Phytopathology* 61, 11–13.

Hornby, D. and Henden, D.R. (1983) A secondary outbreak of take-all. *Rothamsted Experimental Station Report for 1972. Part 1*, pp. 197–198.

Jörg, E. (1986) Komplexer Schaderregerbefall in Winterweizen. *Mitteilungen aus der Biologischen Bundesanstalt für Land- und Forstwirtschaft, Berlin-Dahlem* 232, 122.

King, P.M. (1984) Crop and pasture rotations at Coonalpyn, South Australia: effects on soil-borne diseases, soil nitrogen and cereal production. *Australian Journal of Experimental Agriculture and Animal Husbandry* 24, 555–564.

Korbas, M. (1995) Pathogenic fungi isolated from diseased stem bases of triticale. *Prace Naukowe Instytutu Ochrony Roślin* 36, 146–151.

Lawn, D.A. and Sayre, K.D. (1992) Soilborne pathogens on cereals in a highland location of Mexico. *Plant Disease* 76, 149–154.

Maas, E.M.C. and Kotzé, J.M. (1981) Fungi associated with root diseases of wheat in South Africa. *Phytophylactica* 13, 155–156.

Marshall, D. (1987) First report of take-all of wheat in Texas. *Plant Disease* 71, 850.

Mercer, P.C., McGimpsey, H.C. and Malone, J.P. (1986) Surveys of diseases of spring barley in Northern Ireland. *Record of Agricultural Research, Department of Agriculture, Northern Ireland* 34, 17–27.

Novokhatka, V.G., Doroshenko, N.V. and Zabolotnaya, V.A. (1990) [Distribution of root and rhizosphere rots of winter wheat in the Ukraine.] *Mikologiya i Fitopatologiya* 24, 352–357.

Oliveira, M.A.R., Diehl, J.A. and Reis, E.M. (1986) [Root diseases of wheat in the state of Paraná in 1983.] *Fitopatologia Brasileira* 11, 465–469.

Orr, D.D. (1991) Cereal disease survey in central Alberta, Canada, 1990. *Canadian Plant Disease Survey* 71, 64.

Plumb, R.T. and Hornby, D. (1987) Take-all research not forgotten. Readers' Letters. *Farmers Weekly*, 21 August, 10.

Reis, E.M. (1982) [Survey of cultivated, wild and weed host plants of fungi that cause root rots of winter cereals and other crops.] *Summa Phytopathologica* 8, 134–142.

Rothrock, C.S. (1985) Take-all of wheat in the southern U.S. In: *Proceedings of the Southern Small Grain Workers Conference, 1985*, Tallahassee, Florida, p. 34.

Rothrock, C.S. and Cunfer, B.M. (1984) Prevalence of take-all caused by *Gaeumannomyces graminis* var. *tritici* on wheat in Georgia. *Plant Disease* 68, 351.

Sturz, A.V. and Bernier, C.C. (1985) Incidence of a 'Take-All Like Fungus' from the crowns, stems and roots of winter wheat grown in Manitoba. *Canadian Plant Disease Survey* 65, 53–55.

Takahashi, H. (1989) Present status of occurrence and control of wheat and barley diseases in Japan. *Japan Pesticide Information* No. 54, 10–16.

Vilich-Meller, V. (1990) Vorkommen von Fusskrankheitserregern in gemischten Getreidebeständen. *Mitteilungen aus der Biologischen Bundesanstalt für Land- und Forstwirtschaft Berlin-Dahlem* 266, 148.

Walczak, F. (1986) [The occurrence of the most important diseases and pests of cereals in Poland in 1981.] *Biuletyn Instytuto Ochrony Roślin* 67, 13–48.

Zoschke, M. and Mueller-Wilmes, U. (1986) Study of continuous winter barley *Hordeum vulgare* cultivation in parabrown earth loess loam. *Bayerisches Landwirtschaftliches Jahrbuch* 63, 455–466.

Effect on host and yield

Augustin, C., Jacob, H.J. and Werner, A. (1997) Effects on growth of wheat plants of isolates of *Gaeumannomyces*/*Phialophora*-complex fungi in different conditions of soil moisture, temperature, and photoperiod. *European Journal of Plant Pathology* 103, 417–426.

Brown, M.E., Hepper, C.M., Chandler, M.R., Gibson, D.V. and Hornby, D. (1983) Comparative anatomical studies of wheat roots invaded by *Gaeumannomyces graminis* var. *tritici* (Ggt), *Phialophora radicicola* var. *graminicola* (Prg) and vesicular-arbuscular endophytes (VAM). *Rothamsted Experimental Station Report for 1982. Part 1*, p. 215..

Christensen, N.W., Taylor, R.G., Jackson, T.L. and Mitchell, B.L. (1981) Chloride effects on water potentials and yield of winter wheat infected with take-all root rot. *Agronomy Journal* 73, 1053–1058.

Christensen, N.W., Brett, M.A. and Hart, J.M. (1989) Yield of take-all infested winter wheat as influenced by inhibiting nitrification with dicyandiamide. *Communications in Soil Science and Plant Analysis* 20, 2137–2148

Cook, R.J. (1984) Root health: importance and relationship to farming practices. In: American Society of Agronomy (ASA), Bezdicek, D.F. and Power, J.F. (eds) *Organic Farming: Current Technology and its Role in Sustainable Agriculture*. Special publication No. 46. ASA, Madison, Wisconsin, pp. 111–127.

Cook, R.J. (1986) Wheat management systems in the Pacific Northwest. *Plant Disease* 70, 894–898.

Cook, R.J. (1987) Dynamics of soilborne plant pathogens and root health. *Abstracts of Papers of the 153rd National Annual Meeting of the American Association for the Advancement of Science*, 49.

Darby, R.J., Widdowson, F.V. and Hewitt, M.W. (1984) Comparisons between the establishment, growth and yield of winter wheat on three clay soils, in experiments testing nitrogen fertilizer in combination with aphicide and fungicides, in 1980 to 1982. *Journal of Agricultural Science, Cambridge* 103, 595–611.

Deacon, J.W. (1985) Programmed senescence and root diseases of cereals. *Phytophylactica* 17, 55.

Deacon, J.W. and Lewis, S.J. (1986) Invasion of pieces of sterile wheat root by *Gaeumannomyces graminis* and *Phialophora graminicola*. *Soil Biology and Biochemistry* 18, 167–172.

Hall, R. (1985) Effects of root pathogens on plant water relations. In: Ayres, P.G. and Boddy, L. (eds) *Water, Fungi and Plants*. Cambridge University Press, Cambridge, pp. 241–266.

Henry, C.M. and Deacon, J.W. (1981) Natural nonpathogenic death of the cortex of wheat and barley seminal roots as evidenced by nuclear staining with acridine orange. *Plant and Soil* 60, 255–274.

Kirk, J.J. and Deacon, J.W. (1986) Early senescence of the root cortex of agricultural grasses, and of wheat following root amputation or infection by the take-all fungus. *New Phytologist* 104, 63–75.

Lipps, P.E. (1984) Effect of take-all on yield components of three soft-red winter wheats in Ohio. *Phytopathology* 74, 1270.

Penrose, L. (1987) Influence of weight of seed on hyphal growth of the take-all pathogen *Gaeumannomyces graminis* var. *tritici* in wheat roots grown in sand culture. *Australian Journal of Experimental Agriculture* 27, 559–562.

Penrose, L.D.J. (1992) Interpretive value of symptoms of infection by *Gaeumannomyces graminis* in wheat seedlings grown in sand culture. *Annals of Applied Biology* 121, 545–557.

Rengel, Z., Pedler, J.F. and Graham, R.D. (1994) Control of Mn status in plants and rhizosphere: Genetic aspects of host and pathogen effects in the wheat take-all interaction. In: Manthey, J.A., Crowley, D.E. and Luster, D.G. (eds) *Biochemistry of Metal Micronutrients in the Rhizosphere*. CRC Press, Boca Raton, Florida, pp. 125–145.

Riveros, F., Carvalho, F.I.F. de and Reis, E.M. (1987) Take-all disease in wheat (*Triticum aestivum* L.): phenotypic changes due to seed size and inoculum concentration. *Revista Brasileira de Genética* 10, 87–100.

Roget, D.K. and Rovira, A.D. (1991) The relationship between incidence of infection by the take-all fungus (*Gaeumannomyces graminis* var. *tritici*), rainfall and yield of wheat in South Australia. *Australian Journal of Experimental Agriculture* 31, 509–513.

Rothrock, C. and Hargrove, W.L. (1987) Influence of fertility practices on take-all and yield of wheat. In: *Proceedings of the Eastern Wheat Workers and Southern Small Grain Workers Conference, 1987*, Griffin, Georgia.

Schönhammer, A. and Fischbeck, G. (1987) Untersuchungen an getreidereichen Fruchtfolgen und Getreidemonokulturen. 1. Mitteilung: Die Differenzierung der Ertragsleistung und deren Struktur im Verlauf von 15 Versuchsjahren. *Bayerisches Landwirtschaftliches Jahrbuch* 64, 175–191.

Stephens, P.M. and Davoren, C.W. (1996) Effect of the lumbricid earthworm *Aporrectodea trapezoides* on wheat grain yield in the field, in the presence or absence of *Rhizoctonia solani* and *Gaeumannomyces graminis* var. *tritici*. *Soil Biology and Biochemistry* 28, 561–567.

Sturz, A.V. and Bernier, C.C. (1989) Influence of crop rotations on winter wheat growth and yield in relation to the dynamics of pathogenic crown and root rot fungal complexes. *Canadian Journal of Plant Pathology* 11, 114–121.

Sward, R.J. and Kollmorgen, J.F. (1986) The separate and combined effects of barley yellow dwarf virus and take-all fungus (*Gaeumannomyces graminis* var. *tritici*) on the growth and yield of wheat. *Australian Journal of Agricultural Research* 37, 11–22.

Watt, A.D. and Wratten, S.D. (1984) The effects of growth stage in wheat on yield reductions caused by the rose-grain aphid *Metopolophium dirhodum*. *Annals of Applied Biology* 104, 393–397.

Grass and turf diseases

Baldwin, N.A. (1990) Fungal diseases of sports turf. *Mycologist* 4, 16–19.

Dernoeden, P.H. (1986) Management of take-all patch of creeping bentgrass with nitrogen, sulfur and PMA. *Phytopathology* 76, 1057.

Dernoeden, P.H. (1987) Management of take-all patch of creeping bentgrass with nitrogen, sulfur and phenyl mercury acetate. *Plant Disease* 71, 226–229.

Dernoeden, P.H. and O'Neill, N.R. (1983) Occurrence of Gaeumannomyces patch disease in Maryland and growth and pathogenicity of the causal agent. *Plant Disease* 67, 528–532.

Elliott, M.L. (1990) Pathogenicity studies of fungi associated with bermudagrass decline. *Phytopathology* 80, 978.

Elliott, M.L. (1992) Cultural and chemical control of bermudagrass decline caused by *Gaeumannomyces graminis* var. *graminis*. *Phytopathology* 82, 1123.

Elliott, M.L. (1995) Disease response of bermudagrasses to *Gaeumannomyces graminis* var. *graminis*. *Plant Disease* 79, 699–702.

Elliott, M.L. (1995) Effect of systemic fungicides on a bermudagrass putting green infested with *Gaeumannomyces graminis* var. *graminis*. *Plant Disease* 79, 945–949.

Elliott, M.L. and Landschoot, P.J. (1991) Fungi similar to *Gaeumannomyces* associated with root rot of turf grasses in Florida. *Plant Disease* 75, 238–241.

Elliott, M.L., Hagan, A.K. and Mullen, J.M. (1993) Association of *Gaeumannomyces graminis* var. *graminis* with a St Augustine grass root rot disease. *Plant Disease* 77, 206–209.

Freeman, T.E. and Augustin, B.J. (1986) Association of *Phialophora radicicola* Cain with declining Bemudagrass in Florida. *Phytopathology* 76, 1057.

Furuya, M., Shimokoji, H., Kawamura, K., Nakazumi, H., Ueda, S., Masutani, T., Higuchi, S., Tsutsui, S., Maki, Y., et al. (1988) [New smooth bromegrass *Bromus inermis* Leyss. cultivar Aikappu.] *Bulletin of Hokkaido Prefectural Agriculture Experiment Stations* No. 57, 35–48.

Jackson, N. (1984) A new cool season patch disease of Kentucky bluegrass turf in northeastern USA. *Phytopathology* 74, 812.

Jackson, N. and Landschoot, P.J. (1986) *Gaeumannomyces cylindrosporus* associated with diseased turfgrass in Rhode Island. *Phytopathology* 76, 654.

Kemp, M.L., Clarke, B.B. and Funk, C.R. (1990) The susceptibility of fine fescues to isolates of *Magnaporthe poae* and *Gaeumannomyces incrustans*. *Phytopathology* 80, 978.

Landschoot, P.J. and Jackson, N. (1990) Pathogenicity of some ectotrophic fungi with *Phialophora* anamorphs that infect the roots of turfgrasses. *Phytopathology* 80, 520–526.

Landschoot, P.J., Jackson, N. and Clarke, B.B. (1988) Taxonomy and pathogenicity of a *Gaeumannomyces* sp. from turfgrass roots. *Phytopathology* 78, 1521.

Lucas, P. and Sarniguet, A. (1991) Screening in the greenhouse of some treatments to control take-all patch on turfgrass. *Phytopathology* 81, 1198.

Lucas, P., Sarniguet, A., Cavelier, N. and Lelarge, S. (1992) Preliminary studies on the efficacy of some treatments to control take-all patch on turfgrass. *Agronomie (Paris)* 12, 187–192.

Lucas, P., Sarniguet, A. and Laurent, C. (1992) Occurrence in France of take-all patches on turfgrass caused by *Gaeumannomyces graminis* var. *avenae*. *Agronomie (Paris)* 12, 183–186.

MacDonald, L.S., Deyoung, R. and Ormrod, D.J. (1991) Turfgrass diseases diagnosed at the B. C. Ministry of Agriculture and Fisheries Plant Diagnostic Laboratory in 1990. *Canadian Plant Disease Survey* 71, 128.

Markham, T.D., Rimelspach, J.E. and Boehm, M.J. (1997) Development and use of a bentgrass bioassay to screen rootzone mixes for potential biocontrol activity against *Gaeumannomyces graminis* var. *avenae*. *Phytopathology* 87, June (Suppl.), S63.

Plumley, K.A., Gould, A.B. and Clarke, B.B. (1997) Impact of temperature, osmotic potential, and osmoregulant on the growth of three ectotrophic root-infecting fungi of Kentucky bluegrass. *Plant Disease* 81, 873–879.

Skipp, R.A. and Christensen, M.J. (1989) Fungi invading roots of perennial ryegrass (*Lolium perenne* L.) in pasture. *New Zealand Journal of Agricultural Research* 32, 423–431.

Skipp, R.A. and Christensen, M.J. (1989) Host specificity of root-invading fungi from New Zealand pasture soils. *Australasian Plant Pathology* 18, 101–103.

Smiley, R.W. (1987) The etiologic dilemma concerning patch diseases of bluegrass turfs. *Plant Disease* 71, 774–781.

Smiley, R.W. and Fowler, M.C. (1984) *Leptosphaeria korrae* and *Phialophora graminicola* associated with fusarium blight syndrome of *Poa pratensis* in New York. *Plant Disease* 68, 440–442.

Smiley, R.W. and Fowler, M.C. (1985) Pathogenicity of *Phialophora graminicola* to six cereal species at three temperatures. *Phytopathology* 75, 1290.

Smiley, R.W. and Fowler, M.C. (1985) Techniques for inducing summer patch symptoms on *Poa pratensis*. *Plant Disease* 69, 482–484.

Smiley, R.W. and Giblin, D.E. (1985) Pathogenesis of bluegrass by *Phialophora graminicola* in relation to root cortical death. *Phytopathology* 75, 1290.

Smiley, R.W. and Giblin, D.E. (1986) Root cortical death in relation to infection of Kentucky bluegrass by *Phialophora graminicola*. *Phytopathology* 76, 917–922.

Smiley, R.W., Fowler, M.C. and Kane, R.T. (1984) Characteristics of pathogens causing patch diseases of *Poa pratensis* in New York. *Phytopathology* 74, 811.

Smiley, R.W., Fowler, M.C. and Kane, R.T. (1985) Temperature and osmotic potential effects on *Phialophora graminicola* and other fungi associated with patch diseases of *Poa pratensis*. *Phytopathology* 75, 1160–1167.

Smiley, R.W., Fowler, M.C. and O'Knefski, R.C. (1985) Arsenate herbicide stress and incidence of summer patch on Kentucky bluegrass turfs. *Plant Disease* 69, 44–48.

Tisserat, N., Nus, A. and Pair, J. (1986) Characteristics and pathogenicity of two fungi isolated from Bermudagrass affected with spring dead spot. *Phytopathology* 76, 1130.

Tisserat, N., Hulbert, S. and Nus, A. (1990) Identification of *Leptosphaeria korrae* with cloned DNA probes. *Phytopathology* 80, 979.

Wilkinson, H.T. (1994) First report of root rot on centipedegrass (*Eremochloa ophiuroides*) caused by *Gaeumannomyces graminis* var. *graminis*. *Plant Disease* 78, 1220.

Wilkinson, H.T. and Kane, R.T. (1988) The relatedness of patch causing fungi in the midwest USA. *Phytopathology* 78, 1612–1613.

Wilkinson, H.T. and Pedersen, D. (1993) *Gaeumannomyces graminis* var. *graminis* infecting St Augustine grass selections in southern California. *Plant Disease* 77, 536.

Wong, P.T.W. and Worrad, D.J. (1989) Preventative control of take-all patch of bentgrass turf using triazole fungicides and *Gaeumannomyces graminis* var. *graminis* following soil fumigation. *Plant Protection Quarterly* 4, 70–72.

Wong, P.T.W., Rasmussen-Dykes, C., Perotti, L.E. and Brown, W.M.J. (1982) Occurrence of ophiobolus patch of turf in Colorado. *Phytopathology* 72, 976.

Worf, G.L., Avenius, R.C. and Stewart, J.S. (1982) A *Gaeumannomyces*-like organism associated with diseased bluegrass in Wisconsin. *Phytopathology* 72, 975–976.

Epidemiology (includes environmental and soil factors)

Bell, T.S. and Yarham, D.J. (1983) Aerial photography in the study of take-all (*Gaeumannomyces graminis*). In: *10th International Congress of Plant Protection 1983*. British Crop Protection Council, Croydon, p. 948.

Cotterill, P.J. and Sivasithamparam, K. (1989) Inoculum of the take-all fungus (*Gaeumannomyces graminis* var. *tritici*) in a Mediterranean-type climate: spatial distribution at a field site in Western Australia. *Phytophylactica* 21, 45–48.

Lucas, P. and Nignon, M. (1987) Influence du type de sol et de ces composantes physicochimiques sur les relations contre une variété de blé (*Triticum aestivum* L. var. Rescler) et deux souches, agressive et hypoaggressive de *Gaeumannomyces graminis* (Sacc.) Von Arx et Olivier var. *tritici* Walker. *Plant and Soil* 97, 105–117.

Murray, G.M., Heenan, D.P. and Taylor, A.C. (1991) The effect of rainfall and crop management on take-all and eyespot of wheat in the field. *Australian Journal of Experimental Agriculture* 31, 645–652.

Novotný, J. (1983) [An evaluation of the effect of some soil properties on the take-all of cereals *Gaeumannomyces graminis*.] *Sborník ÚVTIZ, Ochrana Rostlin* 19, 211–219.

Roseman, T.S., Phillips, J.D. and Huber, D.M. (1988) Microelement immobilization predisposes wheat to take-all. *Phytopathology* 78, 1504.

Alternative hosts

Abel, C. (1995) Higher yielding hybrid wheats here by 1998. *Farmers Weekly*, 20 January, 55–56.

Anon. (1989) Triticale fails to live up to promise on take-all. *Farmers Weekly*, 3 February, 49.

Mathematics and statistics

Gilligan, C.A. (1983) A test for randomness of infection by soil-borne pathogens. *Phytopathology* 73, 300–303.

Payne, R.W., Hornby, D. and Bateman, G.L. (1992) Experiments for testing treatments to control take-all. In: *Abstracts. XVIth International Biometrics Conference*, Hamilton, New Zealand, 7–11 December 1992, p. 176.

Fertilizers

Brennan, R.F. (1992) Effect of ammonium chloride, ammonium sulphate and sodium nitrate on take-all and grain yield of wheat in southwestern Australia. *Journal of Plant Nutrition* 15, 2639–2651.

Brennan, R.F. (1992) Effect of superphosphate and nitrogen on yield and take-all of wheat. *Fertilizer Research* 31, 43–49.

Brennan, R.F. (1992) The role of manganese and nitrogen nutrition in the susceptibility of wheat plants to take-all in Western Australia. *Fertilizer Research* 31, 35–41.

Brennan, R.F. (1993) Effect of ammonium chloride, ammonium sulphate and sodium nitrate on take-all and grain yield of wheat grown on soils in south-western Australia. *Journal of Plant Nutrition* 16, 349–358.

Cotterill, P.J. and Sivasithamparam, K. (1988) Effects of ammonium nitrogen on development and infectivity of the take-all fungus. *Annals of Applied Biology* 113, 461–470.

Coventry, D.R. and Kollmorgen, J.F. (1987) An association between lime application and the incidence of take-all symptoms on wheat on an acid soil in north-eastern Victoria. *Australian Journal of Experimental Agriculture* 27, 695–699.

Darby, R.J., Widdowson, F.V., Bird, E. and Hewitt, M.V. (1986) The relationship of soil mineral nitrate–nitrogen concentration and of fertilizer–nitrogen with the amount of nitrogen taken up by winter wheat in experiments testing nitrogen fertilizer in combination with aphicide and fungicides from 1980 to 1982. *Journal of Agricultural Science, Cambridge* 106, 497–508.

Engel, R.E. and Mathre, D.E. (1988) Effect of fertilizer nitrogen source and chloride on take-all of irrigated hard red spring wheat. *Plant Disease* 72, 393–396.

Gutteridge, R.J., Hornby, D. and Bateman, G.L. (1984) Observations on the current interest in chloride and take-all. In: *British Crop Protection Conference 1984. Pests and Diseases*. Vol. 1, pp. 83–84.

Hornby, D. (1991) A dose of manganese may not help take-all. Readers' Letters. *Farmers Weekly* 115, 7.

Huber, D.M. (1987) Immobilization of manganese predisposes wheat to take-all. *Phytopathology* 77, 1715.

Huber, D.M. (1989) The role of nutrition in the take-all disease of wheat and other small grains. In: Englehard, A.W. (ed.) *Management of Diseases with Macro- and Micro-elements*. American Phytopathological Society, St Paul, Minnesota, pp. 46–74.

Huber, D.M. and Dorich, R.A. (1988) Effect of nitrogen fertility on take-all disease in wheat. *Down to Earth* 44, 12–17.

Huber, D.M., McCay-Buis, T.S., Graham, R.D. and Robinson, N. (1993) Cultural conditions affecting take-all and their effect on manganese availability. *Proceedings of the 6th International Congress of Plant Pathology, 27 July-5 Aug., 1993*, Montreal, Canada.

Leggett, M.E., Sivasithamparam, K. and McFarlane, M.J. (1991) Effect of nitrogen supply on rhizosphere interactions and take-all disease of wheat. *Canadian Journal of Microbiology* 37, 42–51.

Lucas, P. and Collet, J.M. (1988) Influence de la fertilisation azotée sur la réceptivité d'un sol au piétin-échaudage, le développement de la maladie au champ et les populations de *Pseudomonas* fluorescents. *EPPO Bulletin* 18, 103–109.

MacNish, G.C. (1988) Changes in take-all (*Gaeumannomyces graminis* var. *tritici*), rhizoctonia root rot (*Rhizoctonia solani*) and soil pH in continuous wheat with annual applications of nitrogenous fertiliser in Western Australia. *Australian Journal of Experimental Agriculture* 28, 333–341.

Mineev, V.G., Belousova, N.A. and Durynina, E.P. (1985) [Effect of increasing doses of mineral fertilizers and liquid and solid manure on the resistance of barley to root rot.] *Agrokhimiya* No. 9, 84–93.

Scheyer, J.M., Christensen, N.W. and Powelson, R.L. (1987) Chloride fertilizer effects on stripe rust development and grain yield of winter wheat. *Plant Disease* 71, 54–57.

Taylor, R.G., Jackson, T.L., Powelson, R.L. and Christensen, N.W. (1983) Chloride, nitrogen form, lime and planting date effects on take-all root rot of winter wheat. *Plant Disease* 67, 1116–1120.

Crop sequence studies

Al-Najjar, A., Basedow, Th. and Schulz, F.A. (1989) Die Auswirkungen abgestufter Produktionsintensitäten im Weizenbau auf Schaderreger, Erträge und Wirtschaftlichkeit. I. Phytomedizinische Aspekte. *Zeitschrift für Pflanzenkrankheiten und Pflanzenschutz* 96, 561–584.

Amein, T.A.M. (1988) [The effect of a forecrop, its irrigation and temperature on wheat root and foot infection by *Gaeumannomyces graminis* var. *tritici* and species of *Fusarium*]. *Roczniki Nauk Rolniczych, Seria E* 18, 131–151.

Anon. (1988) Peas and barley – togetherness pays dividends. *Farmers Weekly*, 4 March, 46–49.

Borisov, G., Dzhumalieva, D., Mitova, T., Tsvetanova, K. and Petkova, M. (1988) Effect of long-term wheat monoculture on some components of soil fertility. *Pochvoznanie i Agrokhimiya* 23, 65–71.

Bowerman, P. and Banfield, C.F. (1982) The effect of break crops on winter wheat. *Experimental Husbandry* 38, 10–19.

Colbach, N. (1994) Influence of crop succession and soil tillage on wheat take-all (*Gaeumannomyces graminis* var. *tritici*). In: Borin, M. and Sattin, M. (eds) *Proceedings of the Third Congress of the European Society for Agronomy*. European Society of Agronomy, Colmar, pp. 672–673.

Colbach, N., Duby, C., Cavelier, A. and Meynard, J.M. (1997) Influence of cropping systems on foot and root diseases of winter wheat: fitting of a statistical model. *European Journal of Agronomy* 6, 61–77.

Cotterill, P.J. and Sivasithamparam, K. (1988) Inoculum of the take-all fungus in rotations of wheat and pasture. Relationships to disease and yield of wheat. *Transactions of the British Mycological Society* 91, 63–72.

Cotterill, P.J. and Sivasithamparam, K. (1988) Reduction of take-all inoculum by rotation with lupins, oats or field peas. *Journal of Phytopathology* 121, 125–134.

Crowe, F.J. and Bockus, W.W. (1984) Effect of soybeans double cropped with wheat on subsequent wheat take-all incidence. *Phytopathology* 74, 853.

Cunfer, B.M. (1997) Effect of twelve crop rotation sequences on take-all of wheat. *Phytopathology* 87, June (Suppl.), S21.

Cunningham, P.C. (1983) Effects of different spring cereal sequences on soil-borne pathogens and grain yields. *Irish Journal of Agricultural Research* 22, 225–242.

Czajka, W., Rogalski, L., Kurowski, T.P. and Cwalina, B. (1994) Crop rotation and health status of stem base of spring barley. *Phytopathologia Polonica* 7, 100–101.

Herman, M. (1986) [Possibilities of the use of cereals in crop rotation and the risks of foot rots.] *Rostlinna Výroba* 32, 991–998.

Herman, M. and Suskevic, M. (1986) [The influence of forecrop and soil treatment on the yield and health condition of winter wheat grown in the region with rainfall shortage.] *Rostlinna Výroba* 32, 247–256.

Hornby, D. and Gutteridge, R.J. (1993) Comparison of take-all in different cereals in monoculture and in cereal rotations. In: *Abstracts. 6th International Congress of Plant Pathology*. Publication Sales and Distribution, National Research Council Canada, Ottawa, Ontario, p. 265.

Kollmorgen, J.F., Griffiths, J.B. and Walsgott, D.N. (1983) The effects of various crops on the survival and carry-over of the wheat take-all fungus *Gaeumannomyces graminis* var. *tritici*. *Plant Pathology* 32, 73–78.

Krüger, W. and Speakman, J.B. (1990) Fruchtfolgeuntersuchungen über die Wurzel- und Halmbasisfäule bei Weizen sowie die Wurzelfäule bei Mais. *Mitteilungen aus der Biologischen Bundesanstalt für Land- und Forstwirtschaft, Berlin-Dahlem* 261, 1–129.

Langdale, G.W., Wilson, R.L. Jr, and Bruce, R.R. (1990) Cropping frequencies to sustain long-term conservation tillage systems. *Soil Science Society of America Journal* 54, 193–198.

Maas, E.M.C. and Kotzé, J.M. (1990) Crop rotation and take-all of wheat in South Africa. *Soil Biology and Biochemistry* 22, 489–494.

McEwen, J., Darby, R.J., Hewitt, M.V. and Yeoman, D.P. (1990) Effects of field beans, fallow, lupins, oats, oilseed rape, peas, ryegrass, sunflowers and wheat on nitrogen residues in the soil and on growth of a subsequent wheat crop. *Journal of Agricultural Science, Cambridge* 115, 209–220.

Olofsson, S. and Wallgren, B. (1984) [Winter wheat in the crop rotation: results from two trial series with various crops preceding winter wheat R-4-1711 and R-4-1712.] *Växtodling (Sveriges lantbruksuniversitet. Institutionen för växtodling)* No. 130, 1–80.

Olsen, C.C. (1984) Winter wheat in crop rotation with other varieties of cereals. *Tidsskrift for Planteavl* 88, 547–556.

Prew, R.D. (1981) The incidence of *Gaeumannomyces graminis* var. *tritici* and *Phialophora radicicola* var. *graminicola* on wheat grown after different cropping sequences. *Soil Biology and Biochemistry* 13, 179–184.

Reeves, T.G., Ellington, A. and Brooke, H.D. (1984) Effects of lupin–wheat rotations on soil fertility, crop disease and crop yields. *Australian Journal of Experimental Agriculture and Animal Husbandry* 24, 595–600.

Reis, E.M., Santos, H.P. dos and Lhamby, J.C.B. (1983) [Crop rotation I. The effect on root diseases of wheat in 1981 and 1982.] *Fitopatologia Brasileira* 8, 431–437.

Rothrock, C.S. and Langdale, G.W. (1988) Influence of soybean and sorghum summer crops on take-all in double-cropped wheat. *Phytopathology* 78, 1526.

Rothrock, C.S. and Langdale, G.W. (1989) Influence of nonhost crops on take-all in double-cropped winter wheat. *Plant Disease* 73, 130–132.

Steinbrenner, K. (1990) Kornertragsentwicklung und Schaderregerauftreten in Fruchtfolgen mit unterschiedlichem Getreideanteil. *Bodenkultur* 41, 23–26.

Steinbrenner, K. and Obenauf, U. (1986) Untersuchungen zum Einfluss der Vorfrucht und Vorvorfrucht auf den Ertrag von Wintergetreidearten und den Befall durch *Gaeumannomyces graminis*. *Archiv für Acker- und Pflanzenbau und Bodenkunde* 30, 773–779.

Steinbrenner, K., Grabert, D., Roth, R. and Obenauf, U. (1990) Fruchtfolgegestaltung: eine grundlegende Massnahme des Umweltschutzes. *Nachrichtenblatt für den Pflanzenschutz in der DDR* 44, 132–141.

Steinbrenner, K. and Obenauf, U. (1988) Untersuchungen zur Anbaupause von Winterweizen. *Archiv für Acker- und Pflanzenbau und Bodenkunde* 32, 57–62.

Sturz, A.V. and Bernier, C.C. (1986) Population dynamics of take-all and the crown root rot pathogens of winter wheat as influenced by crop sequences. *Canadian Journal of Plant Pathology* 8, 354.

Sturz, A.V. and Bernier, C.C. (1987) Incidence of pathogenic fungal complexes in the crowns and roots of winter and spring wheat relative to cropping practice. *Canadian Journal of Plant Pathology* 9, 265–271.

Vechet, L. (1983) The importance of break crops as suppressors of foot rot in continuously grown winter wheat. *Sborník ÚVTIZ, Ochrana Rostlin* 19, 205–210.

Zimmermann, A. (1983) Wurzelfäule an Weizen durch *Gaeumannomyces graminis* var. *tritici* – interaktionen zwischen Fruchtfolgestellung und bodenphysikalischen Parametern. *Zeitschrift für Pflanzenkrankheiten und Pflanzenschutz* 90, 505–514.

Other agronomic factors (e.g. tillage)

Anderson, M.A., Turkington, T.K., Briggs, K.G., Evans, K.G. and Penny, D. (1996) The influence of cultivar and soil pH on take-all of wheat. *Canadian Journal of Plant Pathology* 18, 488.

Bodker, L., Schulz, H. and Kristensen, K. (1990) [Influence of cultural practices on incidence of take-all (*Gaeumannomyces graminis* var. *tritici*) in winter wheat and winter rye.] *Tidsskrift for Planteavl* 94, 201–209.

Bräutigam, V. and Tebrügge, F. (1994) Wirkung langjährig pflugloser Bodenbearbeitungssysteme auf *Pseudocercosporella herpotrichoides* (Fron) Deighton, *Fusarium* spp. und *Gaeumannomyces graminis* (Sacc.) Arx et Olivier. *Mededelingen Faculteit Landbouwkundige en Toegepaste Biologische Wetenschappen Universiteit Gent* 59, 1009–1015.

Cassell, D. and Hering, T.F. (1982) The effect of water potential on soil-borne diseases of wheat seedlings. *Annals of Applied Biology* 101, 367–376.

Christensen, B.T. and Schjonning, P. (1987) Straw incorporation. *Tidsskrift for Planteavl* 91, 297–300.

Cotterill, P.J. and Sivasithamparam, K. (1987) Intermittent wetting of soils at high temperature reduces survival of the take-all fungus. *Plant and Soil* 103, 289–291.

Cotterill, P.J. and Sivasithamparam, K. (1988) Importance of the proportion of grassy weeds within legume crops in the perpetuation of *Gaeumannomyces graminis* var. *tritici*. *Plant Pathology* 37, 337–343.

Cotterill, P.J. and Sivasithamparam, K. (1988) Survival of the take-all fungus in the presence and absence of susceptible grasses. *Australian Journal of Soil Research* 26, 313–322.

Cunfer, B.M. and Rothrock, C.S. (1994) The influence of conservation tillage and double-cropping practices on diseases of wheat in Georgia. *Georgia Agriculture Experiment Station Bulletin* 418, 22 pp.

Dalbiès, A. and Doré, T. (1994) Agronomic consequences of set-aside; preliminary results. In: Borin, M. and Sattin, M. (eds) *Proceedings of the Third Congress of the European Society for Agronomy*. European Society of Agronomy, Colmar, pp. 678–679.

Deadman, M.L., Soleimani, M.J. and Nkemka, P.N. (1996) Wheat disease and cereal clover bicrops – attempting to resolve the enigma. *Aspects of Applied Biology* 47, 173–178.

de Boer, R.F., Steed, G.R. and Macauley, B.J. (1992) Effects of stubble and sowing treatments on take-all of wheat in north-eastern Victoria. *Australian Journal of Experimental Agriculture* 32, 641–644.

de Boer, R.F., Steed, G.R., Kollmorgen, J.F. and Macauley, B.J. (1993) Effects of rotation, stubble retention and cultivation on take-all and eyespot of wheat in northeastern Victoria, Australia. *Soil and Tillage Research* 25, 263–280.

Ellis, F.B., Christian, D.G. and Cannell, R.Q. (1982) Direct drilling shallow tine cultivation and plowing on a silt loam soil. *Soil and Tillage Research* 2, 115–130.

Garrett, S.D. (1985) Effect of soil texture on microbial abbreviation of saprophytic survival by the take-all fungus of wheat. *Proceedings of the Indian Academy of Sciences (Plant Sciences)* 94, 85–90.

Hannukkala, A.O. and Tapio, E. (1990) Conventional and organic cropping systems at Suitia. V: Cereal diseases. *Journal of Agricultural Science in Finland* 62, 339–347.

Harris, R. (1993) Set-aside provides chance for recovery. *Farmers Weekly*, 12 November, 38.

Herman, M. (1984) Micromycetes in the rhizosphere the rhizoplane and the roots of wheat under conventional and zero tillage. *Soil and Tillage Research* 4, 591–598.

Herman, M. (1985) Antagonistic activity of the rhizosphere mycoflora against *Gaeumannomyces graminis* under conventional and zero tillage. *Soil and Tillage Research* 5, 371–380.

Herman, M. (1990) Effect of tillage systems and crop sequence on the rhizosphere microflora of winter wheat. *Soil and Tillage Research* 15, 297–306.

Herzog, R. and Roth, R. (1986) Influence of many years' differentiated primary tillage on infestation with weeds and occurrence of cereal diseases on lightly loamed sandy soil under continuous cereal growing. *Archiv für Acker- und Pflanzenbau und Bodenkunde* 30, 647–653.

Hornby, D., Bateman, G.L. and Gutteridge, R.J. (1987) Winter wheat: factors affecting take-all. *Friends of Rothamsted Newsletter* No. 5, Winter 86/87, 12–13.

Hornby, D., Bateman, G.L. and Gutteridge, R.J. (1987) Take-all: combining factors known to affect take-all. *Rothamsted Report for 1986*, Part 1, 117–118.

Jensen, A. (1988) Plant pathological aspects of ploughless cultivation. Scandinavian Association of Agricultural Scientists on Reduced Cultivation Seminar, Horsens, Denmark, Feb 9–11, 1988. *Sveriges Lantbruksuniversitet, Uppsala, Institutionen för Markvetenskap, Rapport Jordbearbetningsavdelningen* No.77, 117–127.

Kollmorgen, J.F., Ridge, P.E. and de Boer, R.F. (1987) Effects of tillage and straw mulches on take-all of wheat in the Northern Wimmera of Victoria. *Australian Journal of Experimental Agriculture* 27, 419–423.

Lucas, P., Sarniguet, A., Collet, J.M. and Lucas, M. (1989) Soil receptivity to take-all (*Gaeumannomyces graminis* var. *tritici*): the effect of certain cultivation techniques. *Soil Biology and Biochemistry* 21, 1073–1078.

Mekwatanakarn, P. and Sivasithamparam, K. (1987) Effect of certain herbicides on saprophytic survival and biological suppression of the take-all fungus. *New Phytologist* 106, 153–160.

Mielke, H. (1983) Untersuchungen über den Einfluss verschiedener Bodenbearbeitungen auf Fusskrankheiten des Getreides. *Nachrichtenblatt des Deutschen Pflanzenschutzdienstes (Stuttgart)* 35, 33–39.

Olsen, C.C. (1984) Sowing time and sowing rate in winter wheat and winter barley. *Tidsskrift for Planteavl* 88, 557–570.

Palantinus, L. (1985) [The effect of irrigation on spring barley diseases.] *Rostlinna Výroba* 31, 85–94.

Prew, R.D. (1981) The effect of minimum cultivation on the incidence of take-all down the root profile of winter wheat. *Annals of Applied Biology* 98, 217–226.

Rasmussen, K.J. (1988) Ploughing, direct drilling and reduced cultivation for cereals. *Tidsskrift for Planteavl* 92, 233–248.

Reis, E.M., Lhamby, J.C., Santos, H.P. and Kochhmann, R.A. (1982) Efeitos de calcário e de sistemas de semeadura na incidência de *Gaeumannomyces graminis* var. *tritici* e de *Helminthosporium sativum* em raízes de trigo. *Summa Phytopathologica* 8, 56–64.

Roget, D.K., Neate, S.M. and Rovira, A.D. (1996) Effect of sowing point design and tillage practice on the incidence of rhizoctonia root rot, take-all and cereal cyst nematode in wheat and barley. *Australian Journal of Experimental Agriculture* 36, 683–693.

Roseman, T.S. and Huber, D.M. (1990) Manganese seed content – an ameliorating factor for take-all of cereals. *Phytopathology* 80, 970.

Rothrock, C.S. (1985) Effect of tillage on take-all of wheat. In: Hargrove, W.L., Boswell, F.C. and Langdale, G.W. (eds) *Proceedings of the 1985 Southern Region No-till Conference.* University of Georgia, Athens, pp. 211–214.

Rothrock, C.S. (1985) Take-all of wheat as affected by tillage and doublecropping. *Phytopathology* 75, 1291.

Rothrock, C.S. (1987) Take-all of wheat as affected by tillage and wheat–soybean doublecropping. *Soil Biology and Biochemistry* 19, 307–312.

Rothrock, C.S. and Cunfer, B.M. (1984) Effect of management practices on take-all of wheat. In: Johnson, J. W. and Hargrove, W. L. (eds) *Small Grain Field Day. Georgia Agriculture Experiment Station Progress Report* No. 8, 2.

Rovira, A.D. (1987) Tillage and soil-borne root diseases in winter cereals. In: Cornish, P.S. and Pratley, J.E. (eds) *Tillage – New Directions in Australian Agriculture.* Inkata Press, Melbourne, pp. 335–354.

Rovira, A.D. (1990) The impact of soil and crop management practices on soil-borne root diseases and wheat yields. *Soil Use and Management* 6, 195–200.

Rovira, A.D. and McDonald, H.J. (1986) Effects of the herbicide chlorsulfuron on rhizoctonia bare patch and take-all of barley and wheat. *Plant Disease* 70, 879–882.

Santos, H.P.D., Reis, E.M., Lhamby, J.C.B. and Sandini, I. (1995) [Agronomic characteristics and control of root diseases in direct-drilled barley grown in rotation with other crops]. *Pesquisa Agropecuária Brasileira* 30, 1297–1303.

Sivasithamparam, K. and Bolland, M.D.A. (1985) Effect of herbicide sprays on the infectivity and propagule numbers of take-all fungus (*Gaeumannomyces graminis* var. *tritici*) and the effect of propagule size and soil temperature on infectivity. *Australasian Plant Pathology* 14, 22–25.

Steinbrenner, K. and Höflich, G. (1984) Einfluss acker- und pflanzenbaulicher Massnahmen auf den Befall des Getreides durch *Pseudocercosporella herpotrichoides* (Fron) Deighton und *Gaeumannomyces graminis* (Sacc.) Arx et Olivier. *Archiv für Acker- und Pflanzenbau und Bodenkunde* 20, 61–71.

Wong, P.T.W., Dowling, P.M., Tesoriero, L.A. and Nicol, H.I. (1993) Influence of preseason weed management and in-crop treatments on two successive wheat crops 2. Take-all severity and incidence of rhizoctonia root rot. *Australian Journal of Experimental Agriculture* 33, 173–177.

Control (includes management)

Anon. (1997) European trials could bring take-all control. *Farmers Weekly*, 7 March, 71.

Bernier, C.C., Lamari, L. and Stebbing, J.A. (1992) Effects of chloride, triadimenol and *Gliocladium roseum* on take-all root rot of wheat. *Phytopathology* 82, 1129.

Bernier, C.C., Lamari, L., Stebbing, J.A. and Flaten, D. (1995) Effects of *Gliocladium roseum*, ammonium fertilizers and triadimenol seed treatment on take-all root rot of wheat. *Canadian Journal of Plant Pathology* 17, 353.

Brisbane, P.G., Harris, J.R. and Moen, R. (1989) Inhibition of fungi from wheat roots by *Pseudomonas fluorescens* 2-79 and fungicides. *Soil Biology and Biochemistry* 21, 1019–1026.

Butterworth, B. (1997) Cereals. Diseases. Compost can cut chemical costs? *Arable Farming*, 10 June, 12–14.

Campbell, R. and Clor, A. (1985) Soil moisture affects the interaction between *Gaeumannomyces graminis* var. *tritici* and antagonistic bacteria. *Soil Biology and Biochemistry* 17, 441–446.

Campbell, R. and Ephgrave, J.M. (1983) Effect of bentonite clay on the growth of *Gaeumannomyces graminis* var. *tritici* and on its interactions with antagonistic bacteria. *Journal of General Microbiology* 129, 771–778.

Capper, A.L. and Campbell, R. (1985) Biological control of take-all. *Journal of Applied Bacteriology* 59, xvii.

Cook, R.J. (1988) Management of the environment for the control of pathogens. *Philosophical Transactions of the Royal Society of London, Series B, Biological Sciences* 318, 171–182.

Cook, R.J. and Weller, D.M. (1987) Management of take-all in consecutive crops of wheat or barley. In: Chet, I. (ed.) *Nonconventional Methods of Disease Control*. John Wiley & Sons, Chichester, pp. 41–76.

Crozier, J.B. and Stromberg, E.L. (1996) Management for suppression of take-all in soft red winter wheat in Virginia. *Phytopathology* 86, S120.

Cunfer, B. and Rothrock, C. (1984) Control of take-all of wheat by seed treatments. In: Johnson, J.W. and Hargrove, W.L. (eds) *Small Grain Field Day. Georgia Agriculture Experiment Station Progress Report* no. 8, May, 1.

Deacon, J.W. (1982) Inoculum, inoculum potential and the biographical [sic] control of take-all. *Bulletin of the British Mycological Society* 16, 5.

Goos, R.J., Johnson, B.E. and Stack, R.W. (1989) Effect of potassium chloride, imazalil and method of imazalil application on barley infected with common root rot. *Canadian Journal of Plant Science* 69, 437–444.

Green, C.F. and Dawkins, T.C.K. (1986) Influence of nitrogen fertilizer and chlormequat on two spring wheat cultivars. *Crop Research* 25, 89–102.

Hornby, D. (1988) Take-all control: first understand the enemy. *Farmers Weekly*, 108, 24.

Hornby, D. and Bateman, G.L. (1993) Take-all disease patterns in fields. In: *Abstracts. 6th International Congress of Plant Pathology*. Publication Sales and Distribution, National Research Council Canada, Ottawa, Ontario, p. 99.

Krikun, J. (1986) A soil column technique for testing the efficacy of metham-sodium against nonsclerotial fungi. *Canadian Journal of Plant Pathology* 8, 345–436.

MacNish, G.C. (1986) Effects of fumigation on soil nitrogen plant nitrogen and root disease incidence in wheat at Wongan Hills Western Australia. *Australian Journal of Soil Research* 24, 81–94.

Mielke, H. and Hopp, H. (1982) The influence of peracetic acid on foot rot disease of cereals. *Zeitschrift für Pflanzenkrankheiten und Pflanzenschutz* 89, 282–290.

Rovira, A. (1990) Ecology, epidemiology and control of take-all, rhizoctonia bare patch and cereal cyst nematode in wheat. *Australasian Plant Pathology* 19, 101–111.

Steinbrenner, K. and Seidel, D. (1982) Komplexe Butzung acker- und pflanzenbaulicher Massnahmen im Rahmen des Pflanzenschutzes. *Nachrichtenblatt für den Pflanzenschutz in der DDR* 7, 137–140.

Sutton, J.C. (1989) Progress and problems in integrating disease management practices in winter wheat. *Canadian Journal of Plant Science* 69, 267.

Yarham, D. (1988) The contribution and value of cultural practices to control arable crop diseases. In: Clifford, B.C. and Lester, E. (eds) *Control of Plant Disease*. Blackwell Scientific Publications, Oxford, pp. 135–154.

Zhang QingDong, Xu EnJun and Zhou YuBin (1995) [Study on application of phosphogypsum for the control of take-all disease of wheat.] *Soils and Fertilizers (Beijing)* No. 6, 35–38.

Biological: unclassified

Augustin, C., Zaspel, I. and Steinbrenner, K. (1988) Möglichkeiten einer mikrobiologischen Bekämpfung der Schwarzbeinigkeit des Getreides. *FZB Report* 1988, 128–134.

Bockus, W.W. and Daniels, B.A. (1981) Displacement of *Gaeumannomyces graminis* var. *tritici* by *Microdochium bolleyi* in wheat tissues. *Phytopathology* 71, 861.

Bull, C.T. and Weller, D.M. (1987) Colonization of wheat roots by bacteria suppressive or non-suppressive to wheat root pathogens. *Phytopathology* 77, 1688.

Clark, B.L., Reeder, R.B. and Ownley, B.H. (1995) Evaluation of *Bacillus* and *Pseudomonas* isolates from Tennessee soil for biological control of take-all. *Phytopathology* 85, 1191.

Cook, R.J. (1989) Plant management in relation to biological control of pathogens. *Journal of Cellular Biochemistry* (Suppl. 13), Part A, 165.

Cook, R.J., Weller, D.M. and Thomashow, L.S. (1986) Enhancement of root health and plant growth by rhizobacteria. *Journal of Cellular Biochemistry* (Suppl. 10), Part C, 13.

Défago, G., Ahl, P., Berling, C.H., Stutz, E., Voisard, C., Haas, D. and Rella, M. (1984) Characteristics of a *Pseudomonas fluorescens* strain involved in suppression of black root rot of tobacco. *Phytopathology* 74, 799.

Dewan, M.M. and Sivasithamparam, K. (1988) *Pythium* spp. in roots of wheat and rye-grass in Western Australia and their effect on root rot caused by *Gaeumannomyces graminis* var. *tritici*. *Soil Biology and Biochemistry* 20, 801–808.

Dewan, M.M. and Sivasithamparam, K. (1989) Behaviour of a plant growth-promoting sterile fungus on agar and roots of rye-grass and wheat. *Mycological Research* 93, 161–166.

Dewan, M.M. and Sivasithamparam, K. (1989) Occurrence of species of *Aspergillus* and *Penicillium* in roots of wheat and ryegrass and their effect on root rot caused by *Gaeumannomyces graminis* var. *tritici*. *Australian Journal of Botany* 36, 701–710.

Dilantha Fernando, W.G. and Pierson, L.S. III (1994) In vitro mycelial growth inhibition of fungal pathogens by *Pseudomonas aureofaciens* strain 30-84. *Phytopathology* 84, 1136.

Dutrecq, A., Stevaux, J., Debras, P. and Rossignol, E. (1989) Mise en évidence et étude de l'activité de quelques organismes antagonistes du *Gaeumannomyces graminis* var. *tritici*. *Annales de Gembloux* 95, 285–299.

Dutrecq, A., Debras, P., Stevaux, J. and Marlier, M. (1991) Activity of 2,4-diacetylphloroglucinol isolated from a strain of *Pseudomonas fluorescens* to *Gaeumannomyces graminis* var. *tritici*. In: Beemster, A.B.R., Bollen, G.J., Gerlagh, M., Ruissen, M.A., Schippers, B. and Tempel, A. (eds) *Biotic Interactions and Soil-borne Diseases*. Elsevier Science Publishers, Amsterdam, pp. 252–257.

El-Nashaar, E.M., Wagih, E.E. and Gough, F.J. (1987) Common gene products among *Pseudomonas* and *Acinetobacter* isolates antagonistic to *Gaeumannomyces graminis* var. *tritici*. *Phytopathology* 77, 641.

Elsherif, M. and Grossmann, F. (1986) Isolation and characterisation of fluorescent pseudomonads and assessment of their antagonistic activity against *Gaeumannomyces graminis* var. *tritici*. *Mitteilungen aus der Biologischen Bundesanstalt für Land- und Forstwirtschaft, Berlin-Dahlem* 232, 262.

Ghisalberti, E.L. and Sivasithamparam, K. (1990) The nature of secondary metabolites produced by biological control agents. *Planta Medica* 56, 641.

Ghisalberti, E.L., Narbey, M.J. and Rowland, C.Y. (1990) Metabolites of *Aspergillus terreus* antagonistic towards the take-all fungus. *Journal of Natural Products* 53, 520–522.

Gillespie-Sasse, L.-M.J., Almassi, F., Ghisalberti, E.L. and Sivasithamparam, K. (1991) Use of a clean seedling assay to test plant growth promotion by exudates from a sterile red fungus. *Soil Biology and Biochemistry* 23, 95–98.

Ismail, E.A., Wilkinson, H.T. and Fouly, H.M. (1996) Differential colonization of *Gaeumannomyces graminis* var. *tritici* infected wheat roots by suppressive and non-suppressive bacteria. *Phytopathology* 86, S51–S52.

Iswandi, A., Vandenabeele, J. and Verstraete, W. (1986) The *in vitro* inhibitory effect of pseudomonas strains on the growth of fungi. *Mededelingen van de Faculteit Landbouwwetenschappen der Rijksuniversiteit te Gent* 51, 1373–1380.

Keel, C. and Défago, G. (1997) Interactions between beneficial soil bacteria and root pathogens: mechanisms and ecological impact. In: Gange, A.C. and Brown, V.K. (eds) *Multitrophic Interactions in Terrestrial Systems, Symposium of the British Ecological Society, 36*, Blackwell Science, Oxford, pp. 27–46.

Leggett, M.E. and Sivasithamparam, K. (1986) Interaction of fluorescent pseudomonads, hyphae of the take-all fungus and growth of wheat roots in soil. In: *British Crop Protection Conference, Pests and Diseases.* British Crop Protection Council, Thornton Heath, pp. 1177–1184.

Leong, J. (1986) Siderophores their biochemistry and possible role in the biocontrol of plant pathogens. *Annual Review of Phytopathology* 24, 187–210.

Lynch, J.M. and Ebben, M.H. (1986) The use of microorganisms to control plant disease. In: Bateson, M., Benham, C.L. and Skinner, F.A. (eds) *Microorganisms in Agriculture.* Blackwell Scientific Publications, Oxford, pp. 115–126.

Mazzola, M. and Cook, R.J. (1989) Influence of wheat root pathogens on maintenance of populations of *Pseudomonas fluorescens* Q72A-80 and 2-79 in the wheat rhizosphere. *Phytopathology* 79, 1159.

Mazzola, M., Cook, R.J., Thomashow, L.S. and Weller, D.M. (1990) Significance of phenazine biosynthesis in the survival of fluorescent pseudomonads in soil habitats. *Phytopathology* 80, 969.

McCay-Buis, T.S., Schulze, D.G., Sutton, S.R. and Huber, D.M. (1993) An *in situ* technique for studying mineral interactions and biological control of take-all. *Proceedings of the 6th International Congress of Plant Pathology, 27 July-5 August, 1993,* Montreal, Canada, p.267.

Parke, J.L., Moen, R., Rovira, A.D. and Bowen, G.D. (1984) Root tip colonization by a pseudomonad suppressive to the take-all disease of wheat. *Phytopathology* 74, 799.

Parke, J.L., Rovira, A.D. and Bowen, G.D. (1984) Soil matric potential affects colonization of wheat roots by a pseudomonad suppressive to take-all. *Phytopathology* 74, 806.

Pierson, E.A. and Weller, D.M. (1991) Recent work on control of take-all of wheat by fluorescent pseudomonads. In: Keel, C., Koller, B. and Défago, G. (eds) *Plant Growth-promoting Rhizobacteria: Progress and Prospects.* IOBC WPRS, Paris, pp. 96–97.

Poplawsky, A.R. and Ellingboe, A.H. (1986) A bacterial DNA segment capable of inhibiting the expression of antibiosis to the take-all pathogen. *Phytopathology* 76, 1135.

Poplawsky, A.R. and Ellingboe, A.H. (1989) Take-all suppressive properties of bacterial mutants affected in antibiosis. *Phytopathology* 79, 143–146.

Poplawsky, A.R., Peng, Y.F. and Ellingboe, A.H. (1986) Bacterial TN5 mutants affected in antibiosis to *Gaeumannomyces graminis* var. *tritici. Phytopathology* 76, 1069.

Poplawsky, A.R., Peng, Y.F. and Ellingboe, A.H. (1988) Genetics of antibiosis in bacterial strains suppressive to take-all. *Phytopathology* 78, 426–432.

Raaijmakers, J.M., Weller, D.M. and Thomashow, L.S. (1997) Frequency of antibiotic-producing *Pseudomonas* spp. in natural environments. *Applied and Environmental Microbiology* 63, 881–887.

Renwick, A., Campbell, R. and Coe, S. (1991) Assessment of *in vivo* screening systems for potential biocontrol agents of *Gaeumannomyces graminis. Plant Pathology* 40, 524–532.

Schmidt, O., Doube, B.M., Ryder, M.H. and Killham, K. (1997) Population dynamics of *Pseudomonas corrugata* 2140R LUX8 in earthworm food and in earthworm casts. *Soil Biology and Biochemistry* 29, 523–528.

Sherman, H. (1987) New antibiotic helps protect plants. *Agricultural Research, Washington* 35, 4.

Shivanna, M.B., Meera, M.S. and Hyakumachi, M. (1996) Role of root colonization ability of plant growth promoting fungi in the suppression of take-all and common root rot of wheat. *Crop Protection* 15, 497–504.

Steinbrenner, K., Augustin, C. and Zaspel, I. (1988) Ergebnisse zur biotechnologischen Bekämpfung von *Gaeumannomyces graminis. Tagungs-Berichte, Erhöhung der Bodenfruchtbarkeit und der Erträge durch wissenschaft Fortschritt* 5–7, 198–205.

Weller, D.M. and Rovira, A.D. (1984) Suppression of take-all of wheat in South Australian soils by fluorescent pseudomonads. *Phytopathology* 74, 806.

Wessendorf, J. and Lingens, F. (1989) Effect of culture and soil conditions on survival of *Pseudomonas fluorescens* R1 in soil. *Applied Microbiology and Biotechnology* 31, 97–102.

Wildermuth, G.B., Rovira, A.D. and Warcup, J.H. (1985) Mechanism and site of suppression of *Gaeumannomyces graminis* var. *tritici* in soil. *Transactions of the British Mycological Society* 84, 3–10.

Wong, P.T.W. (1985) Survival of fungal antagonists of *Gaeumannomyces graminis* var. *tritici.* In: Parker, C.A., Rovira, A.D., Moore, K.J., Wong, P.T.W. and Kollmorgen, J.F. (eds) *Ecology and Management of Soilborne Plant Pathogens.* APS Press, St Paul, Minnesota, pp. 148–150.

Zunke, U., Wyss, U., Rössner, J. and Nagel, S. (1986) Zum Parasitierungsverhalten des mykophagen Nematoden *Aphelenchoides hamatus* an Pilzen und Wurzelhaaren. *Mitteilungen aus der Biologischen Bundesanstalt für Land- und Forstwirtschaft, Berlin-Dahlem* 232, 421.

Biological: natural

Bowerman, P. (1981) Exploitation of take-all decline in wheat and barley. *Experimental Husbandry* 37, 74–82.

Chakraborty, S. (1985) Survival of wheat take-all fungus in suppressive and non-suppressive soils. *Pedobiologia* 28, 13–18.

Dewan, M.M. and Sivasithamparam, K. (1988) Identity and frequency of occurrence of *Trichoderma* spp. in roots of wheat and rye-grass in Western Australia and their effect on root rot caused by *Gaeumannomyces graminis* var. *tritici*. *Plant and Soil* 109, 93–102.

Höper, H. and Alabouvette, C. (1996) Importance of physical and chemical soil properties in the suppressiveness of soils to plant diseases. *European Journal of Soil Biology* 32, 41–58.

Kiewnick, S., Jacobsen, B.J. and Sikora, R.A. (1997) Structural differences of fungal and bacterial communities in soils with high and low antagonistic potentials. *Phytopathology* 87, June (Suppl.), S51.

Kloepper, J.W., Schroth, M.N., Leong, J. and Teintze, M. (1981) *Pseudomonas* siderophores: a mechanism explaining disease suppressive soils. *Phytopathology* 71, 232.

McQuilken, M.P. (1995) Promoting natural biological control of soil-borne plant pathogens. In: *Integrated Crop Protection: Towards Sustainability?* BCPC, Farnham, pp. 59–66.

Raaijmakers, J.M., Weller, D.M. and Thomashow, L.S. (1996) Frequency of antibiotic-producing pseudomonads in take-all suppressive soils. *Phytopathology* 86, S36–S37.

Rengel, Z. (1997) Decreased capacity of take-all fungus to oxidize manganous ions is associated with take-all decline. *Journal of Plant Nutrition* 20, 455–460.

Rothrock, C. and Cunfer, B.M. (1984) Absence of take-all decline under wheat–soybean doublecropping. *Phytopathology* 74, 799–800.

Rouxel, F. (1991) Natural suppressiveness of soils to plant diseases. In: Beemster, A.B.R., Bollen, G.L., Gerlach, M., Ruissen, B., Schippers, B. and Tempel, A. (eds) *Biotic Interactions and Soil-borne Diseases*. Elsevier, Amsterdam, pp. 287–296.

Weller, D.M. and Cook, R.J. (1981) Pseudomonads from take-all conducive and suppressive soils. *Phytopathology* 71, 264.

Wildermuth, G.B. (1982) Soils suppressive to *Gaeumannomyces graminis* var. *tritici*: induction by other fungi. *Soil Biology and Biochemistry* 14, 569–574.

Zriba, N., Mathre, D.E. and Sherwood, J.E. (1997) Characterization of *Phialophora* spp. from a Montana take-all suppressive soil. *Phytopathology* 87, June (Suppl.), S110.

Biological: introduced organisms

Andreoli, Y.E., Laich, F.S. and Navarro, C.A. (1993) *In vitro* control of *Sclerotinia sclerotiorum* and *Gaeumannomyces graminis* by fluorescent pseudomonas bacteria. *Revista Argentina de Microbiologia* 25, 70–79.

Augustin, C. (1989) Möglichkeiten einer Schadminderung der Schwarzbeinigkeit an Weizen mittels apathogener Pilze. I. Mitt.: Selektion und Testung apathogener Pilze im Pflanzentest unter semi- und unsterilen Bedingungen. *Zentralblatt für Mikrobiologie* 144, 563–570.

Augustin, C. (1990) Möglichkeiten einer Schadminderung der Schwarzbeinigkeit an Weizen mittels apathogener Pilze. II. Mitteilung: Einsatz von Pilzisolaten in Gefäss- und Feldversuchen. *Zentralblatt für Mikrobiologie* 145, 579–584.

Augustin, C. (1993) Untersuchung zur Bekämpfung der Schwarzbeinigkeit des Weizens durch den Einsatz von schwachvirulenten und apathogenen Pilzisolaten. *Mitteilungen der Gesellschaft für Pflanzenbauwissenschaften* 6, 253–256.

Augustin, C. (1994) Untersuchungen zur Bekämpfung der Schwarzbeinigkeit des Weizens durch den Einsatz von schwachvirulenten und apathogenen Pilzisolaten. In: Werner, A. and Augustin, C. (eds) *Landnutzungs- und Pflanzenschutzforschung*. ZALF-Bericht 12, Müncheberg, pp. 70–75.

Bednarova-Civinova, M., Petrikova, V., Staněk, M. and Vančura, V. (1981) The influence of seed bacterization on *Gaeumannomyces graminis* and on the growth and yield of wheat. *Sborník ÚVTIZ, Ochrana Rostlin* 17, 89–96.

Brisbane, P.G., Janik, L.J., Tate, M.E. and Warren, R.F.O. (1987) Revised structure for the phenazine antibiotic from *Pseudomonas fluorescens* 2-79 NRRL B-15132. *Antimicrobial Agents and Chemotherapy* 31, 1967–1971.

Bull, C.T., Weller, D.M. and Thomashow, L.S. (1991) Relationship between root colonization and suppression of *Gaeumannomyces graminis* var. *tritici* by *Pseudomonas fluorescens* strain 2-79. *Phytopathology* 81, 954–959.

Chen, C. and Collins, D.J. (1995) Screening of indigenous isolated bacteria for biological control of take-all of wheat in Alabama. *Phytopathology* 85, 509.

Crozier, J.B. and Stromberg, E.L. (1997) Suppression of take-all with a proprietary seed dressing and a *Bacillus* isolate. *Phytopathology* 87, June (Suppl.), S21.

Dahiya, J.S. and Woods, D.L. (1987) Control of *Rhizoctonia solani*, causal agent of brown girdling root rot of canola-rapeseed by *Pseudomonas fluorescens*. *Canadian Journal of Plant Pathology* 9, 275–276.

Dahiya, J.S., Woods, D.L. and Tewari, J.P. (1988) Control of *Rhizoctonia solani*, causal agent of brown girdling root rot of rapeseed by *Pseudomonas fluorescens*. *Botanical Bulletin of Academia Sinica (Taiwan)* 29, 135–142.

Deacon, J.W. (1988) Biocontrol of soil-borne plant pathogens with introduced inocula. *Philosophical Transactions of the Royal Society of London, Series B, Biological Sciences* 318, 249–264.

Défago, G., Berling, C.H., Burger, U., Haas, D., Kahr, G., Keel, C., Voisard, C., Wirthner, P. and Wüthrich, B. (1990) Suppression of black root rot of tobacco and other root diseases by strains of *Pseudomonas fluorescens*: potential applications and mechanisms. In: Hornby, D. (ed.) *Biological Control of Soil-borne Plant Pathogens*. CAB International, Wallingford, pp. 93–108.

Dewan, M.M. and Sivasithamparam, K. (1989) Efficacy of the treatment with a sterile red fungus for control of take-all in wheat. *New Zealand Journal of Crop and Horticultural Science* 17, 333–336.

Dewan, M.M. and Sivasithamparam, K. (1989) Growth promotion of rotation crop species by a sterile fungus from wheat and effect of soil temperature and water potential on its suppression of take-all. *Mycological Research* 93, 156–160.

Duffy, B.K. (1995) Fungal inhibition and suppression of root disease using clinical isolates of *Pseudomonas aeruginosa*. *Phytopathology* 85, 1188.

Duffy, B.K. and Weller, D.M. (1992) Suppression of take-all by *Trichoderma koningii* used individually and in combination with fluorescent pseudomonas spp. *Phytopathology* 82, 1080.

Duffy, B.K. and Weller, D.M. (1996) Biological control of take-all of wheat in the Pacific Northwest of the USA using hypovirulent *Gaeumannomyces graminis* var. *tritici* and fluorescent pseudomonads. *Journal of Phytopathology* 144, 585–590.

Duffy, B.K., Simon, A. and Weller, D.M. (1996) Combination of *Trichoderma koningii* with fluorescent pseudomonads for control of take-all on wheat. *Phytopathology* 86, 188–194.

Elsherif, M. and Grossmann, F. (1994) Comparative investigations on the antagonistic activity of fluorescent pseudomonads against *Gaeumannomyces graminis* var. *tritici in vitro* and *in vivo*. *Microbiological Research* 149, 371–377.

Hamdan, H., Thomashow, L.S. and Weller, D.M. (1988) Relative importance of fluorescent siderophore and phenazine antibiotic by *Pseudomonas fluorescens* 2-79 in suppression of take-all. *Phytopathology* 78, 1522.

Hamdan, H., Weller, D.M. and Thomashow, L.S. (1991) Relative importance of fluorescent siderophores and other factors in biological control of *Gaeumannomyces graminis* var. *tritici* by *Pseudomonas fluorescens* 2-79 and m4-80r. *Applied and Environmental Microbiology* 57, 3270–3277.

Harrison, L.A., Letendre, L., Kovacevich, P., Pierson, E. and Weller, D. (1993) Purification of an antibiotic effective against *Gaeumannomyces graminis* var. *tritici* produced by a biocontrol agent *Pseudomonas aureofaciens*. *Soil Biology and Biochemistry* 25, 215–221.

Harvey, D.E. and Pierson, L.S. III (1997) Identification of *rpoS* of *Pseudomonas aureofaciens* 30-84. *Phytopathology* 87, June (Suppl.), S40.

Heron, D.S. and Weller, D.M. (1988) Influence of bacterial surface properties on colonization of wheat roots by biocontrol pseudomonads. *Phytopathology* 78, 1522.

Howie, W.J. and Cook, R.J. (1985) The effect of motility on wheat root colonization by fluorescent pseudomonads antagonistic to take-all of wheat. *Phytopathology* 75, 1344.

Huber, D.M., Wagner, J.E., El-Nashaar, H.M. and Moore, L.W. (1986) Interactions of a peat carrier and potential biological control agents. *Phytopathology* 76, 1104–1105.

Huber, D.M., El-Nasshar, H., Moore, L.W., Mathre, D.E. and Wagner, J.E. (1989) Interaction between a peat carrier and bacterial seed treatments evaluated for biological control of the take-all diseases of wheat *Triticum aestivum* L. *Biology and Fertility of Soils* 8, 166–171.

Kim, D.-S., Bonsall, R.F., Thomashow, L.S. and Weller, D.M. (1995) Construction of transgenic *Pseudomonas fluorescens* Q69c-80 for improved biocontrol activity to take-all. *Phytopathology* 85, 1146.

Kim, D.-S., Cook, R.J. and Weller, D.M. (1997) *Bacillus* sp. L324-92 for biological control of three root diseases of wheat grown in reduced tillage. *Phytopathology* 87, 551–558.

Kim, D.-S., Weller, D.M. and Cook, R.J. (1997) Population dynamics of *Bacillus* sp. L324-92R$_{12}$ and *Pseudomonas fluorescens* 2-79RN$_{10}$ in the rhizosphere of wheat. *Phytopathology* 87, 559–564.

Lamers, J.G., Schippers, B. and Geels, F.P. (1988) Soil-borne diseases of wheat in the Netherlands and results of bacterization with pseudomonads against *Gaeumannomyces graminis* var. *tritici*. In: Jorna, M.L. and Slootmaker, L.A.J. (eds) *Cereal Breeding Related to Integrated Cereal Production*. PUDOC, Wageningen, pp. 134–139.

Luz, W.C. da (1993) [Microbiological control of take-all of wheat by seed treatment.] *Fitopatologia Brasileira* 18, 82–85.

Maplestone, P.A. and Campbell, R. (1989) Colonization of roots of wheat seedlings by bacilli proposed as biocontrol agents against take-all. *Soil Biology and Biochemistry* 21, 543–550.

Mazzola, M., Fujimoto, D.K. and Cook, R.J. (1994) Differential sensitivity of *Gaeumannomyces graminis* populations to antibiotics produced by biocontrol fluorescent pseudomonads. *Phytopathology* 84, 1091.

Mazzola, M., Fujimoto, D.K., Thomashow, L.S. and Cook, R.J. (1995) Variation in sensitivity of *Gaeumannomyces graminis* to antibiotics produced by fluorescent *Pseudomonas* spp. and effect on biological control of take-all of wheat. *Applied and Environmental Microbiology* 61, 2554–2559.

Ownley, B.H., Weller, D.M. and Thomashow, L.S. (1989) Effects of soil pH on suppression of take-all by *Pseudomonas fluorescens* 2-79. *Phytopathology* 79, 1159.

Ownley, B.H., Weller, D.M. and Alldredge, J.R. (1990) Influence of soil edaphic factors on suppression of take-all by *Pseudomonas fluorescens* 2-79. *Phytopathology* 80, 995.

Ownley, B.H., Weller, D.M. and Alldredge, J.R. (1991) Relation of soil chemical and physical factors with suppression of take-all by *Pseudomonas fluorescens* 2-79. In: Keel, C., Koller, B. and Défago, G. (eds) *Plant Growth-promoting Rhizobacteria: Progress and Prospects*. IOBC WPRS, Paris, pp. 299–301.

Ownley, B.H., Weller, D.M. and Thomashow, L.S. (1992) Influence of *in situ* and *in vitro* pH on suppression of *Gaeumannomyces graminis* var. *tritici* by *Pseudomonas fluorescens* 2-79. *Phytopathology* 82, 178–184.

Peng, Y. and Ellingboe, A.H. (1990) Mutations in *Pseudomonas fluorescens* improve antibiosis and biocontrol of take-all of wheat. *Phytopathology* 80, 1016.

Pierson, E.A. and Weller, D.M. (1993) The use of bacterial mixtures to improve the biological control of take-all of wheat. *Phytopathology* 83, 1365.

Pierson, E.A. and Weller, D.M. (1994) Use of mixtures of fluorescent pseudomonads to suppress take-all and improve the growth of wheat. *Phytopathology* 84, 940–947.

Pierson, L.S. III and Thomashow, L.S. (1988) Role of phenazine antibiotics produced by *Pseudomonas aureofaciens* 30-84 in take-all suppression. *Phytopathology* 78, 1522.

Reeder, R.B. and Ownley, B.H. (1994) Influence of temperature on production of hydrogen cyanide and 2,4-diacetylphloroglucinol, and on inhibition of *Gaeumannomyces graminis* var. *tritici* by *Pseudomonas aureofaciens* Q2-87. *Phytopathology* 84, 1114.

Ross, I.L. and Ryder, M.H. (1994) Hydrogen cyanide production by a biocontrol strain of *Pseudomonas corrugata*: evidence that cyanide antagonises the take-all fungus *in vitro*. In: Ryder, M.H., Stephens, P.M. and Bowen, G.D. (eds) *Improving Plant Productivity with Rhizosphere Bacteria*. CSIRO Publications, East Melbourne, pp. 131–133.

Rowland, C.Y., Kurtböke, D.I., Shankar, M. and Sivasithamparam, K. (1994) Nutritional and biological activities of a sterile red fungus which promotes plant growth and suppresses take-all. *Mycological Research* 98, 1453–1457.

Ryder, M.H. and Borrett, M.A. (1991) Root colonization by non-fluorescent pseudomonads used for the control of wheat take-all. In: Keel, C., Koller, B. and Défago, G. (eds) *Plant Growth-promoting Rhizobacteria: Progress and Prospects*. IOBC WPRS, Paris, pp. 302–307.

Shankar, M., Kurtböke, D.I., Gillespie-Sasse, L.M.J., Rowland, C.Y. and Sivasithamparam, K. (1994) Possible roles of competition for thiamine, production of inhibitory compounds, and hyphal interactions in suppression of the take-all fungus by a sterile red fungus. *Canadian Journal of Microbiology* 40, 478–483.

Shankar, M., Kurtböke, D.I. and Sivasithamparam, K. (1994) Nutritional and environmental factors affecting growth and antifungal activity of a sterile red fungus against *Gaeumannomyces graminis* var. *tritici*. *Canadian Journal of Botany* 72, 198–202.

Slininger, P.J. and Shea-Wilbur, M.A. (1995) Liquid-culture pH, temperature, and carbon (not nitrogen) source regulate phenazine productivity of the take-all biocontrol agent *Pseudomonas fluorescens* 2-79. *Applied Microbiology and Biotechnology* 43, 794–800.

Slininger, P.J., van Cauwenberge, J.E., Bothast, R.J., Weller, D.M., Thomashow, L.S. and Cook, R.J. (1996) Effect of growth culture physiological state, metabolites, and formulation on the viability, phytotoxicity, and efficacy of the take-all biocontrol agent *Pseudomonas fluorescens* 2-79 stored encapsulated on wheat seeds. *Applied Microbiology and Biotechnology* 45, 391–398.

Tahvonen, R., Hannukkala, A. and Avikainen, H. (1995) Effect of seed dressing treatment of *Streptomyces griseoviridis* on barley and spring wheat in field experiments. *Agricultural Science in Finland* 4, 419–427.

Thomashow, L.S. and Pierson, L.S. III (1991) Genetic aspects of phenazine antibiotic production by fluorescent pseudomonads that suppress take-all disease of wheat. In: Hennecke, H. and Verma, D.P.S. (eds) *Current Plant Science and Biotechnology*, Kluwer Academic, Dordrecht, pp. 443–449.

Thomashow, L.S. and Weller, D.M. (1987) Role of phenazine antibiotic in disease suppression by *Pseudomonas fluorescens* 2-79. *Phytopathology* 77, 1724.

Thomashow, L.S. and Weller, D.M. (1990) Role of antibiotics and siderophores in biocontrol of take-all disease of wheat. *Plant and Soil* 129, 93–100.

Thomashow, L.S., Pierson, L.S. III, Weller, D.M. and Bonsall, R. (1988) Detection of phenazine antibiotics produced by *Pseudomonas* spp. in soil. *Phytopathology* 78, 1522.

Thomashow, L.S., Essar, D.W., Fujimoto, D.K., Pierson, L.S. III, Thrane, C. and Weller, D.M. (1993) Genetic and biochemical determinants of phenazine antibiotic production in fluorescent pseudomonads that suppress take-all disease of wheat. In: Nester, E.W. and Verma, D.P.S. (eds) *Advances in Molecular Genetics of Plant–Microbe Interactions*. Kluwer Academic Publishers, Dordrecht, Netherlands; Norwell, Massachusetts, pp. 535–541.

Voisard, C., Bull, C.T., Keel, C., Laville, J., Maurhofer, M., Schnider, U., Défago, G. and Haas, D. (1994) Biocontrol of root diseases with *Pseudomonas fluorescens* CHA0: current concepts and experimental approaches. In: O'Gara, F., Dowling, D.N. and Boesten, B. (eds) *Molecular Ecology of Rhizosphere Microorganisms. Biotechnology and the Release of GMOs*. VCH, Weinheim, pp. 67–89.

Vraný, J., Vančura, V. and Staněk, M. (1981) Control of microorganisms in the rhizosphere of wheat by inoculation of seeds with *Pseudomonas putida* and by foliar application of urea. *Folia Microbiologica* 26, 45–51.

Warren, G., Loper, J., Mills, D. and Thomashow, L. (1995) Pseudomonas. In: Wang, K., Herrera-Estrella, A. and Van Montagu, M. (eds) *Transformation of Plants and Soil Microorganisms*. Cambridge University Press, Cambridge, pp. 3–9.

Weller, D.M. (1982) Colonization of wheat roots by a take-all suppressive pseudomonad. *Phytopathology* 72, 949.

Weller, D.M. (1984) Distribution of a take-all suppressive strain of *Pseudomonas fluorescens* on seminal roots of winter wheat. *Applied and Environmental Microbiology* 48, 897–899.

Weller, D.M. and Cook, R.J. (1981) Control of take-all of wheat with fluorescent pseudomonads. *Phytopathology* 71, 1007.

Weller, D.M. and Thomashow, L.S. (1988) Biocontrol of take-all of wheat and the potential for use of engineered bacterial antagonists. *Abstracts of Papers of the 154th National Annual Meeting of the American Association for the Advancement of Science*, 55.

Weller, D.M. and Thomashow, L.S. (1989) Antibiotics: evidence for their operation and sites where they might be produced. *Journal of Cellular Biochemistry* (Suppl. 13), Part A, 154.

Weller, D.M. and Thomashow, S. (1994) Current challenges in introducing beneficial microorganisms into the rhizosphere. In: O'Gara, F.O., Dowling, D.N. and Boesten, B. (eds) *Molecular Ecology of Rhizosphere Microorganisms. Biotechnology and the Release of GMOs*. VCH, Weinheim, pp. 1–18.

Weller, D.M., Zhang, B.X. and Cook, R.J. (1985) The application of a rapid screening test for selection of take-all suppressive bacteria. *Plant Disease* 69, 710–713.

Weller, D.M., Howie, W.J. and Cook, R.J. (1985) Relationships of in vitro inhibition of *Gaeumannomyces graminis* var. *tritici* and in vivo suppression of take-all by fluorescent pseudomonads. *Phytopathology* 75, 1301.

Weller, D.M., Howie, W.J. and Cook, R.J. (1988) Relationship between *in vitro* inhibition of *Gaeumannomyces graminis* var. *tritici* and *in vivo* suppression of take-all of wheat by fluorescent pseudomonads. *Phytopathology* 78, 1094–1100.

Wilkinson, H.T., Weller, D.M. and Alldredge, J.R. (1982) Enhanced biological control of wheat take-all when inhibitory *Pseudomonas* strains are introduced on inoculum or seed as opposed to directly into soil. *Phytopathology* 72, 948–949.

Wong, P.T.W. and Baker, R. (1981) Control of wheat take-all and ophiobolus patch of *Agrostis* turfgrass by fluorescent pseudomonads from a fusarium wilt-suppressive soil. *Phytopathology* 71, 1008.

Wong, P.T.W. and Baker, R. (1984) Suppression of wheat take-all and ophiobolus patch by fluorescent pseudomonads from a *Fusarium*-suppressive soil. *Soil Biology and Biochemistry* 16, 397–404.

Wüthrich, B., Haldimann, P. and Défago, G. (1991) Effects of pH, nitrogen source and manganese on suppression of wheat take-all by *Pseudomonas fluorescens* strain CHA0 under gnotobiotic conditions. In: Keel, C., Koller, B. and Défago, G. (eds) *Plant Growth-promoting Rhizobacteria – Progress and Prospects*. Organisation Internationale de Lutte Biologique et Intégrée contre les Animaux et les Plantes Nuisibles, Paris, France, pp. 340–345.

Zaspel, I. (1992) [Influence of a treatment of seeds with bacterial antagonists against *Gaeumannomyces graminis* and the relation to yield and infestation by wheat.] *Zentralblatt für Mikrobiologie* 147, 173–181.

Zhang, B.-X., Weller, D.M. and Cook, R.J. (1983) Mass screening of fluorescent pseudomonads for suppressiveness to take-all. *Phytopathology* 73, 963.

Zriba, N. and Mathre, D.E. (1995) Biological control of take-all using antagonistic sterile fungi. *Phytopathology* 85, 1190.

Zriba, N. and Mathre, D.E. (1996) Suppression of *Gaeumannomyces graminis* var. *tritici* by a *Phialophora* sp. and a *Bacillus* sp. *Phytopathology* 86, S52.

Fungicides

Amein, T.A.M. (1988) [Effect of chemical seed treatment on limiting wheat roots and stem base infection by *Gaeumannomyces graminis* var. *tritici* and *Fusarium* spp.] *Roczniki Nauk Rolniczych, Seria E* 18, 161–169.

Bockus, W.W. (1982) Triadimenol seed treatment reduces fall infections of winter wheat by *Gaeumannomyces graminis* var. *tritici*. *Phytopathology* 72, 972.

Chastagner, G.A. and Staley, J.M. (1989) Fungicidal control of take-all patch. *Phytopathology* 79, 1169.

Cotterill, P.J. and Ballinger, D.J. (1989) Use of guazatine and flutriafol for the control of take-all and rhizoctonia rot of wheat. *Australasian Plant Pathology* 18, 64–70.

Cotterill, P.J. and McLean, L.K. (1992) Evaluation of fungicides to control take-all and rhizoctonia root rot of wheat. *Plant Protection Quarterly* 7, 51–54.

Crombie, L., Crombie, W.M.L. and Whiting, D.A. (1985) The avenacins. Natural fungicides from oat roots active against 'take-all' attack. In: Smith, I.M. (ed.) *Fungicides for Crop Protection: 100 Years of Progress*. BCPC Publications, Croydon, pp. 267–272.

Deas, A.H.B. (1986) Triadimefon: relationship between metabolism and fungitoxicity. *Pesticide Science* 17, 69–70.

Duffy, B.K. (1997) Pencycuron, a Rhizoctonia specific fungicide, controls both *Rhizoctonia* and *Gaeumannomyces graminis* on wheat. *Phytopathology* 87, June (Suppl.), S26.

Gladders, P. (1988) The contribution and value of pesticides to disease control in combinable break crops. In: Clifford, B.C. and Lester, E. (eds) *Control of Plant Disease*. Blackwell Scientific Publications, Oxford, pp. 29–50.

Graneto, M.J., Phillion, D.P., Pratt, J.K. and Wong, S.C. (1994) Control of *Gaeumannomyces* sp. in plants by application of novel substd. aryl fungicides. Europe, Patent, 619297, 12 October 1994.

Maxwell, B.D. (1997) The radiolabelled synthesis of N-ethyl-2-chloro-6-trimethylsilylbenzamide(Ring-14C(U)), a fungicide candidate for wheat take-all disease. *Journal of Labelled Compounds and Radiopharmaceuticals* 39, 268–273.

Sheng, X.L., Xi, X., Bai, B. and Zhu, F.C. (1994) [Study on the optimum method for applying triadimefon against take-all in wheat.] *Plant Protection* 20, 28–29.

Sheng, X.L., Xu, X., Jin, X.L., Zhao, J.L. and Yang T.J. (1995) [Study on controlling take-all of wheat with propiconazole.] *Plant Protection* 21, 37–38.

Sherwin, W.A. (1987) Chemical control of take-all. *Plant Protection Quarterly* 2, 98.

Smiley, R.W., Uddin, W., Ott, S. and Rhinhart, K.E.L. (1990) Influence of flutolanil and tolclofos-methyl on root and culm diseases of winter wheat. *Plant Disease* 74, 788–791.

Smiley, R.W., Wilkens, D.E. and Klepper, E.L. (1990) Impact of fungicide seed treatments on rhizoctonia root rot, take-all, eyespot and growth of winter wheat. *Plant Disease* 74, 782–787.

Sutton, J.C. (1989) Fungicide seed treatments, sowing dates and inoculum densities of *Gaeumannomyces graminis* var. *tritici* in relation to take-all in Ontario winter wheat. *Fitopatologia Brasileira* 14, 253–260.

Sutton, J.C., James, T.D.W. and Turek, R. (1987) Prochloraz and triadimenol suppress take-all of winter wheat in Ontario. *Phytopathology* 77, 122.

White, G.A., Phillips, J.N., Huppatz, J.L., Witrzens, B. and Grant, S.J. (1986) Pyrazole carboxanilide fungicides I. Correlation of mitochondrial electron transport inhibition and anti-fungal activity. *Pesticide Biochemistry and Physiology* 25, 163–168.

Yarham, D.J. (1995) Soil-borne diseases of cereals. In: Hewitt, H.G., Tyson, D., Hollomon, D.W., Smith, J.M., Davies W.P. and Dixon K.R. (eds) *A Vital Role for Fungicides in Cereal Production*. BIOS Scientific Publishers, Oxford, pp. 83–94.

Zhang, B.Y., Chen, H.G., Zhou, T.W., Luo, X.Y., Zhang, H.F. and Yang, F.R. (1994) [Techniques of using propiconazole to control take-all of wheat.] *Phytophylactica Sinica* 21, 121–126.

Host resistance

Anon. (1984) Take-all. *Plant Breeding Institute. Annual Report*, 98.

Heun, M. and Mielke, H. (1982) Resistenzzüchtung gegen *Pseudocercosporella herpotrichoides* und *Gaeumannomyces graminis* beim Weizen. *Mitteilungen aus der Biologischen Bundesanstalt für Land- und Forstwirtschaft, Berlin-Dahlem* 207, 1–50.

Höxter, H., Ludwig, W., Sieber, G., Trube, P., Miedaner, T. and Geiger, H.H. (1990) Immunologische Erfassung von *Microdochium nivale* in Schneeschimmel- und Fusskrankheitsprüfungen von Roggen-Inzuchtlinien. *Mitteilungen aus der Biologischen Bundesanstalt für Land- und Forstwirtschaft Berlin-Dahlem* 266, 155.

Hollins, T.W. and Scott, P.R. (1985) Plant Pathology Department. Take-all. Sharp eyespot. *Plant Breeding Institute. Annual Report*, 101–102.

Ismail, E., Wilkinson, H.T., Pedersen, W.L. and Fouly, H. (1995) Quantification and classification of resistance to take-all among winter wheat varieties: *in vivo* evaluation. *Phytopathology* 85, 1169.

Marshall, D.R. (1991) Alternative approaches and perspectives in breeding for higher yields. *Field Crops Research* 26, 171–190.

Martyniuk, S., Wroblewska, B., Jurzysta, M. and Bialy, Z. (1996) Saponins as inhibitors of cereal pathogens: *Gaeumannomyces graminis* v. *tritici* and *Cephalosporium gramineum*. In: Lyr, H., Russell, P.E. and Sisler, H.D. (eds) *Modern Fungicides and Antifungal Compounds*. Intercept, Andover, pp. 193–197.

Penrose, L. (1985) Evidence for resistance in wheat cultivars grown in sand culture to the take-all pathogen, *Gaeumannomyces graminis* var. *tritici*. *Annals of Applied Biology* 107, 105–108.

Penrose, L. (1987) Thickening and browning of cortical cell walls in seminal roots or wheat seedlings infected with *Gaeumannomyces graminis* var. *tritici*. *Annals of Applied Biology* 110, 463–470.

Penrose, L.D.J. (1991) Disease in wheat genotypes naturally infected with *Gaeumannomyces graminis* var. *tritici*. *Annals of Applied Biology* 118, 513–526.

Penrose, L.D.J. (1994) Resistance to *Gaeumannomyces graminis* in wheat genotypes grown in field environments and sand culture. *Soil Biology and Biochemistry* 26, 719–726.

Penrose, L.D.J. (1995) Two wheat genotypes differ in root disease due to *Gaeumannomyces graminis* without interaction with site. *Soil Biology and Biochemistry* 27, 133–138.

Riveros, F., Carvalho, F.I.F. de, Reis, E.M. and Loch, L.C. (1987) Response of different traits to take-all disease in nine wheat (*Triticum aestivum* L.) genotypes. *Revista Brasileira de Genetica* 10, 101–108.

Seah, S., Sivasithamparam, K. and Turner, D.W. (1996) Salicylic acid does not induce systemic acquired resistance in roots of wheat seedlings and is not effective in reducing susceptibility to the take-all fungus. *Phytopathology* 86, S91.

Wilhelm, N.S., Graham, R.D. and Rovira, A.D. (1987) Manganese suppresses the take-all disease of wheat by increasing the plant's internal resistance to the penetration of the fungal hyphae into the root. *International Botanical Congress Abstracts* 17, 36.

Index

Many entries have a UK component or bias, which is not always identified in the index to avoid repetition. Other countries and regions are indexed as main entries and/or as sub-entries, particularly of *'Phialophora'*, 'Take-all' and 'Wheat'. Disease names are sub-entries of 'Diseases' and microorganism names are listed as main entries, or as sub-entries of 'BCAs'.

α-amylase activity   25, 283, 284
Acremonium   256
ADAS   159
    advice   204–205
    Aerial Photography Unit   120
    Cereal Disease Survey   60
    Disease Intelligence Reports   16, 141
    Plant Disease Clinic   28
    Research Centres   118, 125, 134
        Arthur Rickwood   20, 105, 190
        Boxworth   20, 114, 115, 116, 121, 128, 131, 133, 158, 159, 198
        Bridgets   20, 159
        Drayton   128, 129
        Gleadthorpe   28, 105, 119, 131, 169, 170, 221
        High Mowthorpe   20, 158, 159
        Rosemaund   20, 28, 143, 153, 153–159, 169, 170, 171
    trials   65, 105, 130, 131, 132, 133, 134, 137, 147, 169, 170, 172, 189–190, 283
        combinable plot   261
    work   297
Aegilops (*see* Goatgrass)
Aerial photographs   120, 148, 150, 152, 159, 161
    cereal diseases   257–258
        take-all   122–127, 148–149, 154–157, 160–161, 268
AFRC Institute of Plant Science Research   303–304
Agricultural depression   9, 113
Agricultural Improvement Council for England and Wales   113, 130
Agriculture

arable
    combinable crops, yields, costs and margins   114, 115, 116, 117
    weather   153
high fertility–high input   209
high input–high output farming   15, 165
labour, availability of   113
machinery, modern, availability of   113
set aside   32–34, 117, 118, 128, 256, 292, 301
    natural regeneration   118, 128, 129
    regulations   181
    scheme   117, 118, 125, 128, 129
    take-all   117, 125–129, 131, 132
    timing of cultivation   129
    volunteers   118, 128, 129
sustainable   9, 178
Agrochemicals   113
Agronomic factors (other than alternative hosts and fertilizers)   xxiii, 184
    crop residues   97, 99
    cultivation   19, 34, 64, 94
        damage   120
        direct drilling   147, 276
        gantry, 12 m   150
        method   147–148, 298
        minimal   29, 121, 161
        no tillage   62, 64
        para-plough   150
        ploughing   34, 118, 119, 129, 147, 150, 151, 205, 206
        power-harrowing   145
        quality   148–150

Agronomic factors *continued*
  cultivation *continued*
    reduced 147
    rotary harrow 168
    seedbed 63, 119, 139, 148, 150, 151, 168, 175, 205
    sub-soiling 119
    tine 147, 148, 205
  husbandry 27, 142–153, 164
    practices 165
  irrigation 56, 101, 150
  rotation 6, 19, 23, 27, 28, 31, 36, 112–129, 162, 164
    alternative 114–115
    alternative crops 36, 105, 113, 119, 162
    changes in practice 112–118
    cropping history 78
    cropping patterns 100, 118, 165
    cropping sequence 97, 143, 144, 158, 181
    double-cropping 45, 62, 75, 101, 209
    flexibility 115
    Norfolk four-course 112
    options 119
    phased sequences 69, 70, 71
    practice 117
    row spacing 94
    short 204
    seeding density/seed rate 36, 99, 172, 174, 198, 206
    sowing, air-drill 150
    sowing, date xxi, 22, 36, 60, 79, 89, 90, 99, 142–147, 168, 170, 172, 173, 197–199, 203, 206
Agronomic practices 165–166
Agropyron repens (*see* Common couch)
Alopecurus myosuroides (*see* Black-grass)
Aluminium toxicity 164
Antagonistic organisms 196, 198
  microflora 71, 97, 102, 121, 124, 147, 167, 188
  root-colonizing fungi 192
Antibiosis 76, 186
Antibiotics 74, 185, 186
  phenazine 186, 187
  phloroglucinol 186–187
  pyoluteorin 187
  pyrones 193
  trichodermin-3 193
Appressoria 220 (*see also* Hyphopodia)
Area payments 115, 117
Aspergillus nidulans, mutants 248
Australia xxi, 1, 6
  research in 62, 72–75, 76, 123, 124, 135, 136, 140, 141, 191
  South 1, 75, 109, 194
  Western 73
Autecology 50, 54
Autocorrelation (*see* Spatial dependence)
Avenacin 105, 109, 111, 220, 229, 249, 250–252
  A-1 250–251

Avena longiglumis 251
Avena strigosa 251
Avenacinase 250–252
Avenacins A-2, B-1 and B-2 250
Avirulent
  fungi 62, 109, 111
  root parasites 111

Bacteria
  aerobic 72, 186
  beneficial 188
  cell turgor 188
  deleterious 72, 188
  exopolysaccharides 186
  fluorescent pseudomonads
    in lesions 147
    rhizosphere 142, 185
  microflora 140, 142
    conducive 121
    naturally occurring 195
  multiplication 188
  neutral 188
  populations 72, 110
  rhizobacteria 186
    ecological competence 186
    growth-stimulating 195
  (*see also* Biological control agents)
Bacteriophages 233
Barley 28
  alternative to wheat 204, 113
  bridge 119
  continuous 119
  early maturity 110
  naked 111
  quality 292
  in rotations 63, 112, 114, 161, 164, 306
  secondary roots 107
  set-aside effects 118
  six-rowed 111
  spring 12, 15, 96, 108, 112, 150
    artificial inoculation 275
    bridge crop 108
    continuous 102, 165, 269
    creating uniform infection 274
    cultivars 60
    foliar diseases 254
    losses attributed to take-all 284
    monoculture 60, 102, 118, 120, 135
    sequence 143
    trials 119
    yield 258
  two-rowed 111
  upsurge in UK production around 1960 10
  winter 12, 28, 29, 108
    after spring barley 143
    alternative to wheat 105
    bridge crop 107, 108
    early-sown 141
    frost damage 141
    fungicides 173, 174

grain filling 107
grain malting quality 287–288
losses from take-all 284
monoculture 69
scorch 141
sequence 143
take-all affecting grain quality 287
take-all and grain 199, 289
take-all, whiteheads 111
winter kill 141
yellow dwarf virus (BYDV) 46
BCAs (biological control agents)
　bacteria 185
　　actinomycetes 75
　　asporogenous 189
　　Bacillus cereus var. *mycoides* 190
　　Bacillus pumilus 190
　　Bacillus spp. 182, 186, 189, 190
　　Bacillus subtilis 182
　　bacterization 182, 189
　　genetically engineered 184, 185, 186
　　genetics, global regulator gene *gacA* 186–187
　　impediments 183
　　interactions with fungicides 189, 194
　　introduced xxi, xxiii
　　pseudomonads 185, 188
　　pseudomonads, fluorescent xxiii, 72, 75, 182, 185, 186, 187, 188, 191, 211
　　pseudomonads, non-fluorescent 186
　　pseudomonads, non-fluorescent strain 2140 186
　　pseudomonads, non-fluorescent strain AN5 186
　　Pseudomonas aureofaciens, strain 30–84 186
　　Pseudomonas corrugata 184, 186
　　Pseudomonas fluorescens 111, 183, 184, 185, 186, 187, 189–190
　　Pseudomonas spp. 75, 76, 182, 184
　　Pseudomonas putida 186
　　requirements 183, 184
　　research approaches 183
　　research worldwide 184–187
　fungi 190–194
　　Gaeumannomyces graminis var. *graminis* xx, xxiii, 191
　　Gliomastix murorum 191
　　hypovirulent Ggt 112, 190, 193, 194
　　Idriella bolleyi 191, 192, 193
　　mixtures/combinations 183, 191
　　Phialophora spp. (*see Phialophora*)
　　Phoma sp. 192
　　sterile fungi xxiii, 75, 102, 191, 192
　　Trichoderma harzianum 191, 193
　　Trichoderma koningii 74, 193
　　Trichoderma lignorum 192, 193
　　Trichoderma spp. 74, 191, 211
　other 194
　　earthworms xxiii, 194
　　Aporrectodea trapezoides 184

mycophagous amoebae 75, 194
screening 178, 182, 196
testing 178, 182
　in pots 186, 192
　field, Europe 193–194
Beans 29, 113, 115, 116, 118, 121, 123, 152, 163
Beet 39
　fodder 37
　sugar 37, 39, 113
Biological buffer 209
Biological control 28, 112, 178–196, 259
　agents (*see* BCAs)
　application/delivery systems
　　alginate beads 189, 190
　　colonized oat grains 191
　　colonized ryegrass seeds 191
　　foliar spray 185
　　incorporation in soil 193
　　in-furrow 190
　　peat preparations 186, 190
　　perlite granules 189
　　seed coating 178, 184, 189–190, 196
　　carboxymethyl cellulose 190
　　seed inoculation 186
　　seed treatment 112, 182, 185, 189, 192, 193, 197, 206
　　to soil 195
　　strategies 178
　　'customized' approach 184
　Australia 229
　benefits 183
　commercial reality 304
　definition 178
　experimental results 189–190
　factors affecting 188–189
　introduced organisms 178, 182–196, 303, 307
　interactions with environmental factors 195
　mechanisms 184, 187, 193–194
　　antibiosis 178, 191, 192, 193
　　microbial antagonism 205
　　non-specific 187
　　several functioning 194, 195
　natural pest control systems 123
　natural phenomena xxiii, 67, 179–182, 195, 302–303
　　combining 182
　　cross protection 302
　　'*Phialophora*' xxi, 178, 180, 181–182
　　take-all decline (*see* TAD)
　patents 175, 182, 191
　products 304
　　commercialization 185, 191
　　GUS 4000 182
　　reliability 195
　　shelf-life 196
　prospects 195–196
　rhizosphere 76, 83, 195
Biomass 51, 64, 181
Biotechnology 178
Black-grass 34, 64

Blue baby syndrome 63
Boron 142
Bovine spongiform encephalopathy (*see* BSE)
Brassica
    juncea (*see* Mustard)
    napus (*see* Oilseed rape)
    spp.
        fodder 124
        glucosinolate content 123
        isothiocyanates 123
Break crops 23, 34, 69, 105, 113, 116, 151–152, 181, 191, 256
    'clean' 199
    in relation to take-all 29, 121–129, 135, 161, 204, 205, 206
Breaks
    in cereal sequences 27, 29, 40, 60, 79, 121, 148, 180–181, 192
    oats 114, 284
Bristol University 190
BSE 292
Buergenerula 213

Canola (*see* Oilseed rape)
Capronia 222
Carex 221
Carrots 113
Cattle 292
Central Science Laboratory 7–8
Cephalosporium 256, 301
    gramineum 256, 283
    maydis 232, 256
Cephalosporium stripe disease 256–257
Ceratobasidium sp. 256
Cereals (*see also* Wheat, Barley, etc.)
    alternating with pasture 209
    consecutive 79, 96, 108, 168, 261
    continuous cropping/growing 27, 78, 113, 204, 117, 163, 180
    deficiencies
        manganese, UK 141
        sulphur 136, 137, 292
    effects on diseases and pests 112
    first 36, 64, 205
        flexibility of cropping 103
    fourth 78
    growth stages xviii, 107, 171
    intensification 9, 16, 17, 46, 113
        factors associated with 17
    intensively grown 45, 113, 161
    measures
        degree-days 47, 99
        quintal xviii
    monoculture 69, 72
    production
        historical outline 12, 112, 130
        over-production 113
        system 291–292
        trends 31–34
        UK 9–17, 20, 22, 207–208
        worldwide 12
    proportion of arable acreage 112
    root system 47, 48
        anucleate cells 47
        compensatory growth 105
        cortex 192, 193
        crown (*see* nodal roots, below)
        nodal roots 47
        period of branching 47
        root cortex death 47–48, 51, 105
        roots 193
        seminal roots 47
    in rotations 124, 174
    runs 120, 179
    second 6, 12, 16, 36, 107, 113, 196, 205
    sequences 71, 106, 108, 114, 115, 121, 152
    sixth 78
    soils in UK 207
    spring 180
    third 12, 27, 36, 205
    tillers 261
    UK
        dominating arable agriculture 113, 114
        regions 136
    volunteers 121
    winter
        take-all and grain quality and quantity 289
    worldwide 159–164
Chemiluminescence 238
China 12, 40
Clasterosphaeria 213
Climate 102
    change 292–293
        disease 293
        global warming 292
        monitoring, Germany 293
        remote-sensing 295
        take-all 293, 299
    interacting with other factors 164
    Mediterranean 100, 163
    rotational options 118–119, 153
    temperate 100
'Climax ecosystem' 140
Clover 9, 34, 116, 129
    mixtures with grass 205
Cochliobolus sativus xix, 45, 46, 235
Colonists
    antagonistic micro-organisms 91
    root 111, 184, 185, 186, 188
    secondary
        fungal 229
        inhibitory effects 75
Combine harvesting 259
Common Agricultural Policy (CAP) 10, 13, 117
Common couch 28, 119, 152, 177, 205, 206
    rhizomes 152, 205, 206
    spraying off 205
    wheat (*see Triticum aestivum*)
Companies, agricultural chemical 176, 177
Competition 76

for nitrogen   152
for plant nutrients   177
plant   192, 205
Competitive saprophytic ability   209, 249
Composts   111
Computer
   modelling (*see* Modelling and Models)
   graphics   80
Constraints
   crop production   196
   economic   98
   environmental   98
Controlled environments   176, 188
   experiments   88, 141
   (*see also* Pathogenicity tests)
Copper   141, 142
Correlations
   polygalacturonase with pathogenicity in Ggt   249
   RFLP types with host plant preference   240
   RFLP types with host species of origin   240, 244
   soil infectivity and λ 269
   take-all with cropping and sowing date   256
   take-all with yield   283
   TAR with specific weight of grain   287
Cotton   101
Covariance   98
Covariates   258, 259
Cover crops   34, 117, 118
Crop
   canopy   66
   catch   105, 125
   debris   162
   emergence   148, 197
   naturally infested   168–169
   nutrients   206, 210
      major   129–137
      minor   137–142
   nutrition   129–142, 165
   stubble   206
Cross protection (*see* Immunization)
Cyanides   185
   hydrogen   187, 211
Cyperaceae   219, 220

Data
   sets
      large   80, 179
      parsimonious   90
      unbalanced   97
   storage and availability   299
Decision making   199–203
   criteria   261
Diagnosis   298
   modern serological and molecular biological methods   307
Diaporthales   213, 220
Diffusible fungal inhibitor   179
Disease

cycle   49–56, 77
   classification   53, 55–56
   oligocyclic   56
   polycyclic   53, 55, 56, 78
definition   262
distribution   261
host
   compensation   281, 283
   response, traditional measures   254
intensity (*see also* Take-all, intensity)   77
minimum thresholds   281
patterns   266–268
progress   259
   curve (DPC)   76–82, 88, 90, 179
   curves, 3D representation   Plate 7, 81
   rate   78
pyramid   48–49
sampling methods   254
severity assay   256
suppression, natural   76
tolerance   280
traditional measures   254
triangle   48
weather   299
uniformity tests   259, 261
–yield relationships   255, 277–288, 297–298, 308
   absolute yield   281
   actual yield   281
   AUDPC   279, 280
   correlations   280
   critical point models   279, 280
   grain quality   282, 284
   models   278–281
   multiple regression models   277–278, 279, 280
   non-linear models   281
   potential yield   281
   practical approaches   282–284
   single point models   277, 278, 279
   statistical considerations   281–282
Disease assessment   261–266
   Horsfall–Barratt logarithmic scale   264
   scales   264, 266
   single plant method   255
   Weber–Fechner law   264
Diseases
   bermudagrass decline   221
   black stem rust   46
   brown foot rot   97, 101
   brown spot of rice   213
   bunts   46
   cankers   220
   common root rot   183, 192 (*see also Cochliobolus sativus*)
   clubroot   115
   crown gall   304
   damping-off   178
   die-back   220
   eyespot   31, 97, 112, 170, 177, 254, 256
   foliar   168, 170

Diseases *continued*
    foot and root disease 97
    foot rot complex 113, 115
    fusarium
        crown rot 183
        seedling blight 192
    large patch disease of zoysiagrass 193
    mushroom blotch 304
    onion white rot 196
    patch diseases of turfgrass 213
    phymatotrichum root rot of cotton 6
    pineapple heart rot 268
    post harvest 178
    powdery mildew 280, 294
    pythium root rot 182, 184
    rhizoctonia
        bare patch 183
        root rot 163, 184
    root disease complex 206
    rusts 46, 293
    sharp eyespot 97, 256
    sheath rot in rice 221
    smuts 46
    soil-borne 261
    take-all (*see* Take-all)
Drinking water 63, 130

Ecologically obligate parasite 50, 209
Ecology, pathogens and antagonists 301
Ectotrophic fungi, patch diseases of turf grasses 223
Egypt 46
Electrophoresis 234, 238
    protein 240
    pulsed field gel 247
ELISA 240
Elymus repens (*see* Common couch)
Elytrigia repens (*see* Common couch)
Endophytes 220
English regions 11
    Bedfordshire 150, 151, 152, 155
    Cambridgeshire 118, 135, 152, 157, 161, 190
    Chalk downs, Hampshire 159
    eastern England 32, 39, 40, 113–114, 118, 119, 136, 143, 155
    Essex 118
    Herefordshire 118, 169, 170
    Hertfordshire 118, 148
    Kent 152
    London Basin 118
    Norfolk 13, 305
    Northamptonshire 119, 190
    Nottinghamshire 118, 119, 169
    Somerset 174
    south-eastern 119
    south-west 136
    Suffolk 13, 109, 118, 130, 131, 149, 151, 158
    Weald of Kent 152
    West Midlands 118
    Wiltshire 119
    Yorkshire Wolds 119
Environmental degradation 44
Enzymes 248, 249, 250
Epidemics
    artificial 298
    defined 76
    disease-free period 279
    disease 'onset' 279
    DPC, importance of shape 78, 280
    polyetic 94, 101
    quantitative analysis 256
    root disease 76
    temporal dynamics 76
Epidemiology 300
    mathematics 82
Erysiphe graminis 46
Europe xxi, 12, 40, 102, 106
    northern 207
    western 40
Eutrophication 64

Fallow 101, 128, 162, 181
    bare 125, 128
Farm management terms 115, 117
Farmyard manure 31, 32, 180
Fertilizer Practice Surveys 12
Fertilizer drill 277
Fertilizers xxiii, 31
    artificial 113
    calcium 35
        gypsum 142
        nitrate 134, 170
    chloride-containing 137, 209, 306
        KCl 130, 134
    compound 111
    copper 119
    increased usage 40, 44, 130
    liming 44, 45, 56, 74, 102, 139, 141, 151, 164, 179, 205, 206, 210
    liquid 134
    magnesium 119
    manganese xxi, 110, 137–141 (*see also* Manganese)
        foliar application 119, 140
        seed treatment 138, 139, 140
        soil-applied 306
        soil treatment 138
        spray 141
    NaCl xx, xxi
    nitrogen xxi, 9, 17, 22, 34, 36, 40, 64, 67, 72, 130–134, 165, 188, 147, 179, 198, 206
        ammonium nitrate 134, 170
        ammonium sulphate 73, 74, 134, 140, 187, 188, 198, 205, 210, 211, 306
        calcium ammonium nitrate 134
        chloride 134, 137
        compensation for disease 165
        form 72–74, 99, 134, 139, 188, 209–211
        nitrate 134, 187, 188

Index

nitrate leaching (*see* Soil, nitrate, leaching)
^{15}N-labelled   64
pollution   9, 67
rates   60, 99, 147, 165, 198, 205, 206
residual   29, 132–133
response of modern cereal cvs   130
split application   119, 166
timing   166, 197, 198, 205, 206
urea   119, 134, 170
oxide terminology   xviii
phosphorus   xx, 31, 135–136
control of take-all   136
phosphate   135, 306
superphosphate   44, 135, 136, 169, 176, 211
potassium   31, 136
'special'   141
sprays   119
sulphur-containing   134
units   xviii
Field
experiments   80, 141, 165, 168, 191
artificial inoculum   271–276, 280
cereals   259
design   168, 258–261
factorial   89, 90, 97, 172, 173, 197–199, 261, 284–286, 287
grain yields per plot   259
inadequate sampling techniques   258
incomplete block design   258
inter-plot interference   259
long-running   80, 96, 108, 296
microplots   258, 280
natural inoculum   258–271
nearest-neighbour models   259
nested and fractional replication   258
novel procedures   259
numbers of samples   261
number of sampling units   261
plot size   259, 261
precision   259, 261
randomized block design   258
regional approaches   288, 290
sampling   80, 81, 261
simulated in pots   198
size of sampling units   259
small plots   176, 190
traditional blocking techniques   259
trials   175, 176, 185, 186, 189
non-experimental assessment, monitoring, surveys   254–258
studies, plant diseases   254
techniques   254–290
tests of genetically engineered pseudomonads   303
variation
interactions with weather   258
residual   258
systematic   258, 259
work methodology   300
First International Workshop on Take-all of Cereals   6

Flax   40
Food productivity   xx, xxi
Forage rape   34
Forecasting   199–203
France   12, 32, 65, 72, 76, 96, 97, 102, 112, 192
Frost damage   141
Fumigants   124, 166, 167
Fungicides   xxiii, 27, 31, 167–177, 198, 208, 305, 307
benomyl   191, 206, 246
benzamide, MON 41100   290
carboxin   191, 249
costs   44–45
drenches   167
foliar application   36, 39, 170, 174
phloem translocation   170
triadimefon plus carbendazim (Bayleton BM)   170, 171
flutriafol   44
formulations   174, 176
fungitoxicity   176
in-furrow treatment   169, 170, 206
interaction with nitrogen   169, 174
intrinsic toxicity   174
lipophilicity   168
mobility   167–168, 176
patents   174, 175, 176
persistence   167–168, 170
physico-chemical properties   168, 170, 174
phytotoxicity   168, 172, 189
protection of nodal roots   168
redistribution   168
registration   168
screening procedures   174, 176
seed treatments   39, 166, 167, 171–177
Baytan   22, 172
coating   171, 172
fuberidazole   199
MON41100   65, 66, 174
organomercury   172, 192
triadimenol   162, 170, 171, 172, 173, 176, 177, 189, 198, 199, 205, 282
slow-release formulations   167, 168, 172
soil treatments   166, 167–170, 176, 197, 199, 294, 298, 299
benomyl   167
drenches   167
flutriafol   169, 170, 174
MON 41100   305
MON 65500   305
nuarimol   167–168, 198
surfactants   167
triadimefon   169
triadimenol   168, 305
sprays   166, 168
sterol biosynthesis-inhibiting (SBI)   168, 169 (*see also* Fungicides, soil treatments, nuarimol; Fungicides, triadimenol)
testing   174–176
lack of standardization   174
timing   198

Fungicides *continued*
    triadimefon 206
    triadimenol xx, 162, 168, 186
    triazole 44
Fusarium 101, 256
    spp. xix, 113, 115, 116, 178, 182, 229
    avenaceum 251
    culmorum 152, 235, 250, 251, 256
    graminearum 251
    (*see also* Brown foot rot)

Gaeumannomyces 182
    caricis 213, 219, 220, 238
    cylindrosporus Plate 3, 181, 213, 219, 222
        asci and ascospores 216
        hosts 220
        perithecia 215, 227
    graminis 188, 213
        ascospores 220, 221, 239, 241, 246
        asci 221, 246
        bioassay 72, 79, 84, 144, 202
        classification 213
        continuum of types 241
        distinguishing amongst vars 220, 221
        genes 247, 248
        genetic diversity, southern Australia 245
        genetics 245–248
        glyphosate synergist 152
        heterokaryons 247
        hosts 220, 221
        hyphopodia 213, 221
        identification, DNA methods 234–240, 242–243, 301
        isolates 242–243, 247, 252
        karyotypes 247
        life cycle 53, 72
        molecular biology 247–248
        mutagenesis 246
        mutants 245, 246, 247, 248, 249–250
        nomenclature 213
        oat-attacking strains 206, 221
        oats 250
        occurrence 40, 41–43, 44, 46, 55
        parasitic phase 50
        pathogenicity, factors implicated in 249–250
        perithecia 220, 221, 225, 226, 227, 247
        phenotypes 247, 248
        phialospores 221, 246
        population studies 244–245
        sexual crosses 246–247
        relationships 220, 246, 247
        transformants, antibiotic resistant 247, 248
        variability 151
        varieties 236, 241, 243
        viruses 252–253
    graminis var. *avenae* Plate 1, 105, 111, 229, 232, 235
        biology 212
        in British Isles 220
        culture 220
        on *Gramineae* 72, 204, 220, 221
        isolates 240, 244, 245
        mutants lacking avenacinase 251
        resistance to avenacin *in vitro* 250–251
    graminis var. *graminis* Plate 3, 191, 219, 229, 230, 232, 235
        Australia 301
        China 301
        different hosts, DNA characteristics 241, 244
        on grasses 193, 221
        isolates, grass hosts 240
        lobed hyphopodia 139, 221
        occurrence 46
        oxidation of Mn 139
        pathogen, weak 178
        serological reactions 240
        on soya bean 101, 301
        survival 62
        transformants, hyphopodia 248
        typical wheat isolates 301
    graminis var. *maydis* 212, 221, 301
        description 221
        hosts 221
        soluble protein profiles 240
        symptoms on maize 221
    graminis var. *tritici* Plates 1, 3
        asci 53, 58
        ascospores Plate 1, 53, 54, 58
        biology 212
        carry-over 29, 146, 177, 198, 269
        on cereals 120, 221
        dynamics 84, 88, 89, 91
        enzymes, endopolygalacturonase 249
        forms of 71, 112
        on grasses 100, 249
        growth (*see Gaeumannomyces*, growth)
        hyphae 75, 140, 187
        in vitro xxiii, 139, 140, 188, 192, 220
        inhibition at low pH 151
        inoculum 29, 269, 271 (*see also* Inoculum, Ggt)
        interaction with host and environment 49–50, 58, 104, 138, 209 (*see also* Disease, triangle; Disease, pyramid)
        isolates (*see* Isolates, Ggt)
        lesions on roots 55, 56
        life cycle 51
        manganese 140, 141
        mutants 191
        mycoviruses 250
        natural, survival 80
        naturally infested soil 138, 168
        nomenclature 1, 213
        oats 105, 111, 249
        occurrence 101
        pathogen 55–56, 88, 91
        pathogenicity 244, 248–249, 250
        perithecia Plate 1, 53, 57

Index

phialidic conidia, germinating 54–55
phialospores, non-germinating 53, 54, 57
populations 181
resting phase 50
RFLP types 244–245
runner hyphae 52, 54, 88
saprophytic state (*see Gaeumannomyces*, growth)
 in soil and rhizosphere xxiii, 62
 spread 78, 85, 88, 91, 147, 150
 surface antigens, extracellular polyphenol oxidase 240
 suppression of 67, 72, 75, 76
 trophic response to roots 187
-type 239, 241, 249
virulence 140
viruses 295
growth
 -cessation structures 213
 combative migration 209
 conditions for 61–63
 ectotrophic 34, 76, 140, 205, 209, 213
 saprophytic 72, 76, 174, 206, 209
  colonization 76
  state 51, 75
  survival 29, 40, 62, 76, 84, 100, 121, 125, 130
 through soil
hyphopodia 213
incrustans 193, 213, 220
 heterothallic, perithecia 228
 on R-PDA 229
medullaris 213, 219, 221
methods of study 212
sp. 221
spp., perithecia 227
tax. spec. 3 221
teleomorphs 219
use of name 213
Gaeumannomyces–Phialophora complex Plate 1, 7, 213–223, 296, 300–301
 code names for isolates 225
 colonies Plate 1
  characteristics 223–224
 combined identification methods, Rothamsted 236, 239
 concept 219, 222
 culture collections 225
 holomorphs 213
 identification, conventional methods 212, 226–232, 242
 isolation 223–224
 lobed hyphopodial types 239–240, 241
 maintenance of fungi 224–225
 methods of inducing perithecia 222, 227
 molecular identification methods 232–244
 perithecia, tests for 24
 regional differences 253
 relationships, taxonomic and ecological 241
 on roots Plate 3

Genetics 178
Genome 234, 247–248
Genstat 262
Geostatistics 268
Gga (*see G. graminis* var. *avenae*)
Ggg (*see G. graminis* var. *graminis*)
Ggt (*see G. graminis* var. *tritici*)
Glasshouse experiments 135, 176, 184, 186, 188, 192
Global Positioning System 295
Glomus fasciculatus 136
Gnomoniaceae 220
Gnotobiotic culture 48, 186
Goatgrass 111
Golf courses 220
GPDATA culture collection 225
Gramineae 187, 213, 219, 220
Grasses 28, 92, 222, 223, 235, 244, 256
 Agrostis spp. 179, 220
 annual 253
 coleoptile 213
 crop 307
 Festuca sp. 72
 grassland 64
 Holcus lanatus 180
 Hordeum murinum 164
 host 121, 206
 leys 48, 112, 152, 180, 181
  carriers of Ggt 204
 Lolium perenne 164, 181
 long-term cropping 204
 maintaining Ggt 163
 Poa annua 181
 rangeland 193
 rhizomes 222
 rye-grass 34, 96, 128, 129
  host of Ggt 128
 seed crops 40
 turf grass 72, 179, 220
 Zoysia (zoysiagrass) 192, 193
  tenuifolia 192
 (*see also* under common or scientific names)
Green manure 118, 205
Ground water pollution 67, 130, 167
Groundnuts 102
Growers' associations 204
Growth chambers (*see* Controlled environments)
Growth regulators 106, 199, 208

Hagberg falling number 283, 287, 289
Haynaldia 111
Herbampulla 213
Herbicides 151–153, 177, 206, 208
 2,4-D-amine + propyzamide 271
 benzoic acid derivatives 152
 diallate 177
 diclofop 177
 difenzoquat 177
 dinoseb 177
 diquat 152

Herbicides *continued*
  glyphosate 28, 33, 152–153, 177, 198, 205, 206
  mecoprop 177, 152
  non-selective 118
  paraquat 152
  phytotoxicity 177
  spraytopping 206
  trifluralin 152
Hessian fly 45
Hordeum spp. 111
Horses 113
Host
  ammonium nitrogen 48, 182
  calcium uptake 53
  compensating for pathogen activity 63
  defence-related phenolics 110, 142
  defences 138, 139, 140, 186, 187
  fungal inhibitor 61
  growth 77, 81, 91
    population 79
    root 78, 84, 85, 87, 88, 91
  infection assays, records 226–227
  parameters 88
    grain weight 255
    grain specific weight 107
    number of ears 255, 136
    number of fertile shoots 110
    number of grains 255
    plant height 105
    plants m^{-2} (plant density) 146
    root volume 194
    1000 kernel/1000-grain weight 105, 136
    tillers per metre 105
  lignin 110, 138, 140, 142
  lignitubers 140, 214
  nitrate nitrogen, effects of 48
  nitrogen uptake and take-all 63–67
  nutrition 165–166
  –parasite interactions 129
  phenols 138
  plant population 88, 99
  resistance 27, 103–112, 176
    breeding 28, 36, 111–112
  root
    amino acids 139
    endodermis 178
    exudates 138, 136, 139
    respiration 129
    sampling 200
    stele 178
    washing 200
  secondary metabolism 140
    metabolites 186 (*see also* Cyanides; siderophores)
  shoot : root dry matter ratio 63
  tissue
    osmotic potential 137
    susceptible 91
    turgor 137
  water potential 61, 130
  uptake of ammonium nitrogen 34
  yield 91
Hymenula cerealis 256
Hypovirulence 296

Identification
  conventional 229
  diagnostic kits, serological 240
  DNA analysis 227, 229
  DNA-based techniques 241
  DNA 'fingerprints' 234
  DNA methods 232–240, 241
    amplification 233, 234
    cloning procedures 233
    PCR 233, 235, 236, 238, 243, 245, 301
    primers 233, 234, 235, 236, 243
    probes 233, 241, 245
    quantitative competitive PCR techniques 236, 237
    RAPDs 234, 243
    restriction enzymes 223, 234, 235, 238, 243
    RFLP analysis 234, 235, 236, 238, 239, 243, 245
    sequencing 233, 239, 241, 243
    types of DNA 234
Ilex paraguariensis Plate 6
Image analysis systems 261–262
Immunization 109, 111, 181, 188, 190, 191, 193
Induced resistance 76 (*see also* Immunization)
Industry, agricultural chemical 167
Infection
  alloinfection 55
  autoinfection 55
  structures 226
Infectious fragments of plant residues per volume of soil (λ) 269
Infestation, artificial 298, 307
Inoculum
  artificial 258
    colonized straw 274
    millet seed 273
    oat grains 273, 276, 277, 278, 283
    requisites 273
    ryegrass 273
    tests of chemical controls 288, 289
    types 273
    vermiculite-maizemeal 273
    uniform infestation 274
  density 258, 259
  Ggt
    added 75, 186
    after different cultivars 109
    age 84
    artificial 46, 47, 75, 80, 86, 89, 100, 136, 171, 175, 176, 177 185, 191, 195, 198, 207, 227, 271, 274, 294
    artificial infestation 168

bioassay   200, 201, 202, 266, 269, 276
decay   77–79, 81, 84, 86, 98, 99
density   60, 72, 78, 79, 89, 91, 100, 136, 143, 181, 198
drilled with seed   175
dynamics   77, 84, 87
EDI   89, 90
estimating in field soil   296
food base   89, 192, 193
infectious fragments of plant residues   269
infectivity   84, 91
initial   78, 86
level   86
minimum
   effective particle   89
   threshold size   89
natural   51, 175
natural host residues   51, 81, 192, 209
particulate   80
potential   106, 152
pressure   221
primary   80, 81
quantification   199, 200
root colonization   110
secondary   80, 81
sources   75, 81, 86
survival in soil   62, 84, 100
thermal inactivation   62
natural, patchy distribution   258
placement   273, 276, 277
spatial distribution, methods of study   266–268
INRA   159
La Verrière   146
Le Rheu (Rennes)   30, 146, 198
Insecticides   199, 208
Integrated control   305–306
Intervention   117
Iron   141, 142, 189, 211
Isodams   40
Isolates
  Ggt
    characteristics, associations with TAD   224
    classification of   227
    differing in virulence   110
    DNA characteristics   244–245
    hypovirulent   190
    loss of pathogenicity   224–225
    N-isolates   221, 227, 240, 249
    pathogenicity   227
    perithecia in rotting tests   227
    R-isolates   221, 227, 240, 249
  maintenance   225
Isozymes   245

Juncaceae   219
Juncus
  geradii   221
  roemerianus   221

La Platina Experimental Station   46
Legumes   73, 100, 129
  grain   37, 162
  (*see also* Peas)
Lentils   163, 184
Leptosphaeria korrae   243
Ley   34, 204
  grass   302
  lucerne   129
  ryegrass   256
Line transects   266–267, 268
Linseed   113, 115, 118
Lucerne   34, 97, 101, 102, 129, 162
Lupins   102, 113, 124, 163

Magnaporthaceae   213
Magnaporthe
  spp.   213, 220
  assemblage   213
  rhizophila   219, 243
    wheat, Transvaal   222
  poae   243
Maize   37, 39, 96, 140, 152, 221, 232, 244, 300, 301
  take-all   212
Management costs   261
Manganese
  availability   103, 138, 139, 152, 210, 211
    to the plant   48
  in biological control   48
  Birnessite   139
  in cereal leaves   141
  deficiency symptoms   138, 141
  effects on the plant   137–138
  key factor   103
  nutrition, susceptibility to takeall   110
  oxidizing organisms   110, 138, 139, 152, 211
    in rhizosphere   110, 138, 139
  -reducing organisms   48, 139, 140
  scale for estimating deficiency in wheat   141
  in seed   105, 138
  take-all in oats   109
  in tissues of take-all tolerant wheat   110
  unifying concept   211
  uptake   139
Mapping, yield and weeds   295
Mathematics and statistics   xxiii, 99
Maximum likelihood estimate   269
Media
  antibiotics
    aureomycin   223
    penicillin   223
    rifampicin   229
    streptomycin   223, 228–229
    terramycin (oxytetracycline)   223
  antifungal agents
    cymoxanil   229
    dicloran   229
    HOE 00703   229
    metalaxyl   229

Media *continued*
    antifungal agents *continued*
        tolclofosmethyl 229
    osmotic potential agar
        Gaeumannomyces–Phialophora complex
            229, 232
        sodium chloride 229
    selective in conjunction with surface
        sterilization 223
    semi-selective
        R-PDA 229
        SM-GGT 228, 229, 230
        with avenacin 242
        with L-DOPA 229, 242, 253
    oat
        -extract agar 242
        -leaf agar 229
        -root agar 229
Melanin 229
Microbiota 34, 89
Modelling 76–100, 307
    DPC 76–82
    general 57, 82–100
Models 82, 83
    abstract 82
    analytical mathematical 83
    cereal root sytem 47
    classification of 82
    definition 82
    descriptive 82, 83
    disease progress 77, 82, 88
    empirical statistical 90–91
        orthogonal polynomial contrasts 90–91
    explanatory 100
    exponential 81
    host growth 57
    linear 77, 79, 87, 90, 96, 98
    logistic 78, 79, 81, 85, 87, 99
    mathematical
        spatial 88–89
        temporal 84–87
    mechanistic 82, 83
    monocyclic root disease 78
    monomolecular 81, 84, 87, 99
    non-linear 77, 79, 81, 87, 90, 96–98
    qualitative 83–84, 137
    rhizosphere microbial dynamics 57
    simple interest (*see* Models, monomolecular)
    simulation 83, 57
        'Life' 57, 92
        paths of growth of seminal roots 47
        wheat crop 91
    soilborne pathogens 83, 87, 88
    sulphur deficiency in cereals 137
    synthetic 98
    take-all 83, 100
        cellular automaton 84, 92–95
        conceptual 83
        crop succession 96–99
        cultural practice 96–99
        deterministic 92
        dynamics of development 90
        epidemic 77
        FORTRAN simulation 88
        linear regression 96
        patch development (cellular automaton, above)
        Poisson 86
        prediction 100
        primary infection 49, 84, 85, 184
        primary and secondary infection 87, 91
        probability 85, 86–87, 90
        secondary infection 55, 85, 87, 184
        sigmoid 86
        statistical 90
        stochastic 88
Moisture stress, ability of different cvs to withstand 300
Molecular biology 178, 212
Mollisia 222
Molluscicides 208
Monsanto Co. 303
Morley Research Centre 134
Most probable number (*see* Maximum likelihood estimate)
Multiple component hypothesis of pathogenesis and parasitism 48
Mustard 118
    inhibition of Ggt 123
Mycoparasitism 75, 191
Mycoses 222
Myetiola destructor (*see* Hessian fly)
Myxobacteria 194

Necrotrophic parasites 220
Nematodes 112, 163, 194
Nitrification inhibitors 139, 306
    Didin 134, 170
Nitrobacter 139
Nitrogen 298
    leaching 66, 301
    rates 306
    Sensitive Areas 64–65, 130
    Vulnerable Zones 64, 130
Nitrosomonas 139
Nucleic acids 232

Oat take-all fungus (*see* G. graminis var. *avenae*)
Oat take-all 109
Oats xxi, 28, 40, 63, 105, 108, 111, 113, 114, 141, 181, 192, 200, 204, 206, 220
    grey speck 108–109
    mutagenesis, sodium azide 251
    naked cultivars 114
Oilseed rape 26, 28, 37–40, 113, 115
    carrier of Ggt 206
    colonized by Ggt 121
    disease of 115
    inhibition of Ggt 123, 124
    low erucic acid cvs 113

natural biofumigation 124
root exudates affect pests and diseases 123
rotation with cereals 113, 117, 118, 121, 123, 124, 152, 203, 206
UK area 121
Omnidemptus 213
Oomycetes 170
Ophiobolus 213
   graminis 1, 7, 34, 213
   miyabeanus 213
Organic
   farm 129
   food base 75
   manures 64, 139
Osmotic potential
   growth of avirulent fungi 229, 232
   in wheat tissue 229

p-aminobenzoic acid 249
Paddocks 206
Parasitism 76
   semi-obligate parasite 88
Pasture 73, 100, 101, 124, 163, 177
   acid old 100
   alternating 253
   clover-ley 269
   legume 177, 206, 209
   old 151
   permanent 194
PatchMaker 292 (*see also* Models, take-all, cellular automaton)
Pathogen suppression 73 (*see also Gaeumannomyces graminis*, suppression of)
Pathogenicity tests
   assessment, length of vascular discoloration 226
   controlled environment rooms 226
   conventional 241
   Ggt, dark lesions penetrating stele 226
   gnotobiotic systems 227
   runner hyphae 226
   signs and symptoms 226–227
Pathozone 85
Peanuts 101
Peas 40, 115, 116, 150
Penicillium spp. 192
Pesticides 207–208
Phialophora
   anamorphs 213, 219, 220, 222
   graminicola 48, 62, 128, 181, 192–193, 204, 219, 220, 222, 223, 226, 229, 232, 302, 303, Plate 1
      biological control 180
      distinguished from *G. graminis* 243
      invasion of grass roots 222
      phialides and phialospores 215
      runner hyphae 222
      swollen cells (vesicles) 214, 222, 228
      'typical' isolates 239
      viruses 252
   -like fungi 300–301
      Egypt 300, 301
      India 300, 301
   parasitica 222
   sp. (lobed hyphopodia) 7, 219, 222, 224, 226, 229, 232, Plate 1
      biological control 191, 191, 192
      cold-tolerant isolate KY 191
      DNA probe pMSU315 235
      hyphopodia 218
      Italian millet 222
      phialides and phialospores 219
      runner hyphae 217
      sclerotium-like bodies 216, 224
      serological reactions 240
      state of Ggg 222
      swollen cells (vesicles) 217, 228
   spp. 219, 227, 307
      antagonistic 181
      Australia 222
      avirulent, cereal roots 229, 242
      biological control 181, 191, 193, 205
      cell-wall-degrading enzymes 249
      grass hosts 181, 191
      non-aggressive 191
      non-pathogenic 295
      Poland 222
      protein electrophoresis 240
      South Africa 222
      vesicles 222
   radicicola sensu Cain 232, Plate 1
   tracheiphila 219
   zeicola 219
      form-genus, classification 221–222
      maize roots, South Africa 222
      possible teleomorph 222
Phoma spp. 113, 115
Phytoalexins 186
Phytophthora nicotianae var. *parasitica* 268
Pig production 292
Plant variables often recorded at Rothamsted 262
Plant Breeding Institute 303
Plant Pathology Laboratory (MAFF) 6
Plant growth
   fungi (PGPF) 192
   promotion 48
   regulation 177
   root-stimulating organisms 196
Plasmids 233, 248
Polymyxa graminis 294
Potato 39, 48, 64, 113, 150
Poultry production 292
Precision farming 295
Predation 76
Professional advisors 204
Protein patterns 240
Pseudocercosporella herpotrichoides 46, 256, 250 (*see also* Diseases, eyespot)
Pseudopezicula rhizophila 219
Pseudotracylla 213
Pulses 113, 117, 206

Pyricularia 220
Pythium spp. 152

Q-factors 252–253
Quantal data 269

Research
　future 308
　priorities
　　controlling take-all 301–306
　　forecasting and risk assessment 299–300
　　importance of take-all 297–298
　　present 296–306
　　take-all biology 300–301
　UK
　　grain yield 284–287
　　grain quality 24, 287–288
Resistance 296, 303–305
Rhizoctonia spp. 152, 182
　cerealis 250, 256
　solani (AG22) 193
Rhizopus spp. 229
Risk assessment 202
　soil types 204
Root
　-colonizing fungi 192
　crops 112, 113
　disease complex 253
　-infecting fungi 167, 178
　　ectotrophic 213
　　highly specialized 50
Rotations
　controlling take-all 301–302
　profitability on different soils 298
Rothamsted 1, 6, 9, 16, 17, 23, 26, 31, 34, 44, 101
　archives 180
　Classical Experiments 31
　　Broadbalk 31, 32, 180, 181
　　Exhaustion Land 108
　experiments 64, 69, 77, 80, 108, 121, 125, 128, 129, 134, 135, 144, 145, 147, 159, 166, 167, 168, 169, 176, 172, 173, 176, 179, 180, 181, 182, 196, 197, 198, 202, 207, 259
　Farm 140, 168, 190, 199
　fields 190
Rotting test 301
Runner hyphae 301
Rye 28, 105, 106, 107–108, 113, 221, 227, 301, 306
　alternative to wheat 204, 205
　monoculture 105
　take-all 288

Salicylic acid
　precursor of pyochelin 187
　systemic acquired resistance (*see* SAR)

Saltmarshes 221
Sampling 261
Saponins 252
SAR 294–295
Schering Agriculture 118
Sclerotinia
　sclerotiorum 113, 115
　trifoliorum 113, 115
Scottish regions
　borders 137
　Orkney 220, 221
　south-east 137
Secondary moulds 256
Sedges (*see also Cyperaceae*) 220, 221
Seedling-infection tests (*see also* Pathogenicity tests) 227, 228
Self-inhibitors 296
Self-seeded plants (*see* Volunteers)
Septoria
　spp. 46, 100
　lycopersici 252
Serology 240, 251
Set aside (*see* Agriculture)
Sheep farming 163
Siderophores 185, 187
Soil
　acid 40
　　grey sand 135
　　sandy 100
　acidification 74
　acidity 76
　aeration 40, 61, 102, 139
　alkaline 63, 138, 151, 159
　biological balance 196
　black
　　earth 102
　　fen 113
　　prairie provinces, Canada 45
　　sands 204
　-borne
　　diseases, take-all as model 306
　　fungi, glyphosate synergists 152
　　organisms, plant diseases 254
　　pathogens, selection pressure on plants 186
　boulder clay 150, 152
　calcareous
　　clay loam 159, 204
　　Mallee 105
　　sands 100
　calcium carbonate content 62
　cation exchange capacity 209
　chalk 119
　chalky
　　boulder clays 62, 148
　　thin 119
　clay 62, 100, 101
　　calcareous pelosol 150
　　content 64, 178
　　Dutch 40
　　illite 189, 211

Index

loam 62, 100, 168, 180, 190
   montmorillonite 189
   non-calcareous 204
   vermiculite 211
compaction 141, 150
condition 86
consolidation 151
cores 200, 266, 269
denitrification 61
drainage 118, 141, 148, 150, 205
erosion 161
exchangeable K 135
extractable P 135
fen 141
   peats 113, 119, 204
fumigation 64, 100, 179
   metham-sodium 288
   methyl bromide 100
Gabalong 101
good tilth 119
heavy 113, 114, 119, 142, 143, 147, 150
infectivity 53, 269, 276
   cultivation 269
   'living histogram' 269, 272
irradiated 73
light 40, 63, 101, 105, 141
   alkaline 119, 204
   mineral 113, 119
   organic 107
   sandy 141
loamy
   drift 190
   peat 113, 190
   sands 102
loess 102
London clay 28, 190
management 148, 150, 151
manganese 109, 194
   deficiencies 209
   equilibrium concentration 211
   solubility 211
matric potential 188, 194
microbial activity 74
microflora 196
mineral
   high organic matter 141, 204
   nitrogen 64
moisture 162
   content 40, 79
   deficit 157, 158, 203
   /water 188, 203
monolith 159
naturally suppressive 76
nitrate, leaching 34, 63–67, 105
nitrification 137, 138, 139, 210, 211
   rate of 134
nitrogen
   ammonium 211
   -deficient 130
   mineralization 64
   nitrate 211
   residual 121
   total 209
organic 141
   amendments 182
   matter 142, 151, 209, 211
organisms
   antagonistic to Ggt 293
Oxford clay 190
oxygen 188
   tension 211
pasteurization 75
peat 105, 118
peaty 102, 141
pH 44, 62, 139, 140, 141, 142, 151, 188–189, 195, 210
phosphate
   -deficient 130, 135, 136, 176, 205
   index 135, 205
physicochemical aspects 62–63
podzols, cultivated 141
poorly structured 120, 143
potassium content 194
puffy 141
red duplex 105
redox potential 139, 211
rhizosphere 34, 51, 72
   denitrification 136
   fungicides 176
   NH^+_4 134
   organisms 129, 138, 139, 186, 200
   pH 129, 134, 139
sandland 28
sandy 136, 141
   clay loam 60, 64, 100, 102, 151
   loam 150, 180, 190
   organic 141
   silt loam 204
series
   Hanslope 62, 114, 118, 159
   Ragdale 62
   Bromyard 159
silt 118, 194
   loam 101, 204
siltland reclaimed 39
silty clay loam 192, 204, 272
soluble P 135
structure 118, 150, 150, 159, 205
temperature 110, 182
texture 86, 102
type 62–63, 102, 118–119, 153, 158, 167, 189, 194, 211
   boundaries 150
   Bunter Sandstone 118, 119, 169
   Chalky Boulder Clay 114, 118, 135, 148, 159
   Old Red Sandstone 118, 169
uncropped 75
virgin 45, 55
   bushland 185
waterlogging 159
yellow sand 101

Sorghum 75
Soya bean 45, 75, 101–102, 138
Spatial
    aggregation analyses 268
    correlation 268
    dependence 89, 271
    distribution 89, 298
Statistics 173, 188, 202
Stochastic variation 83, 88
Stomach cancer 64
Straw 34
    burning 147, 148, 256
    incorporation 147, 205, 256, 300
    yields 136
Sugar beet (*see* Beet, sugar)
Sunflower 37, 113
Surface sterilization 223
Survey 297, 307
    aerial, take-all 257
    cereal diseases, ADAS 282
    cropping patterns, farmer survey 118
    value 258
    wheat diseases 254, 255
Swedes 48
Systemic acquired resistance (*see* SAR)

TAD (take-all decline) xxi, 17, 26, 28, 102, 261, 275, 276, 291, 307
    'asporogenous' bacteria 185
    breakdown 179–180
    classical 209
    concepts 71
    definition 67
    development 301
    different cereals 180
    disease patches in 121
    dynamics 91
    in Europe 102
    experimental sites 296
    exploiting 119–129, 204
    expression 179
    extent 181
    grass weeds 121, 152
    induction 179
    manifestations 81, 89, 179
    mechanisms 69, 71, 78, 81, 83
        changes, environmental 179
        changes, Ggt 179, 190–191
        changes, inoculum 179
        microbial antagonism 179
        specific antagonists 179, 182, 185, 186
    as a method of disease control 120, 190
    models 92–97
    naming 178
    natural and artificial epidemics 80, 195
    peak of disease, natural 195
    primary peak of disease 120, 142, 180
    resident organisms 178, 179
    robustness 69, 195, 209
    secondary peaks of disease 120

soil
    conditions 179
    type 180
sowing dates 143
    timing 181
    weather 179
in spring barley 60, 69, 108
in UK 67, 69–71, 179
in USA 101, 161
Take-all (referring mostly to wheat)
Africa 46
alternative hosts xxiii, 69, 71, Plate 8
amount, sequential samples 264
amplifiers 96
Asia 46
assessment 60, 61, 70, 198, 271
    changing 259, 260
    density of infected plants 84
    functional state of roots 266
    keys 255, 264–265
    nodal roots infected, % 81
    numbers of infected plants per plot 81
    numerical classification 256
    plants infected, % 81, 96
    relationship between methods 266, 267
    roots infected, length 88, 96, 97
    roots infected, log transformation of %
        151
    roots infected, % 133
    Rothamsted 262–263
    scale 259, 260, 261, 268
    seminal roots infected, % 81
    terms 256
    take-all rating (TAR) 135, 145, 150, 171, 190, 201, 202, 197, 203
    take-all index 172
    variables 262–266
    variates, derived 263
    visual 264
Australia 44–45, 50, 100–101, 109, 110, 112, 130, 134, 137, 141, 142, 150, 151, 161, 162, 163, 174, 182, 183, 185, 186, 191, 206–209, 211, 221, 256, 271, 274, 283, 306
    New South Wales 44, 124, 191
    South 44, 185
    Victoria 44, 105, 142, 163, 169
    Western 44, 71, 76, 100, 109, 142, 163, 177, 185, 191, 209, 269
Australasia and Oceania 46
barley
    spring 60
    winter 199
Belgium 39, 193
bibliography 1
Canada 45, 141, 171, 182
    Alberta 142
    Nova Scotia 45
cereals
    relative resistances 105, 109–111
    sequences 69, 71, 105–108, 143–144

changes 28–31
China 141, 185, 221
control 27–28
    advice 134
    chemical 166–178, 259
    combining techniques 112, 198
    'damage limitation' 165
    forms of nitrogen 134, 137, 138
    impact 207
    integrated 178, 184, 196–199
    mixtures of cultivars 110
    packages 196–197
    recommendations (*see* Take-all, recommendations)
    rotation 206
    strategies 307
    'total management system' 184
    world-wide context 206–70
    (*see also* Biological control)
couch grass Plate 8
crop maturation 283
Czechoslovakia 147, 281
database, bibliographic xxiii, 1
differences amongst cereals 105–108
durum wheat, spring 101
dynamics 79, 91
effects
    climate, in relation to 153–159
    differences among cereals 107
    on disease–yield relationships 107, 157, 158
    general 24, 298
    on grain yield 23, 24, 25, 26, 44–45, 58, 80, 105, 177, 181, 207
    on host 24, 25–26, 60, 64
    on host and yield xxiii, 65, 105, 113, 119
    on host roots 52–53, 110, 118, 130, 139, 155, 157
    seasonal variation 157
    timing of severe take-all 24, 27, 105
Eire 60, 120, 220, 284
epidemics 55, 71, 77, 78, 80–81, 101, 259
    accelerating 90
    artificial vs. natural 276, 278
    artificially created 80, 274
    decelerating 90
    long duration 81, 112
    monitoring 203
    polyetic changes in population structure of pathogen 244
    static 87
epidemiology 67, 69, 85
escape 86, 107, 112, 119, 130, 136, 142
Europe 38, 46, 134, 177, 182, 184, 186
exploiting differences between cereals 105–109
expression 188
factors affecting 103, 147, 159, 181, 190, 197, 209, 284
    agronomic and edaphic 299
    climate 159
    edaphic 159
    previous cropping 203, 206
    sowing date 143, 299
field experiments, artificial inoculation 271–276
foci 82
France 34–39, 95, 96, 109, 112, 140, 146, 171, 174, 187, 188, 193, 194, 196
Germany 17, 33, 37, 38, 40, 102, 105, 106, 112, 121, 136, 281, 288
'haydie' 256
herbicides 300
host
    compensatory root growth 112
    graded response to nitrogen in wheat cvs 109–110
    growth stage 26, 98, 99, 144, 199
    infection 40, 80, 81, 147, 155
    predisposition 210
    resistance amongst cultivars 209
    tolerance 105
incidence 39, 40, 44, 60, 88, 97, 120, 144, 145, 173
index (*see also* TI) 60, 65, 194
infection
    categories 255, 264, Plate 4
    mechanical barrier to 110
    primary 78–79, 80, 84, 89, 91, 97, 146
    secondary 78, 79, 80, 84, 88, 89, 91, 146
intensity 39, 52, 58, 60, 76, 131, 141, 145, 168, 203
irrigation 161
Israel 109, 111
Italy 40
Japan 102, 110
levels 59, 90, 93, 97
management 98, 103, 106
    strategies 165
mapping 266, 267, 268
    kriging 268, 271
mixed inoculations 283
natural infection 259
naturally occurring 167, 176, 190, 191
Netherlands 39–40, 55
North America 46
oilseed rape Plate 8
patch (in turf grass) 72
patches 9, 11, 28, 31, 38, 39, 60, 64, 83, 129, 152, 179, 202, Plates 5, 6
    acid 151, 205
    in Brazil Plate 6
    in France 35, 38
patterns 84, 92, 93, 180
    changing 259, 260
    methods of study 268
    Rothamsted 268
    scale 259, 260, 261, 268
prediction 157, 176, 209
prevalence 39
rating (TAR) 31, 34, 108

Take-all *continued*
  recommendations 167, 196–197
    Australia 206
    Germany 205, 206
    UK 204–205
    USA 206
  research 208
    differences, regional xix–xxiii, xxi, 207
    generalization, problems xix–xxiii
    global, development xxi
    global, value of a framework xx
    trends 31–34
    trends since 1980 xxii–xxiii
    UK 274–276
  resistance, in species other than wheat 109, 111
  resistant lines, wheat 112
  rye 105
  severity 60, 61, 78, 88, 94, 96, 97, 144, 177
    effect of nitrogen 131, 132, 133
  signs on wheat, rotting test Plate 1
  soil infectivity 53, 60, 79, 200, 144
  soils
    artificially infested 176
    conditions favouring 56, 61–63, 75, 97, 101–102, 211
    conducive to 67–76, 89
    naturally infested 16, 80, 110
    receptivity 187, 188, 198
    suppressive 40, 67–76, 89, 101, 141, 179, 182, 185
  spatial distribution 82, 89, 120
  spread 91, 78, 88
  surveys 23, 35, 60, 101, 124 (*see also* Winter Wheat Disease Survey)
  symptoms 49
    above ground 80
    blackened roots 80, 255
    in the crop 255, 256, 257, 259, 260
    pathogenicity tests Plate 3
    premature ripening 150, 172
    rolled flag leaf Plate 5
    root lesions 53, 75, 101, 134, 136, 142, 185
    seedling blight 49, Plate 2
    seedling disease 162, Plate 2
    stembase blackening 111
    stunted plants Plate 5
    vascular discoloration, root Plate 2
    on wheat 60
    whiteheads 38, 39, 59, 102, 111, 138, 176, 177, 185, 256, 257, Plate 5
  South Africa 102, 163, 207
  South America 46
    Brazil 45, 101, 161, 164
    Chile 46, 283
  suppression 69, 73, 76, 188, 192
    in monoculture 90
  Sweden 40
  Switzerland 102, 211
  tolerance 112

  UK
    biological control 179, 182, 183, 185, 189, 190, 191, 192, 196
    concerns 33
    epidemics 15, 59–61
    fungicides 206
    historical background 1–17
    importance 18–26
      constraints on farmers 19
      estimates elusive 23–24
      losses 18
    issues 6, 7, 29, 32–34, 35
    occurrences 7–9, 16–17, 81, 161
    Northern Ireland 254
  USA 45, 49, 109, 110, 134, 137, 141, 161, 171, 181–183, 189, 209, 211, 283, 306
    Alabama 101
    Columbia Basin 161
    Georgia 75, 105, 171
    Indiana 45, 101
    Kansas 45, 62, 280, 283, 290
    Montana 75, 101, 171, 175
    North Dakota 101
    Oregon 177
    Pacific Northwest 45, 61, 75, 101, 147, 183, 184, 185, 191, 288
    southeastern 101
  USSR, former 40
  variability, small plots 259
  variables, selection 87, 88
  variance 64, 151
  wheat
    fourth 260
    spring 175, 185, 190, 207, 208
    third 260, 261
    winter 8, 185, 255
  yield loss 23, 290
    artificial inoculum 283–284
    estimates, timing assessments 283
    experimental requirements 287
    studies using fungicide 290
  –yield relationships 274, 284
    difficulties 277
    functional roots 281
    split-line model 281
  weather 17, 31, 40, 58–61, 80, 158, 165, 172, 203, 292
    changes 203
    drought 171
    patterns 21, 100
    precipitation 40, 100
    years which favoured take-all 23, 31, 39–40, 58, 97
TAR (take-all rating) 262
  defined 264
  grain–yield relationship 285, 287
  1000-grain weight relationship 286, 287
  grain hectolitre weight relationship 286
  regression analyses 289
Technical representatives 204

Techniques
    acetylene inhibition   136
    DNA probe   71
    genetic manipulation   195
    inoculum and plant chambers   100
    micro-XANES spectroscopy   139
    nitrosoguanidine treatment of protoplasts   191
    'poisoned-food' assays   174, 176
    reverse genetics and gene disruption   250, 251
    ribosomal DNA sequence analysis   220
    soil sandwich   72, 100, 101, 174
    Southern blotting   234
    X-ray CT   62
Teleomorph   220
TI (take-all index), winter wheat yield relationship   287
Toxins   213
Transformations
    logarithmic   259
    logit   262, 278
Trefoil   34
Trichocladium medullare   219, 221
Triticale   11, 28, 44, 69, 105, 106, 108, 109, 111, 113, 301, 306
    alternative to wheat   105, 204
    bridge crop   107, 108
    cultivar, Purdy   107
    monoculture   69
    properties   106
    winter   106
        take-all and grain   289
Triticum
    germplasm   248
    aestivum   12, 111
    tauschii (*see* Wheat, diploid wild)
    turgidum var. *durum* (*see* Wheat, durum)   15
    turgidum var. *dicoccum* (*see* Wheat, emmer)
Turf grass pathogens   253

USA   xxi, 40, 103, 107, 109
    recent research   75–76, 192, 193
    (*see also* Take-all)

Valsaceae   220
Variance ratio comparisons   268
Variogram   268, 271
Venn diagram   xx
Vermiculite   227
Vesicles, swollen terminal   301
Vesicular-arbuscular mycorrhizal fungi   136
Vitaceae   219
Volunteers   184
    carriers of Ggt   121, 145, 147, 198
    cereal   29, 206
    distribution   129
    grasses   206
    populations   145, 147
    rye   28
    spilled grain   145

wheat   28, 54, 62, 146, 206
    soil shading   62

Wales   8, 11, 19, 20
Wheat (referred to frequently throughout text)
    areas, England and Wales   19
    breeding for cross protection ability   194
    climate   14, 19, 23, 37, 101
    consecutive   118, 119, 139
    continuous   114, 118–119, 128, 162, 180, 181, 241
    crown   188
    cultivars   6, 19, 21, 31, 36, 44, 45, 194
        Arminda   198
        Avalon (bread)   109, 189
        Axona   189
        Brimstone   32, 180, 189
        Brock   189
        Cappelle Desprez   10, 13, 32, 55, 56, 114
        Fakta   110
        Fenma   189
        Flanders   32
        Galahad (winter)   107
        Mercia   24, 133, 189
        Mineret   189
        nodal roots   112
        Norman (feed)   109, 189
        Red Club   32
        Red Standard   32
        Red Rostock   32
        Riband (winter)   256
        Squarehead's master   32
        Stacey   106
        Wembley   189
        Virtue (winter)   267
    in different regions
        Brazil, contour banks   Plate 6
        Europe   140, 293, 304
            France   35–39
            Germany   12, 32, 256
            USA   75
                Indiana   138
                Montana   162
                Pacific Northwest   161, 162
    diploid, wild   111
    drilling   119
    'dryland'   161, 162
    durum   40, 101, 208
    emmer   109, 111
    first   18, 22, 181, 193, 199, 200, 205
    fourth   205
    genetic improvement   31, 111
    genotypes   110, 140, 142
    hybrid   304–305
    intensification   39
    irrigation   45, 96, 161–163, 184, 209
    limiting production   32
    management   97–98
    microflora   140, 185

Wheat *continued*
- monoculture 17, 75, 90, 96, 97, 102, 140, 141
  - long-term 119, 140, 142, 191
- phases of plant growth
  - model 24
  - nitrogen needs 36
  - tissue water potential 61
- planting in paired rows 206
- principal types 14, 103, 161
  - classification 21
  - differences among 103, 105
  - hard red winter 103
  - soft red winter 105, 138
  - soft white winter 103, 161
- production 209
  - climatic change 291, 293
  - constraints 161, 163, 164, 165
  - outside Europe 207, 208
  - UK 22
- rhizosphere 140, 188
- roots
  - axenically grown 134
  - development stimulated by P fertilizer 136
  - exudates 140
  - growth 118
  - mycoflora, Canada 192
  - necrosis 72, 188
  - nodal 188
  - nodal, production 157
  - seminal 188
  - surface, pH 188
  - water potential 191
- second 23, 161, 169, 199, 200, 203, 205
- seedling
  - assays 222
  - vigour 138
- sequence 142, 159
- sixth 161
- spring 120, 162, 163, 208
  - creating uniform infection 274, 284
  - irrigated 171
  - losses attributed to take-all 284
  - potassium in grain and straw 136
  - sowing quality cvs in autumn 143
  - take-all–yield relationship 284
- sub-crown internode 162, 176
- successive crops 84, 101, 131
- take-all fungus (*see Gaeumannomyces graminis* var. *tritici*)
- third 181, 205
- uncertainties in future production 23
- varieties (*see* Wheat, cultivars)
- winter 208
  - artificial inoculation 274–276
  - consecutive 69, 74, 78, 79, 80, 90, 91, 101, 180 (*see also* Wheat, winter, monoculture)
  - continuous 16, 20–21, 28, 31 (*see also* Wheat, winter, monoculture)
  - eighth 97
  - fifth 64, 97, 120, 152
  - first 12, 14, 20, 21, 27, 29, 32, 36, 39, 77–79, 90, 91, 97, 101, 120, 121, 128, 130, 131, 144, 145, 148, 150, 152, 153, 180
  - fourth 81, 120, 152, 195
  - monoculture 28, 31, 40, 96 (*see also* Wheat, winter, continuous)
    - long-term 75, 119
  - ninth 28, 97
  - % nitrogen in grain 289
  - root exudation 48
  - root growth 159
  - second 9, 14, 18–23, 26, 28–31, 36, 39, 78–79, 90, 91, 97, 113, 120, 121–128, 130, 131, 134, 144, 145, 148, 150, 153, 155
  - seminal roots 49, 159
  - sixth 97
  - surveys 8, 39
  - tenth 97
  - third 20, 28, 31, 36, 79, 144, 152
  - varieties (*see* Wheat, cultivars)
  - vernalization 163
  - yield
    - average 31
    - components 24
    - in consecutive crops 20–23
    - seasonal variation 22
  - yield responses, soil fumigation 288
Winter Wheat Disease Survey 7, 29 (*see also* Take-all, surveys)
Woburn Experimental Farm (Bedfordshire) 26, 29, 34, 60, 80, 128, 173, 180, 190, 203
- field experiments 259, 269, 269, 275
Woodland 64, 220
World War II 10, 113

'Xenodochus Cerealium' 1

Yield loss
- indicators 257
- prediction 298
- (*see also* Barley; Take-all; Wheat)

'Zenodochius Cerealium' 1
Zinc 141, 142